Mechanics of Elastic Waves and Ultrasonic Nondestructive Evaluation

Mechanics of Elastic Waves and Ultrasonic Nondestructive Evaluation

Dr. Tribikram Kundu

CRC Press
Taylor & Francis Group
Boca Raton London New York

CRC Press is an imprint of the
Taylor & Francis Group, an **informa** business

CRC Press
Taylor & Francis Group
6000 Broken Sound Parkway NW, Suite 300
Boca Raton, FL 33487-2742

Printed and bound by CPI Group (UK) Ltd, Croydon, CR0 4YY on acid-free paper

International Standard Book Number-13: 978-1-4987-6717-0 (Hardback)

Library of Congress Cataloging-in-Publication Data

Names: Kundu, T. (Tribikram) author.
Title: Mechanics of elastic waves and ultrasonic nondestructive evaluation / Tribikram Kundu.
Description: First edition. | Boca Raton, FL : CRC Press/Taylor & Francis Group, 2018. | Includes bibliographical references and index.
Identifiers: LCCN 2018038804| ISBN 9781138035942 (hardback : acid-free paper) | ISBN 9781315226194 (ebook)
Subjects: LCSH: Ultrasonic testing. | Elastic waves.
Classification: LCC TA417.4 .K859 2018 | DDC 620.1/1274--dc23
LC record available at https://lccn.loc.gov/2018038804

Visit the Taylor & Francis Web site at
http://www.taylorandfrancis.com

and the CRC Press Web site at
http://www.crcpress.com

This book is dedicated to my wife Nupur, whose continuous inspiration, support and sacrifice made this book writing project possible.

Contents

Preface

This book presents necessary background knowledge on mechanics to understand and analyze elastic wave propagation in solids and fluids. This knowledge is necessary for elastic wave propagation modeling and for interpreting experimental data generated during ultrasonic nondestructive testing and evaluation (NDT&E). The book covers both linear and nonlinear analyses of ultrasonic NDT&E techniques. The materials presented here also include some exercise problems and solution manual. Therefore, this book can serve as a textbook or reference book for a graduate level course on elastic waves and/or ultrasonic nondestructive evaluation. It will be also useful for instructors who are interested in designing short courses on elastic wave propagation in solids or NDT&E.

The book has following five chapters:

Chapter 1: Mechanics of Elastic Waves – Linear Analysis

Chapter 2: Guided Elastic Waves – Analysis and Applications in Nondestructive Evaluation

Chapter 3: Modeling Elastic Waves by Distributed Point Source Method (DPSM)

Chapter 4: Nonlinear Ultrasonic Techniques for Nondestructive Evaluation

Chapter 5: Acoustic Source Localization

The materials covered in the first two chapters provide the fundamental knowledge on linear mechanics of deformable solids while Chapter 4 covers nonlinear mechanics. Thus, both linear and nonlinear ultrasonic techniques are covered here. Nonlinear ultrasonic techniques are becoming more popular in recent years for detecting very small defects and damages. However, this topic is hardly covered in currently available textbooks. Researchers mostly rely on published research papers and research monographs to learn about nonlinear ultrasonic techniques. Chapter 3 describes elastic wave propagation modeling techniques using DPSM. Chapter 5 is dedicated to an important and very active research field – acoustic source localization – that is essential for structural health monitoring and for localizing crack and other types of damage initiation regions.

Two graduate level courses on elastic waves can be developed from the materials presented in this book. Chapters 1, 2 and some parts of Chapter 5 can be covered in the first introductory graduate level course on elastic waves, while materials in Chapters 3, 4 and parts of Chapter 5 can be offered in a more advanced graduate level second course on elastic waves and ultrasonic NDE. Alternately, instructors can also pick and choose some basic and advanced topics from these five chapters to cover those materials of the instructor's choice in one graduate level course.

Short courses can also be developed from this book. Five modules can be designed from the five chapters to deliver in one short course. This short course duration can be anywhere between five and fifteen hours, depending on in what depth each topic is covered by the instructor.

Author

Professor Tribikram Kundu received his bachelor degree in mechanical engineering from IIT Kharagpur, where he was the winner of the President of India Gold Medal (PGM). After completing his MS and PhD at UCLA and winning the outstanding graduate student award at UCLA he joined the faculty at the University of Arizona where he was promoted to Full Professor and was later distinguished as a Faculty Fellow in the College of Engineering. To date he has supervised 40 PhD students, published nine books, 18 book chapters and 333 technical papers: 166 of those in refereed scientific journals. He has won the Humboldt Research Prize (Senior Scientist Award) in 2003 and Humboldt Fellowship award in 1989 from Germany, 2012 NDE Life Time Achievement Award from SPIE (the International Society for Optics and Photonics), 2015 Research Award for Sustained Excellence from ASNT (American Society for Nondestructive Testing), 2015 Lifetime Achievement Award and 2008 Person of the Year Award from the Structural Health Monitoring Journal. He received a number of invited Professorships from France, Germany, Sweden, Switzerland, Spain, Italy, South Korea, Poland, China, Japan, Singapore and India. He is a Fellow of five professional societies – ASME, ASCE, SPIE, ASNT and ASA.

1 Mechanics of Elastic Waves – Linear Analysis

It is necessary to have a good understanding of the fundamentals of mechanics to appreciate the physics of elastic wave propagation in solid and fluid materials. With this in mind, this chapter is divided into two focus areas. The first part is devoted to the theory of elasticity and continuum mechanics, and in the second part the basic equations of elastic wave propagation in materials are derived. It is necessary to fully comprehend the first chapter before progressing to the rest of the book.

1.1 FUNDAMENTALS OF THE CONTINUUM MECHANICS AND THE THEORY OF ELASTICITY

Relations between the displacement, strain and stress in an elastic body are derived in this section.

1.1.1 Deformation and Strain Tensor

Figure 1.1 shows the reference state R and the current deformed state D of a body in the Cartesian $x_1x_2x_3$ coordinate system. Deformation of the body and displacement of individual particles in the body are defined with respect to this reference state. As different points of the body move, due to applied force or change in temperature, the configuration of the body changes from the reference state to the current deformed state. After reaching equilibrium in one deformed state if the applied force or temperature changes again, the deformed state also changes. The current deformed state of the body is the equilibrium position under current state of loads. Typically, the stress-free configuration of the body is considered as the reference state but it is not necessary for the reference state to always be stress-free. Any possible configuration of the body can be considered as the reference state. For simplicity, if it is not stated otherwise, the initial stress-free configuration of the body, before applying any external disturbance (force, temperature etc.), will be considered as its reference state.

Consider two points P and Q in the reference state of the body. They move to P* and Q* positions after deformation. Displacement of points P and Q are denoted by vectors **u** and **u+du**, respectively. (Note: Here and in subsequent derivations vector quantities will be denoted by bold letters.) Position vectors of P, Q, P* and Q* are **r**, **r+dr**, **r*** and **r*+dr***, respectively. Clearly, displacement and position vectors are related in the following manner:

$$\mathbf{r}^* = \mathbf{r} + \mathbf{u}$$
$$\mathbf{r}^* + d\mathbf{r}^* = \mathbf{r} + d\mathbf{r} + \mathbf{u} + d\mathbf{u} \tag{1.1}$$
$$\therefore d\mathbf{r}^* = d\mathbf{r} + d\mathbf{u}$$

In terms of the three Cartesian components the above equation can be written as

$$\left(dx_1^*\mathbf{e}_1 + dx_2^*\mathbf{e}_2 + dx_3^*\mathbf{e}_3\right) = \left(dx_1\mathbf{e}_1 + dx_2\mathbf{e}_2 + dx_3\mathbf{e}_3\right) + \left(du_1\mathbf{e}_1 + du_2\mathbf{e}_2 + du_3\mathbf{e}_3\right) \tag{1.2}$$

where $\mathbf{e}_1$, $\mathbf{e}_2$ and $\mathbf{e}_3$ are unit vectors in x_1-, x_2- and x_3-directions, respectively.

In index or tensorial notation Eq. (1.2) can be written as

$$dx_i^* = dx_i + du_i \tag{1.3}$$

where the free index i can take values 1, 2 or 3.

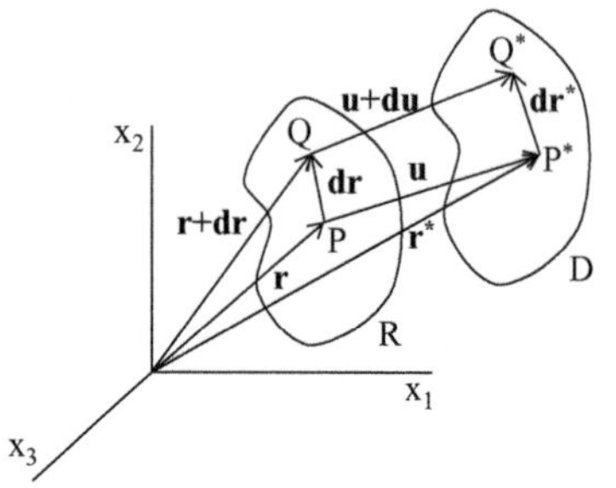

Figure 1.1 Deformation of a body: R is the reference state D is the deformed state.

Applying chain rule Eq. (1.3) can be written as

$$dx_i^* = dx_i + \frac{\partial u_i}{\partial x_1}dx_1 + \frac{\partial u_i}{\partial x_2}dx_2 + \frac{\partial u_i}{\partial x_3}dx_3$$
$$\therefore dx_i^* = dx_i + \sum_{j=1}^{3}\frac{\partial u_i}{\partial x_j}dx_j = dx_i + u_{i,j}dx_j \tag{1.4}$$

In the above equation comma (,) means derivative, and the summation convention (repeated dummy indices mean summation over 1, 2 and 3) has been adopted.

Eq. (1.4) can also be written in matrix notation in the following form:

$$\begin{Bmatrix} dx_1^* \\ dx_2^* \\ dx_3^* \end{Bmatrix} = \begin{Bmatrix} dx_1 \\ dx_2 \\ dx_3 \end{Bmatrix} + \begin{bmatrix} \frac{\partial u_1}{\partial x_1} & \frac{\partial u_1}{\partial x_2} & \frac{\partial u_1}{\partial x_3} \\ \frac{\partial u_2}{\partial x_1} & \frac{\partial u_2}{\partial x_2} & \frac{\partial u_2}{\partial x_3} \\ \frac{\partial u_3}{\partial x_1} & \frac{\partial u_3}{\partial x_2} & \frac{\partial u_3}{\partial x_3} \end{bmatrix} \begin{Bmatrix} dx_1 \\ dx_2 \\ dx_3 \end{Bmatrix} \tag{1.5}$$

In short form, Eq. (1.5) can be written as

$$\{\mathbf{dr}^*\} = \{\mathbf{dr}\} + [\nabla\mathbf{u}]^{\mathrm{T}}\{\mathbf{dr}\} \tag{1.6}$$

If one defines

$$\varepsilon_{ij} = \frac{1}{2}\left(u_{i,j} + u_{j,i}\right) \tag{1.7a}$$

and

$$\omega_{ij} = \frac{1}{2}\left(u_{i,j} - u_{j,i}\right) \tag{1.7b}$$

then Eq. (1.6) takes the following form:

$$\{\mathbf{dr}^*\} = \{\mathbf{dr}\} + [\varepsilon]\{\mathbf{dr}\} + [\omega]\{\mathbf{dr}\} \tag{1.7c}$$

1.1.1.1 Interpretation of ε_{ij} and ω_{ij} for Small Displacement Gradient

Consider the special case when $\mathbf{dr} = dx_1\mathbf{e}_1$. Then, after deformation, three components of $\mathbf{dr}^*$ can be computed from Eq. (1.5).

$$dx_1^* = dx_1 + \frac{\partial u_1}{\partial x_1}dx_1 = (1+\varepsilon_{11})dx_1$$

$$dx_2^* = \frac{\partial u_2}{\partial x_1}dx_1 = (\varepsilon_{21}+\omega_{21})dx_1 \tag{1.8}$$

$$dx_3^* = \frac{\partial u_3}{\partial x_1}dx_1 = (\varepsilon_{31}+\omega_{31})dx_1$$

In this case, the initial length of the element PQ is $dS = dx_1$; the final length of the element P^*Q^* after deformation is

$$dS^* = \left[(dx_1^*)^2 + (dx_2^*)^2 + (dx_3^*)^2\right]^{\frac{1}{2}} = dx_1\left[(1+\varepsilon_{11})^2 + (\varepsilon_{21}+\omega_{21})^2 + (\varepsilon_{31}+\omega_{31})^2\right]^{\frac{1}{2}}$$
$$\approx dx_1\left[1+2\varepsilon_{11}\right]^{\frac{1}{2}} = dx_1(1+\varepsilon_{11}) \tag{1.9}$$

In Eq. (1.9) we have assumed that the displacement gradients $u_{i,j}$ are small. Hence, ε_{ij} and ω_{ij} are small. Because that is true, the second order terms involving ε_{ij} and ω_{ij} could be ignored.

Engineering normal strain (E_{11}) in x_1 direction, from its definition, can be written as

$$E_{11} = \frac{dS^* - dS}{dS} = \frac{dx_1(1+\varepsilon_{11}) - dx_1}{dx_1} = \varepsilon_{11} \tag{1.10}$$

Similarly, one can show that ε_{22} and ε_{33} are engineering normal strains in x_2 and x_3 directions, respectively.

To interpret ε_{12} and ω_{12} consider two mutually perpendicular elements PQ and PR in the reference state. In the deformed state these elements are moved to P^*Q^* and P^*R^* positions, respectively as shown in Figure 1.2.

Let the vectors PQ and PR be $(\mathbf{dr})_{PQ} = dx_1\mathbf{e}_1$ and $(\mathbf{dr})_{PR} = dx_2\mathbf{e}_2$, respectively. Then, after deformation, three components of $(\mathbf{dr}^*)_{PQ}$ and $(\mathbf{dr}^*)_{PR}$ can be written in the forms of Eqs. (1.11) and (1.12), respectively, as given below:

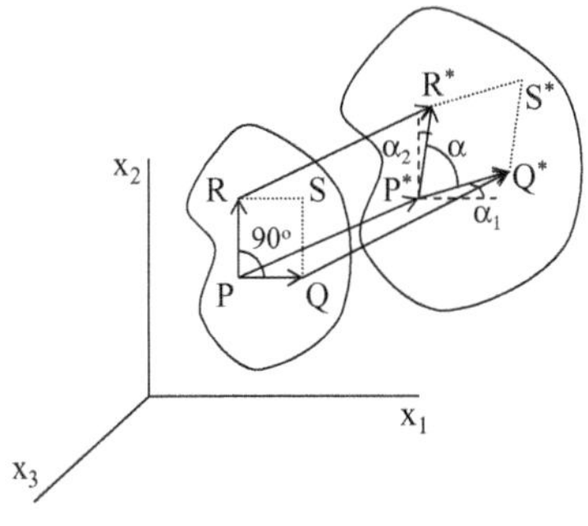

Figure 1.2 Two mutually perpendicular elements, PQ and PR before deformation, no longer remain perpendicular after deformation.

$$\begin{aligned}
(dx_1^*)_{PQ} &= dx_1 + \frac{\partial u_1}{\partial x_1}dx_1 = (1+\varepsilon_{11})dx_1 \\
(dx_2^*)_{PQ} &= \frac{\partial u_2}{\partial x_1}dx_1 = (\varepsilon_{21}+\omega_{21})dx_1 \\
(dx_3^*)_{PQ} &= \frac{\partial u_3}{\partial x_1}dx_1 = (\varepsilon_{31}+\omega_{31})dx_1
\end{aligned} \tag{1.11}$$

$$\begin{aligned}
(dx_1^*)_{PR} &= \frac{\partial u_1}{\partial x_2}dx_2 = (\varepsilon_{12}+\omega_{12})dx_1 \\
(dx_2^*)_{PR} &= dx_2 + \frac{\partial u_2}{\partial x_2}dx_2 = (1+\varepsilon_{22})dx_2 \\
(dx_3^*)_{PR} &= \frac{\partial u_3}{\partial x_2}dx_2 = (\varepsilon_{32}+\omega_{32})dx_1
\end{aligned} \tag{1.12}$$

Let α_1 be the angle between P^*Q^* and the horizontal axis, and α_2 the angle between P^*R^* and the vertical axis as shown in Figure 1.2. Note that $\alpha+\alpha_1+\alpha_2=90°$. From Eqs. (1.11) and (1.12) one can show

$$\begin{aligned}
\tan\alpha_1 &= \frac{(\varepsilon_{21}+\omega_{21})dx_1}{(1+\varepsilon_{11})dx_1} \approx \varepsilon_{21}+\omega_{21} = \varepsilon_{12}+\omega_{21} \\
\tan\alpha_2 &= \frac{(\varepsilon_{12}+\omega_{12})dx_2}{(1+\varepsilon_{22})dx_2} \approx \varepsilon_{12}-\omega_{21}
\end{aligned} \tag{1.13}$$

In the above equation, we have assumed small displacement gradient and hence $1+\varepsilon_{ij}\approx 1$. For small displacement gradient $\tan\alpha_i \approx \alpha_i$ and one can write

$$\begin{aligned}
\alpha_1 &= \varepsilon_{12}+\omega_{21} \\
\alpha_2 &= \varepsilon_{12}-\omega_{21} \\
\therefore \varepsilon_{12} &= \frac{1}{2}(\alpha_1+\alpha_2) \quad \& \quad \omega_{21} = \frac{1}{2}(\alpha_1-\alpha_2)
\end{aligned} \tag{1.14}$$

From Eq. (1.14) it is concluded that $2\varepsilon_{12}$ is the change in the angle between the elements PQ and PR after deformation. In other words, it is the engineering shear strain and ω_{21} is the rotation of the diagonal PS (see Figure 1.2) or the average rotation of the rectangular element PQSR about x_3 axis after deformation.

In summary, ε_{ij} and ω_{ij} are strain tensor and rotation tensor for small displacement gradients.

Example 1.1

Prove that the strain tensor satisfies the relation $\varepsilon_{ij,k\ell}+\varepsilon_{k\ell,ij}=\varepsilon_{ik,j\ell}+\varepsilon_{j\ell,ik}$.

This relation is known as the compatibility condition.

SOLUTION

Left-hand side $=\varepsilon_{ij,k\ell}+\varepsilon_{k\ell,ij}=\frac{1}{2}(u_{i,jk\ell}+u_{j,ik\ell}+u_{k,\ell ij}+u_{\ell,kij})$

$$\text{Right-hand side} = \varepsilon_{ik,j\ell} + \varepsilon_{j\ell,ik} = \frac{1}{2}\left(u_{i,kj\ell} + u_{k,ij\ell} + u_{j,\ell ik} + u_{\ell,jik}\right)$$

Since the order of derivative should not make any difference, $u_{i,jk\ell} = u_{i,kj\ell}$; similarly, the other three terms in the two expressions can be shown as equal. Thus the two sides of the equation are proved to be identical.

Example 1.2

Check if the following strain state is possible for an elasticity problem

$$\varepsilon_{11} = k\left(x_1^2 + x_2^2\right), \quad \varepsilon_{22} = k\left(x_2^2 + x_3^2\right), \quad \varepsilon_{12} = kx_1x_2x_3, \quad \varepsilon_{13} = \varepsilon_{23} = \varepsilon_{33} = 0$$

SOLUTION

From the compatibility condition, $\varepsilon_{ij,k\ell} + \varepsilon_{k\ell,ij} = \varepsilon_{ik,j\ell} + \varepsilon_{j\ell,ik}$ given in Example 1.1 one can write

$\varepsilon_{11,22} + \varepsilon_{22,11} = 2\varepsilon_{12,12}$ by substituting $i = 1, j = 1, k = 2, \ell = 2$.

$$\varepsilon_{11,22} + \varepsilon_{22,11} = 2k + 0 = 2k$$

$$2\varepsilon_{12,12} = 2kx_3$$

Since the two sides of the compatibility equation are not equal, the given strain state is not a possible strain state.

1.1.2 Traction and Stress Tensor

Force per unit area on a surface is called traction. To define traction at a point P (see Figure 1.3), one needs to state on which surface, going through that point, the traction is defined. The traction value at point P will change if the orientation of the surface on which the traction is defined is changed.

Figure 1.3 shows a body in equilibrium under the action of some external forces i f it is cut into two halves by a plane going through point P. In general, to keep each half of the body in equilibrium some force will exist at the cut plane. Force per unit area in the neighborhood of point P is defined as the traction at point P. If the cut plane is changed, then the traction at the same point will change. Hence, to define traction at a point its three components must be given and the plane on which it is defined must be identified. Thus, the traction can be denoted as $\mathbf{T}^{(\mathbf{n})}$, where the superscript **n** denotes the unit vector normal to the

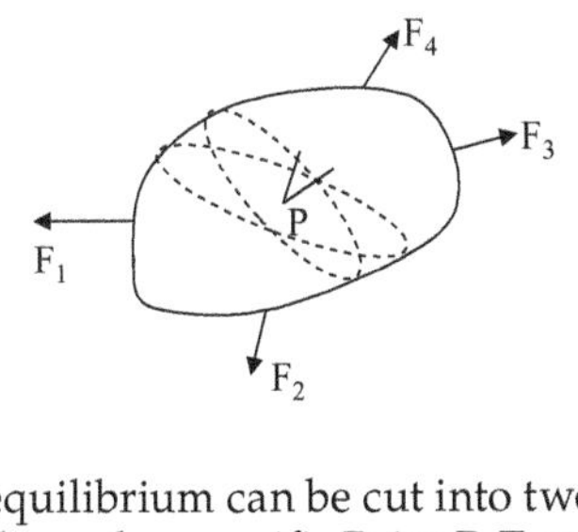

Figure 1.3 A body in equilibrium can be cut into two halves by an infinite number of planes going through a specific Point P. Two such planes are shown in the figure.

plane on which the traction is defined and where $\mathbf{T}^{(n)}$ has three components that correspond to the force per unit area in x_1-, x_2- and x_3-directions, respectively.

Stress is similar to traction – both are defined as force per unit area. The only difference is that the stress components are always defined normal or parallel to a surface while traction components are not necessarily normal or parallel to the surface. A traction $\mathbf{T}^{(n)}$ on an inclined plane is shown in Figure 1.4. Note that neither $\mathbf{T}^{(n)}$ nor its three components T_{ni} are necessarily normal or parallel to the inclined surface, but its two components σ_{nn} and σ_{ns} are perpendicular and parallel to the inclined surface, and are called normal and shear stress components.

Stress components are described by two subscripts. The first subscript indicates the plane (or normal to the plane) on which the stress component is defined and the second subscript indicates the direction of the force per unit area or stress value. Following this convention, different stress components in the $x_1x_2x_3$ coordinate system are defined in Figure 1.5.

Note that on each of the six planes, meaning the positive and negative x_1-, x_2- and x_3-planes, three stress components (one normal and two shear stress components) are defined. If the outward normal to the plane is in the positive direction, then we call the plane a positive plane, otherwise it is a negative plane. If the force direction is positive on a positive plane, or negative on a negative plane, then the stress is positive. All stress components shown on positive x_1-, x_2- and x_3-planes and negative x_1-plane in Figure 1.5 are positive stress components. Stress components on the other two negative planes are not shown to keep the figure simple. Dashed arrows show three of the stress components on the negative x_1-plane while solid arrows show the stress components on positive planes.

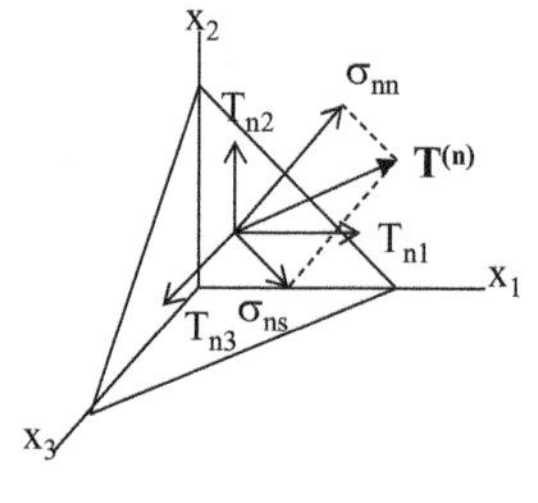

Figure 1.4 Traction $\mathbf{T}^{(n)}$ on an inclined plane can be decomposed into its three components T_{ni}, or into two components – normal and shear stress components (σ_{nn} and σ_{ns}).

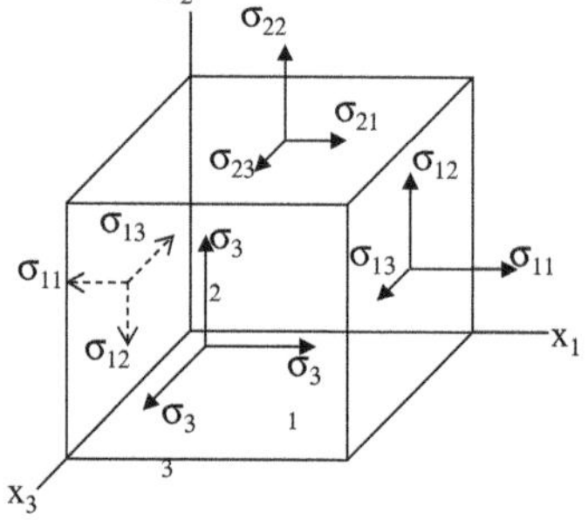

Figure 1.5 Different stress components in the $x_1x_2x_3$ coordinate system.

If the force direction and the plane direction have different signs, one positive and one negative, then the corresponding stress component is negative. Hence, in Figure 1.5, if we change the direction of the arrow of any stress component, then that stress component becomes negative.

1.1.3 Traction–Stress Relation

Let us take a tetrahedron OABC from a continuum body in equilibrium (see Figure 1.6). Forces (per unit area) acting in the x_1-direction on the four surfaces of OABC are shown in Figure 1.6. From its equilibrium in the x_1-direction one can write

$$\sum F_1 = T_{n1}A - \sigma_{11}A_1 - \sigma_{21}A_2 - \sigma_{31}A_3 + f_1V = 0 \tag{1.15}$$

where A is the area of the surface ABC; A_1, A_2 and A_3 are the areas of the other three surfaces OBC, OAC and OAB, respectively. f_1 is the body force per unit volume in the x_1-direction.

If n_j is the j-th component of the unit vector $\mathbf{n}$ that is normal to the plane ABC, then one can write $A_j = n_jA$ and $V = (Ah)/3$, where h is the height of the tetrahedron measured from the apex O. Hence, Eq. (1.15) is simplified to

$$T_{n1} - \sigma_{11}n_1 - \sigma_{21}n_2 - \sigma_{31}n_3 + f_1\frac{h}{3} = 0 \tag{1.16}$$

In the limiting case when the plane ABC passes through point O, the tetrahedron height h vanishes and Eq. (1.16) is simplified to

$$T_{n1} = \sigma_{11}n_1 + \sigma_{21}n_2 + \sigma_{31}n_3 = \sigma_{j1}n_j \tag{1.17}$$

In Eq. (1.17) summation convention (repeated index means summation) has been used.

Similarly, from the force equilibrium in x_2- and x_3-directions one can write

$$\begin{aligned} T_{n2} &= \sigma_{j2}n_j \\ T_{n3} &= \sigma_{j3}n_j \end{aligned} \tag{1.18}$$

Combining Eqs. (1.17) and (1.18), the traction-stress relation is obtained in index notation

$$T_{ni} = \sigma_{ji}n_j \tag{1.19}$$

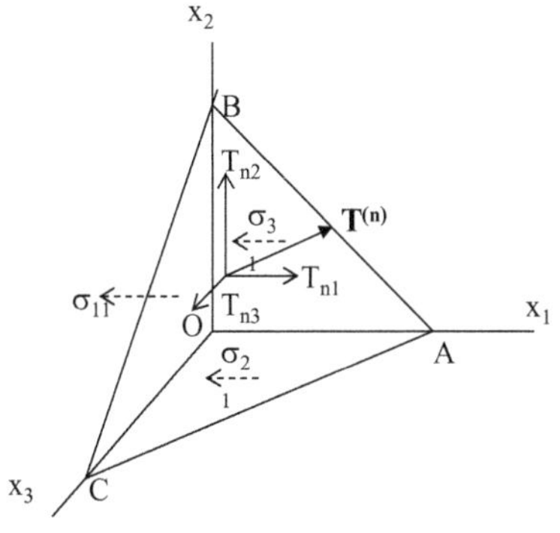

Figure 1.6 A tetrahedron showing traction components on plane ABC, and x_1-direction stress components on planes AOC, BOC and AOB.

where the free index i takes values 1, 2 and 3 to generate three equations, and the dummy index j takes values 1, 2 and 3 and is added inside each equation.

For simplicity, the subscript n of T_{ni} is omitted and is written as T_i. It is implied that the unit normal vector to the surface, on which the traction is defined, is **n**. Hence, Eq. (1.19) can be rewritten as

$$T_i = \sigma_{ji} n_j \tag{1.19a}$$

1.1.4 Equilibrium Equations

If a body is in equilibrium, the resultant force and moment on that body must be equal to zero.

1.1.4.1 *Force Equilibrium*

The resultant forces in the x_1-, x_2- and x_3-directions are equated to zero to obtain the governing equilibrium equations. First, x_1-direction equilibrium is studied. Figure 1.7 shows all forces acting in the x_1-direction on an elemental volume.

Thus, the zero resultant force in the x_1-direction gives

$$-\sigma_{11} dx_2 dx_3 + \left(\sigma_{11} + \frac{\partial \sigma_{11}}{\partial x_1} dx_1\right) dx_2 dx_3$$

$$-\sigma_{21} dx_1 dx_3 + \left(\sigma_{21} + \frac{\partial \sigma_{21}}{\partial x_2} dx_2\right) dx_1 dx_3$$

$$-\sigma_{31} dx_2 dx_1 + \left(\sigma_{31} + \frac{\partial \sigma_{31}}{\partial x_3} dx_3\right) dx_1 dx_2 + f_1\, dx_1 dx_2 dx_3 = 0$$

or,

$$\left(\frac{\partial \sigma_{11}}{\partial x_1} dx_1\right) dx_2 dx_3 + \left(\frac{\partial \sigma_{21}}{\partial x_2} dx_2\right) dx_1 dx_3 + \left(\frac{\partial \sigma_{31}}{\partial x_3} dx_3\right) dx_1 dx_2 + f_1\, dx_1 dx_2 dx_3 = 0$$

or,

$$\frac{\partial \sigma_{11}}{\partial x_1} + \frac{\partial \sigma_{21}}{\partial x_2} + \frac{\partial \sigma_{31}}{\partial x_3} + f_1 = 0$$

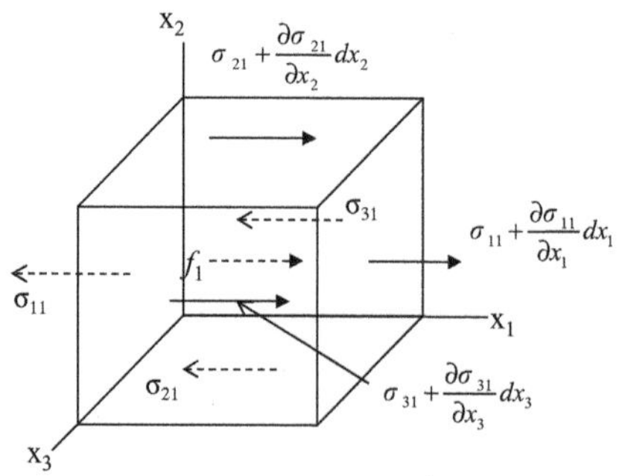

Figure 1.7 Forces acting in the x_1-direction of an elemental volume.

or,

$$\frac{\partial \sigma_{j1}}{\partial x_j} + f_1 = 0 \tag{1.20}$$

In Eq. (1.20), repeated index 'j' indicates summation.

Similarly, equilibrium in x_2- and x_3-directions gives

$$\frac{\partial \sigma_{j2}}{\partial x_j} + f_2 = 0$$
$$\frac{\partial \sigma_{j3}}{\partial x_j} + f_3 = 0 \tag{1.21}$$

Three equations of (1.20) and (1.21) can be combined in the following form:

$$\frac{\partial \sigma_{ji}}{\partial x_j} + f_i = \sigma_{ji,j} + f_i = 0 \tag{1.22}$$

The force equilibrium equations given in Eq. (1.22) are written in index notation, where the free index i takes three values 1, 2 and 3 and corresponds to three equilibrium equations, and comma (,) indicates derivative.

1.1.4.2 Moment Equilibrium

Let us now compute the resultant moment in the x_3-direction (or, in other words, moment about the x_3-axis) for the elemental volume shown in Figure 1.8.

If we calculate the moment about an axis parallel to the x_3 axis and passing through the centroid of the elemental volume shown in Figure 1.8, then only four shear stresses shown on the four sides of the volume can produce moment. Body forces in x_1- and x_2-directions will not produce any moment because the resultant body force passes through the centroid of the volume. Since the resultant moment about this axis should be zero one can write

$$\left(\sigma_{12} + \frac{\partial \sigma_{12}}{\partial x_1} dx_1\right) dx_2 dx_3 \frac{dx_1}{2} + (\sigma_{12}) dx_2 dx_3 \frac{dx_1}{2} - \left(\sigma_{21} + \frac{\partial \sigma_{21}}{\partial x_2} dx_2\right) dx_1 dx_3 \frac{dx_2}{2} - (\sigma_{21}) dx_1 dx_3 \frac{dx_2}{2} = 0$$

Ignoring the higher order terms one gets

$$2(\sigma_{12}) dx_2 dx_3 \frac{dx_1}{2} - 2(\sigma_{21}) dx_1 dx_3 \frac{dx_2}{2} = 0$$

or, $\sigma_{12} = \sigma_{21}$.

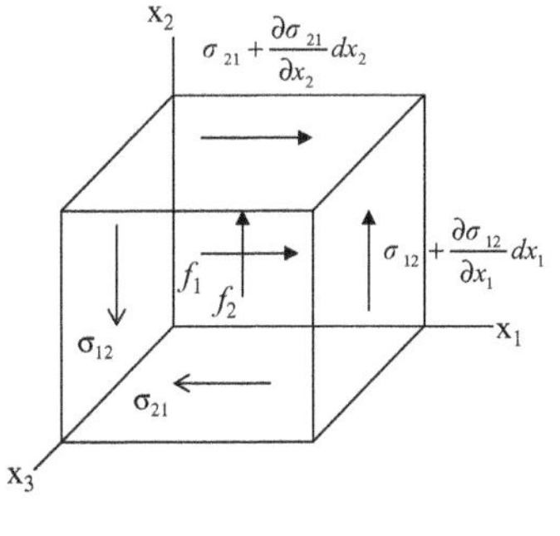

Figure 1.8 Forces on an element that may contribute to the moment in the x_3-direction.

Similarly, applying moment equilibrium about the other two axes, one can show that $\sigma_{13}=\sigma_{31}$ and $\sigma_{32}=\sigma_{23}$. Or, in index notation,

$$\sigma_{ij} = \sigma_{ji} \tag{1.23}$$

Thus, the stress tensor is symmetric. It should be noted here that if the body has internal body couple (or body moment per unit volume), then the stress tensor would not be symmetric.

Because of the symmetry of the stress tensor, Eqs. (1.19a) and (1.22) can also be written in the following form:

$$\begin{aligned} T_i &= \sigma_{ij} n_j \\ \sigma_{ij,j} + f_i &= 0 \end{aligned} \tag{1.24}$$

1.1.5 Stress Transformation

Let us now investigate how the stress components in two Cartesian coordinate systems are related.

Figure 1.9 shows an inclined plane ABC whose normal is in the $x_{1'}$-direction; thus, the $x_{2'}x_{3'}$-plane is parallel to the ABC plane. Traction $\mathbf{T}^{(1')}$ is acting on this plane. Three components of this traction in $x_{1'}$-, $x_{2'}$- and $x_{3'}$-directions are the three stress components $\sigma_{1'1'}$, $\sigma_{1'2'}$ and $\sigma_{1'3'}$, respectively. Note that the first subscript indicates the plane on which the stress is acting and the second subscript gives the stress direction.

From Eq. (1.19) one can write

$$T_{1'i} = \sigma_{ji} n_j^{(1')} = \sigma_{ji} \ell_{1'j} \tag{1.25}$$

where $n_j^{(1')} = \ell_{1'j}$ is the j-th component of the unit normal vector on plane ABC or, in other words, the direction cosines of the $x_{1'}$-axis.

Note that the dot product between $\mathbf{T}^{(1')}$ and the unit vector $\mathbf{n}^{(1')}$ gives the stress component $\sigma_{1'1'}$; therefore,

$$\sigma_{1'1'} = T_{1'i} \ell_{1'i} = \sigma_{ji} \ell_{1'j} \ell_{1'i} \tag{1.26}$$

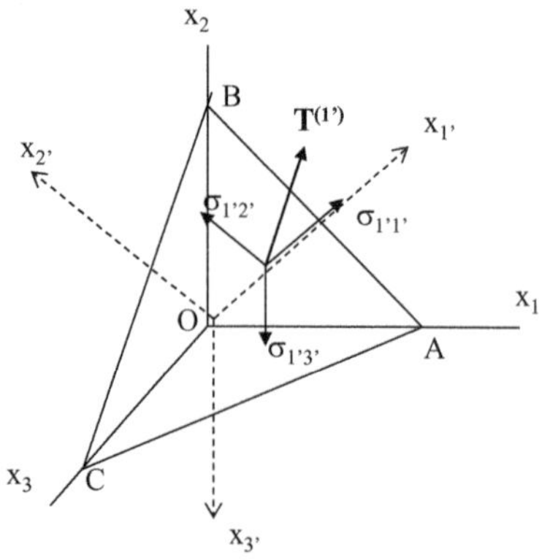

Figure 1.9 Stress components in the $x_{1'}x_{2'}x_{3'}$ coordinate system.

Similarly, the dot product between $\mathbf{T}^{(1')}$ and the unit vector $\mathbf{n}^{(2')}$ gives $\sigma_{1'2'}$ and the dot product between $\mathbf{T}^{(1')}$ and the unit vector $\mathbf{n}^{(3')}$ gives $\sigma_{1'3'}$. Thus we get

$$\begin{aligned}\sigma_{1'2'} &= T_{1'i}\ell_{2'i} = \sigma_{ji}\ell_{1'j}\ell_{2'i} \\ \sigma_{1'3'} &= T_{1'i}\ell_{3'i} = \sigma_{ji}\ell_{1'j}\ell_{3'i}\end{aligned} \tag{1.27}$$

Eqs. (1.26) and (1.27) can be written in index notation in the following form:

$$\sigma_{1'm'} = \ell_{1'j}\sigma_{ji}\ell_{m'i} \tag{1.28}$$

In Eq. (1.28) the free index m' can take values 1′, 2′ or 3′.

Similarly, from the traction vector $\mathbf{T}^{(2')}$ on a plane whose normal is in the $x_{2'}$-direction one can show

$$\sigma_{2'm'} = \ell_{2'j}\sigma_{ji}\ell_{m'i} \tag{1.29}$$

and from the traction vector $\mathbf{T}^{(3')}$ on the $x_{3'}$-plane one can derive

$$\sigma_{3'm'} = \ell_{3'j}\sigma_{ji}\ell_{m'i} \tag{1.30}$$

Eq. (1.28) through Eq. (1.30) can be combined to obtain the following equation in index notation:

$$\sigma_{n'm'} = \ell_{n'j}\sigma_{ji}\ell_{m'i}$$

Note that in the above equation i, j, m' and n' are all dummy indices and can be interchanged to obtain

$$\sigma_{m'n'} = \ell_{m'i}\sigma_{ij}\ell_{n'j} = \ell_{m'i}\ell_{n'j}\sigma_{ij} \tag{1.31}$$

1.1.5.1 *Kronecker Delta Symbol (δ_{ij}) and Permutation Symbol (ε_{ijk})*

In index notation Kronecker Delta Symbol (δ_{ij}) and Permutation Symbol (ε_{ijk}, also known as Levi-Civita symbol and alternating symbol) are often used. They are defined in the following manner:

$$\begin{aligned}\delta_{ij} &= 1 \quad \text{for} \quad i = j \\ \delta_{ij} &= 0 \quad \text{for} \quad i \neq j\end{aligned}$$

and

$\varepsilon_{ijk} =$ 1 for i, j, k having values 1, 2, 3 or 2, 3, 1 or 3, 1, 2.
$\varepsilon_{ijk} =$ −1 for i, j, k having values 3, 2, 1 or 1, 3, 2 or 2, 1, 3.
$\varepsilon_{ijk} =$ 0 for i, j, k not having three distinct values.

1.1.5.1.1 *Examples of Application of δ_{ij} and ε_{ijk}*

Note that

$$\frac{\partial x_i}{\partial x_j} = \delta_{ij}; \qquad \mathbf{e}_i \bullet \mathbf{e}_j = \delta_{ij}$$

$$\text{Det}\begin{vmatrix} a_{11} & a_{12} & a_{13} \\ a_{21} & a_{22} & a_{23} \\ a_{31} & a_{32} & a_{33}\end{vmatrix} = \varepsilon_{ijk}a_{1i}a_{2j}a_{3k}; \qquad \mathbf{b}\times\mathbf{c} = \varepsilon_{ijk}b_j c_k \mathbf{e}_i$$

where $\mathbf{e}_i$ and $\mathbf{e}_j$ are unit vectors in x_i- and x_j-directions, respectively, in the $x_1x_2x_3$ coordinate system. Also note that $\mathbf{b}$ and $\mathbf{c}$ are two vectors, while $[a]$ is a matrix.

One can prove that the following relation exists between these two symbols:

$$\varepsilon_{ijk}\varepsilon_{imn} = \delta_{jm}\delta_{kn} - \delta_{jn}\delta_{km}$$

Example 1.3

Starting from the stress transformation law, prove that $\sigma_{m'n'}\sigma_{m'n'} = \sigma_{ij}\sigma_{ij}$ where $\sigma_{m'n'}$ and σ_{ij} are stress tensors in two different Cartesian coordinate systems.

SOLUTION

$$\sigma_{m'n'}\sigma_{m'n'} = \left(\ell_{m'i}\ell_{n'j}\sigma_{ij}\right)\left(\ell_{m'p}\ell_{n'q}\sigma_{pq}\right) = \left(\ell_{m'i}\ell_{n'j}\right)\left(\ell_{m'p}\ell_{n'q}\right)\sigma_{ij}\sigma_{pq} = \left(\ell_{m'i}\ell_{m'p}\right)\left(\ell_{n'j}\ell_{n'q}\right)\sigma_{ij}\sigma_{pq}$$

$$= \delta_{ip}\delta_{jq}\sigma_{ij}\sigma_{pq} = \sigma_{ij}\sigma_{ij}$$

1.1.6 Definition of Tensor

A Cartesian tensor of order (or rank) 'r' in 'n' dimensional space is a set of n^r numbers (called the elements or components of tensor) that obey the following transformation law between two coordinate systems:

$$t_{m'n'p'q'.....} = \left(\ell_{m'i}\ell_{n'j}\ell_{p'k}\ell_{q'\ell}\ldots\right)\left(t_{ijk\ell....}\right) \tag{1.32}$$

where $t_{m'n'p'q'}\ldots$ and $t_{ijk}\ldots$ each has 'r' number of subscripts. 'r' number of direction cosines $(\ell_{m'i}\ell_{n'j}\ell_{p'k}\ell_{q'\ell}\ldots)$ are multiplied on the right-hand side.

Comparing Eq. (1.31) with the definition of tensor Eq. (1.32), one can conclude that the stress is a second rank tensor.

1.1.7 Principal Stresses and Principal Planes

Planes on which the traction vectors are normal are called principal planes. Shear stress components on the principal planes are equal to zero. Normal stresses on the principal planes are called principal stresses.

In Figure 1.10 let $\mathbf{n}$ be the unit normal vector on the principal plane ABC and λ, the principal stress value on this plane. Hence, the traction vector on plane ABC can be written as

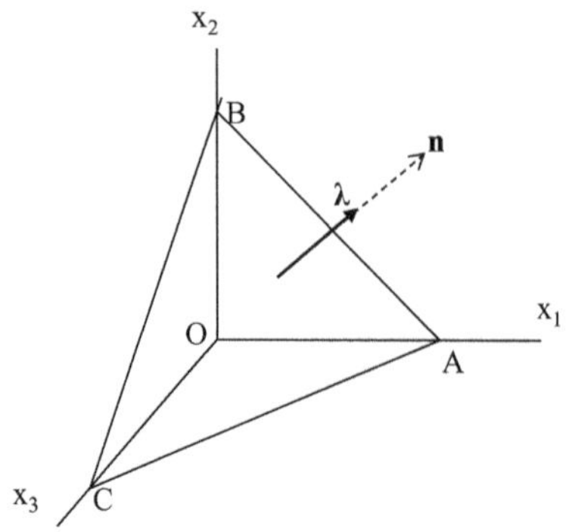

Figure 1.10 Principal stress λ on the principal plane ABC.

$$T_i = \lambda n_i$$

Again from Eq. (1.24) we get the following:

$$T_i = \sigma_{ij} n_j$$

From the above two equations one can write

$$\sigma_{ij} n_j - \lambda n_i = 0 \tag{1.33}$$

The above equation is an eigenvalue problem that can be rewritten as

$$(\sigma_{ij} - \lambda \delta_{ij}) n_j = 0 \tag{1.34}$$

The system of homogeneous equations shown in Eqs. (1.33) and (1.34) gives nontrivial solution for n_j when the determinant of the coefficient matrix is zero. Thus, for a nontrivial solution

$$Det \begin{bmatrix} (\sigma_{11} - \lambda) & \sigma_{12} & \sigma_{13} \\ \sigma_{12} & (\sigma_{22} - \lambda) & \sigma_{23} \\ \sigma_{13} & \sigma_{23} & (\sigma_{33} - \lambda) \end{bmatrix} = 0$$

or,

$$\begin{aligned} &\lambda^3 - (\sigma_{11} + \sigma_{22} + \sigma_{33})\lambda^2 + \left(\sigma_{11}\sigma_{22} + \sigma_{22}\sigma_{33} + \sigma_{33}\sigma_{11} - \sigma_{12}^2 - \sigma_{23}^2 - \sigma_{31}^2\right)\lambda \\ &\quad - \left(\sigma_{11}\sigma_{22}\sigma_{33} + 2\sigma_{12}\sigma_{23}\sigma_{31} - \sigma_{11}\sigma_{23}^2 - \sigma_{22}\sigma_{31}^2 - \sigma_{33}\sigma_{12}^2\right) = 0 \end{aligned} \tag{1.35}$$

In index notation the above equation can be written as

$$\lambda^3 - \sigma_{ii}\lambda^2 + \frac{1}{2}(\sigma_{ii}\sigma_{jj} - \sigma_{ij}\sigma_{ji})\lambda - \varepsilon_{ijk}\sigma_{1i}\sigma_{2j}\sigma_{3k} = 0 \tag{1.36}$$

In Eq. (1.36), ε_{ijk} is the permutation symbol that takes values 1, –1 or 0. If the subscripts *i, j, k* have three distinct values 1, 2 and 3, (or, 2, 3, 1 or, 3, 1, 2), respectively, then its value is 1; if the values of the subscripts are in the opposite order 3, 2, 1, (or, 2, 1, 3, or, 1, 3, 2), then ε_{ijk} is –1; and if *i, j, k* do not have three distinct values, then $\varepsilon_{ijk} = 0$.

Cubic equation (1.36) should have three roots of λ. Three roots correspond to the three principal stress values. After getting λ, the unit vector components n_j can be obtained from Eq. (1.34) and satisfying the constraint condition

$$n_1^2 + n_2^2 + n_3^2 = 1 \tag{1.37}$$

Note that for three distinct values of λ, there are three **n** values corresponding to the three principal directions.

Since the principal stress values should be independent of the starting coordinate system, the coefficients of the cubic equation (1.36) should not change irrespectively of whether we start from the $x_1x_2x_3$ coordinate system or $x'_1x'_2x'_3$ coordinate system. Thus,

$$\begin{aligned} &\sigma_{ii} = \sigma_{i'i'} \\ &\sigma_{ii}\sigma_{jj} - \sigma_{ij}\sigma_{ji} = \sigma_{i'i'}\sigma_{j'j'} - \sigma_{i'j'}\sigma_{j'i'} \\ &\varepsilon_{ijk}\sigma_{1i}\sigma_{2j}\sigma_{3k} = \varepsilon_{i'j'k'}\sigma_{1'i'}\sigma_{2'j'}\sigma_{3'k'} \end{aligned} \tag{1.38}$$

The three equations in Eq. (1.38) are known as the three stress invariants. After some algebraic manipulations, the second and third stress invariants can be further simplified, and the three stress invariants can be written as

$$\begin{aligned}
&\sigma_{ii} = \sigma_{i'i'} \\
&\sigma_{ij}\sigma_{ji} = \sigma_{i'j'}\sigma_{j'i'} \;\text{ or }\; \frac{1}{2}\sigma_{ij}\sigma_{ji} = \frac{1}{2}\sigma_{i'j'}\sigma_{j'i'} \\
&\sigma_{ij}\sigma_{jk}\sigma_{ki} = \sigma_{i'j'}\sigma_{j'k'}\sigma_{k'i'} \;\text{ or }\; \frac{1}{3}\sigma_{ij}\sigma_{jk}\sigma_{ki} = \frac{1}{3}\sigma_{i'j'}\sigma_{j'k'}\sigma_{k'i'}
\end{aligned} \tag{1.39}$$

Example 1.4

1. Obtain the principal values and principal directions for the stress tensor

$$[\sigma] = \begin{bmatrix} 2 & -4 & -6 \\ -4 & 4 & 2 \\ -6 & 2 & -2 \end{bmatrix} \text{MPa}$$

given one value of the principal stress is 9.739 MPa.
2. Compute the stress state in the $x_1'x_2'x_3'$ coordinate system. Direction cosines of $x_1'x_2'x_3'$ axes are given below:

	x_1'	x_2'	x_3'
ℓ_1	0.7285	0.6601	0.1831
ℓ_2	0.4827	−0.6843	0.5466
ℓ_3	0.4861	−0.3098	−0.8171

SOLUTION

1. Characteristic equation is obtained from Eq. (1.35)

$$\lambda^3 - (\sigma_{11} + \sigma_{22} + \sigma_{33})\lambda^2 + \left(\sigma_{11}\sigma_{22} + \sigma_{22}\sigma_{33} + \sigma_{33}\sigma_{11} - \sigma_{12}^2 - \sigma_{23}^2 - \sigma_{31}^2\right)\lambda$$
$$- \left(\sigma_{11}\sigma_{22}\sigma_{33} + 2\sigma_{12}\sigma_{23}\sigma_{31} - \sigma_{11}\sigma_{23}^2 - \sigma_{22}\sigma_{31}^2 - \sigma_{33}\sigma_{12}^2\right) = 0$$

For the given stress tensor it becomes

$$\lambda^3 - 4\lambda^2 - 60\lambda + 40 = 0$$

The above equation can be written as

$$\lambda^3 - 9.739\lambda^2 + 5.739\lambda^2 - 55.892\lambda - 4.108\lambda + 40 = 0$$
$$\Rightarrow (\lambda - 9.739)\left(\lambda^2 + 5.739\lambda - 4.108\right) = 0$$
$$\Rightarrow (\lambda - 9.739)(\lambda + 6.3825)(\lambda - 0.6435) = 0$$

whose three roots are

$$\lambda_1 = -6.3825$$
$$\lambda_2 = 9.739$$
$$\lambda_3 = 0.6435$$

These are the three principal stress values.
Principal directions are obtained from Eq. (1.34)

$$\begin{bmatrix} (\sigma_{11}-\lambda_1) & \sigma_{12} & \sigma_{13} \\ \sigma_{12} & (\sigma_{22}-\lambda_1) & \sigma_{23} \\ \sigma_{13} & \sigma_{23} & (\sigma_{33}-\lambda_1) \end{bmatrix} \begin{Bmatrix} \ell_{1'1} \\ \ell_{1'2} \\ \ell_{1'3} \end{Bmatrix} = 0$$

where $\ell_{1'1}, \ell_{1'2}, \ell_{1'3}$ are direction cosines for the principal direction associated with the principal stress λ_1.
From the above equation one can write

$$\begin{bmatrix} (2+6.3825) & -4 & -6 \\ -4 & (4+6.3825) & 2 \\ -6 & 2 & (-2+6.3825) \end{bmatrix} \begin{Bmatrix} \ell_{1'1} \\ \ell_{1'2} \\ \ell_{1'3} \end{Bmatrix} = 0$$

The second and third equations of the above system of three homogeneous equations can be solved to obtain two direction cosines in terms of the third one, as given below.

$$\ell_{1'2} = 0.1333\ell_{1'1}$$
$$\ell_{1'3} = 1.3082\ell_{1'1}$$

Normalizing the direction cosines, as shown in Eq. (1.37), we get the following:

$$1 = \ell_{1'1}^2 + \ell_{1'2}^2 + \ell_{1'3}^2 = \ell_{1'1}^2\left(1 + 0.1333^2 + 1.3082^2\right)$$
$$\Rightarrow \ell_{1'1} = \pm 0.605$$
$$\Rightarrow \ell_{1'2} = 0.1333\ell_{1'1} = \pm 0.081$$
$$\Rightarrow \ell_{1'3} = 1.3082\ell_{1'1} = \pm 0.791$$

Similarly, for the second principal stress $\lambda_2 = 9.739$ the direction cosines are $\begin{Bmatrix} \ell_{2'1} = \pm 0.657 \\ \ell_{2'2} = \mp 0.612 \\ \ell_{2'3} = \mp 0.440 \end{Bmatrix}$

And for the third principal stress $\lambda_3 = 0.6435$ the direction cosines are $\begin{Bmatrix} \ell_{3'1} = \pm 0.449 \\ \ell_{3'2} = \pm 0.787 \\ \ell_{3'3} = \mp 0.423 \end{Bmatrix}$

2. From Eq. (1.31) $\sigma_{m'n'} = \ell_{m'i}\ell_{n'j}\sigma_{ij}$

   In matrix notation $[\sigma'] = [\ell][\sigma][\ell]^T$

   where

$$[\ell]^T = \begin{bmatrix} \ell_{1'1} & \ell_{2'1} & \ell_{3'1} \\ \ell_{1'2} & \ell_{2'2} & \ell_{3'2} \\ \ell_{1'3} & \ell_{2'3} & \ell_{3'3} \end{bmatrix} = \begin{bmatrix} 0.7285 & 0.6601 & 0.1831 \\ 0.4827 & -0.6843 & 0.5466 \\ 0.4861 & -0.3098 & -0.8171 \end{bmatrix}$$

$$\text{Thus, } [\sigma'] = [\ell][\sigma][\ell]^T = \begin{bmatrix} -4.6033 & -0.8742 & 2.9503 \\ -0.8742 & 9.4682 & 1.6534 \\ 2.9503 & 1.6534 & -0.8650 \end{bmatrix} \text{MPa}$$

1.1.8 Transformation of Displacement and Other Vectors

The vector **V** can be expressed in two coordinate systems in the following manner (see Figure 1.11):

$$V_1e_1 + V_2e_2 + V_3e_3 = V_{1'}e_{1'} + V_{2'}e_{2'} + V_{3'}e_{3'} \tag{1.40}$$

If one adds the projections of V_1, V_2 and V_3 of Eq. (1.40) along the $x_{j'}$ direction, then the sum should be equal to the component $V_{j'}$. Thus,

$$V_{j'} = \ell_{j'1}V_1 + \ell_{j'2}V_2 + \ell_{j'3}V_3 = \ell_{j'k}V_k \tag{1.41}$$

Comparing Eqs. (1.41) and (1.32), one can conclude that vectors are first order tensors, or tensors of rank 1.

1.1.9 Strain Transformation

Eq. (1.7a) gives the strain expression in the $x_1x_2x_3$ coordinate system. In the $x_1'x_2'x_3'$ coordinate system the strain expression is given by $\varepsilon_{i'j'} = \frac{1}{2}\left(u_{i',j'} + u_{j',i'}\right)$. Now,

$$u_{i',j'} = \frac{\partial u_{i'}}{\partial x_{j'}} = \frac{\partial\left(\ell_{i'm}u_m\right)}{\partial x_{j'}} = \ell_{i'm}\frac{\partial\left(u_m\right)}{\partial x_{j'}} = \ell_{i'm}\frac{\partial u_m}{\partial x_n}\frac{\partial x_n}{\partial x_{j'}} = \ell_{i'm}\frac{\partial u_m}{\partial x_n}\ell_{nj'} = \ell_{i'm}\ell_{j'n}\frac{\partial u_m}{\partial x_n} \tag{1.42}$$

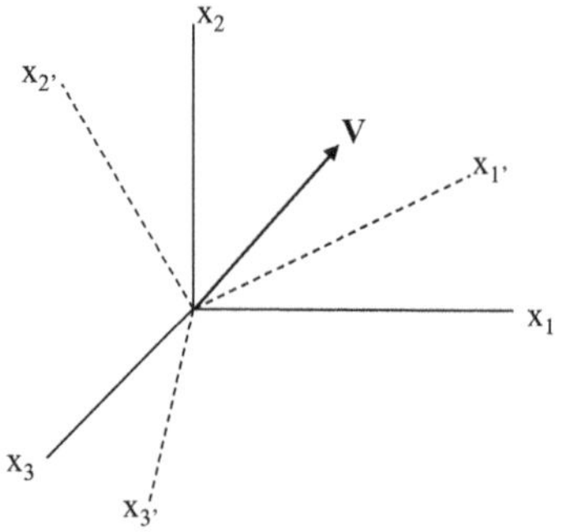

Figure 1.11 A vector **V** and two Cartesian coordinate systems.

Similarly,

$$u_{j',i'} = \ell_{j'n}\ell_{i'm}\frac{\partial u_n}{\partial x_m} \tag{1.43}$$

Hence,

$$\begin{aligned}\varepsilon_{i'j'} &= \frac{1}{2}\left(u_{i',j'} + u_{j',i'}\right) = \frac{1}{2}\left(\ell_{i'm}\ell_{j'n}u_{m,n} + \ell_{j'n}\ell_{i'm}u_{n,m}\right) \\ &= \frac{1}{2}\ell_{i'm}\ell_{j'n}\left(u_{m,n} + u_{n,m}\right) = \ell_{i'm}\ell_{j'n}\varepsilon_{mn}\end{aligned} \tag{1.44}$$

It should be noted here that the strain transformation law [Eq. (1.44)] is identical to the stress transformation law [Eq. (1.31)]. Therefore, strain is also a second rank tensor.

1.1.10 Definition of Elastic Material and Stress–Strain Relation

Elastic or conservative material can be defined in many ways as given below.

1. The material that has one-to-one correspondence between stress and strain is called elastic material.
2. The material that follows the same stress–strain path during loading and unloading is called elastic material.
3. For elastic materials, the strain energy density function (U_0) exists and it can be expressed in terms of the state of current strain only ($U_0 = U_0(\varepsilon_{ij})$), and independent of the strain history or strain path.

If the stress–strain relation is linear, then material is called linear elastic material, otherwise it is nonlinear elastic material. Note that the elastic material does not necessarily mean that the stress–strain relation is linear, and the linear stress–strain relation does not automatically mean that the material is elastic. If the stress–strain path is different during loading and unloading, then the material is no longer elastic even if the path is linear during loading and unloading. Figure 1.12 shows different stress–strain relations and indicates for each plot if the material is elastic or inelastic.

For conservative material, the external work done on the material must be equal to the total increase in the strain energy of the material. If the variation of the external work done on the body is denoted by δW, and the variation of the internal strain energy, stored in the body, is δU, then $\delta U = \delta W$. Note that δU can be expressed in terms of the strain energy density variation (δU_0), and δW can be

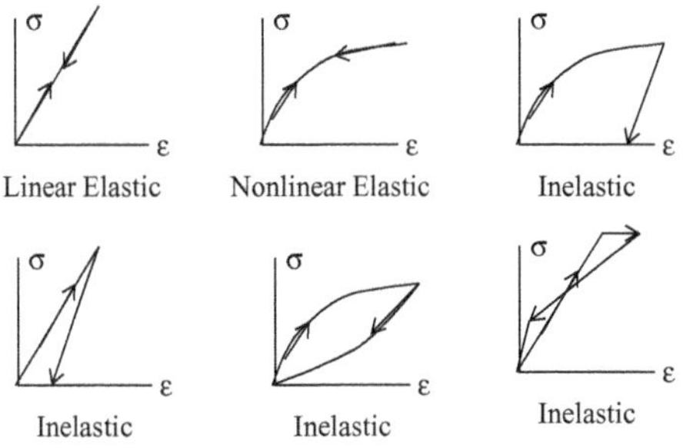

Figure 1.12 Stress–strain relations for elastic and inelastic materials.

expressed in terms of the applied body force (f_i), the surface traction (T_i) and the variation of displacement (δu_i) in the following manner:

$$\begin{aligned} \delta U &= \int_V \delta U_0 dV \\ \delta W &= \int_V f_i \delta u_i dV + \int_S T_i \delta u_i dS \end{aligned} \tag{1.45}$$

In Eq. (1.45) integrals over 'V' and 'S' indicate volume and surface integrals, respectively. From Eq. (1.45) one can write

$$\int_V \delta U_0 dV = \int_V f_i \delta u_i dV + \int_S T_i \delta u_i dS = \int_V f_i \delta u_i dV + \int_S \sigma_{ij} n_j \delta u_i dS = \int_V f_i \delta u_i dV + \int_S \left(\sigma_{ij} \delta u_i\right) n_j dS$$

Applying Gauss divergence theorem on the second integral of the right-hand side one gets

$$\begin{aligned} \int_V \delta U_0 dV &= \int_V f_i \delta u_i dV + \int_V \left(\sigma_{ij} \delta u_i\right)_{,j} dV = \int_V f_i \delta u_i dV + \int_V \left(\sigma_{ij,j} \delta u_i + \sigma_{ij} \delta u_{i,j}\right) dV \\ &= \int_V \left(f_i \delta u_i + \sigma_{ij,j} \delta u_i + \sigma_{ij} \delta u_{i,j}\right) dV = \int_V \left(\left(f_i + \sigma_{ij,j}\right) \delta u_i + \sigma_{ij} \delta u_{i,j}\right) dV \end{aligned} \tag{1.46}$$

After substituting the equilibrium equation [see Eq. (1.24)] the above equation is simplified to

$$\begin{aligned} \int_V \delta U_0 dV &= \int_V \left(\sigma_{ij} \delta u_{i,j}\right) dV = \int_V \frac{1}{2}\left(\sigma_{ij} \delta u_{i,j} + \sigma_{ij} \delta u_{i,j}\right) dV = \int_V \frac{1}{2}\left(\sigma_{ij} \delta u_{i,j} + \sigma_{ji} \delta u_{j,i}\right) dV \\ &= \int_V \sigma_{ij} \frac{1}{2}\left(\delta u_{i,j} + \delta u_{j,i}\right) dV = \int_V \sigma_{ij} \delta \varepsilon_{ij} dV \end{aligned} \tag{1.47}$$

Since Eq. (1.47) is valid for any arbitrary volume V, the integrand of the left- and right-hand sides must be equal to each other. Hence,

$$\delta U_0 = \sigma_{ij} \delta \varepsilon_{ij} \tag{1.48}$$

However, from the definition of elastic materials

$$\begin{aligned} U_0 &= U_0\left(\varepsilon_{ij}\right) \\ \therefore \delta U_0 &= \frac{\partial U_0}{\partial \varepsilon_{ij}} \delta \varepsilon_{ij} \end{aligned} \tag{1.49}$$

For arbitrary variation of $\delta \varepsilon_{ij}$ from Eqs. (1.48) and (1.49) one can write

$$\begin{aligned} \sigma_{ij} \delta \varepsilon_{ij} &= \frac{\partial U_0}{\partial \varepsilon_{ij}} \delta \varepsilon_{ij} \\ \therefore \sigma_{ij} &= \frac{\partial U_0}{\partial \varepsilon_{ij}} \end{aligned} \tag{1.50}$$

From Eq. (1.50) the stress–strain relation can be obtained by assuming some expression of U_0 in terms of the strain components (Green's approach). For example, if one assumes that the strain energy density function is a quadratic function (complete second degree polynomial) of the strain components as shown below

$$U_0 = D_0 + D_{kl}\varepsilon_{kl} + D_{klmn}\varepsilon_{kl}\varepsilon_{mn} \tag{1.51}$$

then,

$$\sigma_{ij} = \frac{\partial U_0}{\partial \varepsilon_{ij}} = D_{kl}\delta_{ik}$$

or,

$$\sigma_{ij} = \frac{\partial U_0}{\partial \varepsilon_{ij}} = D_{kl}\delta_{ik}\delta_{jl} + D_{klmn}\left(\delta_{ik}\delta_{jl}\varepsilon_{mn} + \varepsilon_{kl}\delta_{im}\delta_{jn}\right) = D_{ij} + D_{ijmn}\varepsilon_{mn} + D_{klij}\varepsilon_{kl}$$

$$= D_{ij} + \left(D_{ijkl} + D_{klij}\right)\varepsilon_{kl}$$

Substituting $(D_{ijkl} + D_{klij}) = C_{ijkl}$ and $D_{ij}=0$ (for zero stress if strain is also zero then this assumption is valid) one gets the linear stress–strain relation (or *constitutive* relation) in the following form:

$$\sigma_{ij} = C_{ijkl}\varepsilon_{kl} \tag{1.52}$$

In Cauchy's approach Eq. (1.52) is obtained by relating stress tensor with strain tensor. Note that Eq. (1.52) is a general linear relation between two second order tensors.

In the same manner, for a nonlinear (quadratic) material the stress–strain relation will be

$$\sigma_{ij} = C_{ij} + C_{ijkl}\varepsilon_{kl} + C_{ijklmn}\varepsilon_{kl}\varepsilon_{mn} \tag{1.53}$$

In Eq. (1.53) the first term on the right-hand side is the residual stress (stress for zero strain), the second term is the linear term and the third term is the quadratic term. If one follows Green's approach, then this nonlinear stress–strain relation can be obtained from a cubic expression of the strain energy density function

$$U_0 = D_{kl}\varepsilon_{kl} + D_{klmn}\varepsilon_{kl}\varepsilon_{mn} + D_{klmnpq}\varepsilon_{kl}\varepsilon_{mn}\varepsilon_{pq} \tag{1.54}$$

In this chapter we limit our analysis to linear materials only. Hence, our stress–strain relation is the one given in Eq. (1.52).

Example 1.5

In the $x_1x_2x_3$ coordinate system the stress–strain relation for a general anisotropic material is given by $\sigma_{ij} = C_{ijkm}\varepsilon_{km}$, and in the $x_{1'}x_{2'}x_{3'}$ coordinate system the stress–strain relation for the same material is given by $\sigma_{i'j'} = C_{i'j'k'm'}\varepsilon_{k'm'}$.

1. Starting from the stress and strain transformation laws, obtain a relation between C_{ijkm} and $C_{i'j'k'm'}$.
2. Is C_{ijkm} a tensor? If yes, what is its rank?

SOLUTION

1. Using Eqs. (1.52) and (1.31) one can write

$$\sigma_{i'j'} = C_{i'j'k'm'}\varepsilon_{k'm'}$$

$$\Rightarrow \ell_{i'r}\ell_{j's}\sigma_{rs} = C_{i'j'k'm'}\ell_{k'p}\ell_{m'q}\varepsilon_{pq}$$

$$\Rightarrow (\ell_{i't}\ell_{j'u})\ell_{i'r}\ell_{j's}\sigma_{rs} = (\ell_{i't}\ell_{j'u})C_{i'j'k'm'}\ell_{k'p}\ell_{m'q}\varepsilon_{pq}$$

$$\Rightarrow \delta_{tr}\delta_{us}\sigma_{rs} = (\ell_{i't}\ell_{j'u})C_{i'j'k'm'}\ell_{k'p}\ell_{m'q}\varepsilon_{pq} = \ell_{i't}\ell_{j'u}\ell_{k'p}\ell_{m'q}C_{i'j'k'm'}\varepsilon_{pq}$$

$$\Rightarrow \sigma_{tu} = (\ell_{i't}\ell_{j'u}\ell_{k'p}\ell_{m'q}C_{i'j'k'm'})\varepsilon_{pq}$$

However,

$$\sigma_{tu} = C_{tupq}\varepsilon_{pq}$$

Therefore,

$$C_{tupq} = \ell_{i't}\ell_{j'u}\ell_{k'p}\ell_{m'q}C_{i'j'k'm'} = \ell_{ti'}\ell_{uj'}\ell_{pk'}\ell_{qm'}C_{i'j'k'm'}$$

Similarly, starting with the equation $\sigma_{pq} = C_{pqrs}\varepsilon_{rs}$ and applying stress and strain transformation laws, one can show that $C_{i'j'k'm'} = \ell_{i'p}\ell_{j'q}\ell_{k'r}\ell_{m's}C_{pqrs}$

2. Clearly C_{ijkm} satisfies the transformation law for a fourth order tensor. Therefore, it is a tensor of order or rank = 4.

1.1.11 Number of Independent Material Constants

In Eq. (1.52) the coefficient values C_{ijkl} depend on the material type and are called material constants or elastic constants. Note that *i*, *j*, *k* and *l* can each take three values 1, 2 or 3. Thus, there are a total of 81 combinations possible. However, not all 81 material constants are independent. Since stress and strain tensors are symmetric we can write

$$C_{ijkl} = C_{jikl} = C_{jilk} \tag{1.55}$$

The above relation in Eq. (1.55) reduces the number of independent material constants from 81 to 36, and the stress–strain relation of Eq. (1.52) can be written in the following form:

$$\begin{Bmatrix} \sigma_{11} \\ \sigma_{22} \\ \sigma_{33} \\ \sigma_{23} \\ \sigma_{31} \\ \sigma_{12} \end{Bmatrix} = \begin{bmatrix} C_{1111} & C_{1122} & C_{1133} & C_{1123} & C_{1131} & C_{1112} \\ C_{2211} & C_{2222} & C_{2233} & C_{2223} & C_{2231} & C_{2212} \\ C_{3311} & C_{3322} & C_{3333} & C_{3323} & C_{3331} & C_{3312} \\ C_{2311} & C_{2322} & C_{2333} & C_{2323} & C_{2331} & C_{2312} \\ C_{3111} & C_{3122} & C_{3133} & C_{3123} & C_{3131} & C_{3112} \\ C_{1211} & C_{1222} & C_{1233} & C_{1223} & C_{1231} & C_{1212} \end{bmatrix} \begin{Bmatrix} \varepsilon_{11} \\ \varepsilon_{22} \\ \varepsilon_{33} \\ 2\varepsilon_{23} \\ 2\varepsilon_{31} \\ 2\varepsilon_{12} \end{Bmatrix} \tag{1.56}$$

In the above expression only six stress and strain components are shown. The other three stress and strain components are not independent because of the symmetry of stress and strain tensors. The 6×6 C-matrix is known as the constitutive matrix. For elastic materials the strain energy density function can be expressed as a function of only strain; then its double derivative will have the form

$$\frac{\partial^2 U_0}{\partial \varepsilon_{ij}\partial \varepsilon_{kl}} = \frac{\partial}{\partial \varepsilon_{ij}}\left(\frac{\partial U_0}{\partial \varepsilon_{kl}}\right) = \frac{\partial}{\partial \varepsilon_{ij}}(\sigma_{kl}) = \frac{\partial}{\partial \varepsilon_{ij}}(C_{klmn}\varepsilon_{mn}) = C_{klmn}\delta_{im}\delta_{jn} = C_{klij} \tag{1.57}$$

Similarly,

$$\frac{\partial^2 U_0}{\partial \varepsilon_{kl} \partial \varepsilon_{ij}} = \frac{\partial}{\partial \varepsilon_{kl}}\left(\frac{\partial U_0}{\partial \varepsilon_{ij}}\right) = \frac{\partial}{\partial \varepsilon_{kl}}(\sigma_{ij}) = \frac{\partial}{\partial \varepsilon_{kl}}(C_{ijmn}\varepsilon_{mn}) = C_{ijmn}\delta_{km}\delta_{ln} = C_{ijkl} \tag{1.58}$$

In Eqs (1.57) and (1.58) the order of derivative has been changed. However, since the order of derivative should not change the final results, one can conclude that $C_{ijkl} = C_{klij}$. In other words, the C-matrix of Eq. (1.56) must be symmetric. Then, the number of independent elastic constants is reduced from 36 to 21 and Eq. (1.56) is simplified to

$$\begin{Bmatrix} \sigma_1 \\ \sigma_2 \\ \sigma_3 \\ \sigma_4 \\ \sigma_5 \\ \sigma_6 \end{Bmatrix} = \begin{bmatrix} C_{11} & C_{12} & C_{13} & C_{14} & C_{15} & C_{16} \\ & C_{22} & C_{23} & C_{24} & C_{25} & C_{26} \\ & & C_{33} & C_{34} & C_{35} & C_{36} \\ & & & C_{44} & C_{45} & C_{46} \\ & \text{symm} & & & C_{55} & C_{56} \\ & & & & & C_{66} \end{bmatrix} \begin{Bmatrix} \varepsilon_1 \\ \varepsilon_2 \\ \varepsilon_3 \\ 2\varepsilon_4 \\ 2\varepsilon_5 \\ 2\varepsilon_6 \end{Bmatrix} \tag{1.59}$$

In Eq. (1.59), for simplicity, we have denoted the six stress and strain components with only one subscript (σ_i and ε_i, where i varies from 1 to 6) instead of traditional notation of two subscripts, and the material constants have been written with two subscripts instead of four.

1.1.12 Material Planes of Symmetry

Eq. (1.59) has 21 independent elastic constants in absence of any plane of symmetry. Such material is called general anisotropic material or *triclinic* material. However, if the material response is symmetric about a plane or an axis then the number of independent material constants is reduced.

1.1.12.1 One Plane of Symmetry

Let the material have only one plane of symmetry and this plane is the x_1-plane; in other words, the x_2x_3-plane whose normal is in the x_1 direction is the plane of symmetry. For this material, if the stress states $\sigma_{ij}^{(1)}$ and $\sigma_{ij}^{(2)}$ are mirror images of each other with respect to the x_1 plane then the corresponding strain states $\varepsilon_{ij}^{(1)}$ and $\varepsilon_{ij}^{(2)}$ should be the mirror images of each other with respect to the same plane. Following the notations of Eq. (1.59) we can say that the stress states $\sigma_{ij}^{(1)} = (\sigma_1, \sigma_2, \sigma_3, \sigma_4, \sigma_5, \sigma_6)$ and $\sigma_{ij}^{(2)} = (\sigma_1, \sigma_2, \sigma_3, \sigma_4, -\sigma_5, -\sigma_6)$ are of mirror symmetry with respect to the x_1-plane. Similarly, the strain states $\varepsilon_{ij}^{(1)} = (\varepsilon_1, \varepsilon_2, \varepsilon_3, \varepsilon_4, \varepsilon_5, \varepsilon_6)$ and $\varepsilon_{ij}^{(2)} = (\varepsilon_1, \varepsilon_2, \varepsilon_3, \varepsilon_4, -\varepsilon_5, -\varepsilon_6)$ are also of mirror symmetry with respect to the same plane. One can easily show by substitution that both states $(\sigma_{ij}^{(1)}, \varepsilon_{ij}^{(1)})$ and $(\sigma_{ij}^{(2)}, \varepsilon_{ij}^{(2)})$ can satisfy Eq. (1.59) only when a number of elastic constants of the C-matrix become zero as shown below:

$$\begin{Bmatrix} \sigma_1 \\ \sigma_2 \\ \sigma_3 \\ \sigma_4 \\ \sigma_5 \\ \sigma_6 \end{Bmatrix} = \begin{bmatrix} C_{11} & C_{12} & C_{13} & C_{14} & 0 & 0 \\ & C_{22} & C_{23} & C_{24} & 0 & 0 \\ & & C_{33} & C_{34} & 0 & 0 \\ & & & C_{44} & 0 & 0 \\ & \text{symm} & & & C_{55} & C_{56} \\ & & & & & C_{66} \end{bmatrix} \begin{Bmatrix} \varepsilon_1 \\ \varepsilon_2 \\ \varepsilon_3 \\ 2\varepsilon_4 \\ 2\varepsilon_5 \\ 2\varepsilon_6 \end{Bmatrix} \tag{1.60}$$

Material with one plane of symmetry is called *monoclinic material.* From the stress–strain relation [Eq. (1.60)] of monoclinic materials one can see that the number of independent elastic constants is 13 for such materials.

1.1.12.2 *Two and Three Planes of Symmetry*

In addition to the x_1 plane, if the x_2 plane is also a plane of symmetry, then two stress and strain states that are symmetric with respect to the x_2 plane must also satisfy Eq. (1.59). Note that, the stress states $\sigma_{ij}^{(1)} = (\sigma_1, \sigma_2, \sigma_3, \sigma_4, \sigma_5, \sigma_6)$ and $\sigma_{ij}^{(2)} = (\sigma_1, \sigma_2, \sigma_3, -\sigma_4, \sigma_5, -\sigma_6)$ are states of mirror symmetry with respect to the x_2 plane and the strain states $\varepsilon_{ij}^{(1)} = (\varepsilon_1, \varepsilon_2, \varepsilon_3, \varepsilon_4, \varepsilon_5, \varepsilon_6)$ and $\varepsilon_{ij}^{(2)} = (\varepsilon_1, \varepsilon_2, \varepsilon_3, -\varepsilon_4, \varepsilon_5, -\varepsilon_6)$ are states of mirror symmetry with respect to the same plane. Like before, one can easily show, by substitution, that both states $(\sigma_{ij}^{(1)}, \varepsilon_{ij}^{(1)})$ and $(\sigma_{ij}^{(2)}, \varepsilon_{ij}^{(2)})$ can satisfy Eq. (1.59) only when a number of elastic constants of the C-matrix become zero as shown below:

$$\begin{Bmatrix} \sigma_1 \\ \sigma_2 \\ \sigma_3 \\ \sigma_4 \\ \sigma_5 \\ \sigma_6 \end{Bmatrix} = \begin{bmatrix} C_{11} & C_{12} & C_{13} & 0 & C_{15} & 0 \\ & C_{22} & C_{23} & 0 & C_{25} & 0 \\ & & C_{33} & 0 & C_{35} & 0 \\ & & & C_{44} & C_{45} & 0 \\ & \text{symm} & & & C_{55} & 0 \\ & & & & & C_{66} \end{bmatrix} \begin{Bmatrix} \varepsilon_1 \\ \varepsilon_2 \\ \varepsilon_3 \\ 2\varepsilon_4 \\ 2\varepsilon_5 \\ 2\varepsilon_6 \end{Bmatrix} \tag{1.61}$$

Eq. (1.60) is the constitutive relation when the x_1-plane is the plane of symmetry and Eq. (1.61) is the constitutive relation for the x_2-plane as the plane of symmetry. Therefore, when both x_1- and x_2-planes are planes of symmetry then the C-m–atrix has only nine independent material constants as shown below:

$$\begin{Bmatrix} \sigma_1 \\ \sigma_2 \\ \sigma_3 \\ \sigma_4 \\ \sigma_5 \\ \sigma_6 \end{Bmatrix} = \begin{bmatrix} C_{11} & C_{12} & C_{13} & 0 & 0 & 0 \\ & C_{22} & C_{23} & 0 & 0 & 0 \\ & & C_{33} & 0 & 0 & 0 \\ & & & C_{44} & 0 & 0 \\ & \text{symm} & & & C_{55} & 0 \\ & & & & & C_{66} \end{bmatrix} \begin{Bmatrix} \varepsilon_1 \\ \varepsilon_2 \\ \varepsilon_3 \\ 2\varepsilon_4 \\ 2\varepsilon_5 \\ 2\varepsilon_6 \end{Bmatrix} \tag{1.62}$$

Note that Eq. (1.62) includes the case when all three planes x_1, x_2 and x_3 are planes of symmetry. Thus when two mutually perpendicular planes are planes of symmetry then the third plane automatically becomes a plane of symmetry. Materials having three planes of symmetry are called *orthotropic* (or *orthogonally anisotropic* or *orthorhombic*) materials.

1.1.12.3 *Three Planes of Symmetry and One Axis of Symmetry*

If the material has one axis of symmetry, in addition to the three planes of symmetry, then it is called *transversely isotropic* (*hexagonal*) material. If x_3-axis is the axis of symmetry, then the material response in x_1- and x_2-directions must be identical. In Eq. (1.62), if we substitute $\varepsilon_1 = \varepsilon_0$, and all other strain components $= 0$, then we get the three nonzero stress components,

$\sigma_1 = C_{11}\varepsilon_0$, $\sigma_2 = C_{12}\varepsilon_0$, $\sigma_3 = C_{13}\varepsilon_0$. Similarly, if the strain state has only one non-zero component $\varepsilon_2 = \varepsilon_0$, while all other strain components are zero, then the three normal stress components are $\sigma_1 = C_{12}\varepsilon_0$, $\sigma_2 = C_{22}\varepsilon_0$, $\sigma_3 = C_{23}\varepsilon_0$. Since x_3 axis is an axis of symmetry, σ_3 should be same for both cases, and σ_1 for the first case should be equal to the σ_2 for the second case, and vice-versa. Thus, $C_{13}=C_{23}$ and $C_{11}=C_{22}$. Then consider two more cases – 1) ε_{23} [or ε_4 in Eq. (1.62)] $=\varepsilon_0$, while all other strain components are zero, and 2) ε_{31} [or ε_5 in Eq. (1.62)] $=\varepsilon_0$, while all other strain components are zero. From Eq. (1.62) one gets $\sigma_4 = C_{44}\varepsilon_0$ for case 1 and $\sigma_5 = C_{55}\varepsilon_0$. Since x_3 axis is the axis of symmetry, σ_4 and σ_5 should have equal values; hence $C_{44}=C_{55.}$ Substituting these constraint conditions in Eq. (1.62) one gets

$$\begin{Bmatrix} \sigma_1 \\ \sigma_2 \\ \sigma_3 \\ \sigma_4 \\ \sigma_5 \\ \sigma_6 \end{Bmatrix} = \begin{bmatrix} C_{11} & C_{12} & C_{13} & 0 & 0 & 0 \\ & C_{11} & C_{13} & 0 & 0 & 0 \\ & & C_{33} & 0 & 0 & 0 \\ & & & C_{44} & 0 & 0 \\ & \text{symm} & & & C_{44} & 0 \\ & & & & & C_{66} \end{bmatrix} \begin{Bmatrix} \varepsilon_1 \\ \varepsilon_2 \\ \varepsilon_3 \\ 2\varepsilon_4 \\ 2\varepsilon_5 \\ 2\varepsilon_6 \end{Bmatrix} \tag{1.63}$$

In Eq. (1.63), although there are six different material constants, only five are independent. Considering the isotropic deformation in the x_1x_2-plane, C_{66} can be expressed in terms of C_{11} and C_{12} in the following manner:

$$C_{66} = \frac{C_{11} - C_{12}}{2} \tag{1.64}$$

1.1.12.4 *Three Planes of Symmetry and Two or Three Axes of Symmetry*

If we now add x_1 as an axis of symmetry, then following the same arguments as before one can show that in Eq. (1.63) the following three additional constraint conditions must be satisfied: $C_{12}=C_{13}$, $C_{11}=C_{33}$, and $C_{44}=C_{66}$. Thus, the constitutive matrix is simplified to:

$$\begin{Bmatrix} \sigma_1 \\ \sigma_2 \\ \sigma_3 \\ \sigma_4 \\ \sigma_5 \\ \sigma_6 \end{Bmatrix} = \begin{bmatrix} C_{11} & C_{12} & C_{12} & 0 & 0 & 0 \\ & C_{11} & C_{12} & 0 & 0 & 0 \\ & & C_{11} & 0 & 0 & 0 \\ & & & C_{66} & 0 & 0 \\ & \text{symm} & & & C_{66} & 0 \\ & & & & & C_{66} \end{bmatrix} \begin{Bmatrix} \varepsilon_1 \\ \varepsilon_2 \\ \varepsilon_3 \\ 2\varepsilon_4 \\ 2\varepsilon_5 \\ 2\varepsilon_6 \end{Bmatrix} \tag{1.65}$$

The addition of the third axis of symmetry does not modify the constitutive matrix any more. Therefore, if two mutually perpendicular axes are the axes of symmetry then the third axis must be an axis of symmetry. These materials have same material properties in all directions and are known as *isotropic* materials. From Eqs. (1.65) and (1.64) one can see that isotropic materials have only two independent material constants. This chapter will concentrate on the analysis of the linear, elastic, isotropic materials.

Example 1.6

Consider an elastic orthotropic material for which the stress–strain relations are given by the following:

$$\varepsilon_{11} = \frac{\sigma_{11}}{E_1} - \nu_{21}\frac{\sigma_{22}}{E_2} - \nu_{31}\frac{\sigma_{33}}{E_3}$$

$$\varepsilon_{22} = \frac{\sigma_{22}}{E_2} - \nu_{12}\frac{\sigma_{11}}{E_1} - \nu_{32}\frac{\sigma_{33}}{E_3}$$

$$\varepsilon_{33} = \frac{\sigma_{33}}{E_3} - \nu_{13}\frac{\sigma_{11}}{E_1} - \nu_{23}\frac{\sigma_{22}}{E_2}$$

$$2\varepsilon_{12} = \frac{\sigma_{12}}{G_{12}} \qquad 2\varepsilon_{21} = \frac{\sigma_{21}}{G_{21}}$$

$$2\varepsilon_{13} = \frac{\sigma_{13}}{G_{13}} \qquad 2\varepsilon_{31} = \frac{\sigma_{31}}{G_{31}}$$

$$2\varepsilon_{23} = \frac{\sigma_{23}}{G_{23}} \qquad 2\varepsilon_{32} = \frac{\sigma_{32}}{G_{32}}$$

where E_i is the Young's Modulus in the x_i direction, ν_{ij} and G_{ij} represent Poisson's Ratio and Shear Modulus respectively in different directions for different values of i and j.

1. How many different elastic constants do you see in the above relations?
2. How many of those do you expect to be independent?
3. How many equations or constraint relations must exist among the above material constants?
4. Do you expect G_{ij} to be equal to G_{ji} for $i \neq j$? Justify your 'yes' or 'no' answer.
5. Do you expect ν_{ij} to be equal to ν_{ji} for $i \neq j$? Justify your 'yes' or 'no' answer.
6. Write down all equations (relating the material constants) that must be satisfied.
7. If the above relations are proposed for an isotropic material, how many independent relations among the above material constants must exist? Do not write down those equations.
8. If the material is transversely isotropic, how many independent relations among the above material constants must exist? Do not write down those equations.

SOLUTION

1. 15
2. 9
3. 6
4. Yes, because ε_{ij} and σ_{ij} are symmetric
5. No, symmetry of the constitutive matrix does not require that V_{ij} to be equal to V_{ji}

6. $$\begin{Bmatrix} \varepsilon_{11} \\ \varepsilon_{22} \\ \varepsilon_{33} \\ \varepsilon_{23} \\ \varepsilon_{31} \\ \varepsilon_{12} \end{Bmatrix} = \begin{bmatrix} \frac{1}{E_1} & -\frac{\nu_{21}}{E_2} & -\frac{\nu_{31}}{E_3} & 0 & 0 & 0 \\ -\frac{\nu_{12}}{E_1} & \frac{1}{E_2} & -\frac{\nu_{32}}{E_3} & 0 & 0 & 0 \\ -\frac{\nu_{13}}{E_1} & -\frac{\nu_{23}}{E_2} & \frac{1}{E_3} & 0 & 0 & 0 \\ 0 & 0 & 0 & \frac{1}{G_{23}} & 0 & 0 \\ 0 & 0 & 0 & 0 & \frac{1}{G_{31}} & 0 \\ 0 & 0 & 0 & 0 & 0 & \frac{1}{G_{12}} \end{bmatrix} \begin{Bmatrix} \sigma_{11} \\ \sigma_{22} \\ \sigma_{33} \\ \sigma_{23} \\ \sigma_{31} \\ \sigma_{12} \end{Bmatrix}$$

From symmetry of the above matrix (also known as compliance matrix),

$$\frac{\nu_{12}}{E_1} = \frac{\nu_{21}}{E_2}$$

$$\frac{\nu_{13}}{E_1} = \frac{\nu_{31}}{E_3}$$

$$\frac{\nu_{23}}{E_2} = \frac{\nu_{32}}{E_3}$$

The other three constraint conditions are $G_{12} = G_{21}$, $G_{13} = G_{31}$, $G_{32} = G_{23}$.

7. Thirteen constraint relations must exist because isotropic material has only two independent material constants.
8. Ten relations should exist since the transversely isotropic solid has five independent material constants.

1.1.13 Stress–Strain Relation for Isotropic Materials – Green's Approach

Consider an isotropic material subjected to two states of strain as shown in Figure 1.13. The state of strain for the first case is ε_{ij} in the $x_1x_2x_3$ coordinate system as shown in the left figure and the strain state for the second case is $\varepsilon_{i'j'}$ in the $x_1x_2x_3$ coordinate system as shown in the right figure. Note that $\varepsilon_{i'j'}$ and ε_{ij} are numerically different. The numerical values for $\varepsilon_{i'j'}$ can be obtained from ε_{ij} by transforming the strain components ε_{ij} from the $x_1x_2x_3$ coordinate system to the $x_{1'}x_{2'}x_{3'}$ coordinate system as shown on the left figure of 1.13. If the strain energy density function in the $x_1x_2x_3$ coordinate system is given by $U_0(\varepsilon_{ij})$, then the strain energy density for these two cases are $U_0(\varepsilon_{ij})$ and $U_0(\varepsilon_{i'j'})$. If the material is anisotropic, then these two values can be different since the strain states are different. However, if the material is isotropic then these two values must be the same since, in the two figures of 1.13, identical numerical values of strain components ($\varepsilon_{i'j'}$) are applied in two different directions. For isotropic material equal strain values applied in two different directions should not make any difference in computing the strain energy density. For $U_0(\varepsilon_{ij})$ and $U_0(\varepsilon_{i'j'})$ to be identical U_0

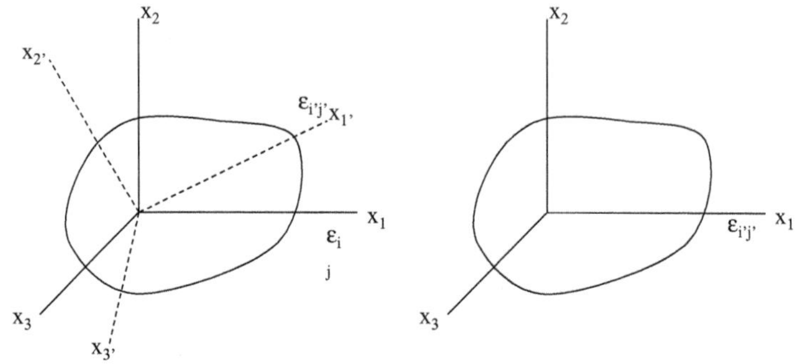

Figure 1.13 Isotropic material subjected to two states of strain.

must be a function of strain invariants because strain invariants are the only parameters that do not change when the numerical values of the strain components are changed from ε_{ij} to $\varepsilon_{i'j'}$.

Three stress invariants have been defined in Eq. (1.39). In the same manner, three strain invariants can be defined as

$$
\begin{aligned}
I_1 &= \varepsilon_{ii} \\
I_2 &= \frac{1}{2}\varepsilon_{ij}\varepsilon_{ji} \\
I_3 &= \frac{1}{3}\varepsilon_{ij}\varepsilon_{jk}\varepsilon_{ki}
\end{aligned}
\tag{1.66}
$$

Note that I_1, I_2 and I_3 are linear, quadratic and cubic functions of strain components, respectively. To obtain linear stress–strain relation from Eq. (1.50) it is clear that the strain energy density function must be a quadratic function of strain as shown below:

$$U_0 = C_1 I_1^2 + C_2 I_2$$

$$\therefore \sigma_{ij} = \frac{\partial U_0}{\partial \varepsilon_{ij}} = 2C_1 I_1 \frac{\partial I_1}{\partial \varepsilon_{ij}} + C_2 \frac{\partial I_2}{\partial \varepsilon_{ij}} = 2C_1 I_1 \delta_{ik}\delta_{kj} + C_2 \frac{1}{2}\left(\delta_{im}\delta_{jn}\varepsilon_{nm} + \varepsilon_{mn}\delta_{in}\delta_{jm}\right) \tag{1.67}$$

$$\therefore \sigma_{ij} = 2C_1 \varepsilon_{kk}\delta_{ij} + C_2 \varepsilon_{ij}$$

In Eq. (1.67), if we substitute $2C_1 = \lambda$, and $C_2 = 2\mu$, then the stress–strain relation takes the following form:

$$\sigma_{ij} = \lambda \delta_{ij}\varepsilon_{kk} + 2\mu\varepsilon_{ij} \tag{1.68}$$

In Eq. (1.68) coefficients λ and μ are known as Lame's first and second constants, respectively. This equation can be expressed in matrix form as in Eq. (1.65) to obtain

$$
\begin{Bmatrix}
\sigma_1 = \sigma_{11} \\
\sigma_2 = \sigma_{22} \\
\sigma_3 = \sigma_{33} \\
\sigma_4 = \sigma_{23} \\
\sigma_5 = \sigma_{31} \\
\sigma_6 = \sigma_{12}
\end{Bmatrix}
=
\begin{bmatrix}
\lambda + 2\mu & \lambda & \lambda & 0 & 0 & 0 \\
 & \lambda + 2\mu & \lambda & 0 & 0 & 0 \\
 & & \lambda + 2\mu & 0 & 0 & 0 \\
 & & & \mu & 0 & 0 \\
 & \text{symm} & & & \mu & 0 \\
 & & & & & \mu
\end{bmatrix}
\begin{Bmatrix}
\varepsilon_1 = \varepsilon_{11} \\
\varepsilon_2 = \varepsilon_{22} \\
\varepsilon_3 = \varepsilon_{33} \\
2\varepsilon_4 = 2\varepsilon_{23} = \gamma_{23} \\
2\varepsilon_5 = 2\varepsilon_{31} = \gamma_{31} \\
2\varepsilon_6 = 2\varepsilon_{12} = \gamma_{12}
\end{Bmatrix}
\tag{1.69}
$$

Note that the shear stress component (σ_{ij}) is simply equal to the engineering shear strain component (γ_{ij}) multiplied by Lame's second constant (μ). Therefore, Lame's second constant is the shear modulus.

Equations (1.68) and (1.69) are also known as generalized Hooke's law in three-dimension, named after Robert Hooke, who first proposed the linear stress–strain model. Eq. (1.68) can be inverted to obtain strain components in terms of the stress components as given below:

In Eq. (1.68) substituting the subscript j by i one can write

$$\sigma_{ii} = \lambda\delta_{ii}\varepsilon_{kk} + 2\mu\varepsilon_{ii} = (3\lambda + 2\mu)\varepsilon_{ii}$$

$$\therefore \varepsilon_{ii} = \frac{\sigma_{ii}}{(3\lambda + 2\mu)} \tag{1.70}$$

Substitution of Eq. (1.70) back into Eq. (1.68) gives

$$\sigma_{ij} = \lambda\delta_{ij}\frac{\sigma_{kk}}{(3\lambda + 2\mu)} + 2\mu\varepsilon_{ij}$$

or,

$$\varepsilon_{ij} = \frac{\sigma_{ij}}{2\mu} - \delta_{ij}\frac{\lambda\sigma_{kk}}{2\mu(3\lambda + 2\mu)} \tag{1.71}$$

1.1.13.1 *Hooke's Law in Terms of Young's Modulus and Poisson's Ratio*

In undergraduate mechanics courses, strains are expressed in terms of stress components, Young's modulus (E), Poisson's ratio (V) and shear modulus or Kirchhoff's modulus (μ) in the following form:

$$\begin{aligned}
\varepsilon_{11} &= \frac{\sigma_{11}}{E} - \frac{\nu\sigma_{22}}{E} - \frac{\nu\sigma_{33}}{E}\\
\varepsilon_{22} &= \frac{\sigma_{22}}{E} - \frac{\nu\sigma_{11}}{E} - \frac{\nu\sigma_{33}}{E}\\
\varepsilon_{33} &= \frac{\sigma_{33}}{E} - \frac{\nu\sigma_{22}}{E} - \frac{\nu\sigma_{11}}{E}\\
2\varepsilon_{12} &= \gamma_{12} = \frac{\sigma_{12}}{\mu} = \frac{2(1+\nu)\sigma_{12}}{E}\\
2\varepsilon_{23} &= \gamma_{23} = \frac{\sigma_{23}}{\mu} = \frac{2(1+\nu)\sigma_{23}}{E}\\
2\varepsilon_{31} &= \gamma_{31} = \frac{\sigma_{31}}{\mu} = \frac{2(1+\nu)\sigma_{31}}{E}
\end{aligned} \tag{1.72}$$

In the above equation, the relation between Young's modulus (E), Poisson's ratio (V) and shear modulus (μ) has been incorporated. Eq. (1.72) can be expressed in index notation:

$$\varepsilon_{ij} = \frac{1+\nu}{E}\sigma_{ij} - \frac{\nu}{E}\delta_{ij}\sigma_{kk} \tag{1.73}$$

Equating the right-hand sides of Eqs. (1.71) and (1.73), Lame's constants can be expressed in terms of Young's modulus and Poisson's ratio. Similarly, the bulk modulus $K = \frac{\sigma_{ii}}{3\varepsilon_{ii}}$ can be expressed in terms of Lame's constants from Eq. (1.70).

Example 1.7

For an isotropic material, obtain the bulk modulus K in terms of (1) E and V, (2) λ and μ.

SOLUTION

1. From Eq. (1.70), $\sigma_{ii} = (3\lambda + 2\mu)\varepsilon_{ii}$; hence, $K = \frac{\sigma_{ii}}{3\varepsilon_{ii}} = \frac{3\lambda + 2\mu}{3}$
2. From Eq. (1.73),

$$\varepsilon_{ii} = \frac{1+\nu}{E}\sigma_{ii} - \frac{\nu}{E}\delta_{ii}\sigma_{kk} = \frac{1+\nu}{E}\sigma_{ii} - \frac{3\nu}{E}\sigma_{kk} = \frac{1-2\nu}{E}\sigma_{ii}$$

Hence, $K = \frac{\sigma_{ii}}{3\varepsilon_{ii}} = \frac{E}{3(1-2\nu)}$

Since the isotropic material has only two independent elastic constants, any of the five commonly used elastic constants (λ, μ, E, ν and K) can be expressed in terms of any other two elastic constants, as shown in Table 1.1.

1.1.14 Navier's Equation of Equilibrium

Substituting the stress–strain relation [Eq. (1.68)] into the equilibrium equation [Eq. (1.24)] one gets

$$\left(\lambda\delta_{ij}\varepsilon_{kk} + 2\mu\varepsilon_{ij}\right)_{,j} + f_i = 0$$

$$\Rightarrow \left(\lambda\delta_{ij}u_{k,k} + 2\mu\frac{1}{2}\left[u_{i,j} + u_{j,i}\right]\right)_{,j} + f_i = 0$$

$$\Rightarrow \lambda\delta_{ij}u_{k,kj} + \mu\left[u_{i,jj} + u_{j,ij}\right] + f_i = 0 \tag{1.74}$$

$$\Rightarrow \lambda u_{k,ki} + \mu\left[u_{i,jj} + u_{j,ji}\right] + f_i = 0$$

$$\Rightarrow (\lambda + \mu)u_{j,ji} + \mu u_{i,jj} + f_i = 0$$

In the vector form the above equation can be written as

$$(\lambda + \mu)\underline{\nabla}(\underline{\nabla} \cdot \mathbf{u}) + \mu\nabla^2\mathbf{u} + \mathbf{f} = 0 \tag{1.75}$$

Because of the vector identity $\nabla^2\mathbf{u} = \underline{\nabla}(\underline{\nabla} \cdot \mathbf{u}) - \underline{\nabla} \times \underline{\nabla} \times \mathbf{u}$, Eq. (1.75) can also be written as

$$(\lambda + 2\mu)\underline{\nabla}(\underline{\nabla} \cdot \mathbf{u}) - \mu\underline{\nabla} \times \underline{\nabla} \times \mathbf{u} + \mathbf{f} = 0 \tag{1.76}$$

In Eqs. (1.75) and (1.76) the dot (•) is used to indicate scalar or dot product and the cross (×) is used to indicate the vector or cross product. In index notations the above two equations can also be written as

Table 1.1: Relations between Different Elastic Constants for Isotropic Materials

	λ	μ	E	ν	K
λ, μ	–	–	$\frac{\mu(3\lambda+2\mu)}{\lambda+\mu}$	$\frac{\lambda}{2(\lambda+\mu)}$	$\frac{3\lambda+2\mu}{3}$
λ, E	–	$\frac{(E-3\lambda)+\sqrt{(E-3\lambda)^2+8\lambda E}}{4}$	–	$\frac{-(E+\lambda)+\sqrt{(E+\lambda)^2+8\lambda^2}}{4\lambda}$	$\frac{(E+3\lambda)+\sqrt{(E+3\lambda)^2-4\lambda E}}{6}$
λ, ν	–	$\frac{\lambda(1-2\nu)}{2\nu}$	$\frac{\lambda(1+\nu)(1-2\nu)}{\nu}$	–	$\frac{\lambda(1+\nu)}{3\nu}$
λ, K	–	$\frac{3(K-\lambda)}{2}$	$\frac{9K(K-\lambda)}{3K-\lambda}$	$\frac{\lambda}{3K-\lambda}$	–
μ, E	$\frac{(2\mu-E)\mu}{E-3\mu}$	–	–	$\frac{E-2\mu}{2\mu}$	$\frac{\mu E}{3(3\mu-E)}$
μ, ν	$\frac{2\mu\nu}{1-2\nu}$	–	$2\mu(1+\nu)$	–	$\frac{2\mu(1+\nu)}{3(1-2\nu)}$
μ, K	$\frac{3K-2\mu}{3}$	–	$\frac{9K\mu}{3K+\mu}$	$\frac{3K-2\mu}{2(3K+\mu)}$	–
E, ν	$\frac{\nu E}{(1+\nu)(1-2\nu)}$	$\frac{E}{2(1+\nu)}$	–	–	$\frac{E}{3(1-2\nu)}$
E, K	$\frac{3K(3K-E)}{9K-E}$	$\frac{3KE}{9K-E}$	–	$\frac{3K-E}{6K}$	–
ν, K	$\frac{3K\nu}{1+\nu}$	$\frac{3K(1-2\nu)}{2(1+\nu)}$	$3K(1-2\nu)$	–	–

$$(\lambda+\mu)u_{j,ji}+\mu u_{i,jj}+f_i=0$$

or, (1.77)

$$(\lambda+2\mu)u_{j,ji}-\mu\varepsilon_{ijk}\varepsilon_{kmn}u_{n,mj}+f_i=0$$

where ε_{ijk} and ε_{kmn} are permutation symbols, defined in Eq. (1.36). The equilibrium equations, expressed in terms of the displacement components [Eqs. (1.75–77)], are known as Navier's equation.

Example 1.8

If a linear elastic isotropic body does not have any body force, prove that

1. The volumetric strain is harmonic $(\varepsilon_{ii,jj}=0)$, and
2. The displacement field is biharmonic $(u_{i,jjkk}=0)$

SOLUTION

1. From Eq. (1.77) for zero body force one can write

$$(\lambda+\mu)u_{j,ji}+\mu u_{i,jj}=0$$

$$\Rightarrow(\lambda+\mu)u_{j,jii}+\mu u_{i,jji}=0$$

Note that

$$u_{i,jji}=u_{i,ijj}=\varepsilon_{ii,jj}$$

and

$$u_{j,jii}=\varepsilon_{jj,ii}=\varepsilon_{ii,jj}$$

The above equation is simplified to $(\lambda+2\mu)\varepsilon_{ii,jj}=0$.
Since $(\lambda+2\mu)\neq 0$, ε_{ii} must be harmonic.

2. Again from Eq. (1.77) for zero body force one can write

$$(\lambda+\mu)u_{j,ji}+\mu u_{i,jj}=0$$

$$\Rightarrow\left((\lambda+\mu)u_{j,ji}+\mu u_{i,jj}\right)_{,kk}=(\lambda+\mu)u_{j,jikk}+\mu u_{i,jjkk}=0$$

From part (a), $\varepsilon_{jj,kk}=u_{j,jkk}=0$

Hence, $u_{j,jkki}=u_{j,jikk}=0$. Substituting it into the above equation one gets

$$(\lambda+\mu)u_{j,jikk}+\mu u_{i,jjkk}=\mu u_{i,jjkk}=0$$

$$\Rightarrow u_{i,jjkk}=0$$

Example 1.9

Obtain the governing equation of equilibrium in terms of displacement for a material whose stress–strain relation is given by

$$\sigma_{ij}=\alpha_{ijkl}\varepsilon_{km}\varepsilon_{ml}+\delta_{ij}\gamma$$

where α_{ijkl} are material properties that are constants over the entire region, and γ is the residual state of stress that varies from point to point.

SOLUTION

Governing equation

$$\sigma_{ij,j}+f_i=0$$

$$\Rightarrow\alpha_{ijkl}\left(\varepsilon_{km}\varepsilon_{ml}\right)_{,j}+\delta_{ij}\gamma_{,j}+f_i=0$$

$$\Rightarrow\alpha_{ijkl}\left(\varepsilon_{km,j}\varepsilon_{ml}+\varepsilon_{km}\varepsilon_{ml,j}\right)+\gamma_{,i}+f_i=0$$

$$\Rightarrow\frac{1}{4}\alpha_{ijkl}\left\{\left(u_{k,mj}+u_{m,kj}\right)\left(u_{m,l}+u_{l,m}\right)+\left(u_{k,m}+u_{m,k}\right)\left(u_{m,lj}+u_{l,mj}\right)\right\}+\gamma_{,i}+f_i=0$$

1.1.15 Fundamental Equations of Elasticity in Other Coordinate Systems

All equations derived so far have been expressed in the Cartesian coordinate system. Although the majority of elasticity problems can be solved in the

Cartesian coordinate system for some problem geometries, such as axisymmetric problems, cylindrical and spherical coordinate systems are better suited for defining the problem and/or solving it. If the equation is given in vector form [Eqs. (1.75) and (1.76)], then it can be used in any coordinate system with appropriate definitions of the vector operators in that coordinate system; however, when it is expressed in index notation in the Cartesian coordinate system [Eq. (1.77)], then that expression cannot be used in cylindrical or spherical coordinate systems. In Table 1.2, different vector operations, strain–displacement relations and equilibrium equations are given in the three coordinate systems that are shown in Figure 1.14.

1.2 TIME DEPENDENT PROBLEMS OR DYNAMIC PROBLEMS

In all equations derived above it is assumed that the body is in static equilibrium. Therefore the resultant force acting on the body is equal to zero. If the body is subjected to a nonzero resultant force, then it will have an acceleration $\ddot{\mathbf{u}}$ (time derivatives are denoted by dots over the variable, two dots mean double derivative) and the equilibrium equation (Eq.1.24) will be replaced by the following governing equation of motion,

$$\sigma_{ij,j} + f_i = \rho \ddot{u}_i \tag{1.78}$$

In the above equation ρ is the mass density.

Therefore, Navier's equation [Eq. (1.76)] for the dynamic case takes the following form:

$$(\lambda + 2\mu)\nabla(\nabla.\mathbf{u}) - \mu\nabla\times\nabla\times\mathbf{u} + \mathbf{f} = \rho\ddot{u} \tag{1.79}$$

1.2.1 Some Simple Dynamic Problems

Example 1.10

Time dependent normal load on the surface of an elastic half-space is shown in Figure 1.15. Compute the displacement and stress field for this problem geometry.

SOLUTION

Since the loading and the problem geometry are independent of x_2 and x_3 coordinates, the solution should be a function of x_1 only. Moreover, the solution should be symmetric about the x_1 axis and this axis can be moved up or down, front or back without violating the symmetry conditions; hence, u_2 and u_3 components of displacement must be equal to zero and the solution must have only one component of displacement, u_1, that will be a function of x_1 only.

Substituting $\mathbf{f} = 0$ (no body force), $u_2 = u_3 = 0$ and $u_1 = u_1(x_1)$ in Eq. (1.79) one gets

$$(\lambda + 2\mu)u_{1,11} = \rho\ddot{u}_1$$

$$\rightarrow u_{1,11} = \frac{\rho}{\lambda + 2\mu}\ddot{u}_1 = \frac{1}{c_P^2}\ddot{u}_1 \tag{1.80}$$

Eq. (1.80) is a one-dimensional wave equation that has a solution of the form

Table 1.2: Important Equations in Different Coordinate Systems (Moon and Spencer, 1965)

Equations	Cartesian Coordinate System	Cylindrical Coordinate System	Spherical Coordinate System
Grad $\phi = \underline{\nabla}\phi$	$\phi_{,i}=\phi_{,1}e_1+\phi_{,2}e_2+\phi_{,3}e_3$	$\frac{\partial\phi}{\partial r}e_r+\frac{\partial\phi}{\partial z}e_z+\frac{1}{r}\frac{\partial\phi}{\partial\theta}e_\theta$	$\frac{\partial\phi}{\partial r}e_r+\frac{1}{r}\frac{\partial\phi}{\partial\beta}e_\beta+\frac{1}{r\sin\beta}\frac{\partial\phi}{\partial\theta}e_\theta$
Div $\boldsymbol{\psi} = \underline{\nabla}\cdot\boldsymbol{\psi}$	$\psi_{i,i}=\psi_{1,1}+\psi_{2,2}+\psi_{3,3}$	$\frac{\partial\psi_r}{\partial r}+\frac{\partial\psi_z}{\partial z}+\frac{1}{r}\frac{\partial\psi_\theta}{\partial\theta}$	$\frac{\partial\psi_r}{\partial r}+\frac{2}{r}\psi_r+\frac{1}{r}\frac{\partial\psi_\beta}{\partial\beta}$ $+\frac{\cot\beta}{r}\psi_\beta+\frac{1}{r\sin\beta}\frac{\partial\psi_\theta}{\partial\theta}$
Curl $\boldsymbol{\psi} = \underline{\nabla}\times\boldsymbol{\psi}$	$\varepsilon_{ijk}\psi_{k,j}=$ $\begin{vmatrix} e_1 & e_2 & e_3 \\ \frac{\partial}{\partial x_1} & \frac{\partial}{\partial x_2} & \frac{\partial}{\partial x_3} \\ \psi_1 & \psi_2 & \psi_3 \end{vmatrix}$	$\frac{1}{r}\begin{vmatrix} e_r & re_\theta & e_z \\ \frac{\partial}{\partial r} & \frac{\partial}{\partial\theta} & \frac{\partial}{\partial z} \\ \psi_r & r\psi_\theta & \psi_z \end{vmatrix}$	$\frac{1}{r^2\sin\beta}\begin{vmatrix} e_r & re_\beta & r\sin\beta e_\theta \\ \frac{\partial}{\partial r} & \frac{\partial}{\partial\beta} & \frac{\partial}{\partial\theta} \\ \psi_r & r\psi_\beta & r\sin\beta\psi_\theta \end{vmatrix}$
Strain-Displacement Relation	Eq. (1.7a) $\varepsilon_{ij}=\frac{1}{2}\left(u_{i,j}+u_{j,i}\right)$	$\varepsilon_{rr}=\frac{\partial u_r}{\partial r}$	$\varepsilon_{rr}=\frac{\partial u_r}{\partial r}$
		$\varepsilon_{\theta\theta}=\frac{1}{r}\frac{\partial u_\theta}{\partial\theta}+\frac{u_r}{r}$	$\varepsilon_{\beta\beta}=\frac{1}{r}\frac{\partial u_\beta}{\partial\beta}+\frac{u_r}{r}$
		$\varepsilon_{zz}=\frac{\partial u_z}{\partial z}$	$\varepsilon_{\theta\theta}=\frac{1}{r\sin\beta}\frac{\partial u_\theta}{\partial\theta}+\frac{u_r}{r}+\frac{u_\beta}{r}\cot\beta$
		$\varepsilon_{rz}=\frac{1}{2}\left(\frac{\partial u_r}{\partial z}+\frac{\partial u_z}{\partial r}\right)$	$2\varepsilon_{r\beta}=\frac{1}{r}\frac{\partial u_r}{\partial\beta}+\frac{\partial u_\beta}{\partial r}-\frac{u_\beta}{r}$
		$\varepsilon_{r\theta}=\frac{1}{2}\left(\frac{1}{r}\frac{\partial u_r}{\partial\theta}-\frac{u_\theta}{r}+\frac{\partial u_\theta}{\partial r}\right)$	$2\varepsilon_{r\theta}=\frac{1}{r\sin\beta}\frac{\partial u_r}{\partial\theta}-\frac{u_\theta}{r}+\frac{\partial u_\theta}{\partial r}$
		$\varepsilon_{z\theta}=\frac{1}{2}\left(\frac{1}{r}\frac{\partial u_z}{\partial\theta}+\frac{\partial u_\theta}{\partial z}\right)$	$2\varepsilon_{\beta\theta}=\frac{1}{r\sin\beta}\frac{\partial u_\beta}{\partial\theta}+\frac{1}{r}\frac{\partial u_\theta}{\partial\beta}-\frac{u_\theta}{r}\cot\beta$

(Continued)

Table 1.2 (Continued): Important Equations in Different Coordinate Systems (Moon and Spencer, 1965)

Equations	Cartesian Coordinate System	Cylindrical Coordinate System	Spherical Coordinate System
Equil-ibrium Equations	Eqs. (1.24 &1.22) $\sigma_{ij,j} + f_i = 0$	$\frac{\partial \sigma_{rr}}{\partial r} + \frac{\partial \sigma_{rz}}{\partial z} + \frac{1}{r}\frac{\partial \sigma_{r\vartheta}}{\partial \theta} + \frac{1}{r}(\sigma_{rr} - \sigma_{\theta\theta}) + f_r = 0$	$\frac{\partial \sigma_{rr}}{\partial r} + \frac{1}{r}\frac{\partial \sigma_{r\beta}}{\partial \beta} + \frac{1}{r\sin\beta}\frac{\partial \sigma_{r\vartheta}}{\partial \theta} + \frac{1}{r}\left[2\sigma_{rr} + (\cot\beta)\sigma_{r\beta} - \sigma_{\beta\beta} - \sigma_{\theta\theta}\right] + f_r = 0$
		$\frac{\partial \sigma_{rz}}{\partial r} + \frac{\partial \sigma_{zz}}{\partial z} + \frac{1}{r}\frac{\partial \sigma_{z\vartheta}}{\partial \theta} + \frac{1}{r}\sigma_{rz} + f_z = 0$	$\frac{\partial \sigma_{r\beta}}{\partial r} + \frac{1}{r}\frac{\partial \sigma_{\beta\beta}}{\partial \beta} + \frac{1}{r\sin\beta}\frac{\partial \sigma_{\beta\vartheta}}{\partial \theta} + \frac{1}{r}\left[3\sigma_{r\beta} + (\cot\theta)(\sigma_{\beta\beta} - \sigma_{\theta\theta})\right] + f_\beta = 0$
		$\frac{\partial \sigma_{r\theta}}{\partial r} + \frac{\partial \sigma_{z\theta}}{\partial z} + \frac{1}{r}\frac{\partial \sigma_{\theta\vartheta}}{\partial \theta} + \frac{2}{r}\sigma_{r\theta} + f_\theta = 0$	$\frac{\partial \sigma_{r\theta}}{\partial r} + \frac{1}{r}\frac{\partial \sigma_{\beta\theta}}{\partial \beta} + \frac{1}{r\sin\beta}\frac{\partial \sigma_{\theta\vartheta}}{\partial \theta} + \frac{2\sigma_{\beta\theta}\cot\beta + 3\sigma_{r\theta}}{r} + f_\theta = 0$

ϕ and $\boldsymbol{\psi}$ are a scalar and a vector function, respectively.
Source: P. Morse and D. E. Spencer, Vectors, D. Van Nostrand Company, Inc., Princeton, NJ, 1965.

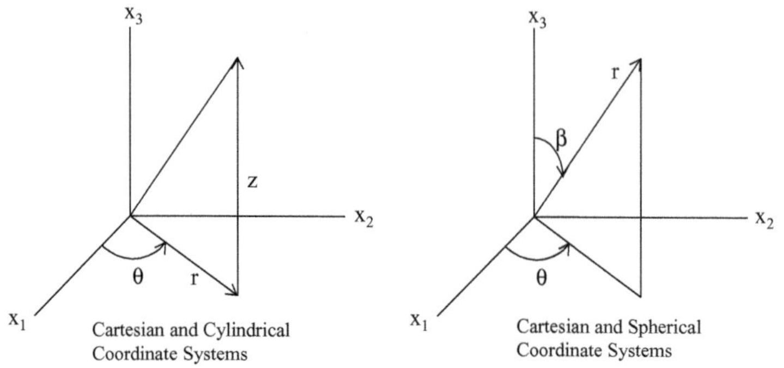

Figure 1.14 Cartesian ($x_1x_2x_3$), cylindrical ($r\theta z$) and spherical ($r\beta\theta$) coordinate systems.

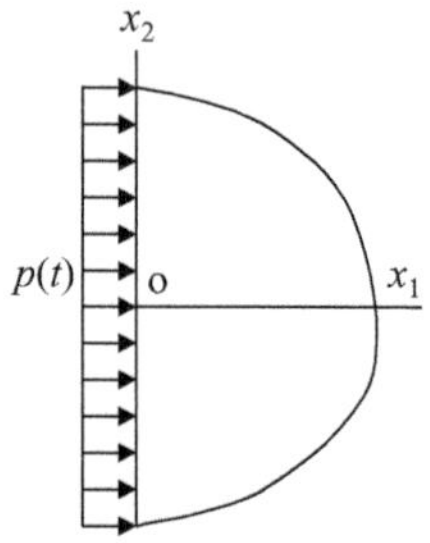

Figure 1.15 A time dependent load $p(t)$ is applied on the surface ($x_1=0$) of an elastic half-space ($x_1>0$).

$$u_1 = F\left(t - \frac{x_1}{c_P}\right) + G\left(t + \frac{x_1}{c_P}\right) \tag{1.81}$$

where

$$c_P = \sqrt{\frac{\lambda + 2\mu}{\rho}} \tag{1.82}$$

Note that when Eq. (1.81) is substituted into Eq. (1.80), both left- and right-hand sides of Eq. (1.80) become equal to

$$u_{1,11} = \frac{1}{c_P^2}\ddot{u}_1 = \frac{1}{c_P^2}\left\{F''\left(t - \frac{x_1}{c_P}\right) + G''\left(t + \frac{x_1}{c_P}\right)\right\}$$

where F'' and G'' represent double derivatives of the functions with respect to their arguments.

From the initial condition of the problem, one can say that at $t=0$, the displacement u_1 and the velocity $\dot{u}_1$ must be equal to zero for all $x_1>0$. Hence, for $x_1>0$

$$u_1\big|_{t=0} = F\left(-\frac{x_1}{c_P}\right) + G\left(\frac{x_1}{c_P}\right) = 0 \tag{1.83}$$

$$\dot{u}_1\Big|_{t=0} = F'\left(-\frac{x_1}{c_P}\right) + G'\left(\frac{x_1}{c_P}\right) \tag{1.84}$$

Taking the derivative of Eq. (1.83) with respect to x_1 gives

$$-F'\left(-\frac{x_1}{c_P}\right) + G'\left(\frac{x_1}{c_P}\right) = 0 \tag{1.85}$$

From Eqs. (1.84) and (1.85) one gets

$$F'\left(-\frac{x_1}{c_P}\right) = G'\left(\frac{x_1}{c_P}\right) = 0 \tag{1.86}$$

Therefore, $F\left(-\frac{x_1}{c_P}\right) = A,\ \ G\left(\frac{x_1}{c_P}\right) = B$, where A and B are two constants.

Since Eq. (1.83) must be satisfied for $x_1 > 0$, constant B must be equal to $-A$. In other words, if the arguments of F is negative and G is positive, then the function values are A and $-A$ respectively, where A can be any positive or negative constant. From Eq. (1.81) one can see that for $t < \frac{x_1}{c_P}$, the argument $\left(t - \frac{x_1}{c_P}\right)$ of function F is negative and the argument $\left(t + \frac{x_1}{c_P}\right)$ of function G is positive. Thus,

$u_1 = F\left(t - \frac{x_1}{c_P}\right) + G\left(t + \frac{x_1}{c_P}\right) = 0$, for $t < \frac{x_1}{c_P}$. For $t > \frac{x_1}{c_P}$, both arguments $\left(t - \frac{x_1}{c_P}\right)$ and $\left(t + \frac{x_1}{c_P}\right)$ are greater than zero. Then the function G should be a constant but the function F is not necessarily a constant. From the initial condition we have proven that if the argument of F is less than zero then only F should be a constant. Absorbing the constant value of $G\left(t + \frac{x_1}{c_P}\right)$, into F and defining a new function f as

$$f(t, x_1) = F\left(t - \frac{x_1}{c_P}\right) + G\left(t + \frac{x_1}{c_P}\right) = F\left(t - \frac{x_1}{c_P}\right) - A = f\left(t - \frac{x_1}{c_P}\right) \tag{1.86a}$$

one can write the displacement field in the following manner:

$$\begin{aligned} u_1 &= 0 && \text{for} \quad t < \frac{x_1}{c_P} \\ u_1 &= f\left(t - \frac{x_1}{c_P}\right) && \text{for} \quad t \geq \frac{x_1}{c_P} \end{aligned} \tag{1.87}$$

To obtain the function f, we utilize the boundary condition at $x_1 = 0$, $\sigma_{11} = -p(t)$. Thus,

$$\sigma_{11} = (\lambda + 2\mu)\varepsilon_{11} + \lambda(\varepsilon_{22} + \varepsilon_{33})$$

$$= (\lambda + 2\mu)u_{1,1} = -\frac{(\lambda + 2\mu)}{c_P} f'(t) = -p(t)$$

$$\Rightarrow f'(t) = \frac{c_P}{(\lambda + 2\mu)} p(t) = \frac{1}{\rho c_P} p(t) \tag{1.88}$$

$$\Rightarrow f(t) = \frac{1}{\rho c_P} \int_0^t p(s) ds$$

The lower limit of the integral is zero since $f(0)=0$, from Eq. (1.87).

Combining Eqs. (1.87) and (1.88) one gets

$$\begin{aligned} u_1 &= f\left(t - \frac{x_1}{c_P}\right) = \frac{1}{\rho c_P} \int_0^{t - \frac{x_1}{c_P}} p(s) ds \quad \text{for} \quad t \geq \frac{x_1}{c_P} \\ u_1 &= 0 \quad \text{for} \quad t < \frac{x_1}{c_P} \end{aligned} \tag{1.89}$$

The stress field can be computed in the following manner:

$$\begin{aligned} \sigma_{11} &= (\lambda + 2\mu)u_{1,1} = -\frac{(\lambda + 2\mu)}{c_P} f'\left(t - \frac{x_1}{c_p}\right) = -\rho c_P f'\left(t - \frac{x_1}{c_p}\right) \\ &= -p\left(t - \frac{x_1}{c_p}\right) \quad \text{for} \quad t \geq \frac{x_1}{c_p} \\ \sigma_{11} &= 0 \quad \text{for} \quad t < \frac{x_1}{c_p} \end{aligned} \tag{1.90}$$

Eqs. (1.89) and (1.90) show that the applied stress field $-p(t)$ at $x_1=0$ takes a time $t = \frac{x_1}{c_P}$ to propagate a distance of x_1. Hence, the propagation velocity of the disturbance is c_P. This wave only generates normal or longitudinal stress in the material; that is why this wave is known as longitudinal or compressional wave. Velocity of the longitudinal wave is greater than that of the shear wave (discussed below); therefore, during an earthquake the longitudinal wave arrives first. That is why it is also known as Primary wave or P-wave.

One can show that if the applied stress field in Figure 1.15 is parallel to the free surface, then the disturbance will propagate with a velocity $c_S = \sqrt{\frac{\mu}{\rho}}$, and only shear stress will be generated in the material. This wave is called a shear wave or secondary wave or S-wave. It is called secondary wave because during an earthquake it arrives second, after the P-wave.

In an infinite space, if a spherical cavity is subjected to an uniform pressure $p(t)$, then one can show that P-waves are generated in the elastic medium that propagates with a velocity c_P away from the cavity.

Some simple problems like the ones described above can be solved directly from Navier's equation, since in these problems only one component of displacement is nonzero, and this nonzero displacement component is a function of only one variable. For the half-space problems, this variable is x_1 and for the spherical cavity problem it is the radial distance r. Thus, effectively those problems are simplified to one-dimensional problems.

If the applied load in Figure 1.15 does not extend to infinity in positive and negative x_2 directions, then the problem is no longer a one-dimensional problem. For example, if the applied load in Figure 1.15 extends to +/− a, in positive and negative x_2 directions, while it extends to infinity in the positive and negative x_3 directions, then the displacement field in the half-space material will have two components of displacement, u_1 and u_2, and both of them will be functions of x_1 and x_2 in general. In other words, now the problem becomes a two-dimensional problem. It is very difficult to solve two- and three-dimensional problems directly from Navier's equation. Stokes-Helmholtz decomposition of the displacement field transforms Navier's governing equation of motion into simple wave equations, as shown in the next section.

1.2.2 Stokes-Helmholtz Decomposition

If ϕ is a scalar function and $\mathbf{A}$ is a vector function, then any displacement field $\mathbf{u}$ can be expressed in the following manner:

$$\mathbf{u} = \underline{\nabla}\phi + \underline{\nabla} \times \mathbf{A} \tag{1.91}$$

The above decomposition is known as the Stokes-Helmholtz decomposition. Since the above vector equation has three parameters (u_1, u_2, u_3) on the left-hand side and four parameters (ϕ, A_1, A_2, A_3) on the right-hand side, one can define an additional relation (known as auxiliary condition or gauge condition)

$$\underline{\nabla} \bullet \mathbf{A} = 0 \tag{1.92}$$

to obtain unique relations between u_1, u_2, u_3 and ϕ, A_1, A_2, A_3.

Substituting Eq. (1.91) in Navier's equation, in absence of a body force, one gets

$$\begin{aligned}
&(\lambda + 2\mu)\underline{\nabla}(\underline{\nabla}.\mathbf{u}) - \mu\underline{\nabla} \times \underline{\nabla} \times \mathbf{u} = \rho\ddot{\mathbf{u}} \\
&\Rightarrow (\lambda + 2\mu)\underline{\nabla}\left(\underline{\nabla}.\{\underline{\nabla}\phi + \underline{\nabla} \times \mathbf{A}\}\right) - \mu\underline{\nabla} \times \underline{\nabla} \times \{\underline{\nabla}\phi + \underline{\nabla} \times \mathbf{A}\} = \rho\left\{\underline{\nabla}\ddot{\phi} + \underline{\nabla} \times \ddot{\mathbf{A}}\right\} \\
&\Rightarrow (\lambda + 2\mu)\underline{\nabla}\left(\nabla^2\phi + \underline{\nabla}.\{\underline{\nabla} \times \mathbf{A}\}\right) - \mu\underline{\nabla} \times \underline{\nabla} \times \{\underline{\nabla}\phi + \underline{\nabla} \times \mathbf{A}\} = \rho\left\{\underline{\nabla}\ddot{\phi} + \underline{\nabla} \times \ddot{\mathbf{A}}\right\}
\end{aligned} \tag{1.93}$$

However, from the vector identity one can write

$$\begin{aligned}
&\underline{\nabla} \bullet (\underline{\nabla} \times \mathbf{A}) = 0 \\
&\underline{\nabla} \times (\underline{\nabla}\phi) = 0 \\
&\underline{\nabla} \times \underline{\nabla} \times \mathbf{A} = \underline{\nabla}(\underline{\nabla} \bullet \mathbf{A}) - \nabla^2\mathbf{A}
\end{aligned} \tag{1.94}$$

Substituting Eqs. (1.94) and (1.92) into Eq. (1.93) one gets

$$\begin{aligned}
&(\lambda + 2\mu)\underline{\nabla}(\nabla^2\phi) - \mu\underline{\nabla} \times \{-\nabla^2\mathbf{A}\} = \rho\left\{\underline{\nabla}\ddot{\phi} + \underline{\nabla} \times \ddot{\mathbf{A}}\right\} \\
&\rightarrow \underline{\nabla}\left[(\lambda + 2\mu)\nabla^2\phi - \rho\ddot{\phi}\right] + \underline{\nabla} \times \left[\mu\nabla^2\mathbf{A} - \rho\ddot{\mathbf{A}}\right] = 0
\end{aligned}$$

Sufficient conditions for the above equation to be satisfied are

$$(\lambda+2\mu)\nabla^2\phi-\rho\ddot{\phi}=0$$

and

$$\mu\nabla^2\mathbf{A}-\rho\ddot{\mathbf{A}}=0$$

Or

$$\begin{aligned}\nabla^2\phi-\frac{\rho}{(\lambda+2\mu)}\ddot{\phi}&=\nabla^2\phi-\frac{1}{c_P^2}\ddot{\phi}=0\\ \nabla^2\mathbf{A}-\frac{\rho}{\mu}\ddot{\mathbf{A}}&=\nabla^2\mathbf{A}-\frac{1}{c_S^2}\ddot{\mathbf{A}}=0\end{aligned}\tag{1.95}$$

Equations (1.95) are wave equations that have solutions in the following form:

$$\begin{aligned}\phi(\mathbf{x},t)&=\phi(\mathbf{n}\bullet\mathbf{x}-c_Pt)\\ \mathbf{A}(\mathbf{x},t)&=\mathbf{A}(\mathbf{n}\bullet\mathbf{x}-c_St)\end{aligned}\tag{1.96}$$

Equations (1.96) represent two waves propagating in the **n** direction with the velocity of c_P and c_S, respectively. Note that **n** is the unit vector in any direction.

When $\mathbf{A}=0$ and $\phi=$nonzero, then from the above solutions one gets

$$\mathbf{u}=\underline{\nabla}\phi=\mathbf{n}\phi'(\mathbf{n}\bullet\mathbf{x}-c_Pt)\tag{1.97}$$

In Eq. (1.97) the prime indicates derivative with respect to the argument. Clearly, here the direction of the displacement vector **u** and the wave propagation direction **n** are the same.

When **A** is not equal to 0 and ϕ is 0, then from the above solutions one gets

$$\mathbf{u}=\underline{\nabla}\times\mathbf{A}=\underline{\nabla}\times\mathbf{A}(\mathbf{n}\bullet\mathbf{x}-c_St)\tag{1.98}$$

Three components of displacement in the Cartesian coordinate system can be written from Eq. (1.98):

$$\begin{aligned}u_1&=n_2A_3'(\mathbf{n}\bullet\mathbf{x}-c_St)-n_3A_2'(\mathbf{n}\bullet\mathbf{x}-c_St)\\ u_2&=n_3A_1'(\mathbf{n}\bullet\mathbf{x}-c_St)-n_1A_3'(\mathbf{n}\bullet\mathbf{x}-c_St)\\ u_3&=n_1A_2'(\mathbf{n}\bullet\mathbf{x}-c_St)-n_2A_1'(\mathbf{n}\bullet\mathbf{x}-c_St)\end{aligned}\tag{1.99}$$

Clearly, the dot product between **n** and **u** [given in Eq. (1.99)] is zero; hence, the direction of the displacement vector **u** is perpendicular to the wave propagation direction **n**. Displacement fields given in Eqs. (1.97) and (1.98) correspond to P- and S-waves, respectively.

1.2.3 Two-Dimensional In-Plane Problems

If the problem geometry is such that $u_1(x_1,x_2)$ and $u_2(x_1,x_2)$ are nonzero while u_3 is equal to zero, then the problem is called an in-plane problem. In Figure 1.15, if the load is applied over a finite length in the x_2 direction but the region of load application is extended to infinity in the x_3 direction, then it will be an in-plane problem. From Eq. (1.99) one can see that if we substitute $A_1=A_2=0$ and $A_3=\psi$, then u_1 and u_2 components survive. To solve two-dimensional in-plane problems in the x_1x_2 plane we take two potential functions, $\phi\ (x_1,x_2,t)$ and $A_3=\psi(x_1,x_2,t)$, to get the displacement components from Eqs. (1.97) and (1.99) in the following form:

$$u_1 = \frac{\partial \phi}{\partial x_1} + \frac{\partial \phi}{\partial x_2}$$
$$u_2 = \frac{\partial \phi}{\partial x_2} - \frac{\partial \psi}{\partial x_1} \tag{1.100}$$

The governing wave equations for this case are

$$\nabla^2\phi - \frac{1}{c_P^2}\ddot{\phi} = \phi_{,11} + \phi_{,22} - \frac{1}{c_P^2}\ddot{\phi} = 0$$
$$\nabla^2\psi - \frac{1}{c_P^2}\ddot{\psi} = \psi_{,11} + \psi_{,22} - \frac{1}{c_S^2}\ddot{\psi} = 0 \tag{1.100a}$$

If the in-plane problem is defined in the x_1x_3-plane, then, to guarantee $u_2=0$, while u_1 and u_3 are nonzero, we need to substitute $A_1=A_3=0$ and $A_2=\psi$. In this case the two potential functions $\phi(x_1,x_3,t)$ and $A_2=\psi(x_1,x_3,t)$ give the following displacement components:

$$u_1 = \frac{\partial \phi}{\partial x_1} - \frac{\partial \psi}{\partial x_3}$$
$$u_3 = \frac{\partial \phi}{\partial x_3} + \frac{\partial \psi}{\partial x_1} \tag{1.101}$$

It should be noted here that Eqs. (1.100) and (1.101) are similar except for their signs.

Combining Eqs. (1.100) and (1.68) one can write

$$\begin{aligned}\sigma_{11} &= (\lambda+2\mu)u_{1,1} + \lambda(u_{2,2}+u_{3,3}) = (\lambda+2\mu)(\phi_{,11}+\psi_{,12}) + \lambda(\phi_{,22}-\psi_{,12})\\ &= (\lambda+2\mu)\{\phi_{,11}+\phi_{,22}\} + 2\mu(\psi_{,12}-\phi_{,22}) = \mu\{\kappa^2(\phi_{,11}+\phi_{,22}) + 2(\psi_{,12}-\phi_{,22})\}\\ &= \mu\{\kappa^2\nabla^2\phi + 2(\psi_{,12}-\phi_{,22})\}\end{aligned} \tag{1.102a}$$

$$\text{where } \kappa^2 = \frac{\lambda+2\mu}{\mu} = \left(\frac{c_P}{c_S}\right)^2 \tag{1.102b}$$

Similarly,

$$\begin{aligned}\sigma_{22} &= (\lambda+2\mu)u_{2,2} + \lambda(u_{1,1}+u_{3,3}) = (\lambda+2\mu)(\phi_{,22}-\psi_{,12}) + \lambda(\phi_{,11}+\psi_{,12})\\ &= (\lambda+2\mu)\{\phi_{,11}+\phi_{,22}\} - 2\mu(\psi_{,12}+\phi_{,11}) = \mu\{\kappa^2(\phi_{,11}+\phi_{,22}) - 2(\psi_{,12}+\phi_{,11})\}\\ &= \mu\{\kappa^2\nabla^2\phi - 2(\phi_{,11}+\psi_{,12})\}\end{aligned} \tag{1.102c}$$

$$\sigma_{12} = 2\mu\times\frac{1}{2}(u_{1,2}+u_{2,1}) = \mu(\phi_{,12}+\psi_{,22}+\phi_{,12}-\psi_{,11}) = \mu(2\phi_{,12}+\psi_{,22}-\psi_{,11})$$

Following similar steps, from Eqs. (1.101) and (1.68) one gets

$$\begin{aligned}\sigma_{11} &= \lambda\nabla^2\phi + 2\mu(\phi_{,11}-\psi_{,13}) = \mu\{\kappa^2\nabla^2\phi - 2(\phi_{,33}+\psi_{,13})\}\\ \sigma_{33} &= \lambda\nabla^2\phi + 2\mu(\phi_{,33}+\psi_{,13}) = \mu\{\kappa^2\nabla^2\phi - 2(\phi_{,11}-\psi_{,13})\}\\ \sigma_{13} &= \mu\{2\phi_{,13}-\psi_{,33}+\psi_{,11}\}\end{aligned} \tag{1.103}$$

The above equations give stress and displacement components in terms of wave potentials for elastic waves propagating in the x_1x_2- and x_1x_3-planes in an infinite isotropic solid medium in absence of any boundary. The effect of a plane boundary on the mechanics of elastic wave propagation is investigated later in this chapter (see Section 1.2.6).

1.2.4 P- and S-Waves

Important results presented above are summarized below.

Elastic waves in an infinite elastic solid can propagate in two different modes: P-wave mode and S-wave mode. When an elastic wave propagates as the P-wave, then only normal stresses (compressional or dilatational) are generated in the solid and the wave propagation speed is $c_P\left(=\sqrt{\frac{\lambda+2\mu}{\rho}}\right)$. When the elastic wave propagates as the S-wave, then only shear stresses are generated in the solid and the propagation speed is $c_S\left(=\sqrt{\frac{\mu}{\rho}}\right)$.

Wave potentials for these two types of waves, propagating in a three-dimensional space in direction **n**, are given by Eq. (1.96). If the problem is simplified to an in-plane problem where the waves propagate in one plane (say the x_1x_2-plane), then the wave potentials, ϕ and ψ, for these two types of waves can be written in the following form:

$$\begin{aligned}\phi(\mathbf{x},t)&=\phi(\mathbf{n}\cdot\mathbf{x}-c_Pt)=\phi(n_1x_1+n_2x_2-c_Pt)=\phi(x_1\cos\theta+x_2\sin\theta-c_Pt)\\ \psi(\mathbf{x},t)&=\psi(\mathbf{n}\cdot\mathbf{x}-c_St)=\psi(n_1x_1+n_2x_2-c_St)=\psi(x_1\cos\theta+x_2\sin\theta-c_St)\end{aligned} \tag{1.104}$$

Eq. (1.104) represent waves propagating in direction **n** in the x_1x_2-plane as shown in Figure 1.16.

Displacement and stress components can be obtained from these wave potential expressions following Eqs. (1.100) and (1.102). Note that in any plane normal to the wave propagation direction **n** the displacement and stress components are identical. In other words, every point on a plane normal to **n** has the same state of motion; these planes are called wave fronts and the propagating P- and S-waves with plane wave fronts are called plane waves.

1.2.5 Harmonic Waves

If the time dependence of the wave motion is $\sin\omega t$, $\cos\omega t$, or $e^{\pm i\omega t}$ (i is the imaginary number $\sqrt{-1}$), then the wave is called a harmonic wave. Any function of time can be expressed as a superposition of harmonic functions by the Fourier series expansion or Fourier integral technique. Therefore, if we have solutions

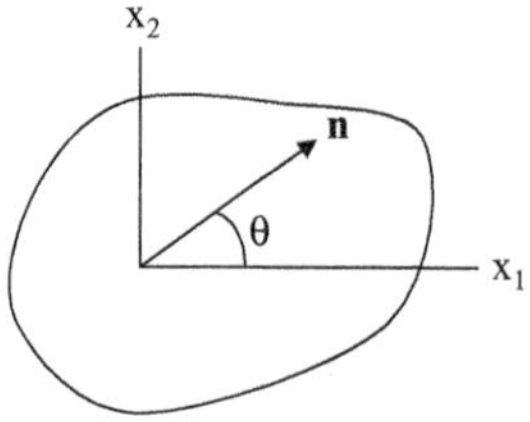

Figure 1.16 Elastic waves propagating in direction **n** in the x_1x_2-plane.

for the harmonic time dependence, then the solution for any other time dependence can be obtained by taking the inverse Fourier transform.

Unless otherwise specified, the time dependence in the subsequent analysis will be taken as $e^{-i\omega t}$. The advantage of this type of time dependence is that the problem becomes much simpler.

Eq. (104) can have any expression for the functions ϕ and ψ. However, for harmonic time dependence $\left(e^{-i\omega t}\right)$ ϕ and ψ must have the following forms:

$$\begin{aligned}\Phi(x_1,x_2,t) &= A\exp(ik_Px_1\cos\theta + ik_Px_2\sin\theta - i\omega t) = \phi(x_1,x_2)e^{-i\omega t}\\ \Psi(x_1,x_2,t) &= B\exp(ik_Sx_1\cos\theta + ik_Sx_2\sin\theta - i\omega t) = \psi(x_1,x_2)e^{-i\omega t}\end{aligned} \tag{1.105}$$

Comparing Eqs. (1.105) and (1.104) one can see that

$$\begin{aligned}k_P &= \frac{\omega}{c_P}\\ k_S &= \frac{\omega}{c_S}\end{aligned} \tag{1.106a}$$

k_P and k_S are called P- and S- wave numbers, respectively. ω is known as the circular frequency (radian per second) and is related to the wave frequency (f in hertz or Hz) in the following manner:

$$\omega = 2\pi f \tag{1.106b}$$

A and B of Eq. (105) are amplitudes of the wave potentials ϕ and ψ, respectively. For the time harmonic waves, the governing wave equations (1.95) and (1.100a) take the following form:

$$\begin{aligned}\nabla^2\phi + \frac{\omega^2}{c_P^2}\phi &= \nabla^2\phi + k_P^2\phi = 0\\ \nabla^2\psi + \frac{\omega^2}{c_S^2}\psi &= \nabla^2\psi + k_S^2\psi = 0\end{aligned} \tag{1.107}$$

Since the time dependence $e^{-i\omega t}$ appears in every term for time harmonic motions, it is customary to ignore it while writing the expressions for potentials, displacements and stresses.

From Eq. (107)

$$\nabla^2\phi = -k_P^2\phi$$

Therefore, from Eq. (1.102b) and the above relation

$$\kappa^2\nabla^2\phi = \left(\frac{c_P}{c_S}\right)^2\nabla^2\phi = \left(\frac{k_S}{k_P}\right)^2\nabla^2\phi = -k_S^2\phi \tag{1.107a}$$

Substituting Eq. (1.107a) into Eqs. (1.102) and (1.103), we get the stress expressions for the harmonic waves in the x_1x_2 coordinate system

$$\begin{aligned}\sigma_{11} &= -\mu\left\{k_S^2\phi + 2(\phi_{,22} - \psi_{,12})\right\}\\ \sigma_{22} &= -\mu\left\{k_S^2\phi + 2(\phi_{,11} + \psi_{,12})\right\}\\ \sigma_{12} &= \mu\left\{2\phi_{,12} + \psi_{,22} - \psi_{,11}\right\}\end{aligned} \tag{1.107b}$$

and in x_1x_3 coordinate system

$$\begin{aligned}\sigma_{11} &= -\mu\left\{k_S^2\phi + 2(\phi_{,33} + \psi_{,13})\right\}\\ \sigma_{33} &= -\mu\left\{k_S^2\phi + 2(\phi_{,11} - \psi_{,13})\right\}\\ \sigma_{13} &= \mu\left\{2\phi_{,13} + \psi_{,11} - \psi_{,33}\right\}\end{aligned} \tag{1.107c}$$

1.2.6 Interaction between Plane Waves and Stress-Free Plane Boundary

So far, we have only talked about the elastic wave propagation in an unbounded elastic solid. Let us now try to understand how the presence of a stress-free surface affects the wave propagation characteristics.

1.2.6.1 P-wave Incident on a Stress-Free Plane Boundary

First consider the effect of a stress-free boundary on the plane P-wave propagating in the direction shown in Figure 1.17. It will be shown here that the reflected wave from the stress-free boundary will have two components: a P-wave component, denoted by PP, and an S-wave component, denoted by PS. In these notations the first letter indicates the type of wave that is incident on the stress-free surface and the second subscript indicates the type of wave generated after the reflection at the surface.

In the absence of reflected waves PP and PS, let us first investigate if only the incident P-wave can satisfy the stress-free boundary conditions at $x_2 = 0$.

Wave potential for the incident P-wave shown in Figure 1.17 is given by

$$\phi = \exp(ik_P x_1 \sin\theta_P - ik_P x_2 \cos\theta_P - i\omega t) = \exp(ikx_1 - i\eta x_2 - i\omega t) \tag{1.108}$$

where $k = k_P \sin\theta_P$ and $\eta = k_P \cos\theta_P$. Amplitude of the incident wave is assumed to be 1. From Eq. (1.102c) one can compute the normal and shear stress components at the interface in the following form:

$$\begin{aligned}\sigma_{22} &= \mu\left\{\kappa^2\nabla^2\phi - 2\phi_{,11}\right\} = \mu\left\{-\kappa^2(k^2 + \eta^2)\phi + 2k^2\phi\right\}\\ \sigma_{12} &= \mu\left\{2\phi_{,12}\right\} = 2\mu k\eta\phi\end{aligned} \tag{1.109}$$

Clearly, σ_{22} and σ_{12} are not equal to zero at $x_2 = 0$. To satisfy the stress-free boundary conditions at $x_2 = 0$, one needs to include two reflected waves PP and PS. Readers can show that inclusion of only PP-waves cannot satisfy the stress-free boundary conditions at $x_2 = 0$.

When both PP- and PS-waves are considered in the reflected field, then the total potential field (incident plus reflected) in the solid is given by

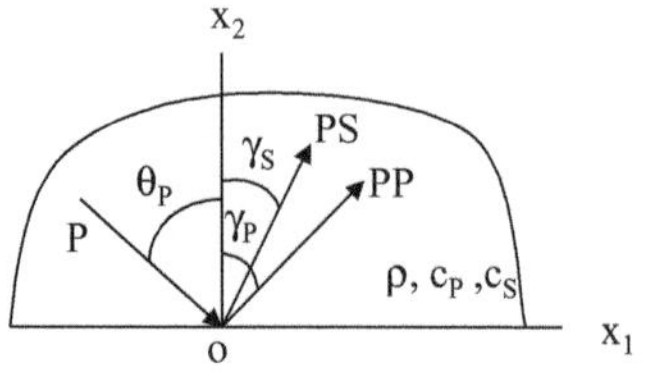

Figure 1.17 Reflection of the plane P-wave by a stress-free plane boundary.

$$\begin{aligned}\phi &= \phi_I + \phi_R \\ &= \exp\left(ik_P x_1 \sin\theta_P - ik_P x_2 \cos\theta_P - i\omega t\right) \\ &\quad + R_{PP}\exp\left(ik_P x_1 \sin\gamma_P + ik_P x_2 \cos\gamma_P - i\omega t\right) \\ \psi &= \psi_R = R_{PS}\exp\left(ik_S x_1 \sin\gamma_S + ik_S x_2 \cos\gamma_S - i\omega t\right)\end{aligned} \tag{1.110}$$

Note that the ϕ expression of the above equation has two terms; the first term (ϕ_I) corresponds to the downward incident P-wave and the second term (ϕ_R) corresponds to the upward reflected P-wave. Amplitude of the incident wave is 1, while that of the reflected P-wave is R_{PP}, and the amplitude of the reflected S-wave is R_{PS}. Inclination angles for these three waves are denoted by θ_P, γ_P and γ_S, respectively, in Figure 1.17.

From Eqs. (1.110) and (1.102) one can compute the stress fields at $x_2=0$.

$$\begin{aligned}\sigma_{22} &= \mu\left\{\kappa^2\nabla^2\phi - 2\left(\psi_{,12} + \phi_{,11}\right)\right\} \\ &= \mu\left\{\kappa^2\left(-k_P^2\left[\sin^2\theta_P + \cos^2\theta_P\right]\phi_I - k_P^2\left[\sin^2\gamma_P + \cos^2\gamma_P\right]\phi_R\right)\right. \\ &\quad \left. + 2k_S^2 \sin\gamma_S \cos\gamma_S \psi + 2k_P^2 \sin^2\theta_P \phi_I + 2k_P^2 \sin^2\gamma_P \phi_R\right\} \\ &= \mu\left\{-\kappa^2 k_P^2\left(\phi_I + \phi_R\right) + 2k_S^2 \sin\gamma_S\cos\gamma_S\psi + 2k_P^2\sin^2\theta_P\phi_I + 2k_P^2\sin^2\gamma_P\phi_R\right\}\end{aligned} \tag{1.111}$$

$$\begin{aligned}\sigma_{12} &= \mu\left\{2\phi_{,12} + \psi_{,22} - \psi_{,11}\right\} \\ &= \mu\left\{2k_P^2\left(\sin\theta_P\cos\theta_P\phi_I - \sin\gamma_P\cos\gamma_P\phi_R\right) - k_S^2\left(\cos^2\gamma_S - \sin^2\gamma_S\right)\psi\right\}\end{aligned}$$

For a stress-free surface at $x_2=0$, both σ_{22} and σ_{12} must be zero at $x_2=0$. Substituting $x_2=0$ in Eq. (1.111) and equating σ_{22} to zero one gets

$$\begin{aligned}\sigma_{22} &= \mu\left\{\left(2k_P^2\sin^2\theta_P - \kappa^2 k_P^2\right)\exp\left(ik_P x_1\sin\theta_P - i\omega t\right)\right. \\ &\quad + \left(2k_P^2\sin^2\gamma_P - \kappa^2 k_P^2\right)R_{PP}\exp\left(ik_P x_1 \sin\gamma_P - i\omega t\right) \\ &\quad \left. + k_S^2 \sin 2\gamma_S R_{PS}\exp(ik_S x_1\sin\gamma_S - i\omega t)\right\} = 0\end{aligned} \tag{1.112}$$

If θ_P, γ_P and γ_S are independent of each other, then the only way one can satisfy the above equation is by equating the coefficient of each of the three terms to zero. This is not possible since the first coefficient cannot be zero for all θ_P. However, if we impose the condition

$$\exp\left(ik_P x_1 \sin\theta_P - i\omega t\right) = \exp\left(ik_P x_1\sin\gamma_P - i\omega t\right) = \exp(ik_S x_1 \sin\gamma_S - i\omega t) \tag{1.113}$$

then Eq. (1.112) will be satisfied if

$$\left(2k_P^2\sin^2\theta_P - \kappa^2 k_P^2\right) + \left(2k_P^2\sin^2\gamma_P - \kappa^2 k_P^2\right)R_{PP} + k_S^2\sin 2\gamma_S R_{PS} = 0 \tag{1.114}$$

Note that Eq. (1.113) implies

$$\begin{aligned}&\gamma_P = \theta_P \\ &k_S\sin\gamma_S = k_P\sin\theta_P \Rightarrow \frac{\sin\gamma_S}{c_S} = \frac{\sin\theta_P}{c_P}\end{aligned} \tag{1.115}$$

Eq. (1.115) is known as Snell's law.

Then we introduce the following symbols:

$$\begin{aligned} k &= k_P \sin\theta_P = k_S \sin\gamma_S \\ \eta &= k_P \cos\theta_P = \sqrt{k_P^2 - k^2} \\ \beta &= k_S \cos\gamma_S = \sqrt{k_S^2 - k^2} \end{aligned} \tag{1.116}$$

in Eq. (1.114) to obtain

$$\left(2k^2 - k_S^2\right) + \left(2k^2 - k_S^2\right) R_{PP} + 2k\beta R_{PS} = 0 \tag{1.117}$$

In Eq. (1.117) we have also used the relation $\kappa^2 k_P^2 = k_S^2$, which is obvious from the definition of κ^2 [Eq. (1.102b)], k_P^2 and k_S^2 [Eq. (1.106)].

Similarly, satisfying $\sigma_{12}=0$ at $x_2=0$ gives rise to the following equation:

$$2k\eta - 2k\eta R_{PP} - (\beta^2 - k^2) R_{PS} = 2k\eta - 2k\eta R_{PP} + (2k^2 - k_S^2) R_{PS} = 0 \tag{1.118}$$

Eqs. (1.117) and (1.118) can be written in matrix form:

$$\begin{bmatrix} \left(2k^2 - k_S^2\right) & 2k\beta \\ -2k\eta & \left(2k^2 - k_S^2\right) \end{bmatrix} \begin{Bmatrix} R_{PP} \\ R_{PS} \end{Bmatrix} = \begin{Bmatrix} -\left(2k^2 - k_S^2\right) \\ -2k\eta \end{Bmatrix} \tag{1.119}$$

Eq. (1.119) is then solved for R_{PP} and R_{PS}:

$$\begin{aligned} R_{PP} &= \frac{4k^2\eta\beta - \left(2k^2 - k_S^2\right)^2}{4k^2\eta\beta + \left(2k^2 - k_S^2\right)^2} \\ R_{PS} &= \frac{-4k\eta\left(2k^2 - k_S^2\right)}{4k^2\eta\beta + \left(2k^2 - k_S^2\right)^2} \end{aligned} \tag{1.120}$$

1.2.6.2 Summary of Plane P-Wave Reflection by a Stress-Free Surface

When a plane P-wave strikes a stress-free surface at an inclination angle θ_P, as shown in Figure 1.17, a plane P-wave (PP) and a plane S-wave (PS) are generated as reflected waves to satisfy the stress-free boundary conditions at the plane surface. The angle of reflection for the P-wave is the same as the incident angle, and the angle of reflection for the S-wave is related to the incident angle by Snell's law [Eq. (1.115)]. Potentials for the incident and reflected waves can be expressed as

$$\begin{aligned} \phi_P &= \phi_I = \exp\left(ikx_1 - i\eta x_2\right) \\ \phi_{PP} &= \phi_R = R_{PP}\exp\left(ikx_1 + i\eta x_2\right) \\ \psi_{PS} &= \psi_R = R_{PS}\exp\left(ikx_1 + i\beta x_2\right) \end{aligned} \tag{1.121}$$

With time dependence, $e^{-i\omega t}$ is implied to appear in every expression and is not shown explicitly. k, η and β are defined in Eq. (1.116). Reflection coefficients R_{PP} and R_{PS} are given in Eq. (1.120).

Example 1.11

Evaluate R_{PP} and R_{PS} for $\theta_P=0$

SOLUTION

From Eq. (1.116) for $\theta_P=0$, $k=0$, $\eta=k_P$ and $\beta=k_S$.

Substituting these in Eq. (1.120) one gets

$$R_{PP}=\frac{4k^2\eta\beta-\left(2k^2-k_S^2\right)^2}{4k^2\eta\beta+\left(2k^2-k_S^2\right)^2}=\frac{0-k_S^4}{0+k_S^4}=-1$$

$$R_{PS}=\frac{-4k\eta\left(2k^2-k_S^2\right)}{4k^2\eta\beta+\left(2k^2-k_S^2\right)^2}=\frac{0}{0+k_S^4}=0$$

Therefore, for normal incidence, the reflected wave does not contain any shear wave. Incident and reflected wave potentials in this case are given by

$$\phi_P=\phi_I=\exp(-i\eta x_2)=\exp(-ik_Px_2)$$

$$\phi_{PP}=\phi_R=-\exp(i\eta x_2)=-\exp(ik_Px_2)$$

Example 1.12

Compute the displacement and stress fields for a plane P-wave striking a stress-free plane boundary at normal incidence ($\theta_P=0$).

SOLUTION

As shown in the previous example for normal incidence of plane P-waves

$$\phi=\phi_P+\phi_{PP}=\exp(-ik_Px_2)-\exp(ik_Px_2)$$

$$\psi=0$$

Hence, from Eq. (1.100)

$$u_1=\phi_{,1}+\psi_{,2}=0$$

$$u_2=\phi_{,2}-\psi_{,1}=-ik_P\left[\exp(-ik_Px_2)+\exp(ik_Px_2)\right]$$

and from Eq. (1.102)

$$\sigma_{11}=\mu\left\{\kappa^2\nabla^2\phi+2\left(\psi_{,12}-\phi_{,22}\right)\right\}$$

$$\sigma_{22}=\mu\left\{\kappa^2\nabla^2\phi-2\left(\psi_{,12}+\phi_{,11}\right)\right\}$$

$$\sigma_{12}=\mu\left\{2\phi_{,12}+\psi_{,22}-\psi_{,11}\right\}$$

From Eqs. (1.107) and (1.102b) one can write

$$\kappa^2\nabla^2\phi=-\kappa^2k_P^2\phi=-\frac{c_P^2}{c_S^2}\frac{\omega^2}{c_P^2}\phi=-k_S^2\phi$$

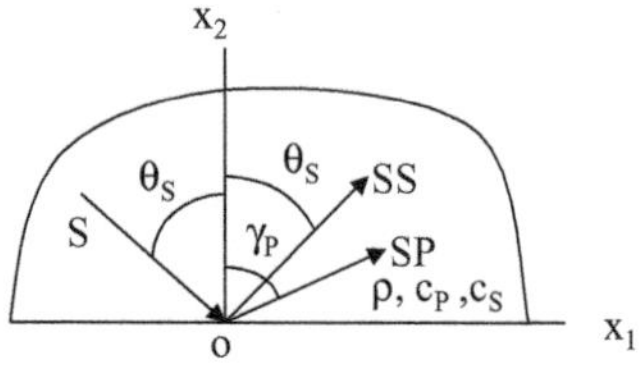

Figure 1.18 Reflection of the plane S-wave by a stress-free plane boundary.

Substituting $\psi = 0$ and the above relation into the expressions for the stress field one gets

$$\sigma_{11} = \mu\{-k_S^2\phi - 2\phi_{,22}\} = -\mu\{(k_S^2 - 2k_P^2)[\exp(-ik_P x_2) - \exp(ik_P x_2)]\} = 2i\mu(k_S^2 - 2k_P^2)\sin(k_P x_2)$$

$$\sigma_{22} = \mu\{-k_S^2\phi - 2\phi_{,11}\} = -\mu k_S^2[\exp(-ik_P x_2) - \exp(ik_P x_2)] = 2i\mu k_S^2 \sin(k_P x_2)$$

$$\sigma_{12} = \mu\{2\phi_{,12}\} = 0$$

Note that all stress components vanish at the boundary at $x_2 = 0$.

1.2.6.3 Shear Wave Incident on a Stress-Free Plane Boundary

Figure 1.18 shows a shear wave incident on a stress-free plane boundary at an angle θ_S. Following a similar analysis as the P-wave incidence case, one can show that for S-wave incidence the reflected wave must also contain both S-wave and P-wave components to satisfy the stress-free boundary conditions. The two reflected waves are denoted by SS and SP, respectively. The wave potentials corresponding to the incident S and reflected SS- and SP-waves are

$$\begin{aligned} \psi_S &= \psi_I = \exp(ikx_1 - i\beta x_2) \\ \phi_{SP} &= \phi_R = R_{SP}\exp(ikx_1 + i\eta x_2) \\ \psi_{SS} &= \psi_R = R_{SS}\exp(ikx_1 + i\beta x_2) \end{aligned} \tag{1.122}$$

In Eq. (1.122) time dependence $e^{-i\omega t}$ is implied and

$$\begin{aligned} k &= k_S \sin\theta_S = k_P \sin\gamma_P \\ \eta &= k_P \cos\gamma_P = \sqrt{k_P^2 - k^2} \\ \beta &= k_S \cos\theta_S = \sqrt{k_S^2 - k^2} \end{aligned} \tag{1.123}$$

θ_S and γ_P satisfy Snell's law,

$$\frac{\sin\theta_S}{c_S} = \frac{\sin\gamma_P}{c_P} \tag{1.124}$$

Note that for $\theta_S = \sin^{-1}\left(\frac{c_S}{c_P}\right)$ the angle γ_P becomes 90°. This angle is called the *critical angle*. Since $c_S < c_P$, the critical angle has a real value. If the angle of incidence exceeds the critical angle value, then the reflected P-wave, as shown in Figure 1.22, is not generated since γ_P then becomes imaginary.

Reflection coefficients R_{SP} and R_{SS} can be obtained by satisfying the stress-free boundary conditions at $x_2 = 0$.

From Eq. (1.102),

$$\sigma_{22} = \mu\{\kappa^2\nabla^2\phi - 2(\psi_{,12} + \phi_{,11})\} = -\mu\{k_S^2\phi + 2(\psi_{,12} + \phi_{,11})\}$$
$$= -\mu\{(k_S^2 - 2k^2)(R_{SP}e^{ikx_1 + i\eta x_2}) - 2k\beta(-e^{ikx_1 - i\beta x_2} + R_{SS}e^{ikx_1 + i\beta x_2})\}$$
$$\sigma_{12} = \mu\{2\phi_{,12} + \psi_{,22} - \psi_{,11}\} \tag{1.125}$$
$$= \mu\{(-2k\eta)R_{SP}e^{ikx_1 + i\eta x_2} + (k^2 - \beta^2)(e^{ikx_1 - i\beta x_2} + R_{SS}e^{ikx_1 + i\beta x_2})\}$$

Since the two stress components of Eq. (1.125) must be zero at $x_2 = 0$,

$$\sigma_{22}\big|_{x_2=0} = -\mu\{(k_S^2 - 2k^2)(R_{SP}e^{ikx_1}) - 2k\beta(-e^{ikx_1} + R_{SS}e^{ikx_1})\}$$
$$= \mu\{(2k^2 - k_S^2)R_{SP} + 2k\beta(-1 + R_{SS})\}e^{ikx_1} = 0$$
$$\sigma_{12}\big|_{x_2=0} = \mu\{(-2k\eta)R_{SP}e^{ikx_1} + (k^2 - \beta^2)(e^{ikx_1} + R_{SS}e^{ikx_1})\}$$
$$= \mu\{-2k\eta R_{SP} + (2k^2 - k_S^2)(1 + R_{SS})\}e^{ikx_1} = 0$$

The two equations above will be satisfied for all x_1 if

$$\begin{bmatrix} (2k^2 - k_S^2) & 2k\beta \\ -2k\eta & (2k^2 - k_S^2) \end{bmatrix} \begin{Bmatrix} R_{SP} \\ R_{SS} \end{Bmatrix} = \begin{Bmatrix} 2k\beta \\ -(2k^2 - k_S^2) \end{Bmatrix} \tag{1.126}$$

Eq. (1.126) can be solved for R_{SP} and R_{SS} to obtain

$$R_{SP} = \frac{4k\beta(2k^2 - k_S^2)}{4k^2\eta\beta + (2k^2 - k_S^2)^2}$$
$$R_{SS} = \frac{4k^2\eta\beta - (2k^2 - k_S^2)^2}{4k^2\eta\beta + (2k^2 - k_S^2)^2} \tag{1.127}$$

Example 1.13

Evaluate R_{SP} and R_{SS} for $\theta_S = 0$

SOLUTION

From Eq. (1.123) for $\theta_S = 0$, we get $k = 0$, $\eta = k_P$ and $\beta = k_S$.

Substituting these in Eq. (1.127) one gets

$$R_{SP} = \frac{4k\beta(2k^2 - k_S^2)}{4k^2\eta\beta + (2k^2 - k_S^2)^2} = 0$$

$$R_{SS} = \frac{4k^2\eta\beta - (2k^2 - k_S^2)^2}{4k^2\eta\beta + (2k^2 - k_S^2)^2} = -1$$

Therefore, for normal incidence the reflected wave does not contain any P-wave. The incident and reflected wave potentials for this case are given by

$$\psi_S = \psi_I = \exp(-ik_S x_2)$$
$$\psi_{SS} = \psi_R = -\exp(ik_S x_2)$$

Example 1.14

Compute displacement and stress fields for the in-plane S-wave striking a stress-free plane boundary at normal incidence ($\theta_S = 0$).

SOLUTION

As shown in the previous example, for normal incidence of the S-wave, the wave potentials are given by

$$\psi = \psi_I + \psi_R = \exp(-ik_S x_2) - \exp(ik_S x_2)$$
$$\phi = 0$$

Hence, from Eq. (1.100)

$$u_1 = \phi_{,1} + \psi_{,2} = -ik_S\left\{e^{-ik_S x_2} + e^{ik_S x_2}\right\} = -2ik_S \cos(k_S x_2)$$
$$u_2 = \phi_{,2} - \psi_{,1} = 0$$

and from Eq. (1.102)

$$\sigma_{11} = \mu\left\{\kappa^2\nabla^2\phi + 2(\psi_{,12} - \phi_{,22})\right\}$$
$$\sigma_{22} = \mu\left\{\kappa^2\nabla^2\phi - 2(\psi_{,12} + \phi_{,11})\right\}$$
$$\sigma_{12} = \mu\left\{2\phi_{,12} + \psi_{,22} - \psi_{,11}\right\}$$

Substituting $\phi = 0$ in the above expressions one gets

$$\sigma_{11} = 2\mu\psi_{,12} = 0$$
$$\sigma_{22} = -2\mu\psi_{,12} = 0$$
$$\sigma_{12} = \mu\left\{\psi_{,22} - \psi_{,11}\right\} = -\mu k_S^2\left\{e^{-ik_S x_2} - e^{ik_S x_2}\right\} = 2i\mu k_S^2 \sin(k_S x_2)$$

Note that all stress components vanish at the boundary at $x_2 = 0$.

1.2.7 Out-of-Plane or Antiplane Motion – SH Wave

In earlier sections we have discussed the in-plane motion, where the displacements and wave propagation directions are confined in the x_1x_2-plane (see section 1.2.3). Hence, x_3-components of displacement and wave velocity are zero. If the waves propagate in the x_1x_2-plane but the particles have only x_3 component of nonzero displacement, then the wave motion is called antiplane or out-of-plane motion. Thus for antiplane problems $u_1 = u_2 = 0$ and $u_3 = u_3(x_1, x_2, t)$.

Substituting the above displacement components into Navier's equation [Eq. (1.79)] and carrying out the dot and cross products (as given in Table 1.2), the following equations are obtained when the body force is absent:

$$(\lambda + 2\mu).0 - \mu(-u_{3,11} - u_{3,22}) = \rho \ddot{u}_3$$

$$\Rightarrow u_{3,11} + u_{3,22} - \frac{\rho}{\mu}\ddot{u}_3 = 0 \qquad (1.128)$$

$$\Rightarrow \nabla^2 u_3 - \frac{1}{c_S^2}\ddot{u}_3 = 0$$

For harmonic waves, like before (see Eq. 1.105), we assume $e^{-i\omega t}$ time dependence and the above equation is simplified to

$$\nabla^2 u_3 + k_S^2 u_3 = 0 \qquad (1.129)$$

Note that the wave Eqs. (1.128) and (1.129) are similar to the Eqs. (1.100a) and (1.107), respectively. The only difference is in the definition of the unknown variable. For Eqs. (1.100a) and (1.107), the variables are the wave potentials, and for Eqs. (1.128) and (1.129) it is the displacement. Eq. (1.128) gives a wave motion with velocity c_S which is the same as the shear wave speed. From the strain–displacement and stress–strain relations (for isotropic material), one can also show that the displacement field $u_1 = u_2 = 0$ and $u_3 = u_3(x_1, x_2, t)$ can only produce shear strain (γ_{13} and γ_{23}) and shear stress (σ_{13} and σ_{23}). It should be noted here that the strain and stress fields corresponding to the in-plane shear wave, discussed earlier, produce nonzero shear strain (γ_{12}) and shear stress (σ_{12}). Thus, the antiplane shear wave and in-plane shear wave have a few things in common – they propagate with the same speed (c_S) and generate only shear stress in the medium. However, they produce different components of nonzero shear stress – because they have different particle displacement or polarization directions as shown in Figure 1.19. When a wave propagates in direction **n** in the x_1x_2-plane and the particle displacement is in the x_3-direction, then this antiplane shear wave is called the shear horizontal or SH-wave; if the particle displacement is normal to the wave propagation direction but lies in the same plane (in this case the x_1x_2-plane), then it is called a shear vertical or SV-wave; and when the particle displacement is parallel to the wave propagation direction, then the wave is a P-wave.

Note that Stokes-Helmholtz decomposition was not necessary for solving the antiplane problem. Direct substitution of $u_1 = u_2 = 0$ and $u_3 = u_3(x_1, x_2, t)$ into Navier's equation simplified it in the form of a wave equation, Eqs. (1.128) and (1.129). Its solution for a harmonic wave is given by, as before,

$$u_3 = A\exp(ikx_1 + i\beta x_2 - i\omega t) \qquad (1.130)$$

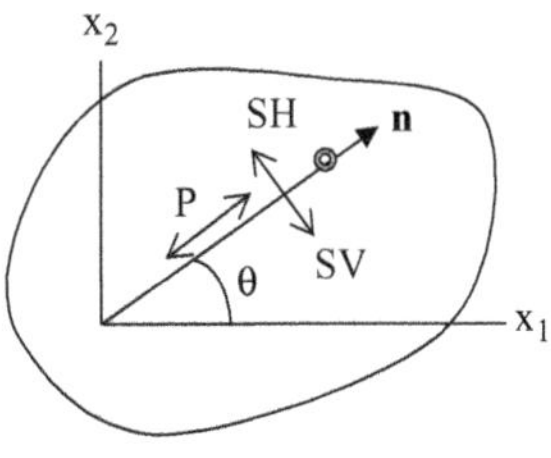

Figure 1.19 Particle displacement directions for P-, SV- and SH-waves propagating in direction **n**. Particle motion for the SH-wave is in the x_3-direction, denoted by a circle.

"A" of Eq. (1.130) represents the amplitude of the wave; k and β must satisfy the relation $\sqrt{k^2+\beta^2}=k_S^2$. In the subsequent analyses the time dependence $e^{-i\omega t}$ will be implied and will not be explicitly written.

1.2.7.1 Interaction of SH-Wave and Stress-Free Plane Boundary

Let us consider a plane SH-wave of unit amplitude striking a stress-free plane boundary at $x_2=0$ at an angle θ_S as shown in Figure 1.20. If the reflected wave amplitude is R and angle of reflection is θ_S then the total displacement field is given by

$$u_3=\exp(ikx_1-i\beta x_2)+\text{R}\exp(ikx_1+i\beta x_2) \tag{1.131}$$

Time dependence $e^{-i\omega t}$ is implied. k and β have been defined in Eq. (1.123).

The stress field can be obtained from this displacement field:

$$\begin{aligned}\sigma_{23}&=2\mu\varepsilon_{23}=2\mu\times\frac{1}{2}(u_{2,3}+u_{3,2})=\mu u_{3,2}\\&=i\beta\mu\left\{-e^{-i\beta x_2}+\text{R}e^{i\beta x_2}\right\}e^{ikx_1}\end{aligned} \tag{1.132}$$

At $x_2=0$, $\sigma_{23}=0$; hence,

$$\sigma_{23}\big|_{x_2=0}=i\beta\mu\{-1+R\}e^{ikx_1}=0$$

$$\Rightarrow R=1$$

The total displacement and stress fields are then given by

$$\begin{aligned}u_3&=\exp(ikx_1-i\beta x_2)+\exp(ikx_1+i\beta x_2)\\&=\left\{e^{-i\beta x_2}+e^{i\beta x_2}\right\}e^{ikx_1}=2\cos(\beta x_2)e^{ikx_1}\end{aligned}$$

$$\sigma_{23}=i\beta\mu\left\{-e^{-i\beta x_2}+e^{i\beta x_2}\right\}e^{ikx_1}=-2\beta\mu\sin(\beta x_2)e^{ikx_1}$$

$$\sigma_{13}=\mu u_{3,1}=ik\mu\left\{e^{-i\beta x_2}+e^{i\beta x_2}\right\}e^{ikx_1}=2ik\mu\cos(\beta x_2)e^{ikx_1}$$

Note that at the boundary σ_{23} is zero but σ_{13} is not.

From the above analysis it is clear that compared to in-plane waves (P and SV) the antiplane wave (SH) is much simpler to analyze, because no mode conversion occurs at the reflecting surface and the reflection coefficient ($R=1$) has a much simpler expression than those given in Eqs. (1.120) or (1.127). Also, the antiplane analysis does not require introduction of Stokes-Helmholtz potential functions, unlike the P-SV case. Because of these simplicities, antiplane problems have been academically popular, although their practical applications are limited.

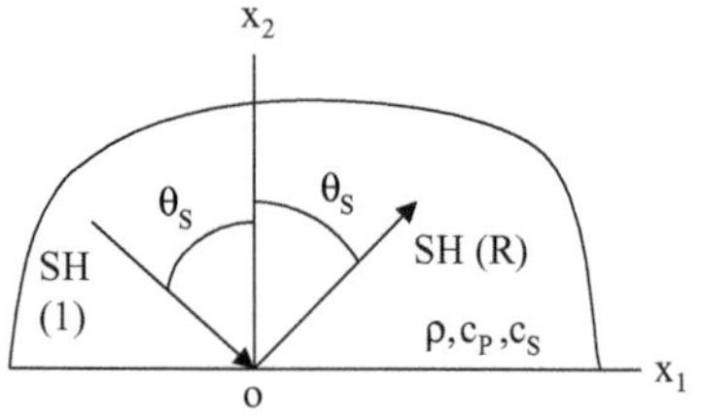

Figure 1.20 SH-wave reflected by a stress-free boundary.

1.2.7.2 Interaction of SH-Wave and a Plane Interface

Let us consider the case where the SH-wave strikes a plane interface between two isotropic elastic solids as shown in Figure 1.21. Material properties (density, P-wave speed and S-wave speed) for the two elastic solids are denoted by ρ_i, c_{Pi} and c_{Si}, where $i=1$ and 2 for the two materials. The amplitudes of the incident, reflected and transmitted waves are 1, R and T, respectively. The angles of incidence and reflection are denoted by θ_1 while the transmission angle is θ_2, as shown in Figure 1.21.

To satisfy the displacement (u_3) and stress (σ_{23}) continuity conditions across the interface at $x_2=0$ for all x_1, the x_1 dependence term for all three waves must be same. This condition gives rise to Snell's law,

$$\frac{\sin\theta_1}{c_{S1}} = \frac{\sin\theta_2}{c_{S2}} \tag{1.133}$$

The displacement fields corresponding to the incident, reflected and transmitted waves then become

$$\begin{aligned} u_{3I} &= e^{ikx_1 - i\beta_1 x_2} \\ u_{3R} &= \mathrm{R}e^{ikx_1 + i\beta_1 x_2} \\ u_{3T} &= Te^{ikx_1 - i\beta_2 x_2} \end{aligned} \tag{1.134}$$

In Eq. (1.134) subscripts I, R and T correspond to incident, reflected and transmitted fields, respectively. Reflected and transmitted wave amplitudes are denoted by R and T, respectively, while the incident wave amplitude is 1.

Note that

$$\begin{aligned} k &= k_{S1}\sin\theta_1 = k_{S2}\sin\theta_2 \\ \beta_1 &= k_{S1}\cos\theta_1 \\ \beta_2 &= k_{S2}\cos\theta_2 \\ k_{S1} &= \frac{\omega}{c_{S1}},\ k_{S2} = \frac{\omega}{c_{S2}} \end{aligned} \tag{1.134a}$$

From displacement and stress continuity at $x_2=0$ one can write

$$\begin{aligned} (1+R)e^{ikx_1} &= Te^{ikx_1} & &\Rightarrow 1+R = T \\ \mu_1\beta_1(-1+R)e^{ikx_1} &= -\mu_2\beta_2 Te^{ikx_1} & &\Rightarrow -1+R = -QT \end{aligned} \tag{1.135}$$

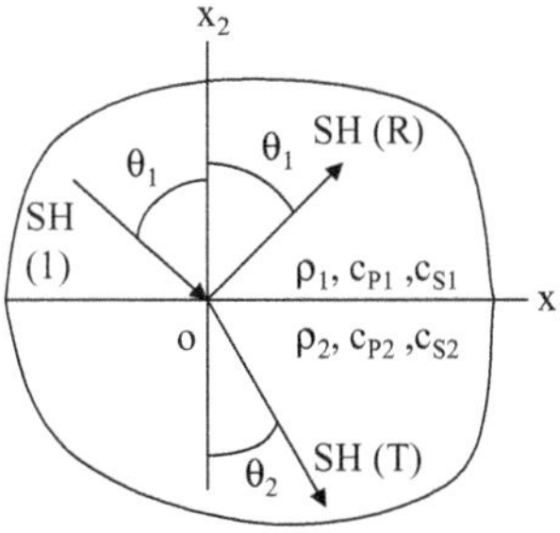

Figure 1.21 Reflection and transmission of the SH-wave at a plane interface.

where

$$Q = \frac{\mu_2\beta_2}{\mu_1\beta_1} \tag{1.136}$$

Eq. (1.135) can be solved for R and T,

$$R = \frac{1-Q}{1+Q}$$
$$T = \frac{2}{1+Q} \tag{1.137}$$

Note that when the two materials are identical, then $Q=1$, $R=0$ and $T=1$. This is expected since in this special case there is no interface.

For normal incidence ($\theta_1=\theta_2=0$) from Eq. (1.134a), $\beta_1=k_{S1}$, $\beta_2=k_{S2}$. Therefore,

$$Q = \frac{\mu_2\beta_2}{\mu_1\beta_1} = \frac{\mu_2 k_{S2}}{\mu_1 k_{S1}} = \frac{\rho_2 c_{S2}^2 \frac{\omega}{c_{S2}}}{\rho_1 c_{S1}^2 \frac{\omega}{c_{S1}}} = \frac{\rho_2 c_{S2}}{\rho_1 c_{S1}} = \frac{Z_{2S}}{Z_{1S}}$$

then

$$R = \frac{1-Q}{1+Q} = \frac{Z_{1S}-Z_{2S}}{Z_{1S}+Z_{2S}}$$
$$T = \frac{2}{1+Q} = \frac{2Z_{1S}}{Z_{1S}+Z_{2S}} \tag{1.137a}$$

1.2.8 Interaction of P and SV-Waves with Plane Interface

These in-plane problems are more complex compared to the SH problem analyzed in the previous section because of the mode conversion, as one can see below. First, the problem of P-wave incidence is solved.

1.2.8.1 P-Wave Striking an Interface

Figure 1.22 shows a plane P-wave of unit amplitude striking a plane interface between two linear elastic isotropic solids. Both reflected and transmitted waves have P- and SV-wave components and their amplitudes are denoted by R_{PP}, R_{PS}, T_{PP} and T_{PS} as shown in the figure.

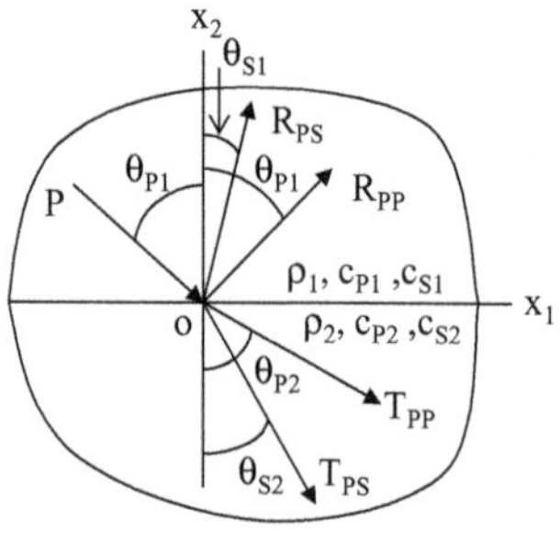

Figure 1.22 Reflected and transmitted waves near an interface for P-wave incidence.

Wave potentials for these five types of wave are given by

$$\begin{aligned} \phi_{IP} &= e^{ikx_1 - i\eta_1 x_2} \\ \phi_{RPP} &= R_{PP} e^{ikx_1 + i\eta_1 x_2} \\ \phi_{TPP} &= T_{PP} e^{ikx_1 - i\eta_2 x_2} \\ \psi_{RPS} &= R_{PS} e^{ikx_1 + i\beta_1 x_2} \\ \psi_{TPS} &= T_{PS} e^{ikx_1 - i\beta_2 x_2} \end{aligned} \tag{1.138}$$

where subscripts I, R and T are used to indicate incident, reflected and transmitted waves, respectively, and

$$\begin{aligned} & k = k_{P1}\sin\theta_{P1} = k_{S1}\sin\theta_{S1} = k_{P2}\sin\theta_{P2} = k_{S2}\sin\theta_{S2} \\ & \eta_1 = k_{P1}\cos\theta_{P1} \\ & \eta_2 = k_{P2}\cos\theta_{P2} \\ & \beta_1 = k_{S1}\cos\theta_{S1} \\ & \beta_2 = k_{S2}\cos\theta_{S2} \\ & k_{P1} = \frac{\omega}{c_{P1}}, \quad k_{P2} = \frac{\omega}{c_{P2}}, \quad k_{S1} = \frac{\omega}{c_{S1}}, \quad k_{S2} = \frac{\omega}{c_{S2}} \end{aligned} \tag{1.139}$$

From the continuity of displacement components u_1, u_2 and stress components σ_{12} and σ_{22} across the interface we get

$$ik(1 + R_{PP}) + i\beta_1 R_{PS} = ikT_{PP} - i\beta_2 T_{PS}$$

$$i\eta_1(-1 + R_{PP}) - ikR_{PS} = -i\eta_2 T_{PP} - ikT_{PS}$$

$$\mu_1\left\{2k\eta_1(1 - R_{PP}) - \beta_1^2 R_{PS} + k^2 R_{PS}\right\} = \mu_2\left\{2k\eta_2 T_{PP} - \beta_2^2 T_{PS} + k^2 T_{PS}\right\}$$

$$-\mu_1\left\{\left(k_{S1}^2 - 2k^2\right)(1 + R_{PP}) - 2k\beta_1 R_{PS}\right\} = -\mu_2\left\{\left(k_{S2}^2 - 2k^2\right)T_{PP} + 2k\beta_2 T_{PS}\right\}$$

The above equations can be written in the following form:

$$\begin{bmatrix} k & \beta_1 & -k & \beta_2 \\ \eta_1 & -k & \eta_2 & k \\ 2k\eta_1 & -\left(2k^2 - k_{S1}^2\right) & 2k\eta_2\mu_{21} & \left(2k^2 - k_{S2}^2\right)\mu_{21} \\ \left(2k^2 - k_{S1}^2\right) & 2k\beta_1 & -\left(2k^2 - k_{S2}^2\right)\mu_{21} & 2k\beta_2\mu_{21} \end{bmatrix} \begin{Bmatrix} R_{PP} \\ R_{PS} \\ T_{PP} \\ T_{PS} \end{Bmatrix} = \begin{Bmatrix} -k \\ \eta_1 \\ 2k\eta_1 \\ -\left(2k^2 - k_{S1}^2\right) \end{Bmatrix} \tag{1.140}$$

In Eq. (1.140) $\mu_{21} = \frac{\mu_2}{\mu_1}$; this equation set can be solved for R_{PP}, R_{PS}, T_{PP} and T_{PS}.

Example 1.15

For normal P-wave incidence ($\theta_{P1} = 0$ in Figure 1.22), calculate the reflected and transmitted wave amplitudes and potential fields in the two solids.

SOLUTION

From Eq. (1.139) one can write for $\theta_{P1} = 0$

$$k = k_{P1}\sin\theta_{P1} = k_{S1}\sin\theta_{S1} = k_{P2}\sin\theta_{P2} = k_{S2}\sin\theta_{S2} = 0$$

$$\Rightarrow \theta_{S1} = \theta_{P2} = \theta_{S2} = 0$$

$$\eta_1 = k_{P1}\cos\theta_{P1} = k_{P1}$$

$$\eta_2 = k_{P2}\cos\theta_{P2} = k_{P2}$$

$$\beta_1 = k_{S1}\cos\theta_{S1} = k_{S1}$$

$$\beta_2 = k_{S2}\cos\theta_{S2} = k_{S2}$$

Eq. (1.140) is then simplified to

$$\begin{bmatrix} 0 & k_{S1} & 0 & k_{S2} \\ k_{P1} & 0 & k_{P2} & 0 \\ 0 & k_{S1}^2 & 0 & -k_{S2}^2\mu_{21} \\ -k_{S1}^2 & 0 & k_{S2}^2\mu_{21} & 0 \end{bmatrix} \begin{Bmatrix} R_{PP} \\ R_{PS} \\ T_{PP} \\ T_{PS} \end{Bmatrix} = \begin{Bmatrix} 0 \\ k_{P1} \\ 0 \\ k_{S1}^2 \end{Bmatrix}$$

From the first and third algebraic equations of the above matrix equation one gets $R_{PS}=T_{PS}=0$; the remaining second and fourth equations form a two-by-two system of equations

$$\begin{bmatrix} k_{P1} & k_{P2} \\ -k_{S1}^2 & k_{S2}^2\mu_{21} \end{bmatrix} \begin{Bmatrix} R_{PP} \\ T_{PP} \end{Bmatrix} = \begin{Bmatrix} k_{P1} \\ k_{S1}^2 \end{Bmatrix}$$

that can be easily solved to obtain

$$R_{PP} = \frac{k_{P1}k_{S2}^2\mu_{21} - k_{P2}k_{S1}^2}{k_{P1}k_{S2}^2\mu_{21} + k_{P2}k_{S1}^2} = \frac{1-\dfrac{k_{P2}k_{S1}^2}{k_{P1}k_{S2}^2\mu_{21}}}{1+\dfrac{k_{P2}k_{S1}^2}{k_{P1}k_{S2}^2\mu_{21}}} = \frac{1-\dfrac{\mu_1 c_{P1} c_{S2}^2}{\mu_2 c_{P2} c_{S1}^2}}{1+\dfrac{\mu_1 c_{P1} c_{S2}^2}{\mu_2 c_{P2} c_{S1}^2}} = \frac{1-\dfrac{\rho_1 c_{P1}}{\rho_2 c_{P2}}}{1+\dfrac{\rho_1 c_{P1}}{\rho_2 c_{P2}}}$$

$$\Rightarrow R_{PP} = \frac{\rho_2 c_{P2} - \rho_1 c_{P1}}{\rho_2 c_{P2} + \rho_1 c_{P1}} = \frac{Z_2 - Z_1}{Z_2 + Z_1}$$

$$T_{PP} = \frac{2k_{P1}k_{S1}^2}{k_{P1}k_{S2}^2\mu_{21} + k_{P2}k_{S1}^2} = \frac{\dfrac{2k_{P1}k_{S1}^2}{k_{P1}k_{S2}^2\mu_{21}}}{1+\dfrac{k_{P2}k_{S1}^2}{k_{P1}k_{S2}^2\mu_{21}}} = \frac{2\dfrac{\mu_1 c_{S2}^2}{\mu_2 c_{S1}^2}}{1+\dfrac{\mu_1 c_{P1} c_{S2}^2}{\mu_2 c_{P2} c_{S1}^2}} = \frac{2\dfrac{\rho_1}{\rho_2}}{1+\dfrac{\rho_1 c_{P1}}{\rho_2 c_{P2}}} \tag{1.141}$$

$$\Rightarrow T_{PP} = \frac{\rho_1}{\rho_2}\frac{2\rho_2 c_{P2}}{\rho_2 c_{P2} + \rho_1 c_{P1}} = \frac{\rho_1}{\rho_2}\frac{2Z_2}{Z_2 + Z_1}$$

In Eq. (1.141) $Z_i = \rho_i c_{Pi}$ is known as the acoustic impedance. Note that when both materials are the same then $R_{PP}=0$ and $T_{PP}=1$.

Therefore, for normal incidence of a plane P-wave at the interface of two materials

$$\phi_1 = \phi_{IP} + \phi_{RPP} = e^{-ik_{P1}x_2} + \frac{Z_2 - Z_1}{Z_2 + Z_1}e^{ik_{P1}x_2}$$

$$\phi_2 = \phi_{TPP} = \frac{\rho_1}{\rho_2}\frac{2Z_2}{Z_2 + Z_1}e^{-ik_{P2}x_2}$$

$$\psi_1 = \psi_2 = 0$$

Example 1.16

Compute stress and displacement fields in the two solids for normal P-wave incidence ($\theta_{P1}=0$ in Figure 1.22).

SOLUTION

Potential fields for this case are shown in the above example. From these potential fields the displacement and stress components are obtained using the following relations (note that in this case $\psi=0$).

$$u_1 = \phi_{,1} + \psi_{,2} = \phi_{,1}$$

$$u_2 = \phi_{,2} - \psi_{,1} = \phi_{,2}$$

$$\sigma_{11} = -\mu\left\{k_S^2\phi + 2\phi_{,22} - 2\psi_{,12}\right\} = -\mu\left\{k_S^2\phi + 2\phi_{,22}\right\}$$

$$\sigma_{22} = -\mu\left\{k_S^2\phi + 2\phi_{,11} + 2\psi_{,12}\right\} = -\mu\left\{k_S^2\phi + 2\phi_{,11}\right\}$$

$$\sigma_{12} = \mu\left\{2\phi_{,12} + \psi_{,22} - \psi_{,11}\right\} = 2\mu\phi_{,12}$$

Hence, for solid 1

$$u_1 = \phi_{,1} = 0$$

$$u_2 = \phi_{,2} = ik_{P1}\left\{-e^{-ik_{P1}x_2} + R_{PP}e^{ik_{P1}x_2}\right\}$$

$$\sigma_{11} = -\mu\left\{k_S^2\phi + 2\phi_{,22}\right\} = \mu_1\left(2k_{P1}^2 - k_{S1}^2\right)\left(e^{-ik_{P1}x_2} + R_{PP}e^{ik_{P1}x_2}\right)$$

$$\sigma_{22} = -\mu\left\{k_S^2\phi + 2\phi_{,11}\right\} = -\mu_1 k_{S1}^2\left(e^{-ik_{P1}x_2} + R_{PP}e^{ik_{P1}x_2}\right)$$

$$\sigma_{12} = 2\mu\phi_{,12} = 0$$

and for solid 2

$$u_1 = \phi_{,1} = 0$$

$$u_2 = \phi_{,2} = -ik_{P2}T_{PP}e^{-ik_{P2}x_2}$$

$$\sigma_{11} = -\mu\left\{k_S^2\phi + 2\phi_{,22}\right\} = \mu_2\left(2k_{P2}^2 - k_{S2}^2\right)T_{PP}e^{-ik_{P2}x_2}$$

$$\sigma_{22} = -\mu\left\{k_S^2\phi + 2\phi_{,11}\right\} = -\mu_2 k_{S2}^2 T_{PP}e^{-ik_{P2}x_2}$$

$$\sigma_{12} = 2\mu\phi_{,12} = 0$$

where R_{PP} and T_{PP} have been defined in Eq. (1.141)

Note that at the interface ($x_2=0$) nonzero displacement and stress components can be computed from the expressions given for solids 1 and 2. Substituting $x_2=0$ in the expressions of the first solid one gets

$$u_2 = ik_{P1}\left\{-1 + R_{PP}\right\} = ik_{P1}\left\{-1 + \frac{Z_2 - Z_1}{Z_2 + Z_1}\right\} = \frac{-2ik_{P1}Z_1}{Z_2 + Z_1} = -\frac{2i\omega\rho_1}{Z_2 + Z_1}$$

$$\sigma_{11} = \mu_1\left(2k_{P1}^2 - k_{S1}^2\right)\left(1 + R_{PP}\right) = \left(\frac{2}{\kappa_1^2} - 1\right)\frac{2\rho_1\omega^2 Z_2}{Z_2 + Z_1}$$

$$\sigma_{22} = -\mu_1 k_{S1}^2\left(1 + R_{PP}\right) = -\frac{2\rho_1\omega^2 Z_2}{Z_2 + Z_1}$$

If $x_2 = 0$ is substituted in the expressions of the second solid, then we obtain

$$u_2 = -ik_{P2}T_{PP} = -i\frac{\omega}{c_{P2}}\frac{\rho_1}{\rho_2}\frac{2Z_2}{Z_2+Z_1} = -\frac{2i\omega\rho_1}{Z_2+Z_1}$$

$$\sigma_{11} = \mu_2\left(2k_{P2}^2 - k_{S2}^2\right)T_{PP} = \left(\frac{2}{\kappa_2^2}-1\right)\frac{2\rho_1\omega^2 Z_2}{Z_2+Z_1}$$

$$\sigma_{22} = -\mu_2 k_{S2}^2 T_{PP} = -\frac{2\rho_1\omega^2 Z_2}{Z_2+Z_1}$$

Note that u_2 and σ_{22} are continuous across the interface, but σ_{11} is not since $\kappa_1\left(=\frac{c_{P1}}{c_{S1}}\right)$ and $\kappa_2\left(=\frac{c_{P2}}{c_{S2}}\right)$ are different.

1.2.8.2 SV-Wave Striking an Interface

Figure 1.23 shows a plane SV-wave of unit amplitude striking a plane interface between two linear elastic isotropic solids. Both reflected and transmitted waves have P and SV-wave components and their amplitudes are denoted by R_{SP}, R_{SS}, T_{SP} and T_{SS}.

Wave potentials for these five types of waves are as follows:

$$\begin{aligned}
\psi_{IS} &= e^{ikx_1 - i\beta_1 x_2}\\
\phi_{RSP} &= R_{SP}e^{ikx_1 + i\eta_1 x_2}\\
\phi_{TSP} &= T_{SP}e^{ikx_1 - i\eta_2 x_2}\\
\psi_{RSS} &= R_{SS}e^{ikx_1 + i\beta_1 x_2}\\
\psi_{TSS} &= T_{SS}e^{ikx_1 - i\beta_2 x_2}
\end{aligned} \tag{1.142}$$

where subscripts *I*, *R* and *T* are used to indicate incident, reflected and transmitted waves, respectively. k, η_i and β_i are defined in Eq. (1.139).

From continuity of displacement components u_1, u_2 and stress components σ_{12} and σ_{22} across the interface we get

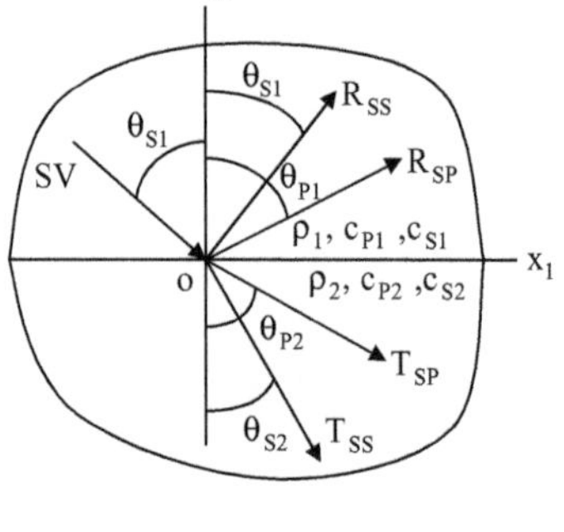

Figure 1.23 Reflected and transmitted waves at an interface for SV-wave incidence.

$$ikR_{SP} + i\beta_1(-1 + R_{SS}) = ikT_{SP} - i\beta_2 T_{SS}$$

$$i\eta_1 R_{SP} - ik(1 + R_{SS}) = -i\eta_2 T_{SP} - ikT_{SS}$$

$$\mu_1\left\{-2k\eta_1 R_{SP} + \left(2k^2 - k_{S1}^2\right)(1 + R_{SS})\right\} = \mu_2\left\{2k\eta_2 T_{SP} + \left(2k^2 - k_{S2}^2\right)T_{SS}\right\}$$

$$-\mu_1\left\{\left(k_{S1}^2 - 2k^2\right)R_{SP} + 2k\beta_1(1 - R_{SS})\right\} = -\mu_2\left\{\left(k_{S2}^2 - 2k^2\right)T_{SP} + 2k\beta_2 T_{SS}\right\}$$

The above equations can be also written in the following form:

$$\begin{bmatrix} k & \beta_1 & -k & \beta_2 \\ \eta_1 & -k & \eta_2 & k \\ 2k\eta_1 & -\left(2k^2 - k_{S1}^2\right) & 2k\eta_2\mu_{21} & \left(2k^2 - k_{S2}^2\right)\mu_{21} \\ \left(2k^2 - k_{S1}^2\right) & 2k\beta_1 & -\left(2k^2 - k_{S2}^2\right)\mu_{21} & 2k\beta_2\mu_{21} \end{bmatrix} \begin{Bmatrix} R_{SP} \\ R_{SS} \\ T_{SP} \\ T_{SS} \end{Bmatrix} = \begin{Bmatrix} \beta_1 \\ k \\ \left(2k^2 - k_{S1}^2\right) \\ 2k\beta_1 \end{Bmatrix} \tag{1.143}$$

In Eq. (1.143) $\mu_{21} = \dfrac{\mu_2}{\mu_1}$; this equation can be solved for R_{SP}, R_{SS}, T_{SP} and T_{SS}.

Example 1.17

For normal SV-wave incidence ($\theta_{S1} = 0$ in Figure 1.23), calculate the reflected and transmitted wave amplitudes and potential fields in the two solids.

SOLUTION

From Eq. (1.139) one can write for $\theta_{S1} = 0$

$$k = k_{P1}\sin\theta_{P1} = k_{S1}\sin\theta_{S1} = k_{P2}\sin\theta_{P2} = k_{S2}\sin\theta_{S2} = 0$$

$$\Rightarrow \theta_{P1} = \theta_{P2} = \theta_{S2} = 0$$

$$\eta_1 = k_{P1}\cos\theta_{P1} = k_{P1} \qquad \eta_2 = k_{P2}\cos\theta_{P2} = k_{P2}$$

$$\beta_1 = k_{S1}\cos\theta_{S1} = k_{S1} \qquad \beta_2 = k_{S2}\cos\theta_{S2} = k_{S2}$$

Hence, Eq. (1.143) is simplified to

$$\begin{bmatrix} 0 & k_{S1} & 0 & k_{S2} \\ k_{P1} & 0 & k_{P2} & 0 \\ 0 & k_{S1}^2 & 0 & -k_{S2}^2\mu_{21} \\ -k_{S1}^2 & 0 & k_{S2}^2\mu_{21} & 0 \end{bmatrix} \begin{Bmatrix} R_{SP} \\ R_{SS} \\ T_{SP} \\ T_{SS} \end{Bmatrix} = \begin{Bmatrix} k_{S1} \\ 0 \\ -k_{S1}^2 \\ 0 \end{Bmatrix}$$

From the second and fourth algebraic equations of the above matrix equation one gets $R_{SP} = T_{SP} = 0$; the remaining first and third equations form a two-by-two system of equations

$$\begin{bmatrix} k_{S1} & k_{S2} \\ k_{S1}^2 & -k_{S2}^2\mu_{21} \end{bmatrix} \begin{Bmatrix} R_{SS} \\ T_{SS} \end{Bmatrix} = \begin{Bmatrix} k_{S1} \\ -k_{S1}^2 \end{Bmatrix}$$

that can be easily solved to obtain

$$R_{SS} = \frac{k_{S1}k_{S2}^2\mu_{21} - k_{S2}k_{S1}^2}{k_{S1}k_{S2}^2\mu_{21} + k_{S2}k_{S1}^2} = \frac{1 - \dfrac{k_{S2}k_{S1}^2}{k_{S1}k_{S2}^2\mu_{21}}}{1 + \dfrac{k_{S2}k_{S1}^2}{k_{S1}k_{S2}^2\mu_{21}}} = \frac{1 - \dfrac{\mu_1 c_{S2}}{\mu_2 c_{S1}}}{1 + \dfrac{\mu_1 c_{S2}}{\mu_2 c_{S1}}} = \frac{1 - \dfrac{\rho_1 c_{S1}}{\rho_2 c_{S2}}}{1 + \dfrac{\rho_1 c_{S1}}{\rho_2 c_{S2}}}$$

$$\Rightarrow R_{SS} = \frac{\rho_2 c_{S2} - \rho_1 c_{S1}}{\rho_2 c_{S2} + \rho_1 c_{S1}} = \frac{Z_{2S} - Z_{1S}}{Z_{2S} + Z_{1S}}$$

$$T_{SS} = \frac{2k_{S1}^3}{k_{S1}k_{S2}^2\mu_{21} + k_{S2}k_{S1}^2} = \frac{\dfrac{2k_{S1}^3}{k_{S1}k_{S2}^2\mu_{21}}}{1 + \dfrac{k_{S2}k_{S1}^2}{k_{S1}k_{S2}^2\mu_{21}}} = \frac{2\dfrac{\mu_1 c_{S2}^2}{\mu_2 c_{S1}^2}}{1 + \dfrac{\mu_1 c_{S2}}{\mu_2 c_{S1}}} = \frac{2\dfrac{\rho_1}{\rho_2}}{1 + \dfrac{\rho_1 c_{S1}}{\rho_2 c_{S2}}} \tag{1.144}$$

$$\Rightarrow T_{SS} = \frac{\rho_1}{\rho_2}\frac{2\rho_2 c_{S2}}{\rho_2 c_{S2} + \rho_1 c_{S1}} = \frac{\rho_1}{\rho_2}\frac{2Z_{2S}}{Z_{2S} + Z_{1S}}$$

In Eq. (1.144) $Z_{iS} = \rho_i c_{Si}$ is the acoustic impedance computed using shear wave speed, instead of the P-wave speed. Note that when both materials are the same then $R_{SS} = 0$ and $T_{SS} = 1$.

Hence, for normal incidence of a plane SV-wave

$$\psi_1 = \psi_{IS} + \psi_{RSS} = e^{-ik_{S1}x_2} + \frac{Z_{2S} - Z_{1S}}{Z_{2S} + Z_{1S}} e^{ik_{S1}x_2}$$

$$\psi_2 = \psi_{TSS} = \frac{\rho_1}{\rho_2}\frac{2Z_{2S}}{Z_{2S} + Z_{1S}} e^{-ik_{S2}x_2}$$

$$\phi_1 = \phi_2 = 0$$

Example 1.18

Compute the stress and displacement fields in the two solids for normal SV-wave incidence ($\theta_{S1} = 0$ in Figure 1.23).

SOLUTION

Potential fields for this case are shown in the above example. From these potential fields, displacement and stress components are obtained using the following relations (in this case $\phi = 0$):

$$u_1 = \phi_{,1} + \psi_{,2} = \psi_{,2}$$

$$u_2 = \phi_{,2} - \psi_{,1} = -\psi_{,1}$$

$$\sigma_{11} = -\mu\left\{k_S^2\phi + 2\phi_{,22} - 2\psi_{,12}\right\} = -2\mu\psi_{,12}$$

$$\sigma_{22} = -\mu\left\{k_S^2\phi + 2\phi_{,11} + 2\psi_{,12}\right\} = -2\mu\psi_{,12}$$

$$\sigma_{12} = \mu\left\{2\phi_{,12} + \psi_{,22} - \psi_{,11}\right\} = \mu\left\{\psi_{,22} - \psi_{,11}\right\}$$

Hence, for solid 1

$$u_1 = \psi_{,2} = ik_{S1}\left(-e^{-ik_{S1}x_2} + R_{SS}e^{ik_{S1}x_2}\right)$$

$$u_2 = -\psi_{,1} = 0$$

$$\sigma_{11} = -2\mu\psi_{,12} = 0$$

$$\sigma_{22} = -2\mu\psi_{,12} = 0$$

$$\sigma_{12} = \mu\{\psi_{,22} - \psi_{,11}\} = -\mu_1 k_{S1}^2\left(e^{-ik_{S1}x_2} + R_{SS}e^{ik_{S1}x_2}\right)$$

and for solid 2

$$u_1 = \psi_{,2} = -ik_{S2}T_{SS}e^{-ik_{S2}x_2}$$

$$u_2 = -\psi_{,1} = 0$$

$$\sigma_{11} = -2\mu\psi_{,12} = 0$$

$$\sigma_{22} = -2\mu\psi_{,12} = 0$$

$$\sigma_{12} = \mu\{\psi_{,22} - \psi_{,11}\} = -\mu_2 k_{S2}^2 T_{SS}e^{ik_{S2}x_2}$$

where R_{SS} and T_{SS} have been defined in Eq. (1.144).

Note that at the interface ($x_2 = 0$) nonzero displacement and stress components can be computed from the expressions given for solids 1 and 2. Substituting $x_2 = 0$ in the expressions for solid 1 one gets

$$u_1 = ik_{S1}\left(-1 + R_{SS}\right) = -\frac{i\omega}{c_{S1}}\frac{2Z_{1S}}{Z_{2S} + Z_{1S}} = \frac{-2i\omega\rho_1}{Z_{2S} + Z_{1S}}$$

$$\sigma_{12} = -\mu_1 k_{S1}^2\left(1 + R_{SS}\right) = -\rho_1\omega^2\frac{2Z_{2S}}{Z_{2S} + Z_{1S}} = \frac{-2\rho_1\omega^2 Z_{2S}}{Z_{2S} + Z_{1S}}$$

If $x_2 = 0$ is substituted in the expressions for solid 2, then we obtain

$$u_1 = -ik_{S2}T_{SS} = -i\frac{\omega}{c_{S2}}\frac{\rho_1}{\rho_2}\frac{2Z_{2S}}{Z_{2S} + Z_{1S}} = \frac{-2i\omega\rho_1}{Z_{2S} + Z_{1S}}$$

$$\sigma_{12} = -\mu_2 k_{S2}^2 T_{SS} = -\rho_2\omega^2\frac{\rho_1}{\rho_2}\frac{2Z_{2S}}{Z_{2S} + Z_{1S}} = \frac{-2\rho_1\omega^2 Z_{2S}}{Z_{2S} + Z_{1S}}$$

Note that u_1 and σ_{12} are continuous across the interface.

1.2.9 Rayleigh Waves in a Homogeneous Half-Space

P-, SV- and SH-type waves, discussed above, propagate inside an elastic body and are known as body waves. When these waves meet a free surface or an interface, they go through reflection and transmission. Some wave motions, however, are confined near a free surface or an interface and are called surface waves or interface waves. A Rayleigh wave is a surface wave.

We have seen earlier that the stress-free boundary conditions at $x_2 = 0$ (Figure 1.17) can be satisfied by the potential expressions given in equation set (1.121). Now let us try to satisfy the stress-free boundary conditions at $x_2 = 0$ with a different set of potential expressions, as given below. If it is possible to satisfy the appropriate boundary conditions with different potential expressions that give a different type of particle motion, then one can logically conclude that this new wave motion can exist in the half-space for that boundary condition. The potential expressions that are tried out now are

$$\begin{aligned} \phi &= A\exp(ikx_1 - \eta x_2) \\ \psi &= B\exp(ikx_1 - \beta x_2) \end{aligned} \tag{1.145}$$

Time dependence exp($-i\omega t$) is implied. Note that as x_2 increases, ϕ and ψ decay exponentially. The displacements and stresses, associated with these potentials, are confined near the free surface at $x_2=0$. Clearly, Eq. (1.145) represents a wave motion that propagates in the x_1 direction with a velocity c, where $c=\omega/k$.

Substituting Eq. (1.145) into the governing wave equations (Eq.1.107) one gets

$$\begin{aligned} -k^2+\eta^2+k_P^2 &= 0 \quad \Rightarrow \eta = \sqrt{k^2-k_P^2} \\ -k^2+\beta^2+k_S^2 &= 0 \quad \Rightarrow \beta = \sqrt{k^2-k_S^2} \end{aligned} \tag{1.146}$$

From Eq. (1.102)

$$\begin{aligned} \sigma_{22} &= -\mu\left\{k_S^2\phi + 2(\psi_{,12}+\phi_{,11})\right\} \\ &= -\mu\left\{\left(k_S^2-2k^2\right)Ae^{ikx_1-\eta x_2} - 2ik\beta Be^{ikx_1-\beta x_2}\right\} \end{aligned}$$

$$\sigma_{12} = \mu\{2\phi_{,12}+\psi_{,22}-\psi_{,11}\} = \mu\left\{(-2ik\eta)Ae^{ikx_1-\eta x_2} + \left(k^2+\beta^2\right)Be^{ikx_1-\beta x_2}\right\}$$

Satisfying the stress-free boundary conditions at $x_2=0$ one gets

$$\left\{\left(2k^2-k_S^2\right)A + 2ik\beta B\right\}e^{ikx_1} = 0$$

$$\left\{-2ik\eta A + \left(2k^2-k_S^2\right)B\right\}e^{ikx_1} = 0$$

The above equations are satisfied if

$$\begin{bmatrix} \left(2k^2-k_S^2\right) & 2ik\beta \\ -2ik\eta & \left(2k^2-k_S^2\right) \end{bmatrix} \begin{Bmatrix} A \\ B \end{Bmatrix} = \begin{Bmatrix} 0 \\ 0 \end{Bmatrix} \tag{1.147}$$

To have nontrivial solutions of A and B, the determinant of the coefficient matrix must be zero

$$\left(2k^2-k_S^2\right)^2 - 4k^2\eta\beta = 0 \tag{1.148}$$

Eq. (1.148) can be also written as

$$\left(2\frac{\omega^2}{c^2}-\frac{\omega^2}{c_S^2}\right)^2 - 4\frac{\omega^2}{c^2}\left(\frac{\omega^2}{c^2}-\frac{\omega^2}{c_P^2}\right)^{\frac{1}{2}}\left(\frac{\omega^2}{c^2}-\frac{\omega^2}{c_S^2}\right)^{\frac{1}{2}} = 0$$

$$\Rightarrow \left(2-\frac{c^2}{c_S^2}\right)^2 - 4\left(1-\frac{c^2}{c_P^2}\right)^{\frac{1}{2}}\left(1-\frac{c^2}{c_S^2}\right)^{\frac{1}{2}} = 0 \tag{1.148a}$$

$$\Rightarrow \left(2-\xi^2\right)^2 - 4\sqrt{\left(1-\xi^2\right)\left(1-\frac{\xi^2}{\kappa^2}\right)} = 0$$

where $\xi = \frac{c}{c_S}$ and $\kappa = \frac{c_P}{c_S}$.

From Eq. (1.146) $\eta = k\sqrt{1-\frac{\xi^2}{\kappa^2}}$, $\beta = k\sqrt{1-\xi^2}$. Since $K>1$, if $\xi<1$, then η and β will have real values. One can show that Eq. (1.148) has only one root between 0 and 1 (Mal and Singh, 1991). To prove it let us remove the radicals from Eq. (1.148a) and, for simplicity, introduce the variable $p=1/K$ to obtain

$$\begin{aligned}
&(2-\xi^2)^4 = 16(1-\xi^2)(1-p^2\xi^2)\\
\Rightarrow\ & 16-32\xi^2+24\xi^4-8\xi^6+\xi^8 = 16-16(1+p^2)\xi^2+16p^2\xi^4\\
\Rightarrow\ & -32+24\xi^2-8\xi^4+\xi^6 = -16(1+p^2)+16p^2\xi^2\\
\Rightarrow\ & \xi^6-8\xi^4+(24-16p^2)\xi^2-16(1-p^2) = 0 = f(\xi^2)
\end{aligned} \tag{1.148b}$$

Note that $f(0) = -16(1-p^2) < 0$ and $f(1) = 1-8+24-16 = 1 > 0$. Eq. (1.148b) is a sixth order polynomial equation of ξ or a third order polynomial equation of ξ^2. Hence, ξ^2 can have a maximum of three roots. Since $f(\xi^2)$ change sign between 0 and 1, as shown above, it must have a minimum of one or a maximum of all three roots between 0 and 1. However, for ξ between 0 and 1 $f'(0) = 24-16p^2 > 0$ and $f''(\xi^2) = 6\xi^2-16 < 0$. Therefore, the slope of $f(\xi^2)$ should have a monotonic variation between 0 and 1. In other words, $f(\xi^2)$ cannot change its trend (either increasing or decreasing) more than once between 0 and 1. So it cannot have three roots between 0 and 1. Therefore, Eq. (1.148b) can have only one root of ξ^2 between 0 and 1. We will denote this root as ξ_R and corresponding c ($=\xi_R.c_S$) as c_R. This wave that is confined near the free surface and has a velocity c_R is called a Rayleigh wave.

Example 1.19

For an elastic solid with Poisson's ratio $\nu=0.25$, calculate c_R.

SOLUTION

From the definitions of p, c_S, c_P and the relations given in Table 1.1, one can write

$$p^2 = \left(\frac{c_S}{c_P}\right)^2 = \frac{\mu}{\lambda+2\mu} = \frac{\mu}{\frac{2\mu\nu}{1-2\nu}+2\mu} = \frac{\mu-2\mu\nu}{-2\mu\nu+2\mu} = \frac{1-2\nu}{2-2\nu} \tag{1.149}$$

For $\nu=0.25$, $p^2=1/3$ and Eq. (1.148b) becomes

$$\xi^6-8\xi^4+\left(24-\frac{16}{3}\right)\xi^2-16\left(1-\frac{1}{3}\right) = 3\xi^6-24\xi^4+56\xi^2-32 = 0$$

Three real roots of the above equation are

$$\xi^2 = 4,\ \left(2+\frac{2}{\sqrt{3}}\right),\ \left(2-\frac{2}{\sqrt{3}}\right)$$

However, only one of these three roots is less than 1, which is necessary to have real values of η and β of Eq. (1.145). Thus,

$$\xi_R^2 = \left(2 - \frac{2}{\sqrt{3}}\right) \Rightarrow \xi_R = 0.91948$$

and $c_R = 0.91948 c_S$.

Similarly, Rayleigh wave speeds for other values of the Poisson's ratio can be computed. Two approximate equations to compute the Rayleigh wave speed from the shear wave speed and Poisson's ratio have been proposed,

$$c_R = \frac{0.862 + 1.14\nu}{1+\nu} c_S \tag{1.150a}$$

$$c_R = \frac{0.87 + 1.12\nu}{1+\nu} c_S \tag{1.150b}$$

Eqs. (1.150a) and (1.150b) are from Schmerr (1998) and Viktorov (1967), respectively.

The ratio of the Rayleigh wave speed and the shear wave speed, computed from the exact and approximate equations for different Poisson's ratios, are given in Table 1.3.

From Eq. (1.147) one can write

$$\frac{B}{A} = \frac{2ik\eta}{\left(2k^2 - k_S^2\right)} = -\frac{\left(2k^2 - k_S^2\right)}{2ik\beta} \tag{1.151}$$

Hence, the displacement field is given by

$$\begin{aligned} u_1 &= \phi_{,1} + \psi_{,2} = \left(ikAe^{-\eta x_2} - \beta Be^{-\beta x_2}\right)e^{ikx_1} = ikA\left(e^{-\eta x_2} + \left[2 - \frac{k_S^2}{2k^2}\right]e^{-\beta x_2}\right)e^{ikx_1} \\ u_2 &= \phi_{,2} - \psi_{,1} = \left(-\eta Ae^{-\eta x_2} + ikBe^{-\beta x_2}\right)e^{ikx_1} = -\eta A\left(e^{-\eta x_2} + \frac{2k^2 - k_S^2}{2\eta\beta}e^{-\beta x_2}\right)e^{ikx_1} \end{aligned} \tag{1.152}$$

In Eq. (1.152) there is only one undetermined constant "A" from which the amplitude of the Rayleigh wave can be computed.

At the surface ($x_2 = 0$)

$$\begin{aligned} u_1 &= ikA\left(3 - \frac{k_S^2}{2k^2}\right)e^{ikx_1} \\ u_2 &= -\eta A\left(1 + \frac{2k^2 - k_S^2}{2\eta\beta}\right)e^{ikx_1} \end{aligned} \tag{1.153}$$

From Eq. (1.153) it is clear that the two displacement components are 90° out-of-phase, because the coefficient of the exponential term of

Table 1.3: Ratio of the Rayleigh Wave Speed and Shear Wave Speed for Different Poisson's Ratios

ν	0	0.05	0.15	0.25	0.30	0.40	0.50
c_R/c_S (exact)	0.87	0.88	0.90	0.92	0.93	0.94	0.95
c_R/c_S (Eq.150a)	0.86	0.88	0.90	0.92	0.93	0.94	0.95
c_R/c_S (Eq.150b)	0.87	0.88	0.90	0.92	0.93	0.94	0.95

the u_1 expression is imaginary while that of the u_2 expression is real. After a detailed analysis of these displacement expressions one can show [Mal and Singh, 1991] that for a propagating Rayleigh wave the particles have an elliptic motion and the particles on the crest of the wave move opposite to the wave propagation direction. This type of motion is called *retrograde* elliptic motion.

1.2.10 Love Wave

In the previous section it has been shown that the Rayleigh wave propagates along a stress-free surface in a homogeneous solid. Let us now investigate if such propagation of antiplane motion is possible. To study this we consider an antiplane wave of the form

$$u_3 = Ae^{ikx_1 - \beta x_2}$$

propagating in the x_1-direction along the stress-free surface (at $x_2=0$) of an elastic half-space. Note that the displacement field decays exponentially as one moves away from the surface.

Shear stress σ_{23} associated with this displacement field is given by

$$\sigma_{23} = \mu u_{3,2} = -\mu A\beta e^{ikx_1 - \beta x_2}$$

Note that at $x_2=0$ this stress component is nonzero unless A is zero. Therefore, SH-waves cannot propagate parallel to a stress-free surface.

Let us now consider antiplane motions in a layered half-space, as shown in Figure 1.24.

Inside the layer of thickness h down-going and up-going SH-waves propagate, and in the substrate an antiplane wave propagates parallel to the interface. The strength of this horizontally propagating wave decays exponentially with x_2. Displacement fields associated with these antiplane waves can be written in the following form:

$$\begin{aligned} u_3^1 &= Ae^{ikx_1 - i\beta_1 x_2} + Be^{ikx_1 + i\beta_1 x_2} \\ u_3^2 &= Ce^{ikx_1 - \beta_2 x_2} \end{aligned} \tag{1.154}$$

In the above equation, A, B, C are the amplitudes of the three waves and the superscripts 1 and 2 indicate whether the displacement is computed in the layer or in the substrate. β_1 and β_2 are real numbers and are defined by

$$\begin{aligned} \beta_1 &= \sqrt{k_{S1}^2 - k^2} \\ \beta_2 &= \sqrt{k^2 - k_{S2}^2} \end{aligned} \tag{1.154a}$$

From the stress-free boundary condition at $x_2=0$ we can write

$$\mu_1 u_{3,2}^1\Big|_{x_2=0} = i\beta_1(-A+B)e^{ikx_1} = 0$$

Therefore, $A=B$.

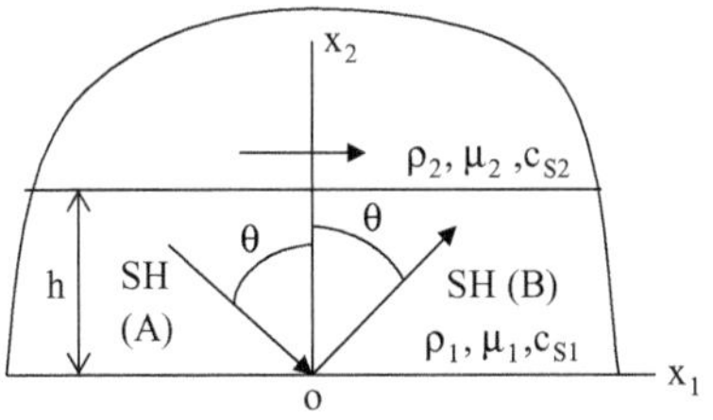

Figure 1.24 Possible antiplane motions in a layered half-space.

From displacement and stress continuity conditions at $x_2 = h$,

$$u_3^1\Big|_{x_3=h} = u_3^2\Big|_{x_3=h} \Rightarrow A\left(e^{-i\beta_1 h} + e^{i\beta_1 h}\right)e^{ikx_1} = Ce^{ikx_1 - \beta_2 h}$$

$$\Rightarrow 2A\cos(\beta_1 h) = Ce^{-\beta_2 h}$$

$$\mu_1 u_{3,2}^1\Big|_{x_3=h} = \mu_2 u_{3,2}^2\Big|_{x_3=h} \Rightarrow i\mu_1\beta_1 A\left(-e^{-i\beta_1 h} + e^{i\beta_1 h}\right)e^{ikx_1} = -\mu_2\beta_2 Ce^{ikx_1 - \beta_2 h}$$

$$\Rightarrow 2A\sin(\beta_1 h) = \frac{\mu_2\beta_2}{\mu_1\beta_1}Ce^{-\beta_2 h}$$

From the above two equations one gets

$$\tan(\beta_1 h) = \frac{\mu_2\beta_2}{\mu_1\beta_1} \Rightarrow \tan\left(h\sqrt{k_{S1}^2 - k^2}\right) = \frac{\mu_2\sqrt{k^2 - k_{S2}^2}}{\mu_1\sqrt{k_{S1}^2 - k^2}} \tag{1.155}$$

Eq. (1.155) can be solved for k, and from the relation $k=\omega/c_{Lo}$ the velocity c_{Lo} of the horizontally propagating wave of Figure 1.24 can be obtained. This wave is known as the *Love wave.* Note that k and c_{Lo} are functions of the wave frequency ω; in other words, this wave is *dispersive.* Eq. (1.155) is known as the *dispersion equation* and it has multiple solutions for k and c_{Lo} for a given value of the frequency ω. For a given frequency ω, the displacement and stress fields inside the solid are different for different solutions of k and c_{Lo}. These displacement and stress profiles are called mode shapes. Different modes are associated with different wave speeds. Wave velocity of a specific mode at a certain frequency is constant, and this velocity is known as the *phase velocity.*

1.2.11 Rayleigh Waves in a Layered Half-Space

Rayleigh wave propagation in an isotropic half-space has been studied in Section 1.2.9. In Section 1.2.10, it was shown that unlike Rayleigh waves, SH-waves cannot propagate parallel to a stress-free surface in an isotropic solid half-space; however, a layered solid half-space can sustain SH motion (Love wave) propagating parallel to the stress-free surface as shown in Figure 1.24. Let us now investigate if the in-plane (P-SV) motion or the Rayleigh wave can propagate parallel to the stress-free surface in a layered half-space as shown in Figure 1.25.

In Figure 1.25 up-going and down-going P-/SV-waves are shown in the layer and a horizontally propagating in-plane wave is shown in the substrate. Potential fields for these different waves are given in Eq. (1.156).

$$\begin{aligned}
\phi_1 &= (a_1 e^{-i\eta_1 x_2} + b_1 e^{i\eta_1 x_2})e^{ikx_1} \\
\psi_1 &= (c_1 e^{-i\beta_1 x_2} + d_1 e^{i\beta_1 x_2})e^{ikx_1} \\
\phi_2 &= b_2 e^{-\eta_2 x_2} e^{ikx_1} \\
\psi_2 &= d_2 e^{-\beta_2 x_2} e^{ikx_1}
\end{aligned} \tag{1.156}$$

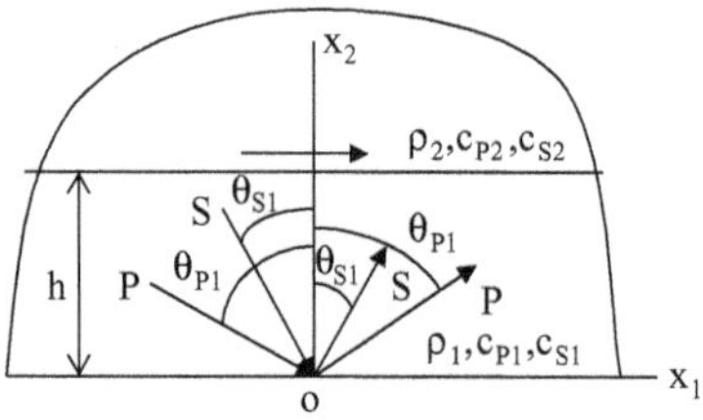

Figure 1.25 Possible in-plane motions in a layered half-space.

where

$$\begin{aligned}
\eta_1 &= \sqrt{k_{P1}^2 - k^2} \\
\beta_1 &= \sqrt{k_{S1}^2 - k^2} \\
\eta_2 &= \sqrt{k^2 - k_{P2}^2} \\
\beta_2 &= \sqrt{k^2 - k_{S2}^2}
\end{aligned} \tag{1.156a}$$

Subscripts 1 and 2 are used for the layer and the substrate, respectively. Note that the potentials decay exponentially in the substrate. Relations between the unknown coefficients a_1, b_1, c_1, d_1, b_2 and d_2 can be obtained from the two boundary conditions at $x_2=0$ and four interface continuity conditions at $x_2=h$. These six conditions are listed below:

$$1)\ \sigma_{22}^1\Big|_{x2=0} = -\mu_1\left\{k_{S1}^2\phi_1 + 2(\psi_{1,12} + \phi_{1,11})\right\}_{x2=0}$$

$$= -\mu_1\left\{\left(k_{S1}^2 - 2k^2\right)(a_1 + b_1) + 2k\beta_1(c_1 - d_1)\right\}e^{ikx_1} = 0$$

$$2)\ \sigma_{12}^1\Big|_{x2=0} = \mu_1\left\{2\phi_{1,12} + \psi_{1,22} - \psi_{1,11}\right\}$$

$$= \mu_1\left\{(2k\eta_1)(a_1 - b_1) + \left(k^2 - \beta_1^2\right)(c_1 + d_1)\right\}e^{ikx_1} = 0$$

$$3)\ \sigma_{22}^1\Big|_{x2=h} = \sigma_{22}^2\Big|_{x2=h} \Rightarrow -\mu_1\left\{k_{S1}^2\phi_1 + 2(\psi_{1,12} + \phi_{1,11})\right\}_{x2=h}$$

$$= -\mu_2\left\{k_{S2}^2\phi_2 + 2(\psi_{2,12} + \phi_{2,11})\right\}_{x2=h}$$

$$\Rightarrow -\mu_1\left\{\left(k_{S1}^2 - 2k^2\right)\left(a_1E_1^{-1} + b_1E_1\right) + 2k\beta_1\left(c_1B_1^{-1} - d_1B_1\right)\right\}e^{ikx_1}$$

$$= -\mu_2\left\{\left(k_{S2}^2 - 2k^2\right)b_2E_2 - 2ik\beta_2 d_2 B_2\right\}e^{ikx_1}$$

$$4)\ \sigma_{12}^1\Big|_{x2=h} = \sigma_{12}^2\Big|_{x2=h} \Rightarrow \mu_1\left\{2\phi_{1,12} + \psi_{1,22} - \psi_{1,11}\right\}_{x2=h}$$

$$= \mu_2\left\{2\phi_{2,12} + \psi_{2,22} - \psi_{2,11}\right\}_{x2=h}$$

$$\Rightarrow \mu_1\left\{(2k\eta_1)\left(a_1E_1^{-1} - b_1E_1\right) + \left(k^2 - \beta_1^2\right)\left(c_1B_1^{-1} + d_1B_1\right)\right\}e^{ikx_1}$$

$$= \mu_2\left\{(-2ik\eta_2)b_2E_2 + \left(k^2 + \beta_2^2\right)d_2B_2\right\}e^{ikx_1}$$

$$5)\ u_1^1\Big|_{x2=h} = u_1^2\Big|_{x2=h} \Rightarrow \left\{\phi_{1,1} + \psi_{1,2}\right\}_{x2=h} = \left\{\phi_{2,1} + \psi_{2,2}\right\}_{x2=h}$$

$$\Rightarrow \left\{ik\left(a_1E_1^{-1} + b_1E_1\right) + i\beta_1\left(-c_1B_1^{-1} + d_1B_1\right)\right\}e^{ikx_1}$$

$$= \left\{ikb_2E_2 - \beta_2 d_2 B_2\right\}e^{ikx_1}$$

$$6)\ u_2^1\Big|_{x2=h} = u_2^2\Big|_{x2=h} \Rightarrow \left\{\phi_{1,2} - \psi_{1,1}\right\}_{x2=h} = \left\{\phi_{2,2} - \psi_{2,1}\right\}_{x2=h}$$

$$\Rightarrow \left\{i\eta_1\left(-a_1E_1^{-1} + b_1E_1\right) - ik\left(c_1B_1^{-1} + d_1B_1\right)\right\}e^{ikx_1}$$

$$= \left\{-\eta_2 b_2E_2 - ikd_2B_2\right\}e^{ikx_1}$$

where

$$E_1 = e^{i\eta_1 h}$$

$$B_1 = e^{i\beta_1 h}$$

$$E_2 = e^{-\eta_2 h}$$

$$B_2 = e^{-\beta_2 h}$$

The above six conditions give the following matrix equation:

$$[A]\begin{Bmatrix} a_1 \\ b_1 \\ c_1 \\ d_1 \\ b_2 \\ d_2 \end{Bmatrix} = \begin{Bmatrix} 0 \\ 0 \\ 0 \\ 0 \\ 0 \\ 0 \end{Bmatrix} \tag{1.157}$$

where the coefficient matrix $[A]$ is given by

$$\begin{bmatrix} (2k^2 - k_{S1}^2) & (2k^2 - k_{S1}^2) & -2k\beta_1 & 2k\beta_1 & 0 & 0 \\ 2k\eta_1 & -2k\eta_1 & (2k^2 - k_{S1}^2) & (2k^2 - k_{S1}^2) & 0 & 0 \\ (2k^2 - k_{S1}^2)\mu_1 E_1^{-1} & (2k^2 - k_{S1}^2)\mu_1 E_1 & -2k\beta_1 B_1^{-1} & 2k\beta_1 B_1 & -(2k^2 - k_{S2}^2)\mu_2 E_2 & -2ik\beta_2\mu_2 B_2 \\ 2k\eta_1\mu_1 E_1^{-1} & -2k\eta_1\mu_1 E_1 & (2k^2 - k_{S1}^2)\mu_1 B_1^{-1} & (2k^2 - k_{S1}^2)\mu_1 B_1 & 2ik\eta_2\mu_2 E_2 & -(2k^2 - k_{S2}^2)\mu_2 B_2 \\ ikE_1^{-1} & ikE_1 & -i\beta_1 B_1^{-1} & i\beta_1 B_1 & -ikE_2 & \beta_2 B_2 \\ -i\eta_1 E_1^{-1} & i\eta_1 E_1 & -ikB_1^{-1} & -ikB_1 & \eta_2 E_2 & ikB_2 \end{bmatrix}$$

Since Eq. (1.157) is a system of homogeneous equations, the determinant of the coefficient matrix $[A]$ must vanish for nontrivial solution of the coefficients a_1, b_1, c_1, d_1, b_2 and d_2. The equation obtained by equating the determinant of $[A]$ to zero is called a dispersion equation because this equation gives the values of k as a function of frequency. From the relation $k = \omega/c_R$ velocity (c_R) of the horizontally propagating wave can be obtained. Here also, like Love waves, the wave velocity is dependent on the frequency and different modes propagate with different speeds. This wave is called a *generalized Rayleigh-Lamb wave* in a layered solid or simply the *Rayleigh wave.* Note that in the homogeneous solid the Rayleigh wave is not dispersive (wave speed is independent of the frequency) but in a layered half-space it is.

The procedure discussed above to solve the wave propagation problem in a layer over a half-space geometry can be generalized for the case when the solid has more than one layer. Note that for the two layers over a half-space, the coefficient matrix $[A]$ will have 10×10 dimension, and for the n layers over a half-space case the dimension of this matrix will be $(4n+2) \times (4n+2)$.

1.2.12 Plate Waves

So far, we have studied the mechanics of elastic wave propagation in a homogeneous or layered half-space and a homogeneous full space. Wave propagation in a plate with two stress-free surfaces is considered in this section. For simplicity, the antiplane (SH) problem will be studied first and then the in-plane (P/SV) problem will be considered.

1.2.12.1 Antiplane Waves in a Plate

Figure 1.26 shows possible SH motions in plate with two stress-free boundaries.

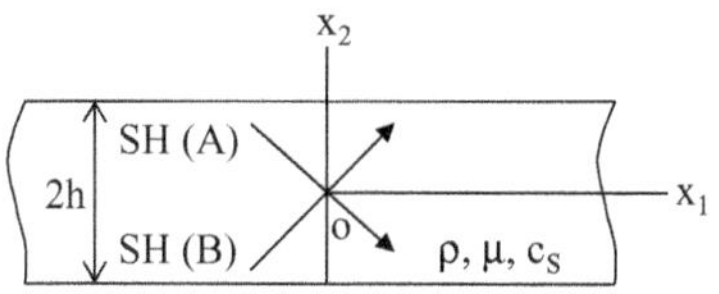

Figure 1.26 Possible SH motions in a plate.

The displacement field in the plate is taken as

$$u_3 = \left(Ae^{-i\beta x_2} + Be^{i\beta x_2}\right)e^{ikx_1}$$
$$\beta = \sqrt{k_S^2 - k^2} \tag{1.158}$$

From the two stress-free boundary conditions at $x_2 = +/-h$,

$$\mu u_{3,2}\big|_{x_2=h} = 0 \;\Rightarrow\; i\beta\mu\left(-Ae^{-i\beta h} + Be^{i\beta h}\right)e^{ikx_1} = 0$$
$$\mu u_{3,2}\big|_{x_2=-h} = 0 \;\Rightarrow\; i\beta\mu\left(-Ae^{i\beta h} + Be^{-i\beta h}\right)e^{ikx_1} = 0 \tag{1.159}$$

From Eq. (1.159) one can write

$$\begin{bmatrix} -e^{-i\beta h} & e^{i\beta h} \\ -e^{i\beta h} & e^{-i\beta h} \end{bmatrix} \begin{Bmatrix} A \\ B \end{Bmatrix} = \begin{Bmatrix} 0 \\ 0 \end{Bmatrix} \tag{1.160}$$

For nontrivial solutions of A and B, the determinant of the coefficient matrix must vanish. Hence,

$$-e^{-2i\beta h} + e^{2i\beta h} = 2i\sin(2\beta h) = 0$$
$$\Rightarrow \sin(2\beta h) = 0 = \sin\{m\pi\}, \quad m = 0,1,2,.... \tag{1.161}$$
$$\Rightarrow \sqrt{k_S^2 - k^2} = \frac{m\pi}{2h}, \quad m = 0,1,2,....$$

k can be computed from the above equation. Clearly, k will be different for different values of m. Let us denote the solution by k_m.

$$k_S^2 - k_m^2 = \left(\frac{m\pi}{2h}\right)^2, \quad m = 0,1,2,....$$
$$\Rightarrow k_m = \sqrt{k_S^2 - \left(\frac{m\pi}{2h}\right)^2}, \quad m = 0,1,2,.... \tag{1.162}$$
$$\Rightarrow c_m = \frac{\omega}{k_m} = \frac{\omega}{\sqrt{k_S^2 - \left(\frac{m\pi}{2h}\right)^2}} = \frac{c_S}{\sqrt{1 - \left(\frac{m\pi}{2h}\right)^2\left(\frac{c_S}{\omega}\right)^2}}, \quad m = 0,1,2,....$$

Note that c_m is a function of ω, making these waves dispersive; however, for $m = 0$, $c_m = c_0 = c_S$. Thus, the 0-th order mode ($m = 0$) is not dispersive but the higher order modes ($m = 1, 2, 3, \ldots$) are (see Figure 1.27).

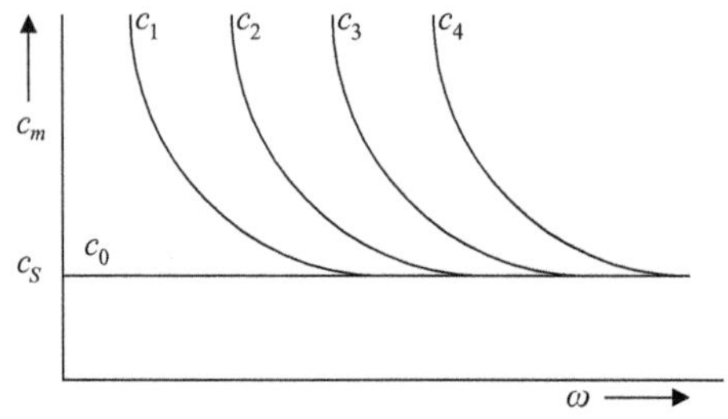

Figure 1.27 Dispersion curves for antiplane plate waves.

1.2.12.1.1 Mode Shapes

Let us now compute mode shapes for $m=0, 1, 2, 3, \ldots$ etc.

For $m=0$, $k_m=k_S$; hence, $\beta=0$, and from Eq. (1.160) $A=B$. Then from Eq. (1.158) the displacement field becomes

$$u_3^0=\left(Ae^{-i\beta x_2}+Be^{i\beta x_2}\right)e^{ikx_1}=2Ae^{ikx_1}$$

Clearly, the displacement field is independent of x_2. Superscript "0" indicates 0-th order mode.

In the same manner for $m=1$

$$k_m=\sqrt{k_S^2-\left(\frac{\pi}{2h}\right)^2}\quad \beta=\sqrt{k_S^2-k_m^2}=\frac{\pi}{2h}$$

From Eq. (1.160), $\dfrac{B}{A}=e^{\pm i\pi}$

Substituting it in Eq. (1.158),

$$\begin{aligned}u_3^1&=\left(Ae^{-i\beta x_2}+Be^{i\beta x_2}\right)e^{ikx_1}\\&=A\left(e^{-\frac{i\pi x_2}{2h}}+e^{\pm i\pi}e^{\frac{i\pi x_2}{2h}}\right)e^{ikx_1}=A\left(e^{-\frac{i\pi x_2}{2h}}-e^{\frac{i\pi x_2}{2h}}\right)e^{ikx_1}\\&=-2iA\sin\left(\frac{\pi x_2}{2h}\right)e^{ikx_1}\end{aligned}$$

For $m=2$

$$k_m=\sqrt{k_S^2-\left(\frac{2\pi}{2h}\right)^2}\quad \beta=\sqrt{k_S^2-k_m^2}=\frac{\pi}{h}$$

From Eq. (1.160), $\dfrac{B}{A}=e^{\pm 2i\pi}$

Substituting it in Eq. (1.158),

$$\begin{aligned}u_3^2&=\left(Ae^{-i\beta x_2}+Be^{i\beta x_2}\right)e^{ikx_1}=A\left(e^{-\frac{i\pi x_2}{h}}+e^{\pm 2i\pi}e^{\frac{i\pi x_2}{h}}\right)e^{ikx_1}\\&=A\left(e^{-\frac{i\pi x_2}{h}}+e^{\frac{i\pi x_2}{h}}\right)e^{ikx_1}=2A\cos\left(\frac{\pi x_2}{h}\right)e^{ikx_1}\end{aligned}$$

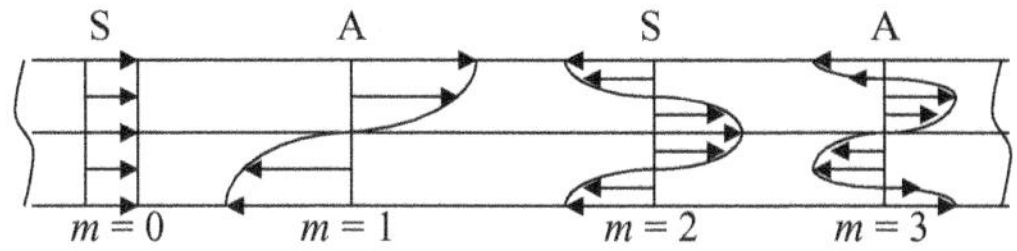

Figure 1.28 Mode shapes for different antiplane plate wave modes.

In this manner, x_2-dependence of the displacement field alternately becomes sine and cosine. Displacement field variation in the x_2-direction is known as the mode shape. Different mode shapes for this problem are plotted in Figure 1.28. Symmetric and antisymmetric modes, relative to the central plane of the plate, are denoted by "S" and "A", respectively. Right and left arrows are used to indicate positive and negative directions of the displacement field, respectively.

1.2.12.2 In-plane Waves in a Plate (Lamb Waves)

Figure 1.29 shows possible in-plane (P/SV) motions in a plate with two stress-free boundaries.

P- and SV-wave potentials in the plate are given by

$$\begin{aligned}
\varphi &= (ae^{-i\eta x_2} + be^{i\eta x_2})e^{ikx_1} \\
\psi &= (ce^{-i\beta x_2} + de^{i\beta x_2})e^{ikx_1} \\
\eta &= \sqrt{k_P^2 - k^2} \\
\beta &= \sqrt{k_S^2 - k^2}
\end{aligned} \tag{1.163}$$

From the stress-free boundary conditions at $x_2 = +/-h$,

$$\begin{aligned}
\sigma_{22}\big|_{x_2=h} &= -\mu\left\{k_S^2\varphi + 2(\psi_{,12} + \varphi_{,11})\right\}_{x_2=h} \\
&= -\mu\left\{\left(k_S^2 - 2k^2\right)\left(aE^{-1} + bE\right) + 2k\beta\left(cB^{-1} - dB\right)\right\}e^{ikx_1} = 0 \\
\sigma_{22}\big|_{x_2=-h} &= -\mu\left\{k_S^2\varphi + 2(\psi_{,12} + \varphi_{,11})\right\}_{x_2=-h} \\
&= -\mu\left\{\left(k_S^2 - 2k^2\right)\left(aE + bE^{-1}\right) + 2k\beta\left(cB - dB^{-1}\right)\right\}e^{ikx_1} = 0 \\
\sigma_{12}\big|_{x_2=h} &= \mu\left\{2\varphi_{,12} + \psi_{,22} - \psi_{,11}\right\}_{x_2=h} \\
&= \mu\left\{(2k\eta)\left(aE^{-1} - bE\right) + \left(k^2 - \beta^2\right)\left(cB^{-1} + dB\right)\right\}e^{ikx_1} = 0 \\
\sigma_{12}\big|_{x_2=-h} &= \mu\left\{2\varphi_{,12} + \psi_{,22} - \psi_{,11}\right\}_{x_2=-h} \\
&= \mu\left\{(2k\eta)\left(aE - bE^{-1}\right) + \left(k^2 - \beta^2\right)\left(cB + dB^{-1}\right)\right\}e^{ikx_1} = 0
\end{aligned} \tag{1.164}$$

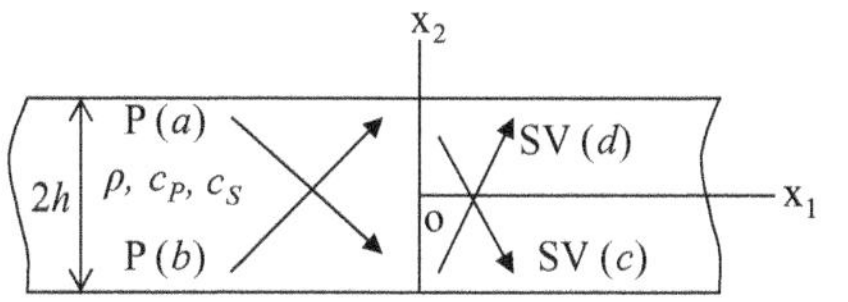

Figure 1.29 Possible in-plane motions in a plate.

where

$$E = e^{i\eta h}$$

$$B = e^{i\beta h}$$

From Eq. (1.164) one can write

$$\begin{bmatrix} (2k^2 - k_S^2)E^{-1} & (2k^2 - k_S^2)E & -2k\beta B^{-1} & 2k\beta B \\ (2k^2 - k_S^2)E & (2k^2 - k_S^2)E^{-1} & -2k\beta B & 2k\beta B^{-1} \\ 2k\eta E^{-1} & -2k\eta E & (2k^2 - k_S^2)B^{-1} & (2k^2 - k_S^2)B \\ 2k\eta E & -2k\eta E^{-1} & (2k^2 - k_S^2)B & (2k^2 - k_S^2)B^{-1} \end{bmatrix} \begin{Bmatrix} a \\ b \\ c \\ d \end{Bmatrix} = \begin{Bmatrix} 0 \\ 0 \\ 0 \\ 0 \end{Bmatrix} \tag{1.165}$$

For nontrivial solutions of a, b, c and d, the determinant of the above 4×4 coefficient matrix must vanish. The equation obtained by equating the determinant of the above matrix to zero is called the dispersion equation because k, as a function of frequency, is obtained from this equation. Like the antiplane problem, multiple values of k are obtained for a given frequency. These multiple values (k_m) correspond to multiple modes of wave propagation. The wave speed c_L is obtained from k_m from the relation $c_L = \omega/k_m$. This wave is called Lamb wave. Since k_m is dispersive, the corresponding wave speed (c_L) is also dispersive (varies with frequency).

Symmetric and Antisymmetric Modes

The above analysis can be simplified significantly by decomposing the problem into symmetric and antisymmetric problems. To this aim, wave potentials in the plate are written in the following manner, instead of Eq. (1.163):

$$\begin{aligned} \phi &= (ae^{\eta x_2} + be^{-\eta x_2})e^{ikx_1} = \{A\sinh(\eta x_2) + B\cosh(\eta x_2)\}e^{ikx_1} \\ \psi &= (ce^{\beta x_2} + de^{-\beta x_2})e^{ikx_1} = \{C\sinh(\beta x_2) + D\cosh(\beta x_2)\}e^{ikx_1} \\ \eta &= \sqrt{k^2 - k_P^2} \\ \beta &= \sqrt{k^2 - k_S^2} \end{aligned} \tag{1.166}$$

From the stress-free boundary conditions at $x_2 = +/-h$,

$$\begin{aligned} &1)\ \sigma_{22}\big|_{x_2=h} = -\mu\{k_S^2\phi + 2(\psi_{,12} + \phi_{,11})\}_{x_2=h} \\ &= -\mu\{(k_S^2 - 2k^2)[A\sinh(\eta h) + B\cosh(\eta h)] \\ &+2ik\beta[C\cosh(\beta h) + D\sinh(\beta h)]\}e^{ikx_1} = 0 \\ &2)\ \sigma_{22}\big|_{x_2=-h} = -\mu\{k_S^2\phi + 2(\psi_{,12} + \phi_{,11})\}_{x_2=-h} \\ &= -\mu\{(k_S^2 - 2k^2)[-A\sinh(\eta h) + B\cosh(\eta h)] \\ &+ 2ik\beta[C\cosh(\beta h) - D\sinh(\beta h)]\}e^{ikx_1} = 0 \end{aligned} \tag{1.167}$$

$$3)\ \sigma_{12}\big|_{x_2=h} = \mu\{2\phi_{,12} + \psi_{,22} - \psi_{,11}\}_{x_2=h}$$

$$= \mu\{(2ik\eta)[A\cosh(\eta h) + B\sinh(\eta h)]$$

$$+ (k^2 + \beta^2)[C\sinh(\beta h) + D\cosh(\beta h)]\}e^{ikx_1} = 0$$

$$4)\ \sigma_{12}\big|_{x_2=-h} = \mu\{2\phi_{,12} + \psi_{,22} - \psi_{,11}\}_{x_2=-h}$$

$$= \mu\{(2ik\eta)[A\cosh(\eta h) - B\sinh(\eta h)]$$

$$+ (k^2 + \beta^2)[-C\sinh(\beta h) + D\cosh(\beta h)]\}e^{ikx_1} = 0$$

Adding the first two equations of Eq. (1.167) and subtracting the fourth equation from the third we get the following two equations:

$$\begin{aligned} &(2k^2 - k_S^2)B\cosh(\eta h) - 2ik\beta C\cosh(\beta h) = 0 \\ &2ik\eta B\sinh(\eta h) + (2k^2 - k_S^2)C\sinh(\beta h) = 0 \end{aligned} \tag{1.168a}$$

Adding the last two equations of Eq. (1.167) and subtracting the second equation from the first one gets the following:

$$\begin{aligned} &(2k^2 - k_S^2)A\sinh(\eta h) - 2ik\beta D\sinh(\beta h) = 0 \\ &2ik\eta A\cosh(\eta h) + (2k^2 - k_S^2)D\cosh(\beta h) = 0 \end{aligned} \tag{1.168b}$$

From Eq. (1.168a) one can write that for nontrivial solutions of B and C, the determinant of the coefficient matrix must vanish:

$$\begin{aligned} &(2k^2 - k_S^2)^2\cosh(\eta h)\sinh(\beta h) - 4k^2\eta\beta\cosh(\beta h)\sinh(\eta h) = 0 \\ &\Rightarrow \frac{\tanh(\eta h)}{\tanh(\beta h)} = \frac{(2k^2 - k_S^2)^2}{4k^2\eta\beta} \end{aligned} \tag{1.169a}$$

Similarly, from Eq. (1.168b), for nontrivial solutions of A and D, the determinant of the coefficient matrix must vanish:

$$\begin{aligned} &(2k^2 - k_S^2)^2\sinh(\eta h)\cosh(\beta h) - 4k^2\eta\beta\sinh(\beta h)\cosh(\eta h) = 0 \\ &\Rightarrow \frac{\tanh(\eta h)}{\tanh(\beta h)} = \frac{4k^2\eta\beta}{(2k^2 - k_S^2)^2} \end{aligned} \tag{1.169b}$$

If A and D vanish, but B and C do not, then Eq. (1.169a) must be satisfied and the potential field is given by

$$\phi = (ae^{\eta x_2} + be^{-\eta x_2})e^{ikx_1} = B\cosh(\eta x_2)e^{ikx_1}$$

$$\psi = (ce^{\beta x_2} + de^{-\beta x_2})e^{ikx_1} = C\sinh(\beta x_2)e^{ikx_1}$$

Therefore, the displacement field is

$$u_1 = \phi_{,1} + \psi_{,2} = \left\{ikB\cosh(\eta x_2) + \beta C\cosh(\beta x_2)\right\}e^{ikx_1}$$

$$u_2 = \phi_{,2} - \psi_{,1} = \left\{\eta B\sinh(\eta x_2) - ikC\sinh(\beta x_2)\right\}e^{ikx_1}$$

Note that u_1 is an even function of x_2, while u_2 is an odd function of x_2. Therefore, in this case the displacement field will be symmetric about the central plane ($x_2 = 0$) of the plate. That is why these modes are called symmetric or extensional modes.

If B and C vanish, but A and D do not, then Eq. (1.169b) must be satisfied. In this case, u_1 will be an odd function of x_2 while u_2 will be an even function of x_2. Therefore, the displacement field will be antisymmetric about the central plane ($x_2 = 0$) of the plate. These modes are known as antisymmetric or flexural modes.

Deformation of the plate for symmetric and antisymmetric modes of wave propagation is shown in Figure 1.30.

Equations (1.169a) and (1.169b) are rather complex and have multiple solutions for k, which can be obtained numerically. From the multiple k values, the Lamb wave speed c_L (= ω/k) for different symmetric and antisymmetric modes of wave propagation can be obtained.

Let us investigate the dispersion equations (1.169a) and (1.169b) more closely for some special cases. For example, at low frequency, since η and β are small,

$$\frac{\tanh(\eta h)}{\tanh(\beta h)} \approx \frac{\eta h}{\beta h} = \frac{\eta}{\beta}$$

Substituting it in Eq. (1.169a) one gets for low frequency:

$$\frac{\eta}{\beta} = \frac{\left(2k^2 - k_S^2\right)^2}{4k^2\eta\beta} \Rightarrow 4k^2\eta^2 = \left(2k^2 - k_S^2\right)^2 \Rightarrow 4k^2\left(k^2 - k_P^2\right) = 4k^4 - 4k^2k_S^2 + k_S^4$$

$$\Rightarrow 4k^2\left(k_S^2 - k_P^2\right) = k_S^4 \Rightarrow k^2 = \frac{k_S^4}{4\left(k_S^2 - k_P^2\right)} \Rightarrow \frac{c_L^2}{\omega^2} = \frac{1}{k^2} = \frac{4\left(k_S^2 - k_P^2\right)}{k_S^4} = 4\frac{c_S^2}{\omega^2}\left(1 - \frac{c_S^2}{c_P^2}\right) \quad (1.170)$$

$$\Rightarrow c_L = 2c_S\sqrt{1 - \frac{c_S^2}{c_P^2}} = c_0$$

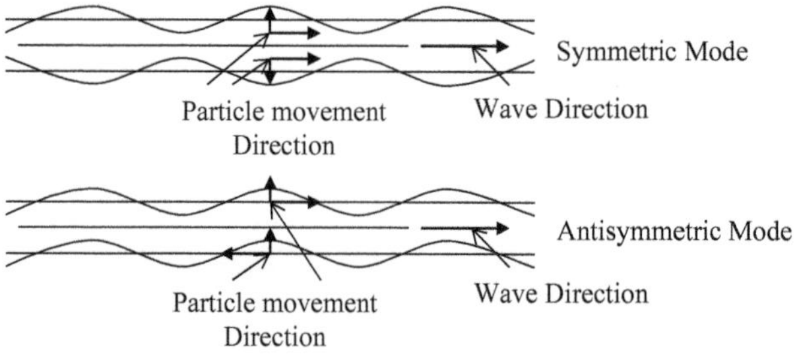

Figure 1.30 Deformation of the plate and particle movement directions for symmetric (top) and antisymmetric (bottom) modes of wave propagation.

Eq. (1.169b) can be solved by the perturbation technique [Bland, 1988; Mal and Singh, 1991] for small frequency to obtain

$$c_L = c_S\left\{\frac{4}{3}\left(1-\frac{c_S^2}{c_P^2}\right)\right\}^{1/4}\left(\frac{\omega h}{c_S}\right)^{1/2} \tag{1.171}$$

Thus, at zero frequency, the phase velocity of the symmetric mode has a finite value [given in Eq. (1.170)] but that of the antisymmetric mode is zero. These modes are called fundamental symmetric and antisymmetric modes and are denoted by the symbols S_0 and A_0, respectively. At high frequencies, A_0 and S_0 modes asymptotically approach the Rayleigh wave speed, as shown in Figure 1.31. Corresponding mathematical proof is given below.

At high frequencies, $\tanh(\eta h) \approx \tanh(\beta h) \approx 1$. Substituting it in Eqs. (1.169a) and (1.169b) one gets

$$1 = \frac{\left(2k^2 - k_S^2\right)^2}{4k^2\eta\beta} \quad \Rightarrow 4k^2\eta\beta = \left(2k^2 - k_S^2\right)^2 \quad \Rightarrow \left(2k^2 - k_S^2\right)^2 - 4k^2\eta\beta = 0$$

The above equation is identical to the Rayleigh wave equation [Eq. (148)]; its solution gives the Rayleigh wave speed. Therefore at high frequencies, Lamb modes attain Rayleigh wave speed.

Although equations (1.169a) and (1.169b) have only one solution each at low frequency, as the signal frequency increases the equations give multiple solutions. Typical plots of the variation of c_L as a function of frequency for an isotropic plate are shown in Figure 1.31.

1.2.13 Phase Velocity and Group Velocity

In Figures 1.27 and 1.31 we see that the wave speeds are function of frequency. Therefore, if any mode of the wave propagates with a single frequency (monochromatic wave), then its velocity can be computed from the dispersion curves like the ones in Figures 1.27 and 1.31. This velocity is called *phase velocity.* The schematic of the monochromatic wave velocity or phase velocity is shown in the top two plots of Figure 1.32.

Wave motions are shown at time 0 (continuous line) and t (dotted line).

In Figure 1.32 note that two waves (1 and 2, top two plots) have slightly different frequencies and phase velocities. Wave 2 has a slightly higher frequency and phase velocity when compared to wave 1. A similar phase velocity–frequency relation (velocity increasing with frequency) is observed for the A_0 mode in Figure 1.31. If both waves, 1 and 2, exist simultaneously, then their

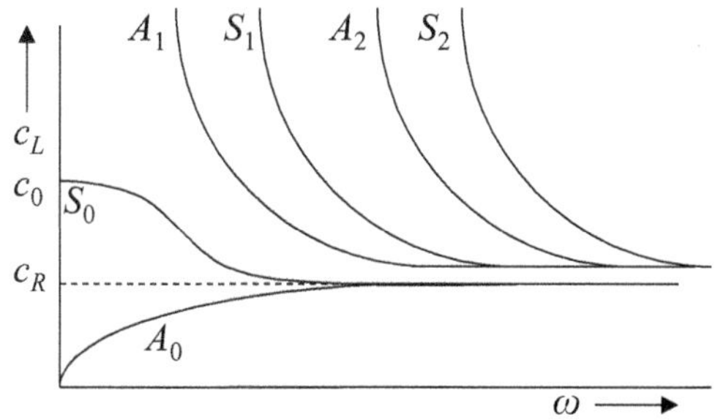

Figure 1.31 Dispersion curves for Lamb wave propagation in a plate.

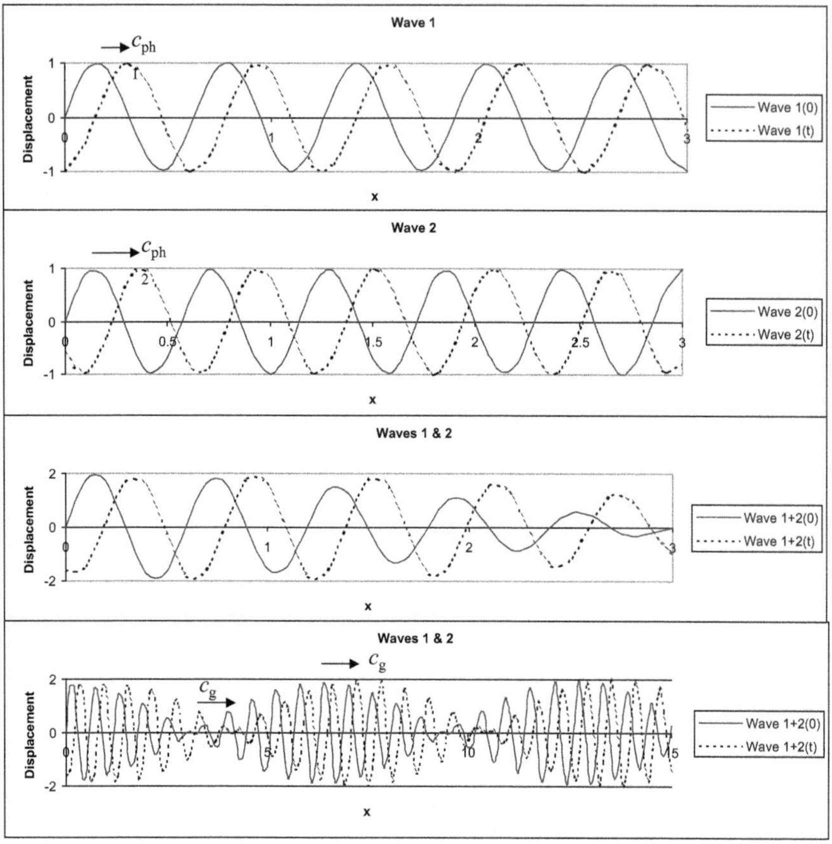

Figure 1.32 Propagation of elastic waves of single frequency (top two figures) and multiple frequencies (bottom two figures). Wave motions are shown at time 0 (continuous line) and t (dotted line).

total response will be as shown in the third and fourth plots of Figure 1.32. The third plot is obtained by simply adding the top two plots. The fourth plot is the same as the third plot; the only difference is that it is plotted in a different scale covering a wider range of x (0 to 15). The fourth plot clearly shows how two waves interfere constructively and destructively to form several groups or packets of waves separated by null regions. In this figure, the continuous lines show the displacement variations as functions of x at time $t=0$, and dotted lines show the displacements at time t (slightly greater than 0). Propagation of these waves is evident in the figure. If the dotted lines are plotted at time $t=1$, then the shift of the peak position represents the phase velocity and is denoted by c_{ph} in the figure. Similarly, the shift in the null position or the peak position of the modulation envelope of the summation of these two waves is the velocity of the envelope and is called the *group velocity*; it is denoted by c_g in the bottom plot of Figure 1.32. The relation between the phase velocity and the group velocity is derived below.

Mathematical representations of two waves of different frequency and phase velocity are

$$u_i^1 = \sin(k_1 x - \omega_1 t)$$
$$u_i^2 = \sin(k_2 x - \omega_2 t) \tag{1.172}$$

where the subscript i can take any value 1, 2 or 3 and the superscript 1 and 2 of u correspond to the first and second waves, respectively. Phase velocities for these two waves are:

$$c_{ph1} = \frac{\omega_1}{k_1}$$
$$c_{ph2} = \frac{\omega_2}{k_2} \tag{1.172a}$$

Superposition of these two waves gives

$$u_i^1 + u_i^2 = \sin(k_1 x - \omega_1 t) + \sin(k_2 x - \omega_2 t)$$
$$= \sin(a+b) + \sin(a-b) = 2\sin(a)\cos(b) \tag{1.173}$$

where

$$a = \frac{1}{2}(k_1 + k_2)x - \frac{1}{2}(\omega_1 + \omega_2)t = k_{av}x - \omega_{av}t$$
$$b = \frac{1}{2}(k_1 - k_2)x - \frac{1}{2}(\omega_1 - \omega_2)t = \left(\frac{\Delta k}{2}x - \frac{\Delta\omega}{2}t\right) \tag{1.174}$$

From Eqs. (1.173) and (1.174),

$$u_i^1 + u_i^2 = 2\cos(b)\sin(a) = 2\cos\left(\frac{\Delta k}{2}x - \frac{\Delta\omega}{2}t\right)\sin(k_{av}x - \omega_{av}t) \tag{1.175}$$

Note that the cosine term shows how the modulation envelope is dependent on x and t. The above analysis shows that when two waves of slightly different wave number and frequency are superimposed, the resultant wave has the average wave number and frequency and is amplitude modulated. The velocity of the amplitude envelope is the *group velocity*, and is given by

$$c_g = \frac{\Delta\omega}{\Delta k} \tag{1.176a}$$

If the two waves that are superimposed have very close frequency and wave number,

$$\omega_1 \approx \omega_2 \approx \omega$$
$$k_1 \approx k_2 \approx k$$

then

$$\omega_{av} \approx \omega$$
$$k_{av} \approx k$$
$$\Delta\omega = d\omega$$
$$\Delta k = dk$$

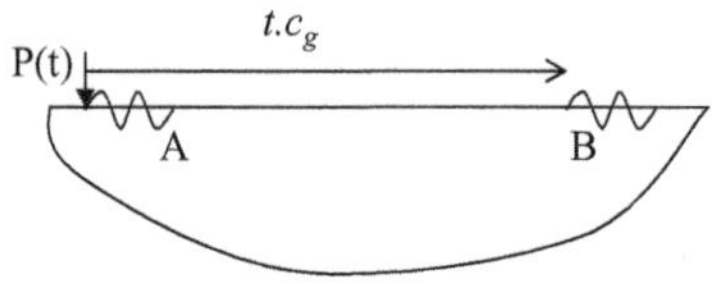

Figure 1.33 Finite pulse propagating with a speed equal to the group velocity.

and we can say that the wave formed by adding the two waves propagates with the same frequency (ω) and wave number (k) as its individual components, but the envelope or group formed in between two successive nulls of the amplitude modulation curve travels with a velocity c_g where

$$c_g = \frac{d\omega}{dk} \tag{1.176b}$$

The concept of group velocity is important because energy travels from one point to another in a solid with this velocity. This concept is illustrated in Figure 1.33.

Let a time dependent force P(t) generate a finite pulse A as shown in Figure 1.33. This pulse can be assumed to be a superposition of a number of waves of different frequencies. Each of these waves can be assumed to exist in the region $-\infty < x < +\infty$. When all these waves are added, a finite pulse like the one shown in Figure 1.33 is generated. This concept is similar to the Fourier series summation concept. Thus, the pulse A of Figure 1.33 is a group of waves and should travel with the group velocity (c_g). Therefore, after a time t, the entire pulse will move to position B at a distance of $t.c_g$.

1.2.14 Point Source Excitation

So far, we have discussed possible wave motions in solids of different geometry – infinite solid, half-space, plate etc. However, except for some simple cases such as the one in Figure 1.15, we did not talk about how the waves are generated. In this section, we will consider the source into our analysis and will calculate the wave motion in a solid due to a point source excitation. The derivation given in Mal and Singh, 1991, for solid full space is followed here.

Let the concentrated time harmonic force $\mathbf{P}\delta(\mathbf{x})e^{-i\omega t}$ act at the origin of a solid full space. Where $\mathbf{P}$ is the force vector acting at the origin, $\delta(\mathbf{x})$ is the three-dimensional delta function that is zero everywhere except at the origin, where it is infinity. When the volume encloses the origin, the volume integral of the delta function is 1. Without the loss of generality we can say that for this harmonic excitation the displacement field will also have the form $\mathbf{u} = \mathbf{u}(\mathbf{x})e^{-i\omega t}$. In the presence of a body force, which in this case is a concentrated force at the origin, Navier's equation [Eq. (1.79)] takes the form

$$\begin{aligned}
&(\lambda + 2\mu)\underline{\nabla}(\underline{\nabla}.\mathbf{u}) - \mu\underline{\nabla}\times(\underline{\nabla}\times\mathbf{u}) + \mathbf{P}\delta(\mathbf{x}) = -\rho\omega^2\mathbf{u} \\
&\Rightarrow c_P^2\underline{\nabla}(\underline{\nabla}.\mathbf{u}) - c_S^2\underline{\nabla}\times(\underline{\nabla}\times\mathbf{u}) + \omega^2\mathbf{u} = -\frac{\mathbf{P}}{\rho}\delta(\mathbf{x})
\end{aligned} \tag{1.177}$$

To solve the above inhomogeneous equation, $\delta(\mathbf{x})$ is represented in the following form:

$$\delta(\mathbf{x}) = -\nabla^2\left(\frac{1}{4\pi r}\right) \tag{1.178}$$

where $r = |\mathbf{x}|$ is the radial distance from the origin. To prove Eq. (1.178), let us consider a function

$$\Phi = \frac{C}{r}$$

where C is a constant. Then one can easily show that

$$\nabla^2\Phi(\mathbf{x}) = 0, \ \mathbf{x} \neq 0$$

Further, by the divergence theorem,

$$\int_V \nabla^2\Phi dV = \int_V \underline{\nabla}.(\underline{\nabla}\Phi)dV = \int_S (\underline{\nabla}\Phi).\mathbf{n}dS = \int_S (\Phi_{,1}n_1 + \Phi_{,2}n_2 + \Phi_{,3}n_3)dS = \int_S \frac{\partial\Phi}{\partial x_i}\frac{\partial x_i}{\partial n}dS$$

$$= \int_S \frac{\partial\Phi}{\partial n}dS = C\int_{\beta=0}^{\pi}\int_{\theta=0}^{2\pi}\left(-\frac{1}{r^2}\right)r^2\sin(\theta)d\theta d\beta = -4\pi C$$

If $C = -1/(4\pi)$, then the above volume integral becomes 1. Hence, $\nabla^2\left(-\frac{1}{4\pi r}\right)$ is zero for $\mathbf{x} \neq 0$ and its volume integral is 1. So it must be the three-dimensional delta function, as given in Eq. (1.178).

Using the vector identity one can write

$$\mathbf{P}\delta(\mathbf{x}) = -\nabla^2\left(\frac{\mathbf{P}}{4\pi r}\right) = -\frac{1}{4\pi}\left[\underline{\nabla}\left(\underline{\nabla}.\frac{\mathbf{P}}{r}\right) - \underline{\nabla}\times\left(\underline{\nabla}\times\frac{\mathbf{P}}{r}\right)\right]$$

Therefore, the equation of motion (1.177) can be rewritten as

$$c_P^2\underline{\nabla}(\underline{\nabla}.\mathbf{u}) - c_S^2\underline{\nabla}\times(\underline{\nabla}\times\mathbf{u}) + \omega^2\mathbf{u} = \frac{1}{4\pi\rho}\left[\underline{\nabla}\left(\underline{\nabla}.\frac{\mathbf{P}}{r}\right) - \underline{\nabla}\times\left(\underline{\nabla}\times\frac{\mathbf{P}}{r}\right)\right] \quad (1.179)$$

Let us now express $\mathbf{u}$ in terms of two scalar functions, ϕ and ψ, in the following manner:

$$\mathbf{u} = \underline{\nabla}(\underline{\nabla}.\mathbf{P}\phi) - \underline{\nabla}\times(\underline{\nabla}\times\mathbf{P}\psi) \quad (1.180)$$

Then,

$$\underline{\nabla}.\mathbf{u} = \nabla^2(\underline{\nabla}.\mathbf{P}\phi) = \underline{\nabla}.(\nabla^2\mathbf{P}\phi)$$

$$\underline{\nabla}\times\mathbf{u} = -\underline{\nabla}\times[\underline{\nabla}\times(\underline{\nabla}\times\mathbf{P}\psi)] = \underline{\nabla}\times(\mathbf{P}\nabla^2\psi)$$

and Eq. (1.179) becomes

$$\underline{\nabla}\underline{\nabla}.\left[\mathbf{P}\left(c_P^2\nabla^2\phi + \omega^2\phi - \frac{1}{4\pi\rho r}\right)\right] - \underline{\nabla}\times\underline{\nabla}\times\left[\mathbf{P}\left(c_S^2\nabla^2\psi + \omega^2\psi - \frac{1}{4\pi\rho r}\right)\right] = 0 \quad (1.181)$$

The above equation is satisfied if ϕ and ψ are the solutions of the following inhomogeneous Helmholtz equations:

$$\nabla^2\phi + k_P^2\phi = \frac{1}{4\pi\rho c_P^2 r}$$

$$\nabla^2\psi + k_S^2\psi = \frac{1}{4\pi\rho c_S^2 r} \quad (1.182)$$

One can easily show that the particular solutions of Eq. (1.182) that are finite at $r=0$ are

$$\phi = \frac{1-e^{ik_P r}}{4\pi\rho\omega^2 r}$$
$$\psi = \frac{1-e^{ik_S r}}{4\pi\rho\omega^2 r} \tag{1.183}$$

From Eqs. (1.183) and (1.180) the displacement field can be written as

$$\mathbf{u} = -\underline{\nabla}\underline{\nabla}.\left[\mathbf{P}\frac{e^{ik_P r}-e^{ik_S r}}{4\pi\rho\omega^2 r}\right] + \frac{\mathbf{P}e^{ik_S r}}{4\pi\rho c_S^2 r} \tag{1.184}$$

or, in index notation,

$$u_i = \frac{1}{4\pi\rho\omega^2}\left[k_S^2\frac{e^{ik_S r}}{r}P_i - \frac{\partial^2}{\partial x_i\partial x_j}\frac{e^{ik_P r}-e^{ik_S r}}{r}P_j\right]$$
$$= \frac{1}{4\pi\rho\omega^2}\left[k_S^2\frac{e^{ik_S r}}{r}\delta_{ij} - \frac{\partial^2}{\partial x_i\partial x_j}\frac{e^{ik_P r}-e^{ik_S r}}{r}\right]P_j = G_{ij}(\mathbf{x};\mathbf{0})P_j \tag{1.185}$$

where $G_{ij}(\mathbf{x};\mathbf{0})\exp(-i\omega t)$ is the i-th component of the displacement produced at $\mathbf{x}$ by a concentrated force $\exp(-i\omega t)$ acting at the origin in the j-th direction. G_{ij} is known as the *steady-state Green's tensor* or in simple words *Green's function* for the infinite isotropic solid.

If the force acts in the j-th direction at $\mathbf{y}$, then the displacement component in the i-th direction is denoted by $G_{ij}(\mathbf{x};\mathbf{y})$ and is obtained by replacing r by $|\mathbf{x}-\mathbf{y}|$; Green's tensor in this case is given by Mal and Singh (1991):

$$G_{ij}(\mathbf{x};\mathbf{y}) = \frac{1}{4\pi\rho\omega^2}\left\{\frac{e^{ik_P r}}{r}\left[k_P^2 R_i R_j + \left(3R_iR_j-\delta_{ij}\right)\left(\frac{ik_P}{r}-\frac{1}{r^2}\right)\right]\right.$$
$$\left. + \frac{e^{ik_S r}}{r}\left[k_S^2\left(\delta_{ij}-R_iR_j\right) - \left(3R_iR_j-\delta_{ij}\right)\left(\frac{ik_S}{r}-\frac{1}{r^2}\right)\right]\right\} \tag{1.186}$$

where

$$R_i = \frac{x_i-y_i}{r} \tag{1.186a}$$

If r is large compared to the wavelength, then $k_P r$ and $k_S r$ are large, and an approximate expression for $G_{ij}(\mathbf{x};\mathbf{y})$ can be obtained by retaining only the terms containing $(k_P r)^{-1}$ and $(k_S r)^{-1}$ but ignoring all higher order terms. Thus, in the *far field*,

$$G_{ij}(\mathbf{x};\mathbf{y}) = \frac{1}{4\pi\rho}\left[\frac{e^{ik_P r}}{c_P^2 r}R_iR_j + \frac{e^{ik_S r}}{c_S^2 r}\left(\delta_{ij}-R_iR_j\right)\right] \tag{1.187}$$

and the displacement vector in the far field is given by

$$\mathbf{u}(\mathbf{x};\mathbf{0}) = \frac{1}{4\pi\rho}\left[\frac{e^{ik_P r}}{c_P^2 r}(\mathbf{e_R}.\mathbf{P})\mathbf{e_R} + \frac{e^{ik_S r}}{c_S^2 r}\left\{\mathbf{P}-(\mathbf{e_R}.\mathbf{P})\mathbf{e_R}\right\}\right] \tag{1.188}$$

1.2.15 Wave Propagation in Fluid

Since perfect fluid does not have any shear stress, the wave propagation analysis in a perfect fluid medium is much simpler and can be considered as a special case of that in the solid. In this book, unless otherwise specified, the fluid is assumed to be perfect fluid. The constitutive relation for a solid [Eq. (1.68)] can be specialized for the fluid case by substituting the shear modulus $\mu = 0$.

$$\sigma_{ij} = \lambda \delta_{ij} \varepsilon_{kk} \tag{1.189}$$

Since fluid can only sustain hydrostatic pressure p, the stress field in a fluid is given by

$$\sigma_{11} = \sigma_{22} = \sigma_{33} = -p \tag{1.190}$$

and all shear stress components are zero. Then, the constitutive relation is simplified to

$$\begin{aligned} -p = \sigma_{11} = \sigma_{22} = \sigma_{33} = \lambda(\varepsilon_{11} + \varepsilon_{22} + \varepsilon_{33}) = \lambda(u_{1,1} + u_{2,2} + u_{3,3}) = \lambda \nabla.\mathbf{u} \\ \Rightarrow \nabla.\mathbf{u} = -\frac{p}{\lambda} \end{aligned} \tag{1.191}$$

The governing equations of motion [Eq. (1.78)] can be specialized in the same manner:

$$\begin{aligned} \sigma_{11,1} + f_1 &= -p_{,1} + f_1 = \rho \ddot{u}_1 \\ \sigma_{22,2} + f_2 &= -p_{,2} + f_2 = \rho \ddot{u}_2 \\ \sigma_{33,3} + f_3 &= -p_{,3} + f_3 = \rho \ddot{u}_3 \\ &\Rightarrow -\nabla p + \mathbf{f} = \rho \ddot{\mathbf{u}} = \rho \dot{\mathbf{v}} \end{aligned} \tag{1.192}$$

Hence,

$$\begin{aligned} -\underline{\nabla}.\underline{\nabla} p + \underline{\nabla}.\mathbf{f} = \underline{\nabla}.\rho \ddot{\mathbf{u}} = \rho \frac{\partial^2 (\underline{\nabla}.\mathbf{u})}{\partial t^2} = -\frac{\rho}{\lambda}\frac{\partial^2 p}{\partial t^2} \\ \Rightarrow -\nabla^2 p + \frac{1}{c_f^2}\frac{\partial^2 p}{\partial t^2} + \underline{\nabla}.\mathbf{f} = 0 \\ \Rightarrow \nabla^2 p - \frac{1}{c_f^2}\frac{\partial^2 p}{\partial t^2} + f = 0 \end{aligned} \tag{1.193}$$

where

$$\begin{aligned} c_f = \sqrt{\frac{\lambda}{\rho}} \\ f = -\nabla.\mathbf{f} \end{aligned} \tag{1.194}$$

In absence of any body force, the above governing equation is simplified to the wave equation or Helmholtz equation

$$\nabla^2 p - \frac{1}{c_f^2}\frac{\partial^2 p}{\partial t^2} = 0 \tag{1.195}$$

Its solution is given by

$$p = f(\mathbf{n}.\mathbf{x} - c_f t) \tag{1.196}$$

It represents a wave propagating in **n** direction with a velocity c_f. Note that this wave produces only normal stress like the P-wave in a solid and its velocity c_f can be obtained from the P-wave speed expression [Eq. (1.82)] by substituting $\mu=0$. Hence, this wave is the compressional wave or P-wave in the fluid. The S-wave cannot be generated in a perfect fluid.

For harmonic time dependence ($e^{-i\omega t}$) and two-dimensional problems p can be expressed as $p(x_1,x_2,t)=p(x_1,x_2)e^{-i\omega t}$ and the governing wave equation is simplified to

$$\nabla^2 p + k_f^2 p = 0 \tag{1.197}$$

where k_f $(=\omega/c_f)$ is the wave number. The solution of Eq. (1.197) is given by

$$p(x_1,x_2) = Ae^{ikx_1+i\eta x_2} \tag{1.198}$$

where

$$\eta = \sqrt{k_f^2 - k^2} \tag{1.199}$$

Because of the plane wave front, the wave defined by Eq. (1.198) is called the plane wave.

1.2.15.1 Relation between Pressure and Velocity

In absence of any body force Eq. (1.192) can be written as

$$-\nabla p = \rho\ddot{\mathbf{u}} = \rho\dot{\mathbf{v}} = \rho\frac{\partial \mathbf{v}}{\partial t}$$

Let us take the dot product on both sides of the equation with **n**, where **n** is some unit vector

$$-\nabla p.\mathbf{n} = \rho\frac{\partial(\mathbf{v}.\mathbf{n})}{\partial t} \tag{1.200}$$

Note that

$$-\nabla p.\mathbf{n} = -\frac{\partial p}{\partial x_i}\frac{\partial x_i}{\partial n} = -\frac{\partial p}{\partial n}$$

$$\rho\frac{\partial(\mathbf{v}.\mathbf{n})}{\partial t} = \rho\frac{\partial(v_n)}{\partial t}$$

Hence,

$$\rho\frac{\partial(v_n)}{\partial t} = -\frac{\partial p}{\partial n} \Rightarrow v_n = -\int\frac{1}{\rho}\frac{\partial p}{\partial n}dt \tag{1.201}$$

1.2.15.2 Reflection and Transmission of Plane Waves at the Fluid–Fluid Interface

Let us now investigate how plane waves in a fluid are reflected and transmitted at an interface between two fluids. Figure 1.34 shows a plane wave p_i of magnitude 1, incident at the interface between two fluids. Reflected and transmitted waves are denoted by p_R and p_T, respectively. Amplitudes of p_R and p_T are R and T, respectively (see Figure 1.34). Since, shear waves cannot be present in a fluid medium reflected and transmitted waves can only have compressional wave or P-wave as shown in the figure.

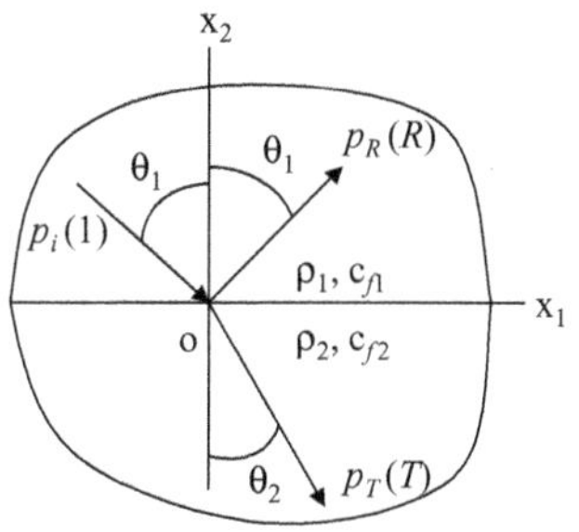

Figure 1.34 Incident, reflected and transmitted waves at a fluid–fluid interface.

Pressure fields corresponding to the incident, reflected and transmitted waves are given by

$$\begin{aligned} p_i &= e^{ikx_1 - i\eta_1 x_2} \\ p_R &= R.e^{ikx_1 + i\eta_1 x_2} \\ p_T &= T.e^{ikx_1 - i\eta_2 x_2} \end{aligned} \tag{1.202}$$

where

$$\begin{aligned} k &= k_{f1}\sin\theta_1 = k_{f2}\sin\theta_2 \\ \eta_j &= k_{fj}\cos\theta_j = \sqrt{k_{fj}^2 - k^2} \\ k_{fj} &= \frac{\omega}{c_{fj}}, \quad j = 1,2 \end{aligned} \tag{1.203}$$

Since the pressure must be continuous across the interface

$$\begin{aligned} &\left[(e^{-i\eta_1 x_2} + R.e^{i\eta_1 x_2})e^{ikx_1}\right]_{x_2=0} = \left[T.e^{-i\eta_2 x_2}e^{ikx_1}\right]_{x_2=0} \\ &\Rightarrow 1 + R = T \end{aligned} \tag{1.204}$$

The second continuity condition that must be satisfied across the interface is the continuity of displacement or its derivative (velocity) normal to the interface. From Eq. (1.201) one can write

$$\begin{aligned} &v_2\Big|_{fluid1, x_2=0} = v_2\Big|_{fluid2, x_2=0} \\ &\Rightarrow \int \frac{1}{\rho_1}\frac{\partial(p_i + p_R)}{\partial x_2}dt\Big|_{x_2=0} = \int \frac{1}{\rho_2}\frac{\partial p_T}{\partial x_2}dt\Big|_{x_2=0} \\ &\Rightarrow \frac{1}{\rho_1}\frac{\partial(p_i + p_R)}{\partial x_2}\Big|_{x_2=0} = \frac{1}{\rho_2}\frac{\partial p_T}{\partial x_2}\Big|_{x_2=0} \\ &\Rightarrow \frac{i\eta_1}{\rho_1}(-1 + R) = -\frac{i\eta_2}{\rho_2}T \end{aligned} \tag{1.205}$$

Eqs. (1.204) and (1.205) can be written in matrix form:

$$\begin{bmatrix} -1 & 1 \\ \frac{\eta_1}{\rho_1} & \frac{\eta_2}{\rho_2} \end{bmatrix} \begin{Bmatrix} R \\ T \end{Bmatrix} = \begin{Bmatrix} 1 \\ \frac{\eta_1}{\rho_1} \end{Bmatrix} \tag{1.206}$$

Eq. (1.206) gives

$$\begin{Bmatrix} R \\ T \end{Bmatrix} = \begin{bmatrix} -1 & 1 \\ \frac{\eta_1}{\rho_1} & \frac{\eta_2}{\rho_2} \end{bmatrix}^{-1} \begin{Bmatrix} 1 \\ \frac{\eta_1}{\rho_1} \end{Bmatrix} = \frac{1}{\left(\frac{\eta_1}{\rho_1}+\frac{\eta_2}{\rho_2}\right)} \begin{bmatrix} -\frac{\eta_2}{\rho_2} & 1 \\ \frac{\eta_1}{\rho_1} & 1 \end{bmatrix} \begin{Bmatrix} 1 \\ \frac{\eta_1}{\rho_1} \end{Bmatrix} = \frac{1}{\left(\frac{\eta_1}{\rho_1}+\frac{\eta_2}{\rho_2}\right)} \begin{Bmatrix} \frac{\eta_1}{\rho_1}-\frac{\eta_2}{\rho_2} \\ 2\frac{\eta_1}{\rho_1} \end{Bmatrix}$$

Hence,

$$\begin{aligned} R &= \frac{\frac{\eta_1}{\rho_1}-\frac{\eta_2}{\rho_2}}{\frac{\eta_1}{\rho_1}+\frac{\eta_2}{\rho_2}} = \frac{\frac{k_{f1}\cos\theta_1}{\rho_1}-\frac{k_{f2}\cos\theta_2}{\rho_2}}{\frac{k_{f1}\cos\theta_1}{\rho_1}+\frac{k_{f2}\cos\theta_2}{\rho_2}} = \frac{\rho_2 c_{f2}\cos\theta_1-\rho_1 c_{f1}\cos\theta_2}{\rho_2 c_{f2}\cos\theta_1+\rho_1 c_{f1}\cos\theta_2} \\ T &= \frac{2\frac{\eta_1}{\rho_1}}{\frac{\eta_1}{\rho_1}+\frac{\eta_2}{\rho_2}} = \frac{2\frac{k_{f1}\cos\theta_1}{\rho_1}}{\frac{k_{f1}\cos\theta_1}{\rho_1}+\frac{k_{f2}\cos\theta_2}{\rho_2}} = \frac{2\rho_2 c_{f2}\cos\theta_1}{\rho_2 c_{f2}\cos\theta_1+\rho_1 c_{f1}\cos\theta_2} \end{aligned} \tag{1.207}$$

1.2.15.3 Plane Wave Potential in a Fluid

So far, plane waves in a fluid have been expressed in terms of the fluid pressure p as given in Eqs. (1.196), (1.198) and (1.202). However, plane waves in a solid were expressed in terms of P- and S-wave potentials. Since the elastic wave in a fluid is the P-wave, one can express the bulk wave in the fluid in terms of the P-wave potential ϕ also, as shown below.

$$\begin{aligned} \phi &= Ae^{ik_f x - i\omega t} \\ \mathbf{u} &= \underline{\nabla}\varphi \end{aligned} \tag{1.208}$$

The above equation is identical for the P-wave propagation in a solid medium.

For two-dimensional time harmonic problems (time dependence $e^{-i\omega t}$ is implied) the wave potential, displacement and stress fields in the x_1x_2 coordinate system are given by

$$\begin{aligned} &\phi = Ae^{ikx_1+i\eta x_2} \\ &u_1 = \frac{\partial\phi}{\partial x_1} \\ &u_2 = \frac{\partial\phi}{\partial x_2} \\ &\sigma_{11} = \sigma_{22} = \lambda\left(u_{1,1}+u_{2,2}\right) = \lambda\left(\frac{\partial^2\phi}{\partial x_1^2}+\frac{\partial^2\phi}{\partial x_1^2}\right) = \lambda\nabla^2\phi = -\lambda k_f^2\phi = -\lambda\frac{\omega^2}{c_f^2}\phi = -\rho\omega^2\phi \\ &\sigma_{12} = 0 \\ &p = -\sigma_{11} = -\sigma_{22} = \rho\omega^2\phi \end{aligned} \tag{1.209}$$

In terms of the P-wave potential, the incident, reflected and transmitted waves shown in Figure 1.34 can be expressed as

$$\begin{aligned}\phi_i &= e^{ikx_1 - i\eta_1 x_2}\\ \phi_R &= R.e^{ikx_1 + i\eta_1 x_2}\\ \phi_T &= T.e^{ikx_1 - i\eta_2 x_2}\end{aligned} \tag{1.210}$$

where k and η_j have been defined in Eq. (1.203); it is assumed here that the incident wave amplitude is 1 while the reflected and transmitted wave amplitudes are R and T, respectively.

From continuity of normal displacement (u_2) and normal stress (σ_{22}) across the interface at $x_2=0$ one can write

$$\begin{aligned} i\eta_1(-1+R) &= -i\eta_2 T\\ -\rho_1\omega^2(1+R) &= -\rho_2\omega^2 T\end{aligned} \tag{1.211}$$

The above two equations can be written in matrix form

$$\begin{bmatrix}\eta_1 & \eta_2\\ -\rho_1 & \rho_2\end{bmatrix}\begin{Bmatrix}R\\ T\end{Bmatrix} = \begin{Bmatrix}\eta_1\\ \rho_1\end{Bmatrix} \tag{1.212}$$

From Eq. (1.212) one gets

$$\begin{aligned}\begin{Bmatrix}R\\ T\end{Bmatrix} &= \begin{bmatrix}\eta_1 & \eta_2\\ -\rho_1 & \rho_2\end{bmatrix}^{-1}\begin{Bmatrix}\eta_1\\ \rho_1\end{Bmatrix} = \frac{1}{(\eta_1\rho_2+\eta_2\rho_1)}\begin{bmatrix}\rho_2 & -\eta_2\\ \rho_1 & \eta_1\end{bmatrix}\begin{Bmatrix}\eta_1\\ \rho_1\end{Bmatrix} = \begin{Bmatrix}\dfrac{\eta_1\rho_2-\eta_2\rho_1}{\eta_1\rho_2+\eta_2\rho_1}\\[2ex] \dfrac{2\eta_1\rho_1}{\eta_1\rho_2+\eta_2\rho_1}\end{Bmatrix}\\ \Rightarrow \begin{Bmatrix}R\\ T\end{Bmatrix} &= \begin{Bmatrix}\dfrac{\rho_2 c_{f2}\cos\theta_1-\rho_1 c_{f1}\cos\theta_2}{\rho_2 c_{f2}\cos\theta_1+\rho_1 c_{f1}\cos\theta_2}\\[2ex] \dfrac{2\rho_1 c_{f2}\cos\theta_1}{\rho_2 c_{f2}\cos\theta_1+\rho_1 c_{f1}\cos\theta_2}\end{Bmatrix}\end{aligned} \tag{1.213}$$

Note that R, defined in Eqs. (1.207) and (1.213), are identical but T has slightly different expressions. The reason for this is that in one case the wave expressions are given in terms of pressure and in another case it is in terms of potentials. The pressure-potential relation is given in Eq. (1.209).

Example 1.20

Obtain Eq. (1.207) from Eq. (1.213)

SOLUTION

From Eq. (1.209) $p=\rho\omega^2\phi$

Hence,

$$\begin{aligned} p_i &= \rho_1\omega^2\phi_i = \rho_1\omega^2 e^{ikx_1-i\eta_1x_2}\\ p_R &= \rho_1\omega^2\phi_R = \rho_1\omega^2 R.e^{ikx_1+i\eta_1x_2}\\ p_T &= \rho_2\omega^2\phi_T = \rho_2\omega^2 T.e^{ikx_1-i\eta_2x_2}\end{aligned}$$

Thus,

$$\frac{|p_R|}{|p_i|} = R = \frac{\rho_2 c_{f2}\cos\theta_1 - \rho_1 c_{f1}\cos\theta_2}{\rho_2 c_{f2}\cos\theta_1 + \rho_1 c_{f1}\cos\theta_2}$$

$$\frac{|p_T|}{|p_i|} = \frac{\rho_2 T}{\rho_1} = \frac{2\rho_2 c_{f2}\cos\theta_2}{\rho_2 c_{f2}\cos\theta_1 + \rho_1 c_{f1}\cos\theta_2} = T_p \tag{1.213a}$$

These expressions are identical to the ones given in Eq. (1.207).

Since the expressions of R are identical in the two cases, it is not necessary to specifically state if the reflection coefficient R is defined for fluid pressure or fluid potential. However, this is not the case for the transmission coefficient T and one should use different symbols for the transmission coefficients for fluid pressure and fluid potential. Similar to the solid material, R and T without any subscript will be used to indicate reflection and transmission coefficients for fluid potentials and the symbol T_p will be used to indicate transmission coefficient for fluid pressure. Relation between T_p and T is

$$T_p = \frac{\rho_2}{\rho_1} T \tag{1.214}$$

Note that for normal incidence

$$R = \frac{Z_2 - Z_1}{Z_2 + Z_1}$$

$$T_p = \frac{2Z_2}{Z_2 + Z_1}$$

1.2.15.4 Point Source in a Fluid

Let us analyze the wave propagation problem in a fluid when the waves are generated by a point source as shown in Figure 1.35. These concentrated pressure sources produce elastic waves with a spherical wave front and are known as spherical waves.

If the time dependence of the point source be $f(t)$, then the governing equation of motion [Eq. (1.193)] can be written as

$$\nabla^2 p - \frac{1}{c_f^2}\frac{\partial^2 p}{\partial t^2} = f(t)\delta(\mathbf{x} - \mathbf{0}) \tag{1.215}$$

where the three-dimensional delta function $\delta(\mathbf{x\text{-}0})$ is zero at all points except at the origin. Because of the axisymmetric nature of the source it will generate an

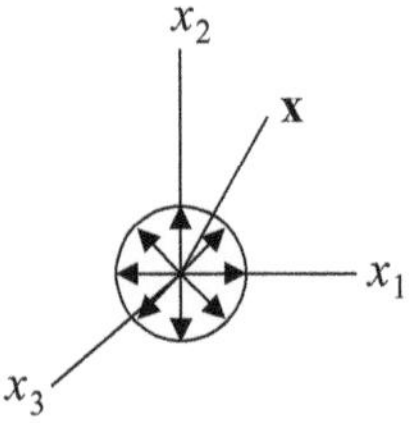

Figure 1.35 A point source at the origin generating spherical wave in a fluid.

axisymmetric wave in the fluid. Since there is no body force in the fluid in the region that excludes the origin, the governing equation of motion in the fluid (excluding the origin) is

$$\nabla^2 p - \frac{1}{c_f^2}\frac{\partial^2 p}{\partial t^2} = 0$$

The problem is axisymmetric; hence, its solution should be axisymmetric. In other words, p should be independent of angles θ and β of the spherical coordinates and be a function of the radial distance r and time t only. For this special case the above governing equation is simplified to

$$\frac{1}{r^2}\left[\frac{\partial}{\partial r}\left(r^2\frac{\partial p}{\partial r}\right)\right] - \frac{1}{c_f^2}\frac{\partial^2 p}{\partial t^2} = 0 \tag{1.216}$$

If we let $p = P(r,t)/r$, then the above equation is simplified to

$$\frac{\partial^2 P}{\partial r^2} - \frac{1}{c_f^2}\frac{\partial^2 P}{\partial t^2} = 0 \tag{1.217}$$

That has a solution in the form

$$P = P_1\left(t - \frac{r}{c_f}\right) + P_2\left(t + \frac{r}{c_f}\right)$$

where P_1 and P_2 represent waves traveling in outward and inward directions, respectively. From physical considerations we cannot have an inward-propagating wave. Hence, the acceptable solution is

$$p(r,t) = \frac{P}{r} = \frac{1}{r}P_1\left(t - \frac{r}{c_f}\right)$$

It can be shown that the function P_1 has the following form [Schmerr, 1998]

$$P_1\left(t - \frac{r}{c_f}\right) = \frac{1}{4\pi}f\left(t - \frac{r}{c_f}\right)$$

Hence, the final solution of Eq. (1.215) is

$$p(r,t) = \frac{1}{4\pi r}f\left(t - \frac{r}{c_f}\right) \tag{1.218}$$

If $f(t)$ is a delta function, then the governing equation and its solution are given by

$$\nabla^2 p - \frac{1}{c_f^2}\frac{\partial^2 p}{\partial t^2} = \delta(t)\delta(\mathbf{x} - \mathbf{0}) \tag{1.219}$$

$$p(r,t) = \frac{1}{4\pi r}\delta\left(t - \frac{r}{c_f}\right) \tag{1.220}$$

The Fourier transforms of Eqs. (1.219) and (1.220) give

$$\begin{aligned} &\nabla^2 G + \frac{\omega^2}{c_f^2}G = \delta(\mathbf{x} - \mathbf{0}) \\ &G(r,\omega) = \frac{e^{i\omega r/c_f}}{4\pi r} \end{aligned} \tag{1.221}$$

where G is the Fourier transform of p.

For harmonic excitation $f(t)=e^{-i\omega t}$ we can assume that the pressure field is also harmonic $p(r,t)=G(r)e^{-i\omega t}$. Then the equation of motion (1.215) becomes

$$\left[\nabla^2 G+\frac{\omega^2}{c_f^2}G\right]e^{-i\omega t}=\delta(\mathbf{x}-\mathbf{0})e^{-i\omega t} \tag{1.222}$$

Comparing Eqs. (1.221) and (1.222), one gets the solution for the harmonic case

$$G(r)=\frac{e^{ik_f r}}{4\pi r} \tag{1.223}$$

where $k_f=\omega/c_f$.

1.2.16 Reflection and Transmission of Plane Waves at a Fluid–Solid Interface

Figure 1.36 shows a plane P-wave of amplitude 1 striking an interface between a fluid half-space (density $=\rho_f$, P-wave speed $=c_f$) and a solid half-space (density $=\rho_S$, P-wave speed $=c_P$, S-wave speed $=c_S$). Reflected and transmitted wave directions in fluid and solid media are shown in the figure. The wave potentials in a fluid and solid are

$$\begin{aligned}\phi_i &= e^{ikx_1-i\eta_f x_2}\\ \phi_R &= R.e^{ikx_1+i\eta_f x_2}\\ \phi_T &= T_P e^{ikx_1-i\eta_S x_2}\\ \psi_T &= T_S e^{ikx_1-i\beta_S x_2}\end{aligned} \tag{1.224}$$

where

$$\begin{aligned}k &= k_f\sin\theta_1 = k_P\sin\theta_{2P} = k_S\sin\theta_{2S}\\ \eta_f &= k_f\cos\theta_1 = \sqrt{k_f^2-k^2}\\ \eta_S &= k_P\cos\theta_{2P} = \sqrt{k_P^2-k^2}\\ \beta_S &= k_S\cos\theta_{2S} = \sqrt{k_S^2-k^2}\\ k_f &= \frac{\omega}{c_f},\quad k_P=\frac{\omega}{c_P},\quad k_S=\frac{\omega}{c_S},\end{aligned} \tag{1.225}$$

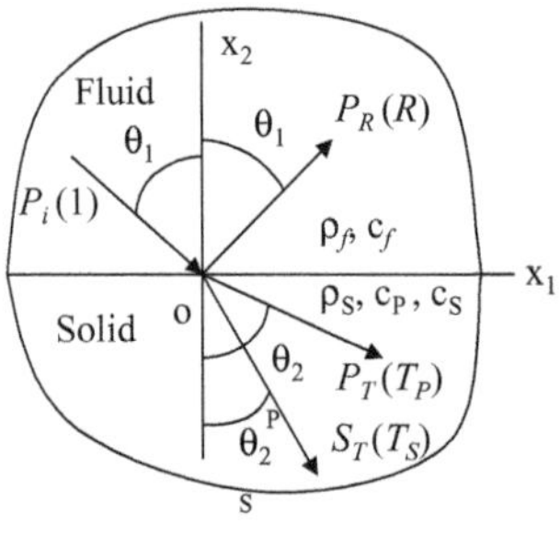

Figure 1.36 Incident, reflected and transmitted waves near an interface between fluid ($x_2 \geq 0$) and solid ($x_2 \leq 0$) half-spaces.

All waves have been expressed in terms of their potentials in Eq. (1.224). From continuity of normal displacement (u_2) and stresses (σ_{22} and σ_{12}) across the interface (at $x_2=0$) one gets

$$\begin{aligned} & i\eta_f(-1+R)=i(-\eta_S T_P - kT_S) \\ & -\rho_f\omega^2(1+R)=-\mu\left\{-\left(2k^2-k_S^2\right)T_P+2k\beta_S T_S\right\} \\ & \mu\left\{2k\eta_S T_P+\left(k^2-\beta_S^2\right)T_S\right\}=0 \end{aligned} \tag{1.226}$$

The above equations can be written in matrix form:

$$\begin{bmatrix} \eta_f & \eta_S & k \\ -\rho_f\omega^2 & -\mu\left(2k^2-k_S^2\right) & 2\mu k\beta_S \\ 0 & 2k\eta_S & \left(2k^2-k_S^2\right) \end{bmatrix} \begin{Bmatrix} R \\ T_P \\ T_S \end{Bmatrix} = \begin{Bmatrix} \eta_f \\ \rho_f\omega^2 \\ 0 \end{Bmatrix} \tag{1.227}$$

Equation (1.227) can be solved to obtain R, T_P and T_S as shown below.

From the third equation in Eq. (1.227) we get the following:

$$T_S=-\frac{2k\eta_S}{\left(2k^2-k_S^2\right)}T_P$$

Substituting this relation in the first two equations of Eq. (1.227), and after some simplification, one gets

$$\begin{aligned} & R-\frac{\eta_S}{\eta_f}\frac{1}{\left(\dfrac{2k^2}{k_S^2}-1\right)}T_P=1 \\ & R+\frac{\rho_S}{\rho_f}\left\{\frac{\left(\dfrac{2k^2}{k_S^2}-1\right)^2+\dfrac{4k^2\eta_S\beta_S}{k_S^4}}{\left(\dfrac{2k^2}{k_S^2}-1\right)}\right\}T_P=-1 \end{aligned} \tag{1.227a}$$

Subtracting the first equation from the second of Eq. (1.227a) it is possible to obtain

$$\left\{\frac{\rho_S}{\rho_f}\frac{\left(\dfrac{2k^2}{k_S^2}-1\right)^2+\dfrac{4k^2\eta_S\beta_S}{k_S^4}}{\left(\dfrac{2k^2}{k_S^2}-1\right)}+\frac{\eta_S}{\eta_f}\frac{1}{\left(\dfrac{2k^2}{k_S^2}-1\right)}\right\}T_P$$

$$=\frac{\dfrac{\rho_S}{\rho_f}\left\{\left(\dfrac{2k^2}{k_S^2}-1\right)^2+\dfrac{4k^2\eta_S\beta_S}{k_S^4}\right\}+\dfrac{\eta_S}{\eta_f}}{\left(\dfrac{2k^2}{k_S^2}-1\right)}T_P=-2$$

or,

$$T_P = \frac{-2\left(\frac{2k^2}{k_S^2}-1\right)}{\frac{\rho_S}{\rho_f}\left\{\left(\frac{2k^2}{k_S^2}-1\right)^2+\frac{4k^2\eta_S\beta_S}{k_S^4}\right\}+\frac{\eta_S}{\eta_f}} \tag{1.227b}$$

Substituting the above expression of T_P into the first equation of (1.227a) one gets

$$R = 1+\frac{\frac{\eta_S}{\eta_f}}{\left(\frac{2k^2}{k_S^2}-1\right)}\left[\frac{-2\left(\frac{2k^2}{k_S^2}-1\right)}{\frac{\rho_S}{\rho_f}\left\{\left(\frac{2k^2}{k_S^2}-1\right)^2+\frac{4k^2\eta_S\beta_S}{k_S^4}\right\}+\frac{\eta_S}{\eta_f}}\right]$$

$$= 1+\frac{-2\frac{\eta_S}{\eta_f}}{\frac{\rho_S}{\rho_f}\left\{\left(\frac{2k^2}{k_S^2}-1\right)^2+\frac{4k^2\eta_S\beta_S}{k_S^4}\right\}+\frac{\eta_S}{\eta_f}}$$

or,

$$R = \frac{\frac{\rho_S}{\rho_f}\left\{\left(\frac{2k^2}{k_S^2}-1\right)^2+\frac{4k^2\eta_S\beta_S}{k_S^4}\right\}-\frac{\eta_S}{\eta_f}}{\frac{\rho_S}{\rho_f}\left\{\left(\frac{2k^2}{k_S^2}-1\right)^2+\frac{4k^2\eta_S\beta_S}{k_S^4}\right\}+\frac{\eta_S}{\eta_f}} \tag{1.227c}$$

After substituting Eq. (1.227b) into the third equation of Eq. (1.227) we get

$$T_S = \frac{\frac{2k\eta_S}{k_S^2}}{\left(\frac{2k^2}{k_S^2}-1\right)}\frac{2\left(\frac{2k^2}{k_S^2}-1\right)}{\frac{\rho_S}{\rho_f}\left\{\left(\frac{2k^2}{k_S^2}-1\right)^2+\frac{4k^2\eta_S\beta_S}{k_S^4}\right\}+\frac{\eta_S}{\eta_f}}$$

$$= \frac{\frac{4k\eta_S}{k_S^2}}{\frac{\rho_S}{\rho_f}\left\{\left(\frac{2k^2}{k_S^2}-1\right)^2+\frac{4k^2\eta_S\beta_S}{k_S^4}\right\}+\frac{\eta_S}{\eta_f}} \tag{1.227d}$$

The above equations [Eqs. (1.227b, c, d)] can be expressed in terms of the angles of incidence and transmission as shown below.

From Eq. (1.225) one can write

$$\left(\frac{2k^2}{k_S^2}-1\right)^2=\left(\frac{2k_S^2\sin^2\theta_{2S}}{k_S^2}-1\right)^2=(-\cos 2\theta_{2S})^2=\cos^2 2\theta_{2S}$$

$$\frac{\eta_S}{\eta_f}=\frac{k_P\cos\theta_{2S}}{k_f\cos\theta_1}=\frac{c_f\cos\theta_{2P}}{c_P\cos\theta_1} \tag{1.227e}$$

$$\frac{4k^2\eta_S\beta_S}{k_S^4}=\frac{4}{k_S^4}(k_S\sin\theta_{2S}.k_P\sin\theta_{2P}.k_S\cos\theta_{2S}.k_P\cos\theta_{2P})$$

$$=\frac{k_P^2}{k_S^2}\sin 2\theta_{2P}.\sin 2\theta_{2S}=\frac{c_S^2}{c_P^2}\sin 2\theta_{2P}.\sin 2\theta_{2S}$$

Substituting the above relations into Eq. (1.227c) we get

$$R=\frac{\frac{\rho_S}{\rho_f}\left\{\left(\frac{2k^2}{k_S^2}-1\right)^2+\frac{4k^2\eta_S\beta_S}{k_S^4}\right\}-\frac{\eta_S}{\eta_f}}{\frac{\rho_S}{\rho_f}\left\{\left(\frac{2k^2}{k_S^2}-1\right)^2+\frac{4k^2\eta_S\beta_S}{k_S^4}\right\}+\frac{\eta_S}{\eta_f}}=\frac{\frac{\rho_S}{\rho_f}\left\{\cos^2 2\theta_{2S}+\frac{c_S^2}{c_P^2}\sin 2\theta_{2P}.\sin 2\theta_{2S}\right\}-\frac{c_f\cos\theta_{2P}}{c_P\cos\theta_1}}{\frac{\rho_S}{\rho_f}\left\{\cos^2 2\theta_{2S}+\frac{c_S^2}{c_P^2}\sin 2\theta_{2P}.\sin 2\theta_{2S}\right\}+\frac{c_f\cos\theta_{2P}}{c_P\cos\theta_1}}$$

or,

$$R=\frac{\frac{\rho_S c_P\cos\theta_1}{\rho_f c_f}\left\{\cos^2 2\theta_{2S}+\frac{c_S^2}{c_P^2}\sin 2\theta_{2P}.\sin 2\theta_{2S}\right\}-\cos\theta_{2P}}{\frac{\rho_S c_P\cos\theta_1}{\rho_f c_f}\left\{\cos^2 2\theta_{2S}+\frac{c_S^2}{c_P^2}\sin 2\theta_{2P}.\sin 2\theta_{2S}\right\}+\cos\theta_{2P}}=\frac{\Delta_2-\Delta_1}{\Delta_2+\Delta_1} \tag{1.227f}$$

where

$$\Delta_1=\cos\theta_{2P}$$
$$\Delta_2=\frac{\rho_S c_P\cos\theta_1}{\rho_f c_f}\left\{\cos^2 2\theta_{2S}+\frac{c_S^2}{c_P^2}\sin 2\theta_{2P}.\sin 2\theta_{2S}\right\} \tag{1.227g}$$

Similarly, from Eq. (1.227b) we get

$$T_P=\frac{-2\left(\frac{2k^2}{k_S^2}-1\right)}{\frac{\rho_S}{\rho_f}\left\{\left(\frac{2k^2}{k_S^2}-1\right)^2+\frac{4k^2\eta_S\beta_S}{k_S^4}\right\}+\frac{\eta_S}{\eta_f}}$$

$$=\frac{2\cos 2\theta_{2S}}{\frac{\rho_S}{\rho_f}\left\{\cos^2 2\theta_{2S}+\frac{c_S^2}{c_P^2}\sin 2\theta_{2P}.\sin 2\theta_{2S}\right\}+\frac{c_f\cos\theta_{2P}}{c_P\cos\theta_1}}$$

or,

$$T_P = \frac{\dfrac{2c_P\cos\theta_1\cos 2\theta_{2S}}{c_f}}{\dfrac{\rho_S c_P\cos\theta_1}{\rho_f c_f}\left\{\cos^2 2\theta_{2S} + \dfrac{c_S^2}{c_P^2}\sin 2\theta_{2P}.\sin 2\theta_{2S}\right\} + \cos\theta_{2P}} \tag{1.227h}$$

$$= \frac{2c_P\cos\theta_1\cos 2\theta_{2S}}{c_f\left(\Delta_2 + \Delta_1\right)}$$

and from Eq. (1.227d) we get

$$T_S = \frac{\dfrac{4k\eta_S}{k_S^2}}{\dfrac{\rho_S}{\rho_f}\left\{\left(\dfrac{2k^2}{k_S^2} - 1\right)^2 + \dfrac{4k^2\eta_S\beta_S}{k_S^4}\right\} + \dfrac{\eta_S}{\eta_f}}$$

$$= \frac{\dfrac{4k_S\sin\theta_1.k_S\cos\theta_1}{k_S^2}}{\dfrac{\rho_S}{\rho_f}\left\{\cos^2 2\theta_{2S} + \dfrac{c_S^2}{c_P^2}\sin 2\theta_{2P}.\sin 2\theta_{2S}\right\} + \dfrac{c_f\cos\theta_{2P}}{c_P\cos\theta_1}} \tag{1.227i}$$

or,

$$T_S = \frac{\dfrac{2c_S^2\sin 2\theta_{2P}}{c_P^2}\dfrac{c_P\cos\theta_1}{c_f}}{\dfrac{\rho_S c_P\cos\theta_1}{\rho_f c_f}\left\{\cos^2 2\theta_{2S} + \dfrac{c_S^2}{c_P^2}\sin 2\theta_{2P}.\sin 2\theta_{2S}\right\} + \cos\theta_{2P}} \tag{1.227j}$$

$$= \frac{2c_S^2\sin 2\theta_{2P}.\cos\theta_1}{c_P c_f\left(\Delta_2 + \Delta_1\right)}$$

If the x_2-axis is defined positive downward, then one can prove that the sign of T_S should change, but the signs of R and T_P remain unchanged.

Example 1.21

In Figure 1.36, suppose the incident and reflected fields in a fluid are defined in terms of the fluid pressure instead of the P-wave potential, while the P- and S-waves in a solid are defined in terms of the wave potentials. Obtain the reflection and transmission coefficients.

SOLUTION

The incident, reflected and transmitted waves shown in Figure 1.36 are written in the following form:

$$\begin{aligned} p_i &= e^{ikx_1 - i\eta_f x_2} \\ p_R &= R^*.e^{ikx_1 + i\eta_f x_2} \\ \phi_T &= T_P^* e^{ikx_1 - i\eta_S x_2} \\ \psi_T &= T_S^* e^{ikx_1 - i\beta_S x_2} \end{aligned} \tag{1.227k}$$

We would like to express R^*, T_P^* and T_S^* in terms of R, T_P and T_S.

Using Eq. (1.209) it is possible to write from Eq. (1.224)

$$\begin{aligned} p_i &= \rho_f\omega^2\phi_i = \rho_f\omega^2 e^{ikx_1 - i\eta_f x_2} \\ p_R &= \rho_f\omega^2\phi_R = \rho_f\omega^2 R.e^{ikx_1 + i\eta_f x_2} \\ \phi_T &= T_P e^{ikx_1 - i\eta_S x_2} \\ \psi_T &= T_S e^{ikx_1 - i\beta_S x_2} \end{aligned} \qquad (1.227l)$$

Therefore, for unit amplitude of the incident pressure field the reflected and transmitted fields are modified in the following manner:

$$\begin{aligned} p_i &= e^{ikx_1 - i\eta_f x_2} \\ p_R &= R.e^{ikx_1 + i\eta_f x_2} \\ \phi_T &= \frac{T_P}{\rho_f\omega^2} e^{ikx_1 - i\eta_S x_2} \\ \psi_T &= \frac{T_S}{\rho_f\omega^2} e^{ikx_1 - i\beta_S x_2} \end{aligned} \qquad (1.227m)$$

Comparing Eqs. (1.227 m) and (1.227k) one can write

$$\begin{aligned} R^* &= R. \\ T_P^* &= \frac{T_P}{\rho_f\omega^2} \\ T_S^* &= \frac{T_S}{\rho_f\omega^2} \end{aligned} \qquad (1.227n)$$

R, T_P and T_S expressions are given in Eqs. (1.227b to j)

1.2.17 Reflection and Transmission of Plane Waves by a Solid Plate Immersed in a Fluid

Figure 1.37 shows an acoustic pressure wave of amplitude 1 striking the top surface of a solid plate of thickness h immersed in a fluid. Clearly, the solid plate divides the fluid space into two half-spaces. The top half-space contains the incident and reflected waves while the bottom half-space contains the transmitted wave. The solid plate will have P- and S-waves generated in it. P- and S-wave potentials are denoted by ϕ and ψ, respectively. Subscripts U and D are used to denote up-going and down-going wave potentials, respectively.

Since the fluid properties are the same in the top and bottom half-spaces the incident, reflected and transmitted wave angles in the fluid are equal (θ). The angles corresponding to the P- and S-waves in the solid plate are denoted by θ_P and θ_S, respectively.

As mentioned earlier, elastic waves in fluid can be mathematically expressed in terms of the pressure field or the potential field. In the previous section we have expressed it in terms of the potential field. Here, let us express it in terms of the pressure field. Waves in the solid will be expressed in terms of the potential field.

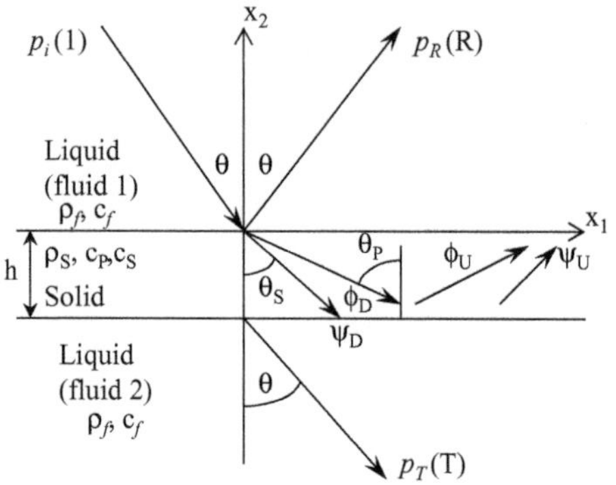

Figure 1.37 Incident, reflected and transmitted pressure waves in fluid half-spaces separated by a solid plate containing up-going and down-going P- and S- waves.

Then, the seven waves shown by seven arrows in Figure 1.37 can be expressed in the following manner:

$$\begin{aligned}
p_i &= e^{ikx_1 - i\eta_f x_2} \\
p_R &= R.e^{ikx_1 + i\eta_f x_2} \\
p_T &= T.e^{ikx_1 - i\eta_f x_2} \\
\phi_D &= P_D.e^{ikx_1 - i\eta_S x_2} \\
\phi_U &= P_U.e^{ikx_1 + i\eta_S x_2} \\
\psi_D &= S_D.e^{ikx_1 - i\beta_S x_2} \\
\psi_U &= S_U.e^{ikx_1 + i\beta_S x_2}
\end{aligned} \tag{1.228}$$

where

$$\begin{aligned}
k &= k_f \sin\theta = k_P \sin\theta_P = k_S \sin\theta_S \\
\eta_f &= k_f \cos\theta = \sqrt{k_f^2 - k^2} \\
\eta_S &= k_P \cos\theta_P = \sqrt{k_P^2 - k^2} \\
\beta_S &= k_S \cos\theta_S = \sqrt{k_S^2 - k^2} \\
k_f &= \frac{\omega}{c_f}, \quad k_P = \frac{\omega}{c_P}, \quad k_S = \frac{\omega}{c_S},
\end{aligned} \tag{1.229}$$

From Eq. (1.209) one can write $u_2 = \dfrac{\partial\phi}{\partial x_2} = \dfrac{1}{\rho\omega^2}\dfrac{\partial p}{\partial x_2}$ and $\sigma_{22} = -p$.

From continuity of normal displacement (u_2) and stresses (σ_{22} and σ_{12}) across the interface (at $x_2 = 0$)

$$\frac{i\eta_f}{\rho_f\omega^2}(-1+R)=i\left(-\eta_S P_D+\eta_S P_U-kS_D-kS_U\right)$$
$$-(1+R)=-\mu\left\{-\left(2k^2-k_S^2\right)\left(P_D+P_U\right)+2k\beta_S\left(S_D-S_U\right)\right\} \tag{1.230a}$$
$$\mu\left\{2k\eta_S\left(P_D-P_U\right)+\left(k^2-\beta_S^2\right)\left(S_D+S_U\right)\right\}=0$$

and from continuity of normal displacement (u_2) and stresses (σ_{22} and σ_{12}) across the interface (at $x_2=h$) one gets

$$-\frac{i\eta_f}{\rho_f\omega^2}TQ_f^{-1}=i\left(-\eta_S Q_P^{-1}P_D+\eta_S Q_P P_U-kQ_S^{-1}S_D-kQ_S S_U\right)$$
$$-TQ_f^{-1}=-\mu\left\{-\left(2k^2-k_S^2\right)\left(Q_P^{-1}P_D+Q_P P_U\right)+2k\beta_S\left(Q_S^{-1}S_D-Q_S S_U\right)\right\} \tag{1.230b}$$
$$\mu\left\{2k\eta_S\left(Q_P^{-1}P_D-Q_P P_U\right)+\left(k^2-\beta_S^2\right)\left(Q_S^{-1}S_D+Q_S S_U\right)\right\}=0$$

In Eq. (1.230) μ is the shear modulus of the solid, and

$$Q_f=e^{i\eta_f h}$$
$$Q_P=e^{i\eta_S h} \tag{1.231}$$
$$Q_S=e^{i\beta_S h}$$

The six equations given in Eqs. (1.230a) and (1.230b) can be written in matrix form:

$$[A]\{X\}=\{b\} \tag{1.232}$$

where

$$[A]=\begin{bmatrix} \frac{\eta_f}{\rho_f\omega^2} & 0 & \eta_S & -\eta_S & k & k \\ 1 & 0 & \mu\left(2k^2-k_S^2\right) & \mu\left(2k^2-k_S^2\right) & -2\mu k\beta_S & 2\mu k\beta_S \\ 0 & 0 & 2k\eta_S & -2k\eta_S & \left(2k^2-k_S^2\right) & \left(2k^2-k_S^2\right) \\ 0 & -\frac{\eta_f Q_f^{-1}}{\rho_f\omega^2} & \eta_S Q_P^{-1} & -\eta_S Q_P & kQ_S^{-1} & kQ_S \\ 0 & Q_f^{-1} & \mu\left(2k^2-k_S^2\right)Q_P^{-1} & \mu\left(2k^2-k_S^2\right)Q_P & -2\mu k\beta_S Q_S^{-1} & 2\mu k\beta_S Q_S \\ 0 & 0 & 2k\eta_S Q_P^{-1} & -2k\eta_S Q_P & \left(2k^2-k_S^2\right)Q_S^{-1} & \left(2k^2-k_S^2\right)Q_S \end{bmatrix} \tag{1.232a}$$

$$\{X\}=\begin{Bmatrix} R \\ T \\ P_D \\ P_U \\ S_D \\ S_U \end{Bmatrix},\quad \{b\}=\begin{Bmatrix} \frac{\eta_f}{\rho_f\omega^2} \\ -1 \\ 0 \\ 0 \\ 0 \\ 0 \end{Bmatrix}$$

Eq. (1.232) can be solved for unknowns R, T etc.

If the x_2-axis in Figure 1.37 is defined as positive downward, then $[A]$ matrix and $\{b\}$ vector of Eq. (1.232) can be written in the following form after using the relations given in Eq. (1.229):

$$[A]=\begin{bmatrix} 0 & 0 & -\dfrac{\sin 2\theta_P}{c_P^2} & \dfrac{\sin 2\theta_P}{c_P^2} & -\dfrac{\cos 2\theta_S}{c_S^2} & -\dfrac{\cos 2\theta_S}{c_S^2} \\ 0 & 0 & -\dfrac{Q_P \sin 2\theta_P}{c_P^2} & \dfrac{\sin 2\theta_P}{c_P^2 Q_P} & -\dfrac{Q_S \cos 2\theta_S}{c_S^2} & -\dfrac{\cos 2\theta_S}{Q_S c_S^2} \\ -1 & 0 & \rho_S\omega^2\left(1+2\dfrac{c_S^2}{c_P^2}\sin^2\theta_P\right) & \rho_S\omega^2\left(1+2\dfrac{c_S^2}{c_P^2}\sin^2\theta_P\right) & -\rho_S\omega^2\sin 2\theta_S & \rho_S\omega^2\sin 2\theta_S \\ 0 & -Q_f & \rho_S\omega^2 Q_P\left(1+2\dfrac{c_S^2}{c_P^2}\sin^2\theta_P\right) & \dfrac{\rho_S\omega^2}{Q_P}\left(1+2\dfrac{c_S^2}{c_P^2}\sin^2\theta_P\right) & -\rho_S\omega^2 Q_S\sin 2\theta_S & \dfrac{\rho_S\omega^2\sin 2\theta_S}{Q_S} \\ \dfrac{\cos\theta}{\rho_f\omega^2 c_f} & 0 & \dfrac{\cos\theta_P}{c_P} & -\dfrac{\cos\theta_P}{c_P} & -\dfrac{\sin\theta_S}{c_S} & -\dfrac{\sin\theta_S}{c_S} \\ 0 & -\dfrac{\cos\theta}{\rho_f\omega^2 c_f}Q_f & \dfrac{Q_P\cos\theta_P}{c_P} & -\dfrac{\cos\theta_P}{c_P Q_P} & -\dfrac{Q_S\sin\theta_S}{c_S} & -\dfrac{\sin\theta_S}{c_S Q_S} \end{bmatrix} \tag{1.232b}$$

$$\{b\}=\begin{Bmatrix} 0 \\ 0 \\ 1 \\ 0 \\ \dfrac{\cos\theta}{\rho_f\omega^2 c_f} \\ 0 \end{Bmatrix}$$

1.2.18 Elastic Properties of Different Materials

The elastic wave speed and density of a number of materials are given in Table 1.4. These values have been collected from a number of references given in the reference list and from other sources. Three wave speeds – P-wave speed (c_P), shear wave speed (c_S), Rayleigh wave speed (c_R) and density (ρ) – are given for a number of materials. For a group of materials some information – shear wave speed, Rayleigh wave speed, or density – is missing.

1.3 Concluding Remarks

In this chapter a brief review of the fundamentals of the mechanics of elastic wave propagation in solid and fluid media are given, starting with the derivation of basic equations of the theory of elasticity and continuum mechanics. Only the basic linear analysis relevant to linear ultrasonic NDE is covered in Chapter 1. Nonlinear analysis is presented in Chapter 4.

Elastic wave propagation in isotropic solids and fluids without material attenuation is discussed in detail in this chapter. Materials of this chapter can be covered in the graduate level first course on elastic wave propagation. Some of the more advanced topics such as elastic wave propagation in multilayered anisotropic plates and pipes with and without material attenuation, ultrasonic field modeling by semi-analytical techniques and nonlinear ultrasonic analysis are covered in the following chapters.

A number of good books on the mechanics of elastic wave propagation in solids are available in the literature and the interested readers are referred to those books for further study. These books have been authored by Brekhovskikh (1960), Kolsky (1963), Achenbach (1973), Auld (1990), Graff (1991), Mal (1991), Schmerr (1998)Rose (1999).

Table 1.4: Elastic Wave Speeds (P-Wave Speed [c_P], S-Wave Speed [c_S], Rayleigh Wave Speed [c_R]) and Density [ρ] of Different Materials

Material	c_P (km/s)	c_S [c_R] (km/s)	ρ (gm/cc)
Acetate, Butyl (n)	1.27		0.871
Acetate, Ethyl	1.18		0.900
Acetate, Methyl ($C_3H_6O_2$)	1.15–1.21		0.928–0.934
Acetate, Propyl	1.18		0.891
Acetone (C_3H_6O)	1.17		0.790
Acetonyl Acetone ($C_6H_{10}O_2$)	1.40		0.729
Acrylic Resin	2.67	1.12	1.18
Air (0°C)	0.33		0.0004
Air (20°C)	0.34		
Air (100°C)	0.39		
Air (500°C)	0.55		
Alcohol, Butyl ($C_4H_{10}O$)	1.24		0.810
Alcohol, Ethyl	1.18		0.789
Alcohol, Methyl	1.12		0.792
Alcohol, Propyl (i)	1.17		0.786
Alcohol, Propyl (n)	1.22		0.804
Alcohol, t-Amyl ($C_5H_{12}O$)	1.20		0.810
Alcohol Vapor (0°C)	0.231		
Alumina	10.82	6.16 [5.68]	3.97
Aluminum	6.25–6.5	3.04–3.13 [2.84–2.95]	2.70–2.80
Ammonia	0.42		
Analine ($C_6H_5NH_2$)	1.69		1.02
Argon	0.319		0.00178
Argon, Liquid (−186°C to −189°C)	0.837–0.863		1.404–1.424
Bakelite	1.59		1.40
Barium Titanate	4.00		6.02
Benzene (C_6H_6)	1.30		0.87
Benzol	1.33		0.878
Benzol, Ethyl	1.34		0.868
Beryllium	12.7–12.9	8.71–8.88 [7.84–7.87]	1.82–1.85
Bismuth	2.18–2.20	1.10 [1.03]	9.80
Boron Carbide	11.00		2.40
Bone	3.0–4.0	1.97–2.25 [1.84–2.05]	1.90
Brass (70% Cu, 30% Zn)	4.28–4.44	2.03–2.12 [1.96]	8.56
Brass, Half Hard	3.83	2.05	8.10
Brass, Naval	4.43	2.12 [1.95]	8.42
Bronze	3.53	2.23 [2.01]	8.86
Butyl Rubber	1.70		1.11
Cadmium	2.78	1.50 [1.39]	8.64
Carbon Bisulfide	1.16		
Carbon Dioxide (CO_2)	0.258		
Curbon Disulfide (CS_2)	1.15		1.26

(Continued)

Table 1.4 (Continued): Elastic Wave Speeds (P-Wave Speed [c_P], S-Wave Speed [c_S], Rayleigh Wave Speed [c_R]) and Density [ρ] of Different Materials

Material	c_P (km/s)	c_S [c_R] (km/s)	ρ (gm/cc)
Carbon Disulfate	0.189		
Carbon Monoxide (CO)	0.337		
Carbon Tetrachloride (CCl_4)	0.93		1.60
Carbon, Vitreous	4.26	2.68 [2.43]	1.47
Castor Oil	1.48		0.969
Cesium (28.5°C)	0.967		1.88
Chlorine	0.205		
Chocolate (dark)	2.58	0.96	1.302
Choroform ($CHCl_3$)	0.987		1.49
Chromium	6.61	4.01 [3.66]	7.19
Columbium	4.92	2.10	8.57
Constantan	5.185.24	2.64 [2.45]	8.88–8.90
Copper	4.66–5.01	2.26–2.33 [1.93–2.17]	8.93
Duraluminium	6.40	3.12 [2.92]	2.80
Cork	0.051		0.24
Diesel Oil	1.25		
Ether Vapor (0°C)	0.179		
Ethyl Ether ($C_4H_{10}O$)	0.986		0.713
Ethylene	0.314		
Flint	4.26	2.96	3.60
Fused Quartz	5.96	3.76	2.20
Gasoline	1.25		0.803
Gallium (29.5°C)	2.74		5.95
Germanium	5.41		5.47
Glass, Crown	5.26–5.66	3.26–3.52 [3.12]	2.24–3.6
Glass, Heavy Flint	5.26	2.96 [2.73]	3.60
Glass, Quartz	5.57–5.97	3.43–3.77 [3.41]	2.2–2.60
Glass, Window	6.79	3.43	
Glass, Plate	5.71–5.79	3.43	2.75
Glass, Pyrex	5.56–5.64	3.28–3.43 [3.01]	2.23
Glycerine ($C_3H_8O_3$)	1.92		1.26
Gold	3.24	1.20 [1.13]	19.3–19.7
Granite	3.90		
Hafnium	3.84		13.28
Helium	0.97		0.00018
Helium, Liquid (−269°C)	0.18		0.125
Helium, Liquid (−271.5°C)	0.231		0.146
Hydrogen	1.28		0.00009
Hydrogen, Liquid (−252.7°C)	1.13		0.355
Ice	3.99	1.99	1.0
Inconel	5.70	3.0 [2.79]	8.25–8.39

(Continued)

Table 1.4 (Continued): Elastic Wave Speeds (P-Wave Speed [c_P], S-Wave Speed [c_S], Rayleigh Wave Speed [c_R]) and Density [ρ] of Different Materials

Material	c_P (km/s)	c_S [c_R] (km/s)	ρ (gm/cc)
Indium	2.22–2.56		7.30
Invar (63.8% Fe, 36%Ni, 0.2%Cu)	4.66	2.66 [2.45]	8.00
Iron, Soft	5.90–5.96	3.22 [2.79–2.99]	7.7
Iron, Cast	4.50–4.99	2.40–2.81 [2.3–2.59]	7.22–7.80
Kerosene	1.32		0.81
Kidney	1.54		1.05
Lead	2.16	0.70 [0.63–0.66]	11.34–11.4
Lead Zirconate Titanate (PZT)	3.79		7.65
Linseed Oil	1.77		0.922
Liver	1.54		1.07
Lucite	2.68–2.70	1.05–1.10	1.15–1.18
Magnesium	5.47–6.31	3.01–3.16 [2.93]	1.69–1.83
Manganese	4.60–4.66	2.35	7.39–7.47
Marble	6.15		9.5
Mercury (20°C)	1.42–1.45		13.5–13.8
Methane	0.43		0.00074
Methylene Iodide	0.98		
Molybdenum	6.30–6.48	3.35–3.51 [3.11–3.25]	10.2
Monel	5.35–6.04	2.72 [1.96–2.53]	8.82–8.83
Monochlorobenzene (C_6H_5Cl)	1.27		1.107
Morpholine (C_4H_9NO)	1.44		1.00
Motor Oil (SAE 20)	1.74		0.87
Mylar	2.54		1.18
n-Hexanol ($C_6H_{14}O$)	1.30		0.819
Neon	0.43		
Neoprene	1.56		1.31
Nickel	5.61–5.81	2.93–3.08 [2.64–2.86]	8.3–8.91
Niobium	5.07	2.09 [1.97]	8.59
Nitric Oxide	0.325		
Nitrobenzene ($C_6H_5NO_2$)	1.46		1.20
Nitrogen (0–20°C)	0.33–0.35		0.00116–0.00125
Nitrogen, Liquid (−197°)	0.869		0.815
Nitrogen, Liquid (−203°C)	0.929		0.843
Nitromethane (CH_3NC_2)	1.33		1.13
Nitrous Oxide	0.26		
Nylon	2.62	1.10 [1.04]	1.11–1.14
Oil	1.38–1.5		0.92–0.953
Olive Oil	1.43		0.948
Oxygen (0–20°C)	0.32–0.33		0.00132–0.00142
Oxygen, Liquid (−183.6°)	0.971		1.143
Oxygen, Liquid (−210°)	1.13		1.272

(Continued)

Table 1.4 (Continued): Elastic Wave Speeds (P-Wave Speed [c_P], S-Wave Speed [c_S], Rayleigh Wave Speed [c_R]) and Density [ρ] of Different Materials

Material	c_P (km/s)	c_S [c_R] (km/s)	ρ (gm/cc)
Parrafin (15°C)	1.30		
Parracin Oil	1.42		0.835
Peanut Oil	1.46		0.936
Pentane	1.01		0.621
Perspex (PMMA)	2.70	1.33 [1.24]	1.19
Petroleum	1.29		0.825
Plastic, Acrylic Resin	2.67	1.12	1.18
Platinum	3.26–3.96	1.67–1.73 [1.60]	21.4
Plexiglas	2.67–2.77	1.12–1.43	1.18–1.27
Plutonium	1.79		15.75
Polycarbonate	2.22		1.19
Polyester Casting Resin	2.29		1.07
Polyethylene	2.0–2.67		0.92–1.10
Polythelyne (Low Density)	1.95	0.54 [0.51]	0.92
Polyethylene (TCI)	1.60		
Polypropylene	2.74		0.904
Polystyrene (Styron 666)	2.40	1.15 [1.08]	1.05
Polystyrene	2.67	1.10	2.80
Polyvinyl Chloride	2.30		1.35
Polyvinyl Chloride Acetate	2.25		
Polyvinyl Formal	2.68		
Polyvinylidene Chorlide	2.40		1.70
Potassium (100°C)	1.86		0.818
Potassium (200°C)	1.81		0.796
Potassium (400°C)	1.71		0.751
Potassium (600°C)	1.60		0.707
Potassium (800°C)	1.49		0.662
Quartz	5.66–5.92	3.76	2.65
Refrasil	3.75		1.73
Rock Salt (x direction)	4.78		
Rochelle Salt	5.36	3.76	2.20
Rubber	1.26–1.85		
Salt Solution (10%)	1.47		
Salt Solution (15%)	1.53		
Salt Solution (20%)	1.60		
Sandstone	2.92	1.84 [1.68]	
Sapphire	9.8–11.15		3.98
Silica (fused)	5.96	3.76 [3.41]	2.15
Silicon Carbide	12.10	7.49 [6.81]	3.21
Silicone Oil (25°C, Dow 710 fluid)	1.35		
Silicon Nitride	10.61	6.20 [5.69]	3.19

(Continued)

Table 1.4 (Continued): Elastic Wave Speeds (P-Wave Speed [c_P], S-Wave Speed [c_S], Rayleigh Wave Speed [c_R]) and Density [ρ] of Different Materials

Material	c_P (km/s)	c_S [c_R] (km/s)	ρ (gm/cc)
Silver	3.60–3.70	1.70 [1.59]	10.5
Silver-18 Ni	4.62	2.31	
Silver Epoxy, e-solder	1.90	0.98 [0.91]	2.71
Sodium (100°C)	2.53		0.926
Sodium (200°C)	2.48		0.904
Sodium (400°C)	2.37		0.857
Sodium (600°C)	2.26		0.809
Sodium (800°C)	2.15		0.759
Spleen	1.50		1.07
Stainless Steel	5.98	3.3 [3.05]	7.80
Steel	5.66–5.98	3.05–3.3 [2.95–3.05]	7.80–7.93
Tantalum	4.16	2.04 [1.90]	16.67
Teflon	1.35–1.45		2.14–2.20
Thorium	2.40	1.56 [1.40]	11.73
Tin	3.32–3.38	1.59–1.67 [1.49]	7.3
Tissue (Beef)	1.55		1.08
Tissue (Brain)	1.49		1.04
Tissue (Human)	1.47		1.07
Titanium	6.10–6.13	3.12–3.18 [2.96]	4.51–4.54
Titanium Carbide	8.27	5.16 [4.68]	5.15
Titanium Carbide with 6% Co	6.66	3.98 [3.64]	15.0
Tungsten	5.18–5.41	2.64–2.89 [2.46–2.67]	19.25
Transformer Oil	1.39		0.92
Uranium	3.37	1.94 [1.78]	19.05
Uranium Dioxide	5.18		10.96
Vanadium	6.02	2.77 [2.60]	6.09
Water (20°C)	1.48		1.00
Water (Sea)	1.53		1.025
Water Vapor (0°C)	0.401		
Water Vapor (100°C)	0.405		
Water Vapor (130°C)	0.424		
Wood (Oak)	4.47–4.64	1.75	0.4615
Wood (Pine, along the fiber)	3.32		
Wood (Poplar, along the fiber)	4.28		
Zinc	4.17–4.19	2.42 [2.22]	7.1–7.14
Zinc Oxide (c-axis)	6.40	2.95 [2.77]	5.61
Zircaloy	4.72	2.36 [2.20]	9.36
Zirconium	4.65	2.25 [2.10]	6.51

Note: For some Materials instead of one value, Ranges (Upper and Lower Bounds) for the Material Properties Collected from Different Sources are Given

EXERCISE PROBLEMS

PROBLEM 1.1

Simplify the following expressions (note that δ_{ij} is the Kronecker delta, repeated index means summation and comma represents derivative):

(a) δ_{mm} (b) $\delta_{ij}\delta_{kj}$ (c) $\delta_{ij}u_{k,kj}$ (d) $\delta_{ij}\delta_{ij}$

(e) $\delta_{mm}\delta_{ij}x_j$ (f) $\delta_{km}u_{i,jk}u_{i,jm}$ (g) $\dfrac{\partial x_m}{\partial x_k}\dfrac{\partial x_m}{\partial x_k}$

PROBLEM 1.2

Express the following mathematical operations in index notation.

(a) $\underline{\nabla}\bullet\mathbf{u}$ (b) $\nabla^2\phi$ (c) $\nabla^2\mathbf{u}$ (d) $\underline{\nabla}\phi$

(e) $\underline{\nabla}\times\mathbf{u}$ (f) $\underline{\nabla}\times(\underline{\nabla}\times\mathbf{u})$ (g) $[C]=[A][B]$ (h) $[A]^T[B]\neq[A][B]^T$

(i) $\{c\}=[A]^T\{b\}$ (j) $\underline{\nabla}\bullet(\underline{\nabla}\times\mathbf{u})$ (k) $\underline{\nabla}\times(\underline{\nabla}\phi)$ (l) $\underline{\nabla}(\underline{\nabla}\bullet\mathbf{u})$

(m) $\underline{\nabla}\bullet(\underline{\nabla}\phi)$

where **u** is a vector quantity and ϕ is a scalar quantity; *A*, *B* and *C* are 3×3 matrices, *c* and *b* are 3×1 vectors.

PROBLEM 1.3

Consider two surfaces passing through the point P (see Figure 1.38). The unit normal vectors on these two surfaces at point P are m_j and n_j, respectively. The traction vectors on the two surfaces at point P are denoted by $\overset{m}{T_i}$ and $\overset{n}{T_i}$, respectively. Check if the dot product between $\overset{m}{\boldsymbol{T}}$ and $\underline{n}$ is the same or different from the dot product between $\overset{n}{\boldsymbol{T}}$ and $\underline{m}$.

PROBLEM 1.4

1. A thin triangular plate is fixed along the boundary OA and is subjected to a uniformly distributed horizontal load p_0 per unit area along the boundary AB as shown in Figure 1.39. Give all boundary conditions in terms of displacement or stress components in the x_1x_2 coordinate system.
2. If p_0 acts normal to the boundary AB, what will be the stress boundary conditions along line AB?

PROBLEM 1.5

The quarter disk of radius *a*, shown in Figure 1.40, is subjected to a linearly varying shear stress which varies from 0 to T_0 along boundaries AO and CO and

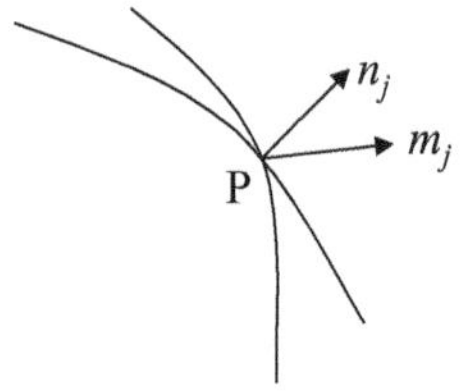

Figure 1.38 Figure for Problem 1.3.

a uniform pressure P_0 along the boundary ABC. Assume that all out-of-plane stress components are zero.

1. Give all stress boundary conditions along the boundaries OA and OC in terms of stress components σ_{11}, σ_{22} and σ_{12} in the Cartesian coordinate system.
2. Give all stress boundary conditions along the boundaries OA and OC in terms of stress components σ_{rr}, $\sigma_{\theta\theta}$ and $\sigma_{r\theta}$ in the cylindrical coordinate system.
3. Give all stress boundary conditions at point B in terms of stress components σ_{11}, σ_{22} and σ_{12} in the Cartesian coordinate system.
4. Give all stress boundary conditions along the boundary ABC in terms of stress components σ_{rr}, $\sigma_{\theta\theta}$ and $\sigma_{r\theta}$ in the cylindrical coordinate system.

PROBLEM 1.6

1. A dam made of isotropic material has two different water heads on two sides, as shown in Figure 1.41. Define all boundary conditions along the boundaries AB and CD in terms of stress components σ_{xx}, σ_{yy} and τ_{xy}.
2. If the dam is made of orthotropic material, what changes, if any, should be in your answer to part (a)?

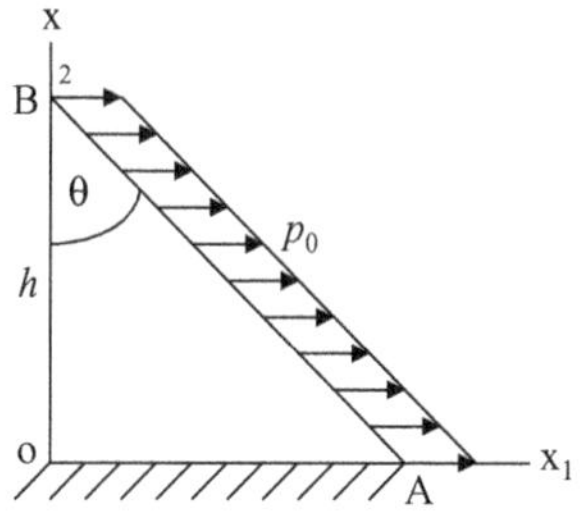

Figure 1.39 Figure for Problem 1.4.

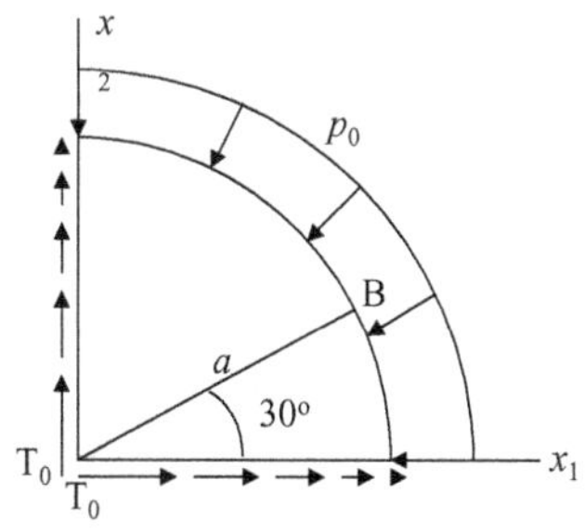

Figure 1.40 Figure for Problem 1.5.

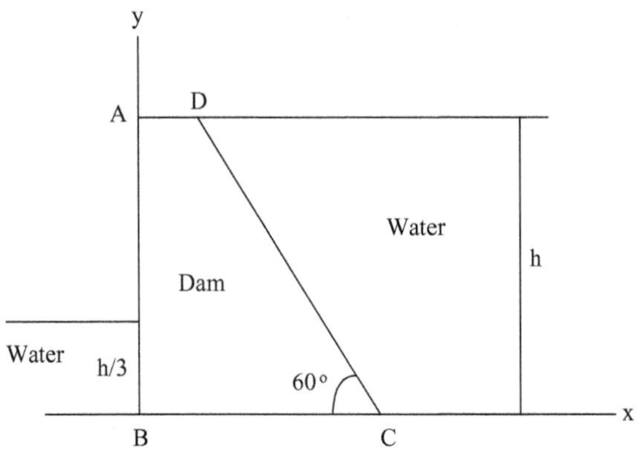

Figure 1.41 Figure for Problem 1.6.

PROBLEM 1.7

Express the surface integral $\oint_S x_i n_j dS$ in terms of the volume V bounded by the surface S (see Figure 1.42). n_j is the j-th component of the outward unit normal vector on the surface.

PROBLEM 1.8

1. Obtain the principal stresses and their directions for the following stress state by solving the appropriate eigenvalue problem:

$$[\sigma]=\begin{bmatrix} 5 & -3 & 0 \\ -3 & 2 & 0 \\ 0 & 0 & 10 \end{bmatrix}$$

2. Solve the same problem by using Mohr's circle technique and compare your results with those obtained in part (a). (Mohr's circle technique is not covered in this chapter. However, since this technique is covered in undergraduate mechanics of materials class, readers are expected to be familiar with it.)

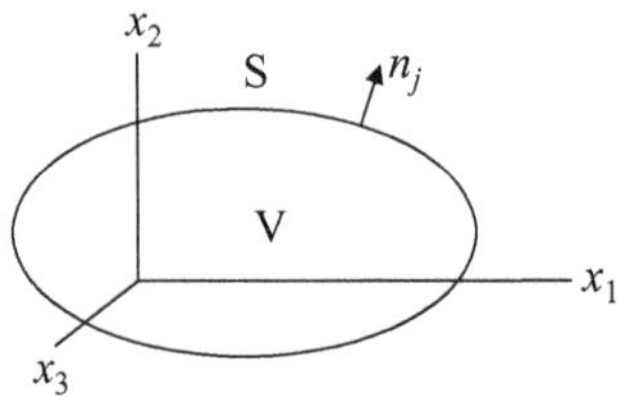

Figure 1.42 Figure for Problem 1.7.

PROBLEM 1.9

An anisotropic elastic solid is subjected to some load that gives a strain state ε_{ij} in the $x_1x_2x_3$ coordinate system. In a different (rotated) $x_1'x_2'x_3'$ coordinate system the strain state is transformed to $\varepsilon_{m'n'}$.

1. Do you expect the strain energy density function U_0 to be a function of strain invariants only?
2. Do you expect the same or different expressions of U_0 when it is expressed in terms of ε_{ij} or $\varepsilon_{m'n'}$?
3. Do you expect the same or different numerical values of U_0 when you compute it from its expression in terms of ε_{ij} and from its expression in terms of $\varepsilon_{m'n'}$?
4. Justify your answers.

Answer parts (a), (b) and (c) if the material is isotropic.

PROBLEM 1.10

Stress–strain relation for a linear elastic material is given by $\sigma_{ij} = C_{ijkl}\varepsilon_{kl}$. Starting from the stress–strain relation for an isotropic material, prove that C_{ijkl} for the isotropic material is given by $\lambda\delta_{ij}\delta_{k\ell} + \mu\left(\delta_{ik}\delta_{j\ell} + \delta_{i\ell}\delta_{jk}\right)$.

PROBLEM 1.11

Obtain the governing equation of equilibrium in terms of displacement for a material whose stress–strain relation is given by

$$\sigma_{ij} = \alpha_{ijkl}\varepsilon_{km}\varepsilon_{ml} + \beta_{ijkl}\varepsilon_{kl} + \delta_{ij}\gamma$$

where α_{ijkl} and β_{ijkl} are material properties that are constants over the entire region, and γ is the residual hydrostatic state of stress that varies from point to point.

PROBLEM 1.12

Starting from the three-dimensional stress transformation law $\sigma_{i'j'}=\lambda_{i'm}\lambda_{j'n}\sigma_{mn}=l_{i'm}l_{j'n}\sigma_{mn}$ prove that for two-dimensional stress transformation the following equations hold good. (Note that direction cosine $\lambda_{i'm}=l_{i'm}=\cos(\theta_{i'm})$, where $\theta_{i'm}$ is the angle between $x_{i'}$ and x_m axes of prime and non-prime coordinate systems, as shown in Figure 1.43.)

$$\sigma_{1'1'} = \sigma_{11}\cos^2\theta + \sigma_{22}\sin^2\theta + 2\sigma_{12}\sin\theta\cos\theta$$

$$\sigma_{2'2'} = \sigma_{11}\sin^2\theta + \sigma_{22}\cos^2\theta - 2\sigma_{12}\sin\theta\cos\theta$$

$$\sigma_{1'2'} = (-\sigma_{11} + \sigma_{22})\sin\theta\cos\theta + \sigma_{12}(\cos^2\theta - \sin^2\theta)$$

PROBLEM 1.13

Consider a bar (one-dimensional structure with zero body force) for which the stress–strain relation is given by $\sigma_{11} = E\varepsilon_{11}$.

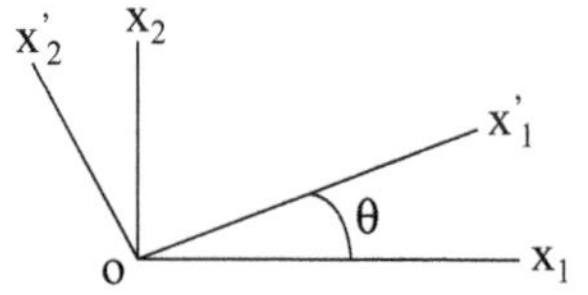

Figure 1.43 Figure for Problem 1.12.

1. Applying Newton's law (force = mass times acceleration) to an elemental segment of the bar, derive the governing equation of motion for the bar in the form of the wave equation. You can assume that σ_{11} and ε_{11} in the bar are functions of x_1 only.
2. What should be the elastic wave speed in the bar?

PROBLEM 1.14

An infinite plate of thickness $2h$ is subjected to a constant normal pressure p_0 at its two surfaces at time $t \geq 0$ (see Figure 1.44). Calculate the displacement field **u** at point P(δ,0,0), where $0<\delta<h$, at time $t=h/c_P$, where c_P is the P-wave speed in the material.

PROBLEM 1.15

1. A half-space is subjected to a time dependent shear stress field p(t) at $x_1=0$ as shown in Figure 1.45. Find the stress and displacement fields at a point (x_1, x_2, x_3) at time t>0.
2. Let p(t) be 1 for 5<t<15 and 0 for other values of t, then plot u_1, u_2, σ_{11} and σ_{12} as a function of time at $x_1 = 5\sqrt{\frac{\mu}{\rho}}$, $x_2 = 0$, $x_3 = 0$ w here μ and ρ are shear modulus and Poisson's ratio, respectively.

PROBLEM 1.16

A linear elastic half-space is subjected to a time dependent inclined loading $p(t)$ on its surface as shown in Figure 1.46. Duration time of the load is T and its

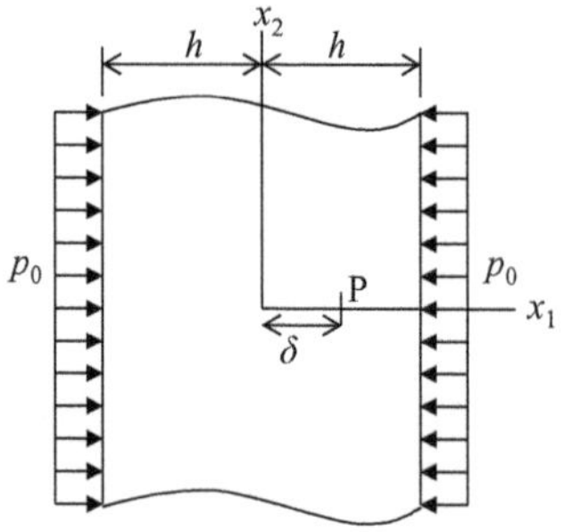

Figure 1.44 Figure for Problem 1.14.

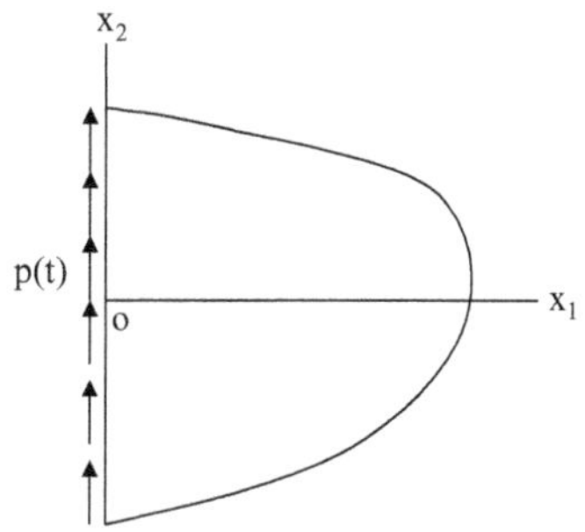

Figure 1.45 Figure for Problem 1.15.

magnitude is P. The problem geometry and applied load are independent of the x_2 and x_3 coordinates. Plot the variation of two displacement components u_1 (by solid line) and u_2 (by dashed line) at point Q ($x_1 = X_0$, $x_2 = Y_0$) on the same plot as a function of time, as shown in the figure. Plot the variation of the two stress components σ_{11} (by solid line) and σ_{12} (by dashed line) as a function of time at the same point Q. Give important values in your plots. Lame's first and second constants and density of the solid are λ, μ and ρ, respectively. You can directly plot the results without showing the intermediate steps. For both displacement and stress components plot positive quantities above the t-axis and negative quantities below the t-axis.

PROBLEM 1.17

1. For a vertically propagating SH-wave in a half-space, as shown in Figure 1.47, calculate the total displacement and stress fields on a semi-circle of radius 'a' in terms of α, a and k_S (ω/c_S).
2. Check if the stress-free boundary conditions are satisfied at the points, where the semi-circle meets the surface at $x_3 = 0$.
3. Find the amplitude of maximum displacement on the semi-circle and locate the point where the displacement amplitude is maximum.

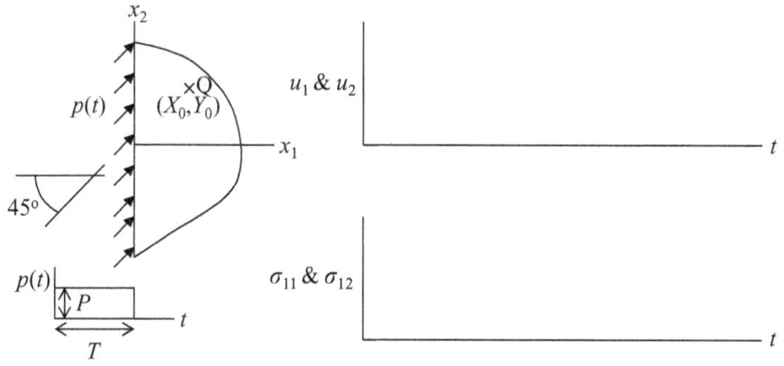

Figure 1.46 Figure for Problem 1.16.

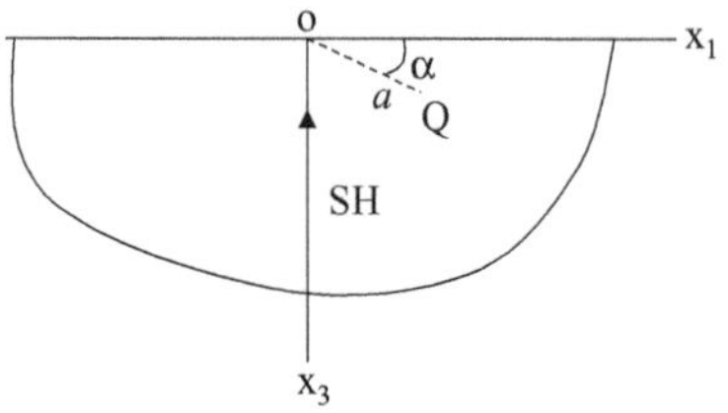

Figure 1.47 Figure for Problem 1.17.

PROBLEM 1.18

A uniform half-space is excited by a time harmonic incident plane P-wave, as shown in Figure 1.48. If the incident field potential is given by

$$\Phi_i(x_1, x_3, t) = \phi_i(x_1, x_3)e^{-i\omega t} = Ae^{i(kx_1 - \eta x_3)}e^{-i\omega t}$$

calculate the displacement field u_1 and u_3 at a point Q as a function of a, α and θ. α varies from 0 to π and θ varies from 0 to $\pi/2$. Assume that the reflection coefficients of the half-space are $R_{PP}(\theta)$ and $R_{PS}(\theta)$ for reflected P- and S-waves, respectively. Assume P- and S-wave speeds to be c_P and c_S, respectively. Express your final results in terms of $R_{PP}(\theta)$, $R_{PS}(\theta)$, a, α, ω, c_P and c_S.

PROBLEM 1.19

1. Prove that the transfer coefficient (T) for an incident SH-wave in a layered half-space (as shown in Figure 1.49) is given by

$$T = \frac{1}{\cos\left\{\dfrac{\omega h}{c_{S1}c_{S2}}\sqrt{c_{S2}^2 - c_{S1}^2 \sin^2\theta}\right\}}$$

   where the transfer coefficient is defined as $T = \dfrac{u_2(x_1, 0)}{u_2(x_1, h)}$ (see Figure 1.49).

   The equation of the incident wave is given by

$$u_{2i} = A\exp\left\{i\frac{\omega}{c_{S2}}(x_1 \sin\theta - x_3 \cos\theta - c_{S2}t)\right\}$$

2. What is the value of the transfer function for (i) $c_{S1} = c_{S2}$, (ii) $h = 0$?

PROBLEM 1.20

Obtain the expressions of the reflected wave amplitudes for a plane P-wave striking a plane rigid boundary at an inclination angle θ, after propagating

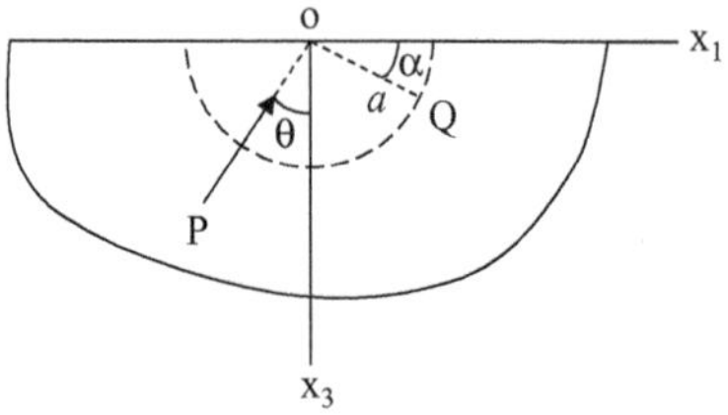

Figure 1.48 Figure for Problem 1.18.

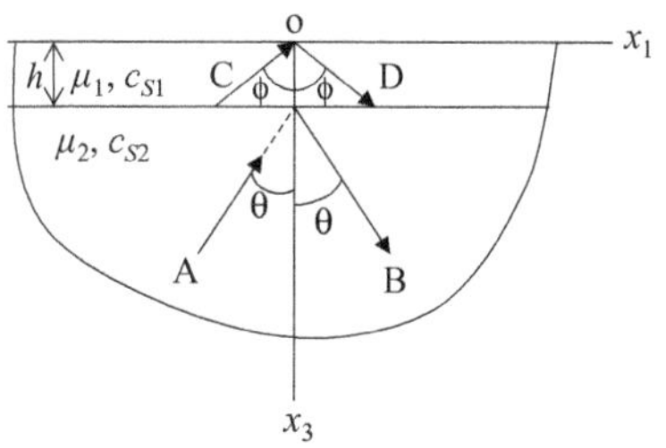

Figure 1.49 Figure for Problem 1.19.

through an elastic half-space. The inclination angle θ is measured from the axis normal to the rigid boundary.

PROBLEM 1.21

For two-dimensional wave propagation problems, P and S-wave potentials are given by

$$\phi = \phi(n_1x_1 + n_2x_2 - c_Pt), \qquad \psi = \psi(n_1x_1 + n_2x_2 - c_St)$$

1. Prove that the wave propagation direction and the wave front are perpendicular to each other for both these waves.
2. When the waves propagate in the x_1 direction, show that for

   (i) $\phi \neq 0, \psi = 0, u_2 = \sigma_{12} = 0$, and for (ii) $\phi = 0, \psi \neq 0, u_1 = \sigma_{11} = \sigma_{22} = 0$

PROBLEM 1.22

A plane SH-wave propagating in a solid medium strikes the stress-free boundaries at $x_1=0$ and $x_2=0$ in the quarter space, as shown in Figure 1.50. The displacement field associated with the incident SH-wave is given by $u_{3i} = e^{-ikx_1 - i\beta x_2}$; the time dependence $e^{-i\omega t}$ is implied. When the incident SH-wave encounters the two stress-free surfaces, reflected plane waves are generated to satisfy the boundary conditions.

Obtain the complete displacement field (considering incident as well as all reflected waves) in this quarter space, and show that the stress-free boundary conditions at the two surfaces are satisfied when the total field is considered.

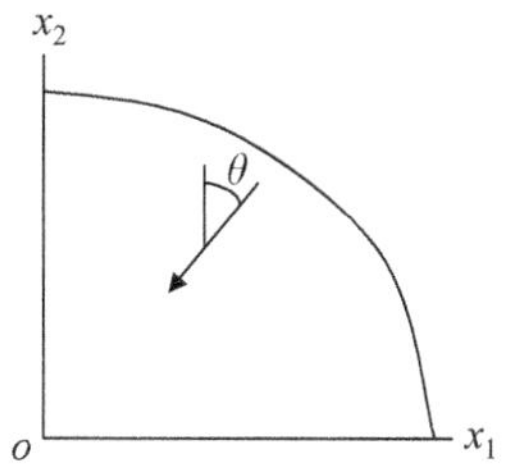

Figure 1.50 Figure for Problem 1.22.

PROBLEM 1.23

$$\left(2-\frac{c_R^2}{c_s^2}\right)^2-4\left(1-\frac{c_R^2}{c_p^2}\right)^{1/2}\left(1-\frac{c_R^2}{c_s^2}\right)^{1/2}=0$$

From the Rayleigh wave dispersion equation prove that when the Rayleigh wave speed is close to the shear wave speed (that is true for most materials) it is possible to approximately obtain the Rayleigh wave speed c_R from the relation

$$\frac{c_R}{c_s}=\frac{0.875+1.125\nu}{1+\nu}$$

where V is Poisson's ratio. (Hint: substitute $c/c_s = 1+\Delta$, where Δ is a small number, and then ignore higher order terms in Δ.)

PROBLEM 1.24

Consider the horizontal shearing motions of a plate of thickness $2h$. Plate surfaces at $x_2 = +h$ and $-h$ are rigidly fixed.

1. Determine the dispersion relation (guided wave speed as a function of frequency) for this problem geometry and plot the dispersion curves.
2. Compute and plot the mode shapes for first few modes.
3. Discuss the differences and similarities between the wave propagation characteristics (dispersion relation and mode shapes) of this problem and the stress-free plate problem discussed in section 1.2.12.1.

PROBLEM 1.25

P-wave is normally incident at the interface of two solids, as shown in Figures 1.51a,b. Incident P-wave amplitude is 1 and reflected wave amplitude is R. Material properties are shown in the figures.

1. What is the value of R for the problem geometry shown in Figure 1.51a? You do not need to derive it; simply give its expression, using the material properties given.

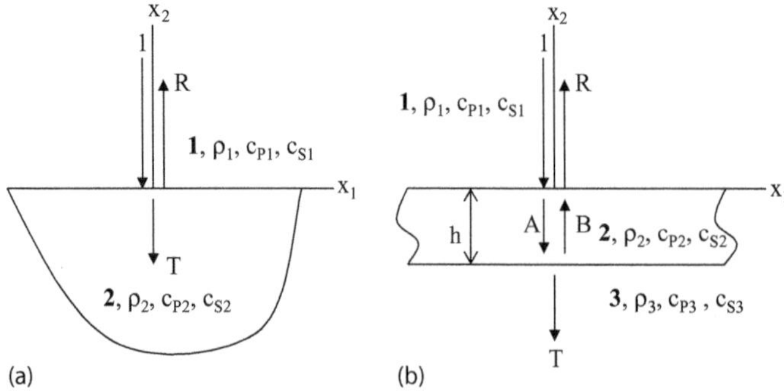

Figure 1.51 Figure for Problem 1.25.

2. For the material properties shown in the two figures, should the two R be same? Note that there are a total of three materials in Figure 1.51b but only two in Figure 1.51a.
3. Should the value of R in Figure 1.51b change when properties of material 3 are changed but those for materials 1 and 2 are unchanged?
4. At what interface of Figure 1.51b (top, bottom, or both) do you need to satisfy the stress and displacement continuity conditions to solve for R?
5. At what interface of Figure 1.51b (top, bottom or both) do you need to satisfy the stress and displacement continuity conditions to solve for T?
6. Write all necessary equations in terms of T and other unknown constants from which you can solve for T of Figure 1.51b. You do not need to give the final expression of T, only write the necessary equations.

PROBLEM 1.26

Consider the phase velocity (c_{ph}) variation with frequency ω, as shown by the continuous line in the top plot of Figure 1.52. Obtain and plot the group velocity (c_g) variations, as a function of frequency ω, in three ranges (for ω between 0 and 1, between 1 and 2, and greater than 2). Give the group velocity values at ω=0.5, 1.5 and 2.5. If you use any relation other than the definitions of the phase velocity and group velocity, then you must derive it.

(Hint: from the definitions of phase velocity and group velocity $c_{ph} = \frac{\omega}{k}$, $c_g = \frac{d\omega}{dk}$, first obtain the group velocity as a function of the phase velocity (c_{ph}) and its derivative $\frac{dc_{ph}}{d\omega}$ and then use that relation to solve the problem.)

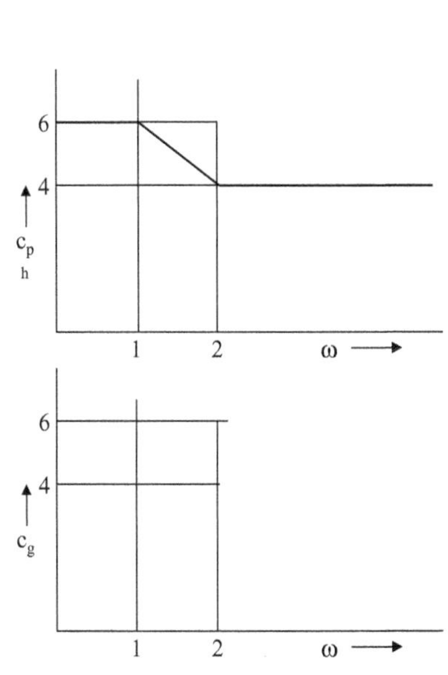

Figure 1.52 Figure for Problem 1.26.

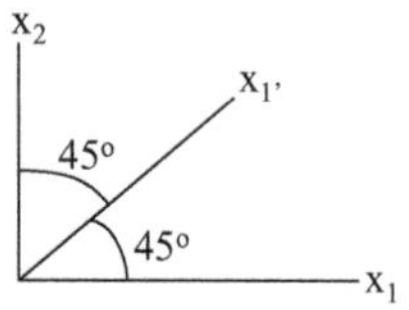

Figure 1.53 Figure for Problem 1.27.

PROBLEM 1.27

If a vector (such as force, velocity) has components 'A' in both x_1- and x_2-directions, then the resultant vector has a magnitude $A\sqrt{2}$ and acts in $x_{1'}$-direction (see Figure 1.53).

Two plane P-waves with harmonic time dependence, propagate in a material in x_1 and x_2 directions with a velocity c_P. Wave potentials for these two waves are given by

$$\phi_1 = Ae^{ik_P x_1 - i\omega t}$$

$$\phi_2 = Ae^{ik_P x_2 - i\omega t}$$

where $k_P = \dfrac{\omega}{c_P}$.

1. In this material, find the total particle displacement (magnitude and direction) at two points – (i) the origin, $x_1 = x_2 = 0$, and (ii) at a second point for which $x_1 = x_2 = h$.
2. It is suggested by a student that the combined effect of these two harmonic waves is a resultant wave propagating in the $x_{1'}$ direction with a wave amplitude $A\sqrt{2}$. Do you agree with this statement? Mathematically justify your 'yes' or 'no' answer.
3. It is suggested by another student that the combined effect of these two harmonic waves is a resultant wave propagating in the $x_{1'}$ direction with a velocity $c_P\sqrt{2}$. Do you agree with this statement? Mathematically justify your 'yes' or 'no' answer.

PROBLEM 1.28

It is well known that when a plane P- or SV-wave strikes a stress-free plane boundary of a solid at an angle then both reflected P- and SV-waves are generated. This phenomenon is known as the mode conversion.

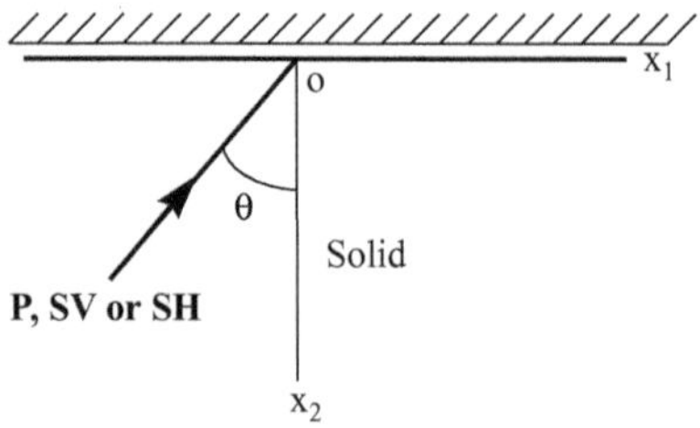

Figure 1.54 Figure for Problem 1.28.

If the solid half-space is pressed against a smooth rigid surface such that the points on the solid boundary cannot move vertically but can freely move horizontally (since the surface is frictionless), then check if the mode conversion occurs for P- and SV-wave incidence and compute all reflection coefficients for the inclined incidence of (i) P-, (ii) SV- and (iii) SH-waves, as shown in Figure 1.54.

REFERENCES

Achenbach, J. D., *Wave Propagation in Elastic Solids,* North-Holland Publishing Company/American Elsevier, Amsterdam (1973).

Auld, B. A., *Acoustic Fields and Waves in Solids,* 2nd ed., Vols. 1 and 2, Kreiger, Malabar, Florida (1990).

Bland, D. R., *Wave Theory and Applications,* Oxford University Press, New York (1988).

Brekhovskikh, L. M., *Waves in Layered Media,* Academic Press, New York/London (1960).

Graff, K. F., *Wave Motion in Elastic Solids,* Dover Publications (1991).

Kolsky, H., *Stress Waves in Solids,* Dover, New York (1963).

Mal, A. K. and S. J. Singh, *Deformation of Elastic Solids,* Prentice Hall, New Jersey (1991).

Moon, P. and D. E. Spencer, *Vectors,* D. Van Nostrand Company, Inc., Princeton, New Jersey (1965).

Rayleigh, J. W. S., *The Theory of Sound,* Dover, New York (1945).

Rose, J. L., *Ultrasonic Waves in Solid Media,* Cambridge University Press, (1999).

Schmerr, L. W., *Fundamentals of Ultrasonic Nondestructive Evaluation – A Modeling Approach,* Plenum Press, New York (1998).

Viktorov, I. A., *Rayleigh and Lamb Waves – Physical Theory and Applications,* Plenum Press, New York (1967).

2 Guided Elastic Waves – Analysis and Applications in Nondestructive Evaluation

In recent years guided waves have been successfully used for detecting defects in pipe and plate type structures such as composite plates, concrete slabs and pipes, metal plates and pipes, etc. This chapter gives the theoretical background for such inspection, lists advantages of the guided wave inspection technique over conventional ultrasonic techniques, and presents results from a number of plate and pipe inspection experiments. With this goal in mind, recent developments by the author and other investigators on plate and pipe inspection techniques that use guided waves are presented here.

2.1 GUIDED WAVES AND WAVE-GUIDES

A wave-guide is a structure with boundaries and/or interfaces that help elastic waves to propagate from one point to another. An elastic wave that propagates through a wave-guide is called a *guided wave*. An elastic full space cannot be considered as a wave-guide since it does not have any boundary or interface, but an elastic half-space with a stress-free boundary can act as a wave-guide if the elastic wave propagates along this boundary. The stress-free boundary can help elastic waves to propagate from one point of the boundary to another point as shown in Chapter 1, Section 1.2.9. A guided wave that propagates along the stress-free boundary of an elastic half-space is called a Rayleigh wave. A single layer over a half-space can be a wave-guide for Love waves (antiplane motions, Section 1.2.10) or Rayleigh waves, also known as generalized Rayleigh-Lamb waves (in-plane motions, Section 1.2.11). Wave-guides can be of any shape or size. Common types of wave-guides are plates, pipes, cylindrical rods (solid or hollow) and bars (of rectangular cross sections or other geometric shapes). Even though the cross-section of a wave-guide often remains constant, it can also vary. Figure 2.1 shows different types of wave-guides – (a) a plate, (b–d) bars with rectangular, circular and I cross-section, (e–f) rods with varying cross sections. In some of these wave-guides the elastic wave can easily propagate while in other cases the propagating guided waves decay fast.

The *Lamb wave* is the *guided wave* that propagates in a plate as shown in Figure 2.1a. The two traction-free boundary surfaces of the plate help the Lamb wave modes to propagate. A Lamb wave, observed in a plate, is also known as the *plate wave*. Similarly, a guided wave propagating through a rectangular or cylindrical rod is known as the *rod wave* and guided waves in a pipe or cylindrical rod are known as *cylindrical guided waves*. All these guided waves have one thing in common; they propagate through a wave-guide satisfying the boundary conditions of the wave-guide. Rayleigh (1885) and Lamb (1917) first solved the guided wave propagation problems in an elastic half-space and elastic plate, respectively. Guided waves in these two wave-guides are named after them. One big difference between these two guided waves is that the Rayleigh wave speed in an elastic half-space is non-dispersive, meaning it is independent of the frequency (see Section 1.2.9), while the Lamb wave speed is dispersive, or dependent on the frequency (see Section 1.2.12). Another difference is that at a given frequency a Rayleigh wave in a half-space can propagate with only one speed while the Lamb wave can propagate with multiple speeds. Different plate wave modes, associated with various types of particle motion in the plate (see Figure 1.28), propagate with different wave speeds at the same frequency (see Figures 1.27 and 1.31). Like Lamb waves, the guided waves through a rectangular bar or a cylindrical rod also show multiple modes and the wave speeds for

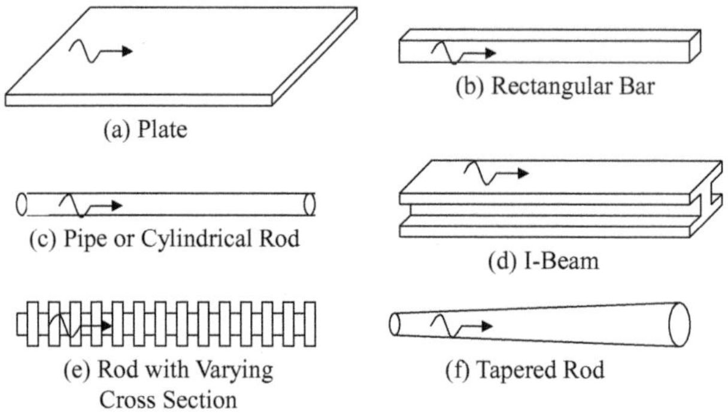

Figure 2.1 Different wave-guides.

these modes are generally dispersive. Because of these similarities, sometimes the guided waves propagating through a pipe, a rectangular plate of finite width, or a rectangular bar are also called Lamb waves. However, strictly speaking those waves are not Lamb waves and should be identified as guided waves but not Lamb waves.

2.1.1 Lamb Waves and Leaky Lamb Waves

When the plate is immersed in a liquid then the surfaces at the liquid-solid interface are not traction-free and the energy of the propagating wave leaks into the surrounding liquid; then the propagating wave is called the *Leaky Lamb Wave*. Strictly speaking, Lamb waves without any leaky energy are observed only in a plate in vacuum. When the plate is immersed in air Lamb waves propagate in the plate leaking energy into the air; thus, strictly speaking, a propagating wave in a plate immersed in air is also a leaky Lamb wave. However, since the intensity of the energy leaking into the surrounding air is very small it is generally ignored, and the wave is not called a leaky Lamb wave when the plate is in the air. For a plate immersed in a liquid the intensity of energy leaking into the surrounding liquid is not negligible and should not be ignored.

2.2 BASIC EQUATIONS – HOMOGENEOUS ELASTIC PLATES IN A VACUUM

Fundamental equations of the Lamb wave propagation in a solid plate have been derived in Chapter 1, Section 1.2.12.2. Those equations are briefly reviewed here. For a linear, elastic, isotropic plate of thickness 2h (see Figure 1.29), the phase velocity dispersion curves are obtained by solving the following dispersion equations [see Eq. (1.169)]:

$$\frac{\tanh(\eta h)}{\tanh(\beta h)} = \frac{\left(2k^2 - k_S^2\right)^2}{4k^2\eta\beta} \tag{2.1a}$$

$$\frac{\tanh(\eta h)}{\tanh(\beta h)} = \frac{4k^2\eta\beta}{\left(2k^2 - k_S^2\right)^2} \tag{2.1b}$$

Equations (2.1a) and (2.1b) give phase velocity dispersion curves for symmetric and anti-symmetric modes, respectively. In the above equations

$$
\begin{aligned}
k &= \frac{\omega}{c_L}\\
\eta &= \sqrt{k^2 - k_P^2}\\
\beta &= \sqrt{k^2 - k_S^2}\\
k_P &= \frac{\omega}{c_P}\\
k_S &= \frac{\omega}{c_S}
\end{aligned}
\tag{2.2}
$$

where:

c_L is the Lamb wave speed (phase velocity)
c_P is the P-wave speed
c_S is the S-wave speed of the plate material
ω is the circular frequency (rad/sec, $\omega = 2\pi f$) of the propagating wave

The potential field, displacement field and stress field for the symmetric and anti-symmetric Lamb modes can be obtained from Eqs. (1.166) and (1.167).

Symmetric Modes:

$$
\begin{aligned}
\phi &= B\cosh(\eta x_2)e^{ikx_1}\\
\psi &= C\sinh(\beta x_2)e^{ikx_1}
\end{aligned}
\tag{2.3}
$$

$$
\begin{aligned}
u_1 &= \phi_{,1} + \psi_{,2} = \{ikB\cosh(\eta x_2) + \beta C\cosh(\beta x_2)\}e^{ikx_1}\\
u_2 &= \phi_{,2} - \psi_{,1} = \{\eta B\sinh(\eta x_2) - ikC\sinh(\beta x_2)\}e^{ikx_1}
\end{aligned}
\tag{2.4}
$$

$$
\begin{aligned}
\sigma_{22} &= -\mu\{k_S^2\phi + 2(\psi_{,12} + \phi_{,11})\} = -\mu\{(k_S^2 - 2k^2)B\cosh(\eta x_2) + 2ik\beta C\cosh(\beta x_2)\}e^{ikx_1}\\
\sigma_{12} &= \mu\{2\phi_{,12} + \psi_{,22} - \psi_{,11}\} = \mu\{(2ik\eta)B\sinh(\eta x_2) + (k^2 + \beta^2)C\sinh(\beta x_2)\}e^{ikx_1}
\end{aligned}
\tag{2.5}
$$

Anti-symmetric Modes:

$$
\begin{aligned}
\phi &= A\sinh(\eta x_2)e^{ikx_1}\\
\psi &= D\cosh(\beta x_2)e^{ikx_1}
\end{aligned}
\tag{2.6}
$$

$$
\begin{aligned}
u_1 &= \phi_{,1} + \psi_{,2} = \{ikA\sinh(\eta x_2) + \beta D\sinh(\beta x_2)\}e^{ikx_1}\\
u_2 &= \phi_{,2} - \psi_{,1} = \{\eta A\cosh(\eta x_2) - ikD\cosh(\beta x_2)\}e^{ikx_1}
\end{aligned}
\tag{2.7}
$$

$$
\begin{aligned}
\sigma_{22} &= -\mu\{k_S^2\phi + 2(\psi_{,12} + \phi_{,11})\} = -\mu\{(k_S^2 - 2k^2)A\sinh(\eta x_2) + 2ik\beta D\sinh(\beta x_2)\}e^{ikx_1}\\
\sigma_{12} &= \mu\{2\phi_{,12} + \psi_{,22} - \psi_{,11}\} = \mu\{(2ik\eta)A\cosh(\eta x_2) + (k^2 + \beta^2)D\cosh(\beta x_2)\}e^{ikx_1}
\end{aligned}
\tag{2.8}
$$

From Eq. (1.168)

$$\frac{B}{C}=\frac{2ik\beta}{\left(2k^2-k_S^2\right)}\frac{\cosh(\beta h)}{\cosh(\eta h)}=-\frac{\left(2k^2-k_S^2\right)}{2ik\eta}\frac{\sinh(\beta h)}{\sinh(\eta h)} \tag{2.9a}$$

and

$$\frac{A}{D}=\frac{2ik\beta}{\left(2k^2-k_S^2\right)}\frac{\sinh(\beta h)}{\sinh(\eta h)}=-\frac{\left(2k^2-k_S^2\right)}{2ik\eta}\frac{\cosh(\beta h)}{\cosh(\eta h)} \tag{2.9b}$$

Substituting Eqs. (2.9a) and (2.9b) in the displacement and stress expressions, we obtain, for symmetric modes

$$u_1=\left\{ikB\cosh(\eta x_2)+\beta C\cosh(\beta x_2)\right\}e^{ikx_1}=B\left\{ik\cosh(\eta x_2)+\frac{\left(2k^2-k_S^2\right)}{2ik}\frac{\cosh(\eta h)}{\cosh(\beta h)}\cosh(\beta x_2)\right\}e^{ikx_1}$$

$$u_2=\left\{\eta B\sinh(\eta x_2)-ikC\sinh(\beta x_2)\right\}e^{ikx_1}=B\left\{\eta\sinh(\eta x_2)-\frac{\left(2k^2-k_S^2\right)}{2\beta}\frac{\cosh(\eta h)}{\cosh(\beta h)}\sinh(\beta x_2)\right\}e^{ikx_1} \tag{2.10}$$

$$\begin{aligned}\sigma_{22}&=-\mu\left\{\left(k_S^2-2k^2\right)B\cosh(\eta x_2)+2ik\beta C\cosh(\beta x_2)\right\}e^{ikx_1}\\&=-\mu B\left\{\left(k_S^2-2k^2\right)\cosh(\eta x_2)+2ik\beta\frac{\left(2k^2-k_S^2\right)}{2ik\beta}\frac{\cosh(\eta h)}{\cosh(\beta h)}\cosh(\beta x_2)\right\}e^{ikx_1}\\&=\mu B\left(2k^2-k_S^2\right)\left\{\cosh(\eta x_2)-\frac{\cosh(\eta h)}{\cosh(\beta h)}\cosh(\beta x_2)\right\}e^{ikx_1}\end{aligned}$$

$$\begin{aligned}\sigma_{12}&=\mu\left\{(2ik\eta)B\sinh(\eta x_2)+\left(k^2+\beta^2\right)C\sinh(\beta x_2)\right\}e^{ikx_1}\\&=\mu B\left\{(2ik\eta)\sinh(\eta x_2)+\left(2k^2-k_S^2\right)\frac{C}{B}\sinh(\beta x_2)\right\}e^{ikx_1}\\&=\mu B\left\{(2ik\eta)\sinh(\eta x_2)-\left(2k^2-k_S^2\right)\frac{2ik\eta}{\left(2k^2-k_S^2\right)}\frac{\sinh(\eta h)}{\sinh(\beta h)}\sinh(\beta x_2)\right\}e^{ikx_1}\\&=2ik\eta\mu B\left\{\sinh(\eta x_2)-\frac{\sinh(\eta h)}{\sinh(\beta h)}\sinh(\beta x_2)\right\}e^{ikx_1}\end{aligned} \tag{2.11}$$

For anti-symmetric modes

$$u_1=\left\{ikA\sinh(\eta x_2)+\beta D\sinh(\beta x_2)\right\}e^{ikx_1}=A\left\{ik\sinh(\eta x_2)+\frac{\left(2k^2-k_S^2\right)}{2ik}\frac{\sinh(\eta h)}{\sinh(\beta h)}\sinh(\beta x_2)\right\}e^{ikx_1}$$

$$u_2=\left\{\eta A\cosh(\eta x_2)-ikD\cosh(\beta x_2)\right\}e^{ikx_1}=A\left\{\eta\cosh(\eta x_2)-\frac{\left(2k^2-k_S^2\right)}{2\beta}\frac{\sinh(\eta h)}{\sinh(\beta h)}\cosh(\beta x_2)\right\}e^{ikx_1} \tag{2.12}$$

$$\begin{aligned}
\sigma_{22} &= -\mu\left\{\left(k_S^2-2k^2\right)A\sinh(\eta x_2)+2ik\beta D\sinh(\beta x_2)\right\}e^{ikx_1}\\
&= -\mu A\left\{\left(k_S^2-2k^2\right)\sinh(\eta x_2)+2ik\beta\frac{\left(2k^2-k_S^2\right)}{2ik\beta}\frac{\sinh(\eta h)}{\sinh(\beta h)}\sinh(\beta x_2)\right\}e^{ikx_1}\\
&= \mu A\left(2k^2-k_S^2\right)\left\{\sinh(\eta x_2)-\frac{\sinh(\eta h)}{\sinh(\beta h)}\sinh(\beta x_2)\right\}e^{ikx_1}
\end{aligned}$$

$$\begin{aligned}
\sigma_{12} &= \mu\left\{(2ik\eta)A\cosh(\eta x_2)+\left(k^2+\beta^2\right)D\cosh(\beta x_2)\right\}e^{ikx_1}\\
&= \mu A\left\{(2ik\eta)\cosh(\eta x_2)+\left(2k^2-k_S^2\right)\frac{D}{A}\cosh(\beta x_2)\right\}e^{ikx_1}\\
&= \mu A\left\{(2ik\eta)\cosh(\eta x_2)-\left(2k^2-k_S^2\right)\frac{2ik\eta}{2k^2-k_S^2}\frac{\cosh(\eta h)}{\cosh(\beta h)}\cosh(\beta x_2)\right\}e^{ikx_1}\\
&= 2ik\eta\mu A\left\{\cosh(\eta x_2)-\frac{\cosh(\eta h)}{\cosh(\beta h)}\cosh(\beta x_2)\right\}e^{ikx_1}
\end{aligned} \tag{2.13}$$

The time dependence $e^{-i\omega t}$ is implied in the above expressions.

2.2.1 Dispersion Curves and Mode Shapes

As explained in Chapter 1, variation of the wave speed as a function of the frequency is known as the dispersion curve. Displacement and stress field variations across the plate thickness are called mode shapes. The steps involved in solving the dispersion equation and mode shapes are discussed in this section.

2.2.1.1 Dispersion Curves

It is necessary to solve Eq. (2.1) to obtain the dispersion curves. It can be solved in one of two ways – 1) fix the frequency (ω) and then try to get the Lamb wave speed (c_L) by satisfying the dispersion equation (2.1), or 2) fix the Lamb wave speed (c_L) and then investigate for what values of frequency ($f=\omega/2\pi$) the dispersion equation is satisfied. If the first method is followed, the frequency is fixed and a value for the Lamb wave speed c_L is assumed. Using these f and c_L values k, η and β are computed from Eq. (2.2) since P- and S-wave speeds (c_P and c_S) in the plate material are known. The values of k, η, β, k_S and h (plate thickness) are substituted into Eq. (2.1), and the left- and right-hand sides of the equation are compared. If the computed values on the two sides of the equation are different, then a new estimate of c_L is made. In other words, the nonlinear equation (2.1) is solved for c_L for a given frequency using the standard techniques for solution of nonlinear equations, such as bisection method, secant method, Newton-Raphson method etc. The complications arise in this case due to the fact that for a single frequency the transcendental dispersion equations (2.1a) and (2.1b) have multiple roots. Thus, it is possible to miss some roots during the root-searching step. However, in principle, by taking a very small step size it is possible to capture all roots at a given frequency. After capturing all roots of the dispersion equation at one frequency, the frequency value is changed and then, following the same root finding technique, all roots for the new frequency are captured. In this manner, a number of roots of Eq. (2.1a) can be found and plotted in the frequency-phase velocity space, as shown in Figure 2.2. Here, roots are found along the vertical grid lines at 0.5 MHz intervals.

Alternately, instead of fixing frequency, it is possible to fix the Lamb wave speed (c_L) and vary the frequency (f) to capture all roots for a given c_L value. Then the c_L value is changed and, again by varying f, all roots are found for the new c_L value. Figure 2.3 shows all roots of Eq. (2.1a) captured in this manner. Roots are found along the horizontal grid lines for the c_L interval of 0.5 km/s. Note that some roots of Figure 2.2 appearing slightly below the c_L value of 3 km/s and 5.5 km/s are not captured in Figure 2.3.

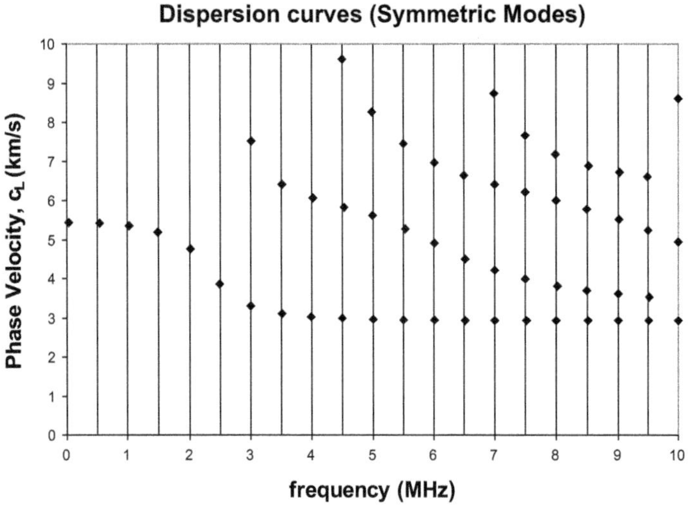

Figure 2.2 Roots of Eq. (2.1a) are plotted at 0.5 MHz frequency interval for a 1 mm thick aluminum plate (c_P=6.32 km/s, c_S=3.13 km/s, ρ=2.7 gm/cc).

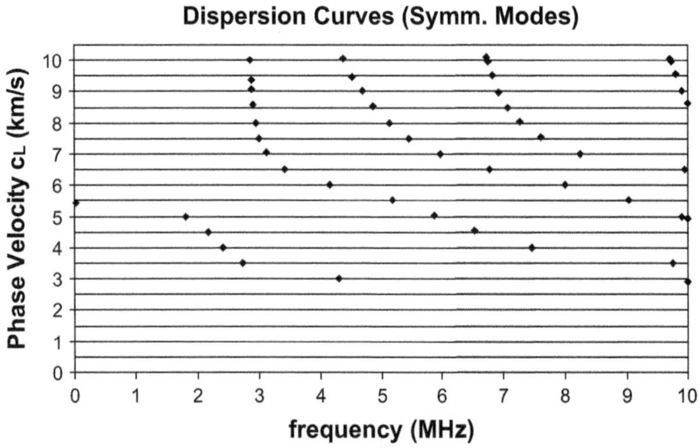

Figure 2.3 Roots of Eq. (2.1a) are plotted at 0.5 km/s interval of c_L for a 1 mm thick aluminum plate. Material properties are given in the caption of Figure 2.2.

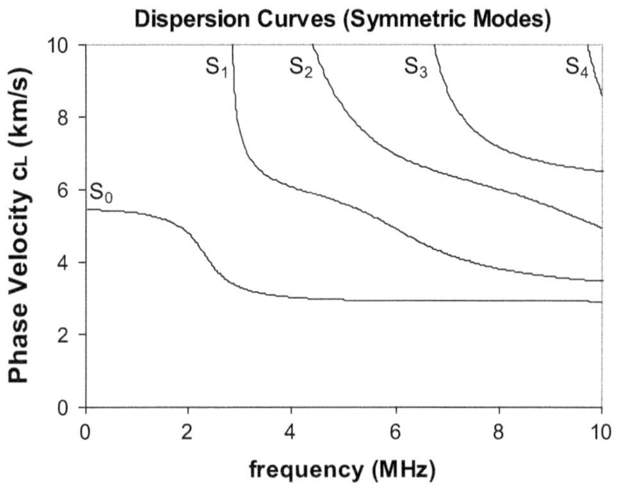

Figure 2.4 Symmetric modes for the Lamb wave propagation in a 1 mm thick aluminum plate – material properties are given in Figure 2.2. Dispersion curves are obtained after solving Eq. (2.1a) and then connecting the roots by continuous lines.

Connecting the neighboring roots in Figure 2.2 (and Figure 2.3), dispersion curves for the symmetric modes are obtained as shown in Figure 2.4. These modes are denoted as S_0, S_1, S_2, S_3 etc. Here, the letter S refers to the symmetric mode and the subscripts 0, 1, 2, 3, ... are numbered from left to right starting with the lowest frequency mode. These numbers are called the order of the modes. Note that the S_0 mode starts at zero frequency, but the higher order modes (S_1, S_2, S_3 etc.) start at nonzero frequencies. The frequency value below which a specific mode is not observed is called the cut-off frequency for that mode. It should be pointed out that there is no cut-off frequency for the S_0 mode; however, the higher order modes have a nonzero cut-off frequency – as the order of the mode becomes higher, the corresponding cut-off frequency also becomes greater.

In the same manner, the anti-symmetric modes can be computed by solving Eq. (2.1b). Anti-symmetric modes for the same plate are shown in Figure 2.5. The anti-symmetric modes are denoted as A_0, A_1, A_2, A_3 etc. The zero-th order anti-symmetric mode A_0 does not have any cut-off frequency. But the higher order modes (A_1, A_2, A_3 etc.) have a nonzero cut-off frequency that increases with the increasing order of the mode.

After superimposing these two sets of modes, the complete set of dispersion curves are obtained. Dispersion curves for aluminum, copper and steel plates of 1 mm thickness are shown in Figure 2.6. This figure shows the effect of the change in material properties on the dispersion curves.

The effect of the variation of the plate thickness on the dispersion curves is then investigated. This effect can be explicitly seen when Eqs. (2.1a) and (2.1b) are rewritten in the following form:

$$\frac{\tanh\left(\omega h\sqrt{\dfrac{1}{c_L^2}-\dfrac{1}{c_P^2}}\right)}{\tanh\left(\omega h\sqrt{\dfrac{1}{c_L^2}-\dfrac{1}{c_S^2}}\right)}=\frac{\left(\dfrac{2}{c_L^2}-\dfrac{1}{c_S^2}\right)^2}{\dfrac{4}{c_L^2}\sqrt{\dfrac{1}{c_L^2}-\dfrac{1}{c_P^2}}\sqrt{\dfrac{1}{c_L^2}-\dfrac{1}{c_S^2}}} \tag{2.14a}$$

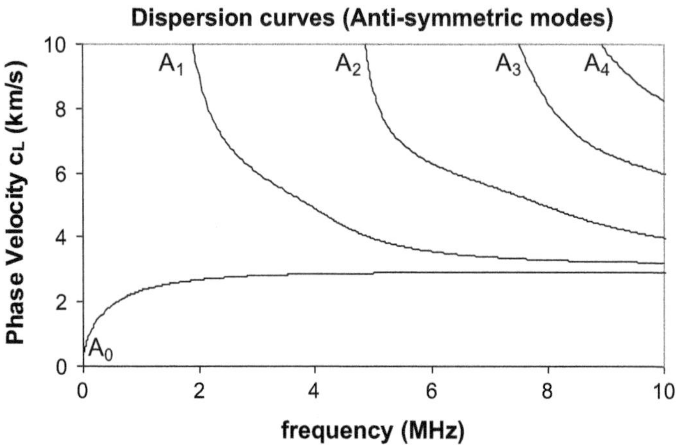

Figure 2.5 Anti-symmetric modes for the Lamb wave propagation in a 1 mm thick aluminum plate – material properties are given in Figure 2.2. Dispersion curves are obtained after solving Eq. (2.1b) and then connecting the roots by continuous lines.

$$\frac{\tanh\left(\omega h\sqrt{\frac{1}{c_L^2}-\frac{1}{c_P^2}}\right)}{\tanh\left(\omega h\sqrt{\frac{1}{c_L^2}-\frac{1}{c_S^2}}\right)}=\frac{\frac{4}{c_L^2}\sqrt{\frac{1}{c_L^2}-\frac{1}{c_P^2}}\sqrt{\frac{1}{c_L^2}-\frac{1}{c_S^2}}}{\left(\frac{2}{c_L^2}-\frac{1}{c_S^2}\right)^2} \tag{2.14b}$$

In Eq. (2.14), note that the half-thickness (h) of the plate and the wave frequency ω (= $2\pi f$) appear in the equation as a product term. Thus from Eq. (2.14) it is possible to plot the variations of c_L as a function of fh (=$\omega h/2\pi$), instead of a function of f only. The main advantage of this plot is that one plot can represent dispersion curves of plates of different thickness. From Eq. (2.14), it is easy to see that the c_L values for a 1 mm thick plate at 5 MHz should be identical to those for a 5 mm thick plate at 1 MHz or 2.5 mm thick plate at 2 MHz since for all these combinations the product between the plate thickness and the frequency is equal to 5 MHz-mm. Three sets of dispersion curves of Figure 2.6 are given for the 1 mm thick plate for the frequency range varying from 0 to 10 MHz. The same curves will represent the dispersion curves for a 2 mm thick plate in the frequency range varying from 0 to 5 MHz, or for a 5 mm thick plate in the frequency range 0 to 2 MHz. In other words, in Figure 2.6, if the frequency axis is changed to the frequency times the plate thickness (or $2fh$), as shown in Figure 2.7, then that plot represents dispersion curves for plates of different thickness. For a smaller plate thickness the frequency value should be greater, and for a thicker plate it should be smaller to obtain the same $2fh$ value along the horizontal axis of the dispersion curve plot, shown in Figure 2.7.

2.2.1.2 Mode Shapes

Any point on the dispersion curves of Figures 2.6 and 2.7 give the frequency–phase velocity combinations for which the Lamb wave can propagate. For the given frequency and phase velocity values, displacement and stress fields inside the plate can be obtained from equations (2.10) and (2.11) for

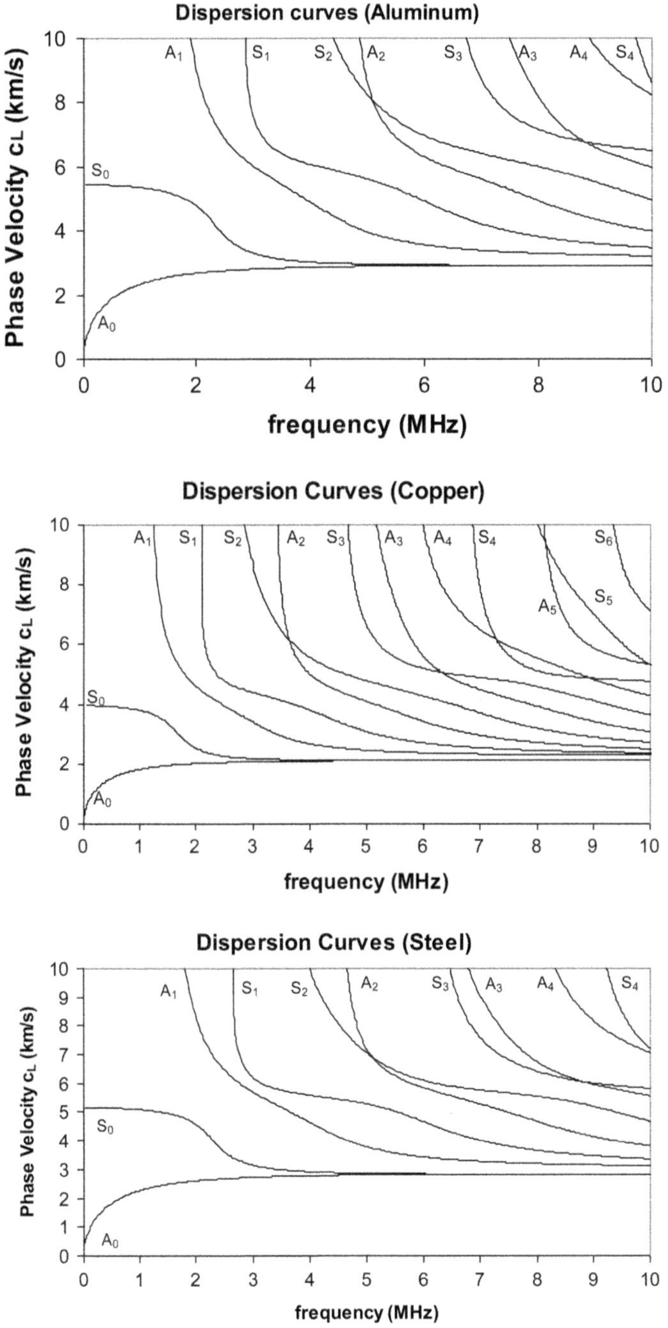

Figure 2.6 Dispersion curves for 1 mm thick aluminum, copper and steel plates. Material properties – (Aluminum: c_P=6.32 km/s, c_S=3.13 km/s, ρ=2.7 gm/cc), (Copper: c_P=4.7 km/s, c_S=2.26 km/s, ρ=8.9 gm/cc), (Steel: c_P=5.96 km/s, c_S=3.26 km/s, ρ=7.9 gm/cc).

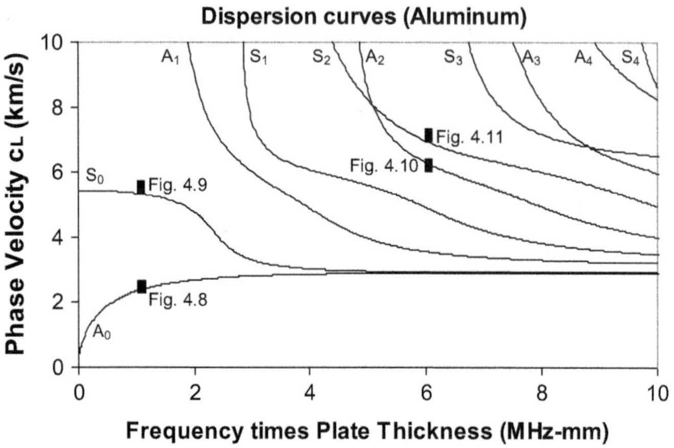

Figure 2.7 Dispersion curves for aluminum plates of different thickness. Displacement and stress variations inside the plate for four phase velocity–frequency combinations, marked by black squares, are shown in Figures 2.8 through 2.11.

symmetric modes, and from (2.12) and (2.13) for anti-symmetric modes. Often, we like to compute the displacement and stress field variations inside the plate. In the wave propagation direction (x_1 direction), the field variation is sinusoidal for a given time and x_2 value, since the x_1 dependence is e^{ikx_1} [see Eqs. (2.10 throuh 2.13)]. The x_2 dependence of the displacement and stress fields is called the mode shapes. Mode shapes for different displacement and stress components are shown in Figures 2.8 through 2.11 for four modes (A_0, S_0, A_2 and S_2) at four different frequency–phase velocity combinations. For all four figures, the aluminum plate thickness is 1 mm and the field values are normalized with respect to their maximum values. The wave frequency is 1 MHz for Figures 2.8 and 2.9, and 6 MHz for Figures 2.10 and 2.11. Rectangular black markers in the dispersion curve plot (Figure 2.7) show the frequency–phase velocity combinations for which the mode shapes are generated.

As expected, the anti-symmetric modes A_0 (Figure 2.8) and A_2 (Figure 2.10) show u_1, σ_{11}, σ_{22} as odd functions of x_2, while u_2 and σ_{12} are even functions. From Figures 2.9 and 2.11, one can see that the symmetric modes (S_0 and S_2) generate u_1, σ_{11}, σ_{22} as even functions of x_2, while u_2 and σ_{12} are odd functions. It should also be noted that the oscillations in the mode shapes increase as the frequency and the order of the mode increase.

What happens to the mode shape if the frequency is changed but the mode order is not? To investigate it, the displacement and stress amplitudes along the plate thickness are plotted for A_2 and S_2 modes in Figures 2.12a,b and 2.13a,b, respectively, for different phase velocities. Note that for both these modes, as the phase velocity increases, the frequency decreases. Details of the oscillation pattern vary with the phase velocity (and frequency) but the general nature of the variation remains approximately similar over a wide frequency and phase velocity range for most field variables. One interesting thing to note is that the shear stress becomes almost equal to zero through the entire plate thickness near the phase velocity of 4.4 km/s for both modes.

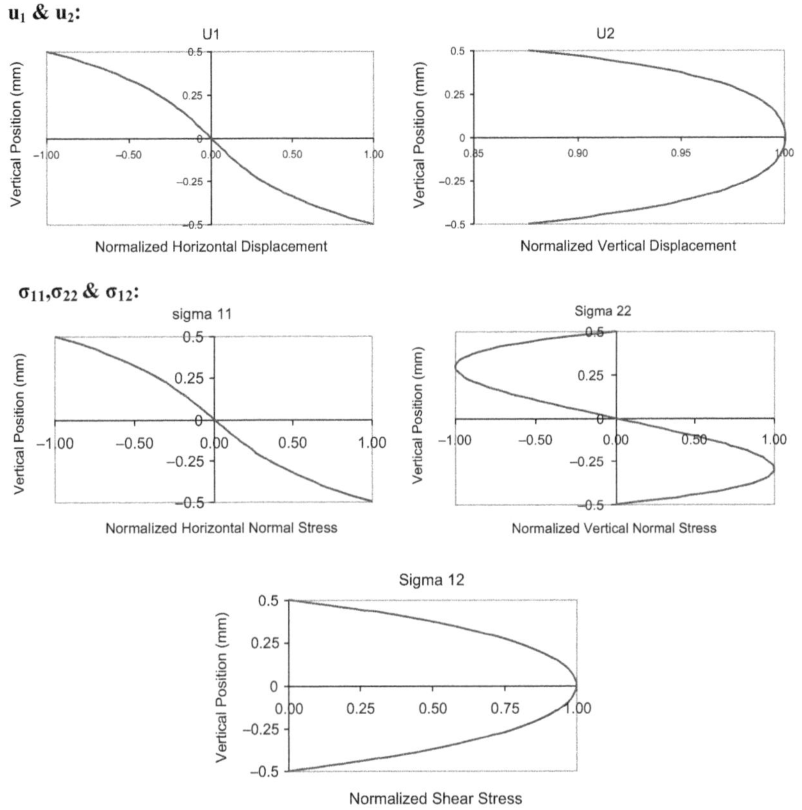

Figure 2.8 Displacement and stress variations along the plate thickness in a 1 mm thick aluminum plate for the A_0 mode of Lamb wave propagation at 1 MHz signal frequency. The corresponding point on the dispersion curve is shown in Figure 2.7.

2.3 HOMOGENEOUS ELASTIC PLATES IMMERSED IN A FLUID

In the previous section we learned how different modes of Lamb wave produce particle displacement in the plate placed in a vacuum. Let us now study the effect of two fluid half-spaces, placed above and below an infinite plate. Unlike the previous case (a plate in a vacuum), here the acoustic energy is no longer trapped inside the plate; it can leak into the surrounding fluid medium as shown in Figure 2.14.

From Eq. (1.166) the potential field in the solid plate is given by

$$
\begin{aligned}
\phi &= (ae^{\eta x_2} + be^{-\eta x_2})e^{ikx_1} = \left\{A\sinh(\eta x_2) + B\cosh(\eta x_2)\right\}e^{ikx_1} \\
\psi &= (ce^{\beta x_2} + de^{-\beta x_2})e^{ikx_1} = \left\{C\sinh(\beta x_2) + D\cosh(\beta x_2)\right\}e^{ikx_1} \\
\eta &= \sqrt{k^2 - k_P^2} \\
\beta &= \sqrt{k^2 - k_S^2}
\end{aligned}
\tag{2.15}
$$

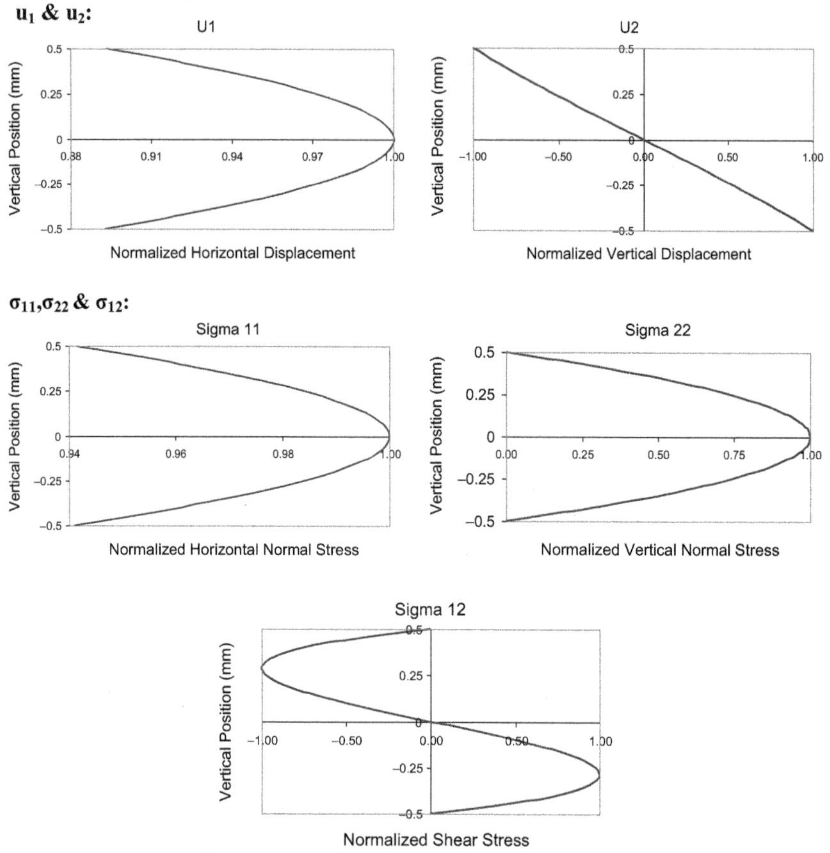

Figure 2.9 Displacement and stress variations along the plate thickness in a 1 mm thick aluminum plate for the S_0 mode of Lamb wave propagation at 1 MHz signal frequency. The corresponding point on the dispersion curve is shown in Figure 2.7.

and the potential field in the fluid is

$$\begin{aligned} \phi_{fL} &= me^{\eta_f x_2} e^{ikx_1} = m\left\{\sinh\left(\eta_f x_2\right) + \cosh\left(\eta_f x_2\right)\right\} e^{ikx_1} \\ \phi_{fU} &= ne^{-\eta_f x_2} e^{ikx_1} = n\left\{\cosh\left(\eta_f x_2\right) - \sinh\left(\eta_f x_2\right)\right\} e^{ikx_1} \\ \eta_f &= \sqrt{k^2 - k_f^2} \end{aligned} \tag{2.16}$$

where ϕ_{fL} and ϕ_{fU} correspond to the wave potentials in the lower and upper fluid half-spaces, respectively.

Instead of writing the wave potentials in the above form, we can separate the symmetric and anti-symmetric components in the following manner:

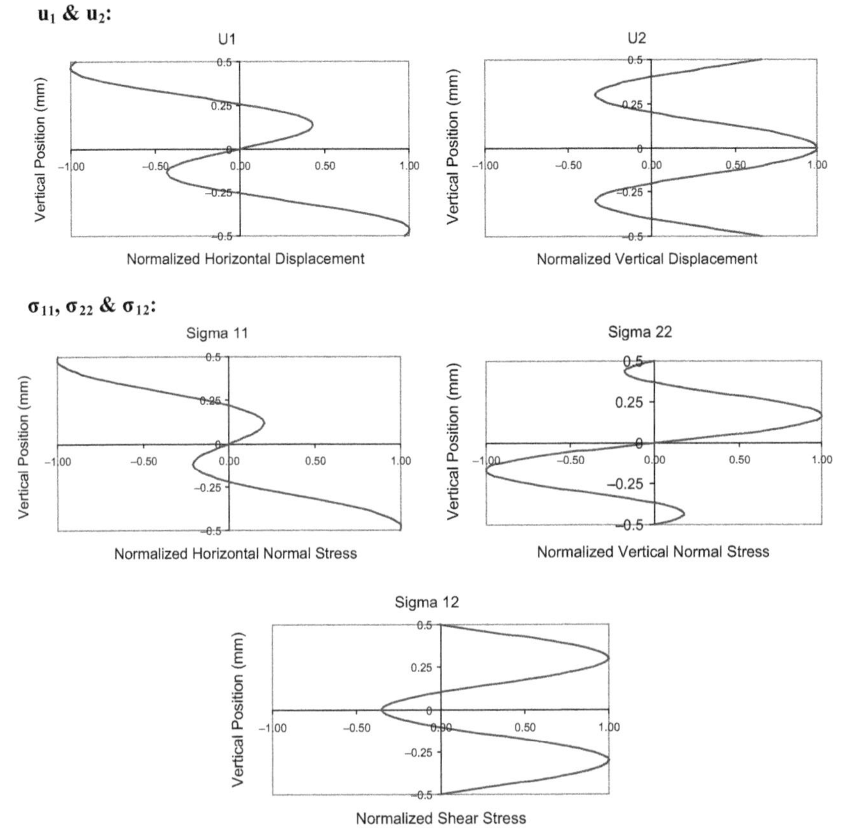

Figure 2.10 Displacement and stress variations along the plate thickness in a 1 mm thick aluminum plate for the A_2 mode of Lamb wave propagation at 6 MHz signal frequency. The corresponding point on the dispersion curve is shown in Figure 2.7.

Symmetric Motion

$$\begin{aligned} \phi &= B\cosh(\eta x_2)e^{ikx_1} \\ \psi &= C\sinh(\beta x_2)e^{ikx_1} \\ \phi_{fL} &= Me^{\eta_f x_2}e^{ikx_1} \\ \phi_{fU} &= Me^{-\eta_f x_2}e^{ikx_1} \end{aligned} \tag{2.17}$$

Anti-symmetric Motion

$$\begin{aligned} \phi &= A\sinh(\eta x_2)e^{ikx_1} \\ \psi &= D\cosh(\beta x_2)e^{ikx_1} \\ \phi_{fL} &= Ne^{\eta_f x_2}e^{ikx_1} \\ \phi_{fU} &= -Ne^{-\eta_f x_2}e^{ikx_1} \end{aligned} \tag{2.18}$$

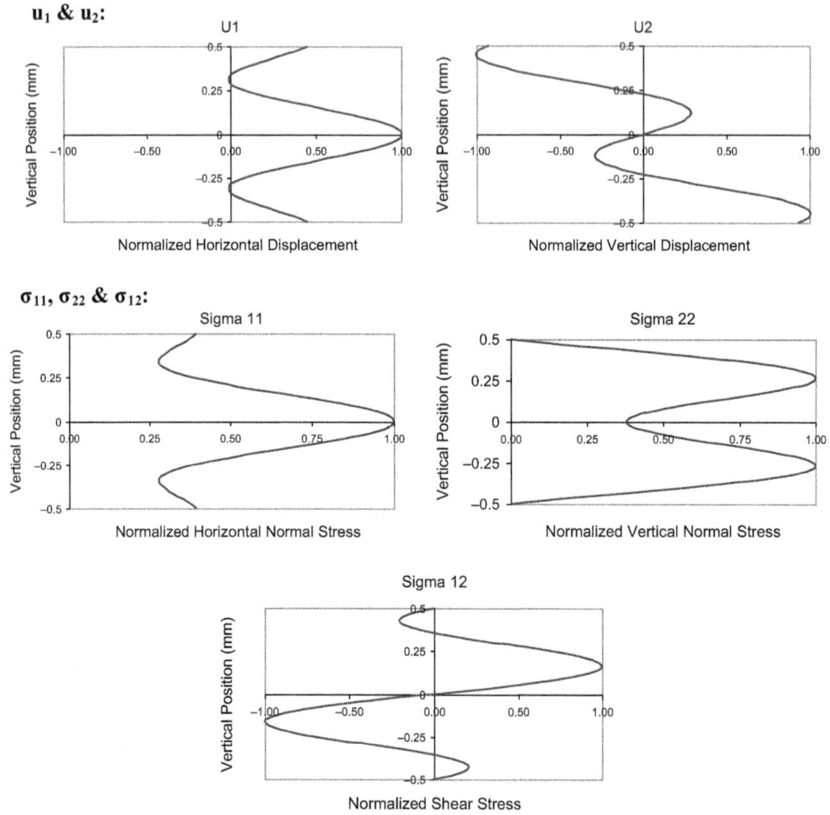

Figure 2.11 Displacement and stress variations along the plate thickness in a 1 mm thick aluminum plate for the S_2 mode of Lamb wave propagation at 6 MHz signal frequency. The corresponding point on the dispersion curve is shown in Figure 2.7.

Then the symmetric and anti-symmetric motions are analyzed separately as given below. ϕ_{fL} and ϕ_{fU} in Eq. (2.16) are defined differently from those given in Eqs. (2.17) and (2.18). However, the two definitions are equivalent (see Exercise Problem 2.7).

2.3.1 Symmetric Motion

At the fluid–solid interface the shear stress component should be zero, and normal stress and normal displacement components should be continuous across the interface. Therefore we can write

$$\begin{aligned} &\sigma_{12}\big|_{x_2=\pm h} = \mu\left\{(2ik\eta)\left[\pm B\sinh(\eta h)\right] + \left(2k^2 - k_S^2\right)\left[\pm C\sinh(\beta h)\right]\right\}e^{ikx_1} = 0 \\ &\Rightarrow 2ik\eta B\sinh(\eta h) + \left(2k^2 - k_S^2\right)C\sinh(\beta h) = 0 \end{aligned} \quad (2.19)$$

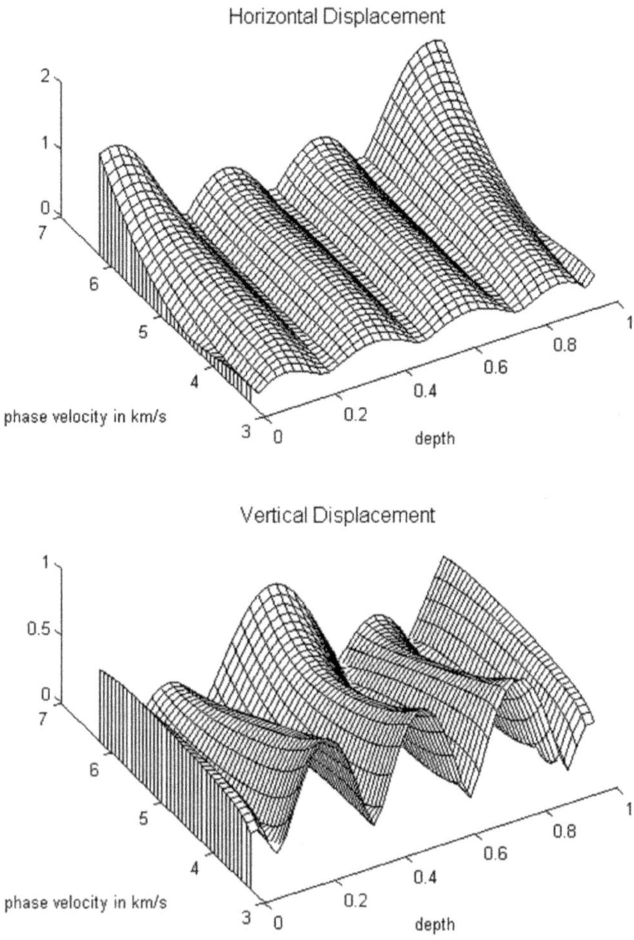

Figure 2.12a Amplitude variations of horizontal (u_1 component, top figure) and vertical (u_2 component, bottom figure) displacements through the plate thickness for the A_2 mode, as the phase velocity varies from 3.4 to 6.3 km/s; depth is 0 for the top of the plate and 1 for the bottom of the plate. Dispersion curves for the aluminum plate are shown in Figure 2.7.

$$\sigma_{22}\big|_{x_2=\pm h} = +\mu\left\{\left(2k^2 - k_S^2\right)\left[B\cosh(\eta h)\right] - 2ik\beta\left[C\cosh(\beta h)\right]\right\}e^{ikx_1} = -\rho_f\omega^2 M e^{-\eta_f h}e^{ikx_1}$$

$$\Rightarrow \left(2k^2 - k_S^2\right)B\cosh(\eta h) - 2ik\beta C\cosh(\beta h) + \frac{\rho_f k_S^2}{\rho}Me^{-\eta_f h} = 0$$

$$u_2\big|_{x_2=\pm h} = \phi_{,2} - \psi_{,1} = \left\{\pm\eta B\sinh(\eta h) \mp ikC\sinh(\beta h)\right\}e^{ikx_1} = \mp\eta_f Me^{-\eta_f h}e^{ikx_1}$$

$$\Rightarrow \eta B\sinh(\eta h) - ikC\sinh(\beta h) + \eta_f Me^{-\eta_f h} = 0$$

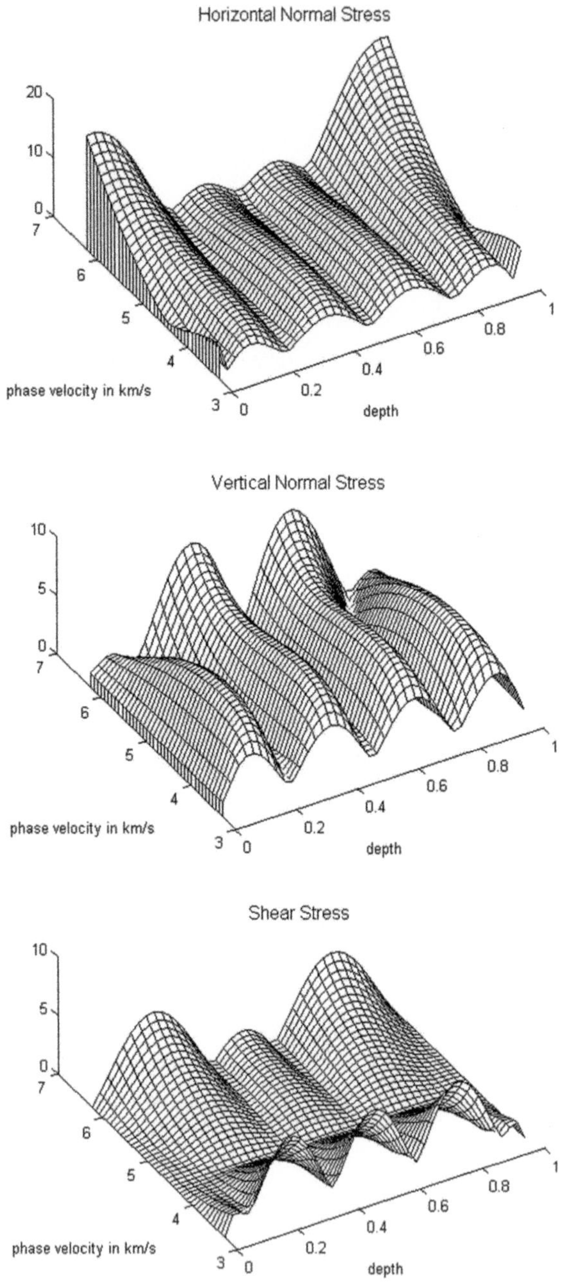

Figure 2.12b Amplitude variations of horizontal normal stress (σ_{11} component, top figure), vertical normal stress (σ_{22} component, middle figure) and shear stress (σ_{12} component, bottom figure) through the plate thickness for the A_2 mode, as the phase velocity varies from 3.4 to 6.3 km/s; depth is 0 for the top of the plate and 1 for the bottom of the plate. Dispersion curves for the aluminum plate are shown in Figure 2.7.

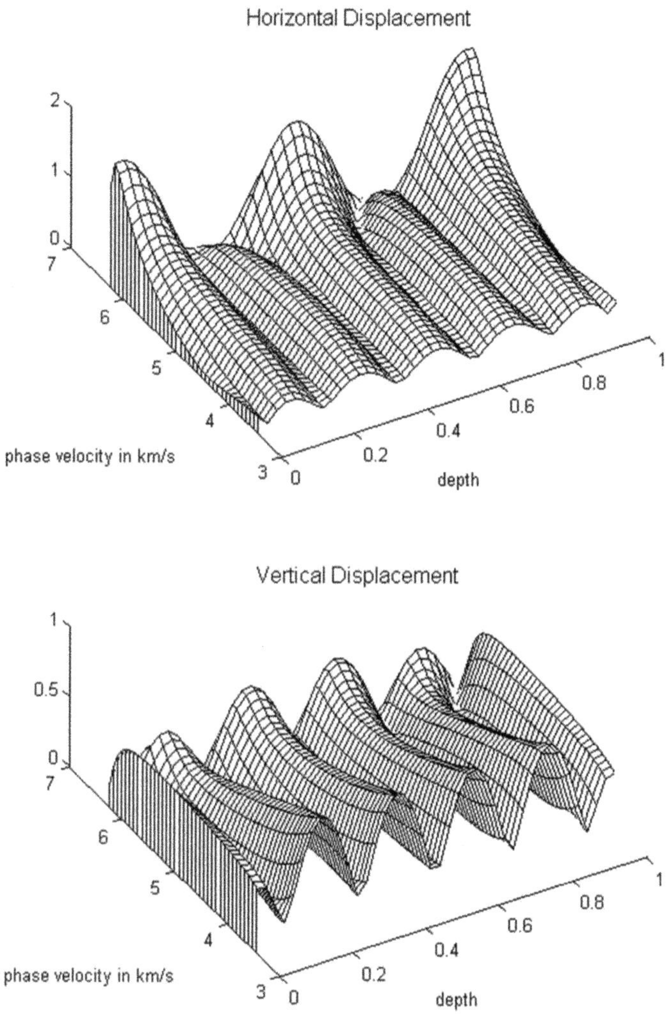

Figure 2.13a Amplitude variations of horizontal (u_1 component, top figure) and vertical (u_2 component, bottom figure) displacements through the plate thickness for the S_2 mode, as the phase velocity varies from 3.5 to 6.3 km/s; depth is 0 for the top of the plate and 1 for the bottom of the plate. Dispersion curves for the aluminum plate are shown in Figure 2.7.

The three algebraic equations given in Eq. (2.19) can be written as a matrix equation:

$$\begin{bmatrix} 2ik\eta\sinh(\eta h) & (2k^2-k_S^2)\sinh(\beta h) & 0 \\ (2k^2-k_S^2)\cosh(\eta h) & -2ik\beta\cosh(\beta h) & \dfrac{\rho_f k_S^2}{\rho}e^{-\eta_f h} \\ \eta\sinh(\eta h) & -ik\sinh(\beta h) & \eta_f e^{-\eta_f h} \end{bmatrix} \begin{Bmatrix} B \\ C \\ M \end{Bmatrix} = \begin{Bmatrix} 0 \\ 0 \\ 0 \end{Bmatrix} \quad (2.20)$$

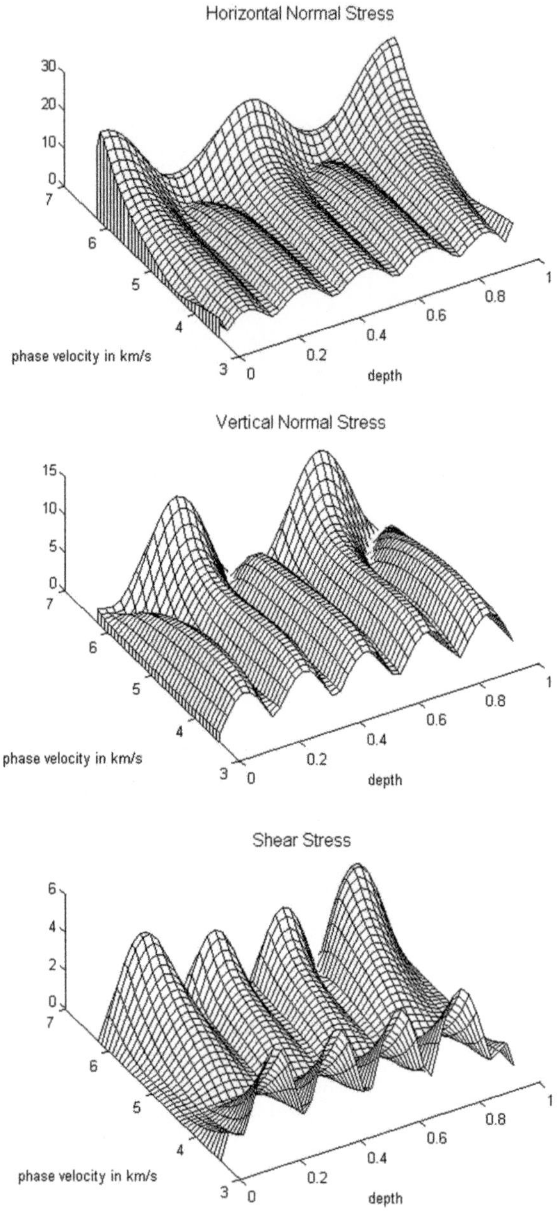

Figure 2.13b Amplitude variations of horizontal normal stress (σ_{11} component, top figure), vertical normal stress (σ_{22} component, middle figure) and shear stress (σ_{12} component, bottom figure) through the plate thickness for the S_2 mode, as the phase velocity varies from 3.5 to 6.3 km/s; depth is 0 for the top of the plate and 1 for the bottom of the plate. Dispersion curves for the aluminum plate are shown in Figure 2.7.

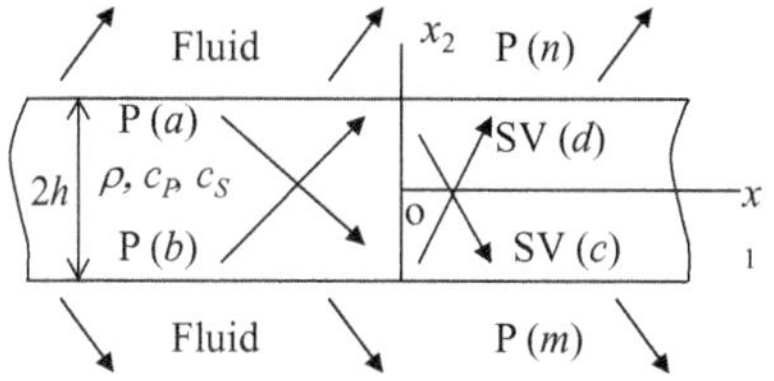

Figure 2.14 The Lamb wave is propagating in the positive x_1 direction while the acoustic energy is leaking into the surrounding fluid medium, thus giving rise to the leaky Lamb wave propagation.

For nonzero solutions of B, C and M, the determinant of the coefficient matrix must vanish. Therefore

$$\eta_f e^{-\eta_f h}\left\{4k^2\eta\beta\sinh(\eta h)\cosh(\beta h)-\left(2k^2-k_S^2\right)^2\cosh(\eta h)\sinh(\beta h)\right\}$$

$$-\frac{\rho_f k_S^2}{\rho}e^{-\eta_f h}\left\{2k^2\eta\sinh(\eta h)\sinh(\beta h)-\eta\left(2k^2-k_S^2\right)\sinh(\eta h)\sinh(\beta h)\right\}=0$$

$$\Rightarrow \eta_f\left\{4k^2\eta\beta\sinh(\eta h)\cosh(\beta h)-\left(2k^2-k_S^2\right)^2\cosh(\eta h)\sinh(\beta h)\right\}$$

$$-\frac{\rho_f\eta k_S^4}{\rho}\sinh(\eta h)\sinh(\beta h)=0 \tag{2.21}$$

$$\Rightarrow 4k^2\eta\beta\sinh(\eta h)\cosh(\beta h)-\left(2k^2-k_S^2\right)^2\cosh(\eta h)\sinh(\beta h)$$

$$=\frac{\rho_f\eta k_S^4}{\rho\eta_f}\sinh(\eta h)\sinh(\beta h)$$

2.3.2 Anti-Symmetric Motion

For the potential field given in Eq. (2.18), the plate motion should be anti-symmetric with respect to the central plane of the plate. As in the previous section, if we apply the vanishing shear stress condition at the fluid–solid interfaces and the continuity of the normal stress and normal displacement components across the two interfaces we get

$$\sigma_{12}\big|_{x_2=\pm h}=\mu\left\{2ik\eta A\cosh(\eta h)+\left(2k^2-k_S^2\right)D\cosh(\beta h)\right\}e^{ikx_1}=0$$

$$\Rightarrow 2ik\eta A\cosh(\eta h)+\left(2k^2-k_S^2\right)D\cosh(\beta h)=0$$

$$\sigma_{22}\big|_{x_2=\pm h}=+\mu\left\{\pm\left(2k^2-k_S^2\right)A\sinh(\eta h)-2ik\beta\left[\pm D\sinh(\beta h)\right]\right\}e^{ikx_1}$$

$$=\pm\rho_f\omega^2 Ne^{-\eta_f h}e^{ikx_1} \tag{2.22}$$

$$\Rightarrow\left(2k^2-k_S^2\right)A\sinh(\eta h)-2ik\beta D\sinh(\beta h)-\frac{\rho_f k_S^2}{\rho}Ne^{-\eta_f h}=0$$

$$u_2\big|_{x_2=\pm h}=\phi_{,2}-\psi_{,1}=\left\{\eta A\cosh(\eta h)-ikD\cosh(\beta h)\right\}e^{ikx_1}=\eta_f Ne^{-\eta_f h}e^{ikx_1}$$

$$\Rightarrow \eta A\cosh(\eta h)-ikD\cosh(\beta h)-\eta_f Ne^{-\eta_f h}=0$$

The three equations of (2.22) can be written in the following matrix form:

$$\begin{bmatrix} 2ik\eta\cosh(\eta h) & (2k^2-k_S^2)\cosh(\beta h) & 0 \\ (2k^2-k_S^2)\sinh(\eta h) & -2ik\beta\sinh(\beta h) & -\dfrac{\rho_f k_S^2}{\rho}e^{-\eta_f h} \\ \eta\cosh(\eta h) & -ik\cosh(\beta h) & -\eta_f e^{-\eta_f h} \end{bmatrix} \begin{Bmatrix} A \\ D \\ N \end{Bmatrix} = \begin{Bmatrix} 0 \\ 0 \\ 0 \end{Bmatrix} \quad (2.23)$$

For nontrivial solutions of A, D and N, the determinant of the coefficient matrix must vanish. Therefore

$$-\eta_f e^{-\eta_f h}\left\{4k^2\eta\beta\cosh(\eta h)\sinh(\beta h)-(2k^2-k_S^2)^2\sinh(\eta h)\cosh(\beta h)\right\}$$

$$+\frac{\rho_f k_S^2}{\rho}e^{-\eta_f h}\left\{2k^2\eta\cosh(\eta h)\cosh(\beta h)-\eta(2k^2-k_S^2)\cosh(\eta h)\cosh(\beta h)\right\}=0$$

$$\Rightarrow -\eta_f\left\{4k^2\eta\beta\cosh(\eta h)\sinh(\beta h)-(2k^2-k_S^2)^2\sinh(\eta h)\cosh(\beta h)\right\} \quad (2.24)$$

$$+\frac{\rho_f\eta k_S^4}{\rho}\cosh(\eta h)\cosh(\beta h)=0$$

$$\Rightarrow 4k^2\eta\beta\cosh(\eta h)\sinh(\beta h)-(2k^2-k_S^2)^2\sinh(\eta h)\cosh(\beta h)$$

$$=\frac{\rho_f\eta k_S^4}{\rho\eta_f}\cosh(\eta h)\cosh(\beta h)$$

Equations (2.21) and (2.24) give the dispersion equations for symmetric and anti-symmetric modes, respectively, of leaky Lamb waves. Leaky Lamb wave dispersion curves for an aluminum plate immersed in the water are shown in Figure 2.15. Similarities and differences between the Lamb wave dispersion curves (Figure 2.6) and leaky Lamb wave dispersion curves (Figure 2.15) should be mentioned here. Note that in Figure 2.15 one additional mode appears for the c_L value below 2 km/s. Viktorov (1967) has shown that for a solid plate immersed in a liquid one additional real root of the dispersion equation appears at every frequency giving a phase velocity value that is less than the P-wave speed in the fluid or the shear wave speed in the solid. This wave mode at the fluid–solid interface was first identified by Scholte (1942) and is considered to be a special case of a *Stonely* wave mode (Stonely, 1924) that is observed at the interface of two solids. This wave mode is known as *Scholte* wave mode or *Stonely-Scholte* wave mode. In absence of the fluid–solid or solid–solid interface, such interface wave modes (Stonely and Scholte) are not observed in a free plate (Figure 2.6). Other than this additional mode, the presence of water does not significantly alter the other Lamb modes of Figure 2.6. Note that when the plate is in a vacuum, the right-hand side of equations (2.21) and (2.24) vanish. This simplifies the dispersion equations to the forms given in Eq. (2.1), and as a result the interface wave mode disappears.

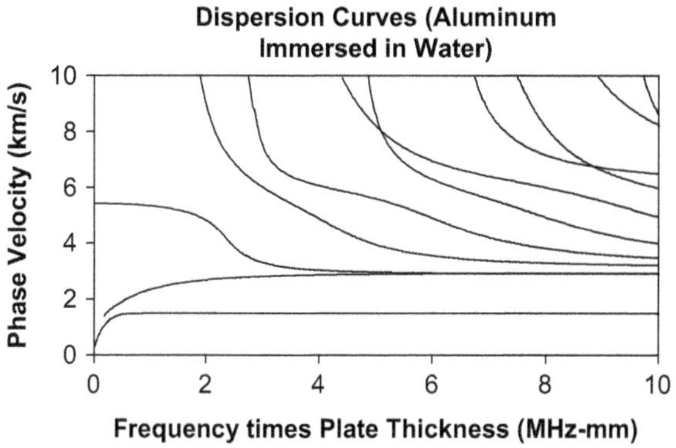

Figure 2.15 Dispersion curves for a 1 mm thick aluminum plate immersed in water. The mode, which is almost horizontal and has a phase velocity less than 2 km/s, corresponds to the Scholte wave. Other curves correspond to the Lamb modes.

Example 2.1

Without separating the symmetric and anti-symmetric components of the plate motion, obtain the dispersion equation for the leaky Lamb wave propagation in a homogeneous isotropic solid plate immersed in a liquid.

SOLUTION

For the problem geometry shown in Figure 2.14, the potential fields corresponding to the down-going and up-going P- and SV-waves are given by Eqs. (2.15) and (2.16).

Wave potentials inside the solid plate:

$$\phi = (ae^{\eta x_2} + be^{-\eta x_2})e^{ikx_1}$$

$$\psi = (ce^{\beta x_2} + de^{-\beta x_2})e^{ikx_1}$$

$$\eta = \sqrt{k^2 - k_P^2}$$

$$\beta = \sqrt{k^2 - k_S^2}$$

Wave potentials in the lower and upper fluid half-spaces:

$$\phi_{fL} = me^{\eta_f x_2} e^{ikx_1}$$

$$\phi_{fU} = ne^{-\eta_f x_2} e^{ikx_1}$$

$$\eta_f = \sqrt{k^2 - k_f^2}$$

If we apply the vanishing shear stress condition at the fluid–solid interfaces and continuity of the normal stress and displacement components across the two interfaces, we get

$$
\begin{aligned}
\sigma_{12}\big|_{x_2=h} &= \mu\{2\phi_{,12}+\psi_{,22}-\psi_{,11}\}_{x_2=h} \\
&= \mu\left\{(2ik\eta)\left(aE-bE^{-1}\right)+\left(2k^2-k_S^2\right)\left(cB+dB^{-1}\right)\right\}e^{ikx_1}=0 \\
&\Rightarrow (2ik\eta)\left(aE-bE^{-1}\right)+\left(2k^2-k_S^2\right)\left(cB+dB^{-1}\right)=0 \\
\sigma_{12}\big|_{x_2=-h} &= \mu\{2\phi_{,12}+\psi_{,22}-\psi_{,11}\}_{x_2=-h} \\
&= \mu\left\{(2ik\eta)\left(aE^{-1}-bE\right)+\left(2k^2-k_S^2\right)\left(cB^{-1}+dB\right)\right\}e^{ikx_1}=0 \\
&\Rightarrow (2ik\eta)\left(aE^{-1}-bE\right)+\left(2k^2-k_S^2\right)\left(cB^{-1}+dB\right)=0 \\
\sigma_{22}\big|_{x_2=h} &= -\mu\left\{k_S^2\phi+2\left(\psi_{,12}+\phi_{,11}\right)\right\}_{x_2=h} \\
&= -\mu\left\{\left(k_S^2-2k^2\right)\left(aE+bE^{-1}\right)+2ik\beta\left(cB-dB^{-1}\right)\right\}e^{ikx_1}=-\rho_f\omega^2 nE_f^{-1}e^{ikx_1} \\
&\Rightarrow \left(2k^2-k_S^2\right)\left(aE+bE^{-1}\right)-2ik\beta\left(cB-dB^{-1}\right)+\frac{\rho_f k_S^2}{\rho}nE_f^{-1}=0 \\
\sigma_{22}\big|_{x_2=-h} &= -\mu\left\{k_S^2\phi+2\left(\psi_{,12}+\phi_{,11}\right)\right\}_{x_2=-h} \\
&= -\mu\left\{\left(k_S^2-2k^2\right)\left(aE^{-1}+bE\right)+2ik\beta\left(cB^{-1}-dB\right)\right\}e^{ikx_1}=-\rho_f\omega^2 mE_f^{-1}e^{ikx_1} \\
&\Rightarrow \left(2k^2-k_S^2\right)\left(aE^{-1}+bE\right)-2ik\beta\left(cB^{-1}-dB\right)+\frac{\rho_f k_S^2}{\rho}mE_f^{-1}=0 \\
u_2\big|_{x_2=h} &= \{\phi_{,2}-\psi_{,1}\}_{x_2=h} \\
&= \left\{\eta\left(aE-bE^{-1}\right)-ik\left(cB+dB^{-1}\right)\right\}e^{ikx_1}=-\eta_f nE_f^{-1}e^{ikx_1} \\
&\Rightarrow \eta\left(aE-bE^{-1}\right)-ik\left(cB+dB^{-1}\right)+\eta_f nE_f^{-1}=0 \\
u_2\big|_{x_2=-h} &= \{\phi_{,2}-\psi_{,1}\}_{x_2=-h} \\
&= \left\{\eta\left(aE^{-1}-bE\right)-ik\left(cB^{-1}+dB\right)\right\}e^{ikx_1}=\eta_f mE_f^{-1}e^{ikx_1} \\
&\Rightarrow \eta\left(aE^{-1}-bE\right)-ik\left(cB^{-1}+dB\right)-\eta_f mE_f^{-1}=0
\end{aligned}
\tag{2.25}
$$

where

$$
\begin{aligned}
E &= e^{i\eta h} \\
B &= e^{i\beta h} \\
E_f &= e^{i\eta_f h}
\end{aligned}
\tag{2.26}
$$

Above six continuity conditions across the two fluid–solid interfaces can be written in the following matrix form:

$$\begin{bmatrix} 2ik\eta E & -2ik\eta E^{-1} & \left(2k^2-k_S^2\right)B & \left(2k^2-k_S^2\right)B^{-1} & 0 & 0 \\ 2ik\eta E^{-1} & -2ik\eta E & \left(2k^2-k_S^2\right)B^{-1} & \left(2k^2-k_S^2\right)B & 0 & 0 \\ \left(2k^2-k_S^2\right)E & \left(2k^2-k_S^2\right)E^{-1} & -2ik\beta B & 2ik\beta B^{-1} & 0 & \frac{\rho_f k_S^2}{\rho}E_f^{-1} \\ \left(2k^2-k_S^2\right)E^{-1} & \left(2k^2-k_S^2\right)E & 2ik\beta B^{-1} & 2ik\beta B & \frac{\rho_f k_S^2}{\rho}E_f^{-1} & 0 \\ \eta E & -\eta E^{-1} & -ikB & -ikB^{-1} & 0 & \eta_f E_f^{-1} \\ \eta E^{-1} & -\eta E & -ikB^{-1} & -ikB & -\eta_f E_f^{-1} & 0 \end{bmatrix} \begin{Bmatrix} a \\ b \\ c \\ d \\ m \\ n \end{Bmatrix} = \begin{Bmatrix} 0 \\ 0 \\ 0 \\ 0 \\ 0 \\ 0 \end{Bmatrix} \quad (2.27)$$

For nonzero wave amplitudes (a, b, c, d, m, n), the determinant of the above 6×6 coefficient matrix must vanish. By equating the determinant of this matrix to zero, the dispersion equation for the leaky Lamb wave propagation is obtained. Note that this dispersion equation gives both symmetric and anti-symmetric modes. However, compared to Eqs. (2.21) and (2.24), the determinant of the above 6×6 matrix is much more complicated. For this reason, symmetric and anti-symmetric modes are computed separately from relatively simpler dispersion equations for symmetric and anti-symmetric modes whenever possible.

2.4 PLANE P-WAVES STRIKING A SOLID PLATE IMMERSED IN A FLUID

In Sections 2.2 and 2.3, Lamb wave propagation in a plate, which is in a vacuum or immersed in a fluid, has been studied by decomposing the particle motions into their symmetric and anti-symmetric components. These Lamb waves are generated in the plate by some external excitations – such as a time dependent force or an elastic wave field striking the plate. In this section, the problem of a plane P-wave striking the plate at an angle θ, as shown in Figure 1.37, is studied.

The pressure fields in the fluid and the potential fields in the solid are given in Eq. (1.228) for a plane P-wave striking a solid plate immersed in a fluid:

$$p_i = e^{ikx_1 - i\eta_f x_2}$$

$$p_R = R.e^{ikx_1 + i\eta_f x_2}$$

$$p_T = T.e^{ikx_1 - i\eta_f x_2}$$

$$\phi_D = P_D.e^{ikx_1 - i\eta_S x_2}$$

$$\phi_U = P_U.e^{ikx_1 + i\eta_S x_2}$$

$$\psi_D = S_D.e^{ikx_1 - i\beta_S x_2}$$

$$\psi_U = S_U.e^{ikx_1 + i\beta_S x_2}$$

Equations (2.15) and (2.16) are similar to Eq. (1.228); the only difference is that in Eq. (1.228), one additional term p_i appears that corresponds to the incoming P-wave striking the plate. p_R and p_T of Eq. (1.228) are similar to ϕ_{fU} and ϕ_{fL} of Eq. (2.16); however, p_R and p_T are pressure fields while ϕ_{fU} and ϕ_{fL} are wave potentials in the fluid. The relation between the pressure (p) and wave potential (ϕ) in a fluid is given in Chapter 1, Eq. (1.209), $p = \rho\omega^2\phi$. It should also be mentioned that η_f, η_s and β_s of Eq. (1.228), and η_f, η and β of Eqs. (2.15 and 2.16), have slightly different definitions. See Eqs. (1.229) and (2.15 and 2.16) for their definitions.

The notations for wave amplitudes have also been changed in Eqs. (2.15) and (2.16) from Eq. (1.228).

Six equations that are obtained from the stress and displacement continuity conditions across the fluid–solid interfaces are given in Eqs. (1.232) and (1.232a). Note that Eq. (1.232) is a system of non-homogeneous equations while Eq. (2.27) gives a system of homogeneous equations. Nonzero entries of the right-hand side vector of Eq. (1.232) are due to the presence of the striking P-wave. The 6×6 coefficient matrix of Eq. (1.232) and (2.27) are similar; however, these two are not identical because the definitions of the wave potentials are slightly different in these two cases.

To obtain the identical coefficient matrix for the above two cases (in presence and in absence of the striking P-wave) the wave potentials in the solid plate and in the two fluid half-spaces can be defined in the following manner.

Wave potentials inside the solid plate:

$$\begin{aligned}
\phi &= (ae^{\eta x_2} + be^{-\eta x_2})e^{ikx_1} \\
\psi &= (ce^{\beta x_2} + de^{-\beta x_2})e^{ikx_1} \\
\eta &= \sqrt{k^2 - k_P^2} = -i\sqrt{k_P^2 - k^2} \\
\beta &= \sqrt{k^2 - k_S^2} = -i\sqrt{k_S^2 - k^2}
\end{aligned} \tag{2.28a}$$

Wave potentials in the lower and upper fluid half-spaces:

$$\begin{aligned}
\phi_{fL} &= me^{\eta_f x_2}e^{ikx_1} \\
\phi_{fU} &= \left(e^{\eta_f x_2} + ne^{-\eta_f x_2}\right)e^{ikx_1} \\
\eta_f &= \sqrt{k^2 - k_f^2} = -i\sqrt{k_f^2 - k^2} = -ik_f\cos\theta; \quad k = k_f\sin\theta
\end{aligned} \tag{2.28b}$$

The extra term in the expression of ϕ_{fU} in Eq. (2.28b) in comparison to that in Eq. (2.16) represents the incoming P-wave that strikes the plate. The reflected and transmitted P-wave amplitudes in the upper and lower fluid half-spaces are denoted by n and m, respectively. Applying the continuity conditions across the two fluid–solid interfaces, the following equations are obtained:

$$\begin{aligned}
&\sigma_{12}\big|_{x_2=h} = 0 \quad \Rightarrow (2ik\eta)\left(aE - bE^{-1}\right) + \left(2k^2 - k_S^2\right)\left(cB + dB^{-1}\right) = 0 \\
&\sigma_{12}\big|_{x_2=-h} = 0 \quad \Rightarrow (2ik\eta)\left(aE^{-1} - bE\right) + \left(2k^2 - k_S^2\right)\left(cB^{-1} + dB\right) = 0 \\
&\sigma_{22}\big|_{x_2=h} = -\mu\left\{\left(k_S^2 - 2k^2\right)\left(aE + bE^{-1}\right) + 2ik\beta\left(cB - dB^{-1}\right)\right\}e^{ikx_1} = -\rho_f\omega^2\left(E_f + nE_f^{-1}\right)e^{ikx_1} \\
&\Rightarrow \left(2k^2 - k_S^2\right)\left(aE + bE^{-1}\right) - 2ik\beta\left(cB - dB^{-1}\right) + \frac{\rho_f k_S^2}{\rho}\left(E_f + nE_f^{-1}\right) = 0 \\
&\sigma_{22}\big|_{x_2=-h} = -\mu\left\{\left(k_S^2 - 2k^2\right)\left(aE^{-1} + bE\right) + 2ik\beta\left(cB^{-1} - dB\right)\right\}e^{ikx_1} = -\rho_f\omega^2 mE_f^{-1}e^{ikx_1} \\
&\Rightarrow \left(2k^2 - k_S^2\right)\left(aE^{-1} + bE\right) - 2ik\beta\left(cB^{-1} - dB\right) + \frac{\rho_f k_S^2}{\rho}mE_f^{-1} = 0
\end{aligned} \tag{2.29}$$

$$u_2\big|_{x_2=h} = \{\phi_{,2} - \psi_{,1}\}_{x_2=h} = \left\{\eta\left(aE - bE^{-1}\right) - ik\left(cB + dB^{-1}\right)\right\}e^{ikx_1} = \eta_f\left(E_f - nE_f^{-1}\right)e^{ikx_1}$$

$$\Rightarrow \eta\left(aE - bE^{-1}\right) - ik\left(cB + dB^{-1}\right) - \eta_f\left(E_f - nE_f^{-1}\right) = 0$$

$$u_2\big|_{x_2=-h} = \{\phi_{,2} - \psi_{,1}\}_{x_2=-h} = \left\{\eta\left(aE^{-1} - bE\right) - ik\left(cB^{-1} + dB\right)\right\}e^{ikx_1} = \eta_f mE_f^{-1}e^{ikx_1}$$

$$\Rightarrow \eta\left(aE^{-1} - bE\right) - ik\left(cB^{-1} + dB\right) - \eta_f mE_f^{-1} = 0$$

Definitions of E, B and E_f are given in Eq. (2.26). Putting these six equations in a matrix form,

$$\begin{bmatrix} 2ik\eta E & -2ik\eta E^{-1} & \left(2k^2 - k_S^2\right)B & \left(2k^2 - k_S^2\right)B^{-1} & 0 & 0 \\ 2ik\eta E^{-1} & -2ik\eta E & \left(2k^2 - k_S^2\right)B^{-1} & \left(2k^2 - k_S^2\right)B & 0 & 0 \\ \left(2k^2 - k_S^2\right)E & \left(2k^2 - k_S^2\right)E^{-1} & -2ik\beta B & 2ik\beta B^{-1} & 0 & \frac{\rho_f k_S^2}{\rho}E_f^{-1} \\ \left(2k^2 - k_S^2\right)E^{-1} & \left(2k^2 - k_S^2\right)E & 2ik\beta B^{-1} & 2ik\beta B & \frac{\rho_f k_S^2}{\rho}E_f^{-1} & 0 \\ \eta E & -\eta E^{-1} & -ikB & -ikB^{-1} & 0 & \eta_f E_f^{-1} \\ \eta E^{-1} & -\eta E & -ikB^{-1} & -ikB & -\eta_f E_f^{-1} & 0 \end{bmatrix} \begin{Bmatrix} a \\ b \\ c \\ d \\ m \\ n \end{Bmatrix} = \begin{Bmatrix} 0 \\ 0 \\ -\frac{\rho_f k_S^2}{\rho}E_f \\ 0 \\ \eta_f E_f \\ 0 \end{Bmatrix} \tag{2.30}$$

Note that the coefficient matrices in Eqs. (2.27) and (2.30) are identical. The only difference between these two sets of equations is in the definition of their right-hand side vectors. Two problems, defined by equations (2.27) and (2.30), are analogous to the free vibration and forced vibration problems in structural dynamics, where the right-hand side vector represents the forcing function. Note that the wave number k in Eq. (2.27) is obtained by satisfying the dispersion equation; in other words, by equating the determinant of the coefficient matrix to zero. However, in Eq. (2.30) k is no longer a variable; it is determined from the striking angle of the incident P-wave [see Eq. (2.28b)].

For generating a Lamb wave in the plate, the wave number k in Eq. (2.30) must be the solution of the Lamb wave dispersion equation. In other words, if the striking angle of the incident P-wave is such that k becomes equal to ω / c_L (c_L is the Lamb wave speed in the plate), then the Lamb wave will be generated. For a given frequency, a number of wave speeds are observed in the Lamb wave dispersion curve plot (see Figure 2.6). Which Lamb mode is generated in the plate depends on what k value corresponds to the striking angle of the P-wave, $k = k_f \sin\theta$. To generate a Lamb mode with the phase velocity c_L, it is necessary to satisfy the following condition:

$$k_f \sin\theta = k = \frac{\omega}{c_L}$$

$$\Rightarrow \sin\theta = \frac{\omega}{c_L k_f} = \frac{c_f}{c_L} \tag{2.31}$$

$$\Rightarrow \theta = \sin^{-1}\left(\frac{c_f}{c_L}\right)$$

Equation (2.31) gives the incident angle (θ) necessary for generating a Lamb mode with a phase velocity c_L in a plate immersed in a fluid that has an acoustic wave speed, c_f. Equation (2.31) is also known as Snell's law.

Figure 2.16 shows a schematic diagram for the Lamb wave generation in a plate by striking the plate with a plane P-wave. If a plane P-wave strikes the left

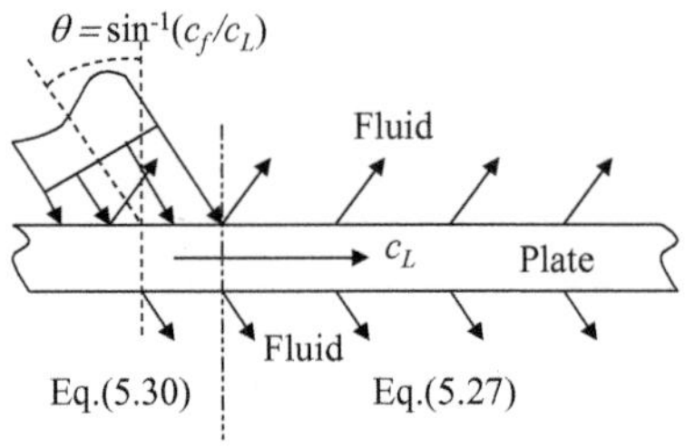

Figure 2.16 Lamb wave generation in a plate immersed in a fluid. Eqs. (2.30) and (2.27) govern the left side (forced vibration) and right side (free vibration) of the plate. The dashed-dotted line separates the left and right sides of the plate.

side of the plate with an incident angle (θ) that is appropriate for generating a Lamb mode, then the Lamb mode is generated in the plate. The generated Lamb wave then propagates to the right with a velocity c_L (=ω/k). Note that Eqs. (2.30) and (2.27) govern motions on the left side (forced vibration) and right side (free vibration) of the plate, respectively. From Eq. (2.31) it is obvious that only the Lamb modes that have phase velocities higher than the acoustic wave speed in the surrounding fluid can be generated in this manner. The acoustic wave speed in water is 1.48 km/s, while in alcohol it is between 1.12 and 1.24 km/s (see Table 1.4). Therefore, to generate a Lamb mode with a phase velocity between 1.2 and 1.48 km/s, the plate must be immersed in alcohol instead of water.

2.4.1 Plate Inspection by Lamb Waves

Above we have learned how different Lamb modes can be generated in a plate by adjusting the striking angle and frequency of the incident P-wave. By placing an ultrasonic transducer at an angle relative to the plate, as shown in Figure 2.16, the required condition can be satisfied. Ultrasonic transducers generate P-waves of certain frequency. The signal frequency is determined by the natural frequency of the ceramic crystal in the transducer. The ultrasonic transducer frequency varies from 500 kHz to 10 MHz for conventional ultrasonic inspection applications. However, specially built ultrasonic transducers can generate ultrasonic signals with a resonance frequency as low as 50 kHz or even lower, and as high as 100 MHz or higher. In acoustic microscopy applications the signal frequency can be as high as 1 to 2 GHz, and in some applications even higher.

2.4.1.1 Generation of Multiple Lamb Modes by Narrowband and Broadband Transducers

Ultrasonic transducers that have well-defined natural frequency and always vibrate close to that frequency are called narrowband transducers. For example, a 5 MHz narrowband transducer can only generate ultrasonic waves with signal frequency equal to 5 MHz or very close to 5 MHz – for example, between 4.8 and 5.2 MHz. On the other hand, a broadband transducer can generate ultrasonic waves over a wide frequency range. For example, a 5 MHz broadband transducer can generate ultrasonic waves of between 1 and 9 MHz. For generating different Lamb modes with a narrowband transducer, since it is not possible to change the signal frequency, it is necessary to change the angle of strike (θ of Figure 2.16). Note that changing θ implies changing the phase velocity, c_L, because θ and c_L are related by Eq. (2.31). Keeping the signal frequency constant and varying the incident angle implies moving along the vertical axis of the dispersion curve plots (see Figure 2.2). Lamb waves will be generated for the incident angle θ, if $\theta = \sin^{-1}\left(\frac{c_f}{c_L}\right)$, where c_f is the P-wave speed in the fluid and c_L is the Lamb wave speed. If we place two ultrasonic transducers, one acting as the transmitter (T) and the second one as the receiver (R) as shown in Figure 2.17a, and record the received signal strength as a function of the incident angle θ, then we should get a curve as shown in Figure 2.17b. Peaks correspond to the incident angles for which a Lamb mode is generated. If the incident angle does not correspond to a Lamb mode-generating angle, then no ultrasonic energy reaches the receiver because no Lamb wave is generated and the energy of the directly reflected beam is mostly confined between the lines AB and CD (see Figure 2.17a). This directly reflected beam is also called a "specularly reflected" beam.

If the transducers are broadband type, then the signal frequency can be changed while maintaining a constant inclination angle of the transducers. Fixed inclination angle means constant phase velocity. Therefore, in this case, as we change frequency we move horizontally in the dispersion curve plot (see Figure 2.3). In this manner, received signal voltage versus frequency curve or $V(f)$ curve is generated, as shown in Figure 2.17c. In this case, peaks will be observed at frequencies for which Lamb modes are generated. If the signal frequency does not correspond to a Lamb wave generation frequency, then the received signal strength should be close to zero since the reflected signal energy in this case remains mostly confined between lines AB and CD and does not reach the receiver.

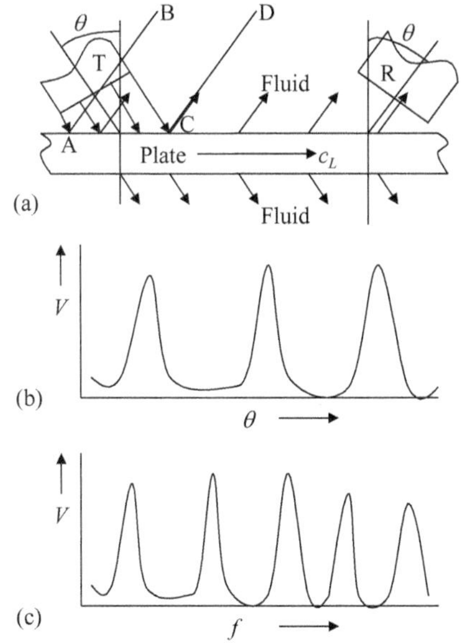

Figure 2.17 (a) Transmitter (T) and receiver (R) for generating and receiving Lamb waves in a plate; (b) Received signal amplitude voltage as a function of the incident angle – peaks indicate generation of Lamb modes; (c) Received signal amplitude voltage as a function of the signal frequency – peaks indicate generation of Lamb modes.

2.4.1.2 *Nondestructive Inspection of Large Plates*

Following the technique presented above, it is possible to generate multiple Lamb modes in a plate. In this section the interaction between Lamb modes and internal defects in a plate is investigated. Ghosh and Kundu (1998) have shown that strong and consistent Lamb modes can be generated in large plates by placing the transducers in small conical water containers attached to the plate, as shown in Figure 2.18, instead of immersing the entire plate in a big water tank. The bottomless but watertight containers are attached to the plate such that the water is in direct contact with the plate without being leaked out. Alleyne and Cawley (1992) also avoided immersing the entire plate in a water tank, and generated Lamb waves in the plate by placing inclined transducers in cylindrical holes filled with water. In both of these cases water coupling is provided between the transducers and the plate – in one case a cylindrical water column, and in the other case a conical water volume. Between these two, the conical water volume is found to produce more consistent signals.

Using the transmitter–receiver arrangement shown in Figure 2.18, Lamb waves are generated in a 0.39 in. (9.9 mm) thick steel plate. Details of this experiment with steel and aluminum plates are given in Ghosh et al. (1998). Here, some experimental results generated for the steel plate are given. Plate length and width are 60 in. (1524 mm) and 3.5 in. (88.9 mm), respectively. A hole of 2 in. (50.8 mm) length and 0.125 in. (3.175 mm) diameter is drilled into the central plane of the plate parallel to its surfaces to artificially produce an internal defect, as shown in Figure 2.19.

Figure 2.20 shows the $V(f)$ curves for the specimen of Figure 2.19 over the non-defective region (region that does not contain the hole) for 26° angle of incidence. The distance between the transmitter and the receiver for all experimental results is 12 in. (304.8 mm). This incident angle corresponds to a phase velocity of 3.4 km/s, derived from Snell's law, Eq. (2.31), with $c_f = 1.49$ km/s.

The four curves of Figure 2.20 correspond to four different experiments carried out on four regions of the plate at four different times. They show the experimental variability. These differences are due to the fact that at low frequencies the signal is not well collimated. Directivity of the signal is controlled by the

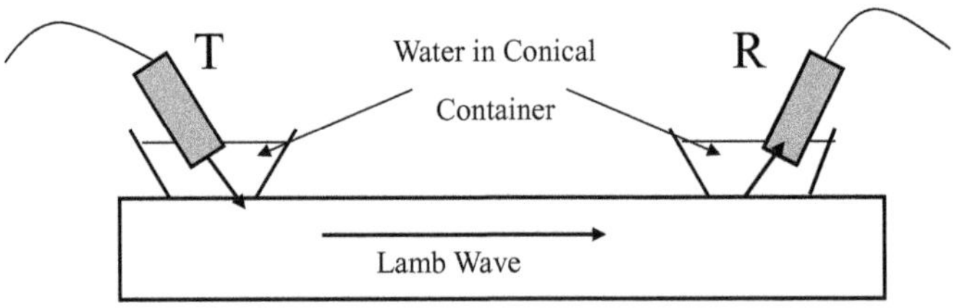

Figure 2.18 Transducers in conical water containers [from Ghosh and Kundu (1998)].

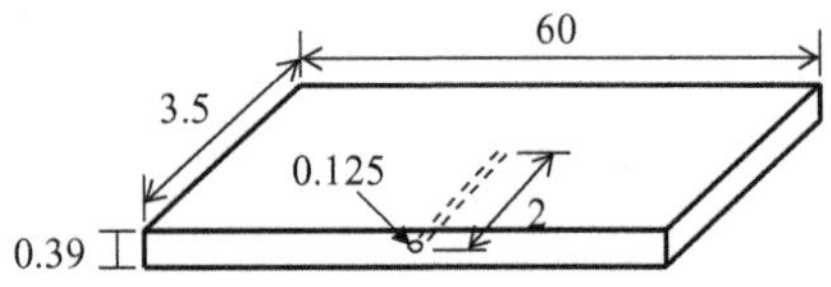

Figure 2.19 Steel plate with a hole, all dimensions are in inch [from Ghosh et al. (1998)].

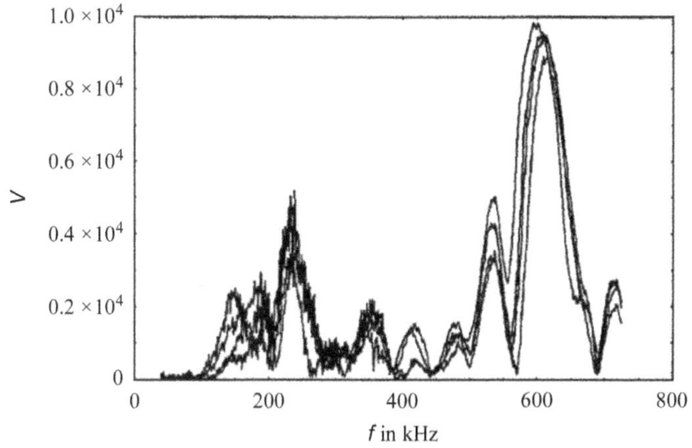

Figure 2.20 Four *V*(*f*) curves generated by the setup of Figure 2.18 over the defect-free region of the plate of Figure 2.19, for 26° incident angle [from Ghosh et al. (1998)].

inclination of the transducer and its relative position with respect to the bottom section of the conical container. During different experiments, the transducer position relative to the bottom section of the conical container slightly changes, giving rise to the variations in the *V*(*f*) curves even when the plate dimension and its properties do not change. One can also see that the noise level is higher at lower frequencies because the signal is less collimated; in other words, its directivity is poorer at lower frequencies.

However, in spite of the existence of noise and experimental variability in *V*(*f*) curves, it is easy to observe two strong and distinguishable peaks – one near 230 kHz and the second near 600 kHz (Figure 2.20). Another peak near 540 kHz is also observed in the *V(f)* plot; however, it is much weaker than the peak near 600 kHz.

Theoretical dispersion curves for this specimen are computed and plotted in Figure 2.21. The experimental points (frequencies at which peaks occur in Figure 2.20) are plotted along with the theoretical curves in Figure 2.21 to identify the Lamb modes that correspond to the two peaks of Figure 2.20. The strong peaks of the *V*(*f*) curves have been denoted by the "*" symbol while the weak peaks are denoted by the "+" symbol in Figure 2.21. Since peak frequencies change from experiment to experiment, a frequency range (instead of a single value) is shown in Table 2.1 and on the dispersion curve plot (Figure 2.21) for every mode. After plotting the experimental points on the theoretical dispersion curves in Figure 2.21, it can be clearly seen that the two peaks of Figure 2.20 correspond to the zero-th order symmetric (S_0) and first order anti-symmetric (A_1) modes.

V(*f*) curves for the defective region are shown in Figure 2.22. For these plots the transmitter and the receiver are placed on opposite sides of the hole. Therefore, the Lamb wave generated in the plate by the transmitter must propagate through the plate region containing the hole before reaching the receiver. The four curves of Figure 2.22 are for four different experiments carried out at four different locations while the defect (the hole) is located between the transmitter and the receiver in all four cases. Note that the S_0 mode has been strongly affected (it has almost disappeared) by the defect (hole) but that the A_1 mode has not changed significantly, indicating that it is not very sensitive to this defect.

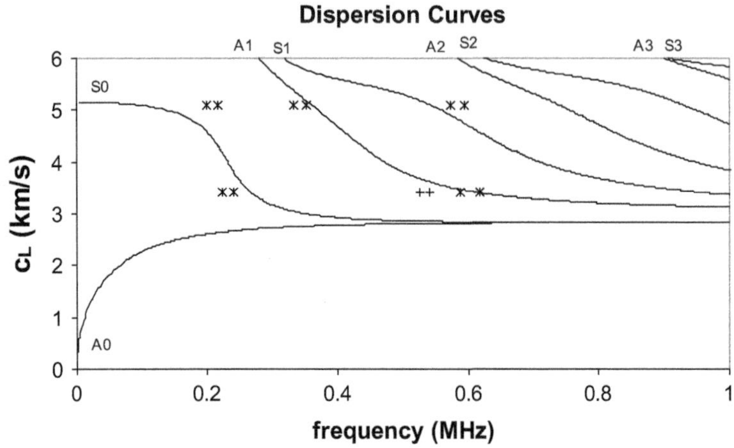

Figure 2.21 Theoretical dispersion curve for a 0.39 inch thick steel plate (c_P=5.72 km/s, c_S=3.05 km/s, ρ=7.9 gm/cc). Experimental points are shown by '*' and '+' symbols, for strong and weak peaks, respectively [from Ghosh et al. (1998)].

Results from Figures 2.20 and 2.22 are summarized in Table 2.1. In this table the row marked as "Freq. Range" describes the frequency range in which the Lamb wave peaks are observed for all four experiments. The rows denoted as "Min", "Max" and "Average Amp." show the minimum, maximum and average values of the four peaks corresponding to the four experimental curves over the non-defective (corresponding column is marked as "Non-def") and defective regions (corresponding column is marked as "Def"). The percentage change of the average amplitude of the S_0 mode due to the presence of the defect is –78.8%, where the minus sign implies a reduction in the peak amplitude. The change of the A_1 mode due to the defect is only −0.89%.

The computed stress patterns inside the plate for phase velocities varying from 3 to 5 km/s for the S_0 and A_1 modes are shown in Figures 2.23 and 2.24, respectively. The depth dimension is normalized with respect to the total plate thickness. In these figures σ_{11}, σ_{22} and σ_{13} stand for the horizontal normal stress (σ_{11}), the vertical normal stress (σ_{22}) and the shear stress component (σ_{12}), respectively. The x_1 and x_2 axes are located parallel and perpendicular to the plate surface, respectively. For the S_0 mode (Figure 2.23) at 3.4 km/s phase velocity, the normal stress σ_{22} is large near the central plane of the plate (depth=0.5) but σ_{12} is

Table 2.1: Lamb Mode Peaks for 26° Incidence Angle (see Figures 2.20 and 2.22)

Mode	S_0		A_1	
Freq. Range (kHz)	**224–241**		**588–617**	
Amplitude Information	**Non-def.**	**Def.**	**Non-def.**	**Def.**
Min.	3333	203	8782	8637
Max.	4782	1534	9797	9536
Average amp.	3713	785	9116	9034
Percentage chang		–78.8		–0.89

Source: Reprinted from Ghosh et al. (1998) with permission from Elsevier.

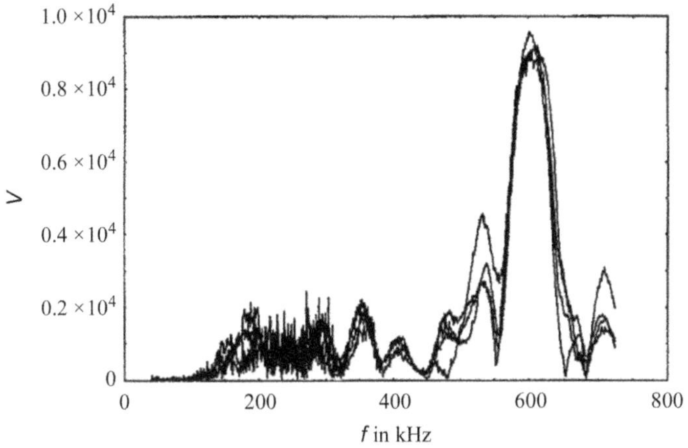

Figure 2.22 *V(f)* curves over the defective region of the steel plate, for 26° angle of incidence [from Ghosh et al. (1998)].

zero there. In presence of the hole, these stress components are altered since the normal and shear stress components at the free surface of the hole must vanish; as a result, the propagating mode is affected. For the A_1 mode (Figure 2.24), normal stresses (σ_{11} and σ_{22}) are zero at the central plane of the plate and σ_{12} has a nonzero value at the central plane of the plate (also, σ_{12} is zero in the entire plate near the phase velocity of 4 km/s). However, since the two normal stress components are zero at the central plane, the defect has a smaller effect on the A_1 mode.

For a 17° incident angle (phase velocity from Snell's law is 5.1 km/s) strong peaks in the *V(f)* curves for the plate without any defect are observed near 210, 340 and 580 kHz (Figure 2.25). These correspond to S_0, A_1 and S_1 modes, respectively (see Figure 2.21). In Figure 2.21 one can notice that theoretical A_1 and S_1 modes are very close to the experimental values. However, the experimental points corresponding to the S_0 mode are not very close to the theoretical S_0 curve. This is due to the fact that the transducer response below 200 kHz is very weak (the transducers have a central frequency of 500 kHz). Secondly, although the transducer is inclined at 17°, not all the energy of the 200 kHz signal that generated the Lamb wave in the plate struck the plate exactly at the 17° inclination. Some ultrasonic energy struck the plate at a slightly larger angle making the corresponding phase velocity smaller [from Snell's law, Eq. (2.31)] and thus bringing the experimental point closer to the theoretical curve.

The four *V(f)* curves for the defective region are shown in Figure 2.26. One can clearly see that due to the presence of the defect, S_0 and S_1 modes change significantly, but that the A_1 mode is again not very sensitive to the defect. This result is summarized in Table 2.2. The amplitude of the S_0 mode is reduced by more than 72% due to the defect; this reduction for the S_1 mode is more than 64%, while the reduction of the A_1 mode is only 3.5%. The justification for the symmetric modes being more sensitive to the defect is the same as before. Figure 2.23 shows that for the S_0 mode, for 5.1 km/s phase velocity σ_{11} is very large and σ_{22} is moderately large at the central plane. As a result, this mode has been significantly affected (–72.79%) by the presence of the defect at the central plane of the plate. Figure 2.27 shows that, for the S_1 mode, for 5.1 km/s phase velocity σ_{11} is small but σ_{22} is large at the central plane of the plate. As a result, this mode has also been strongly affected (–64.48%) by the defect, although to a lesser degree

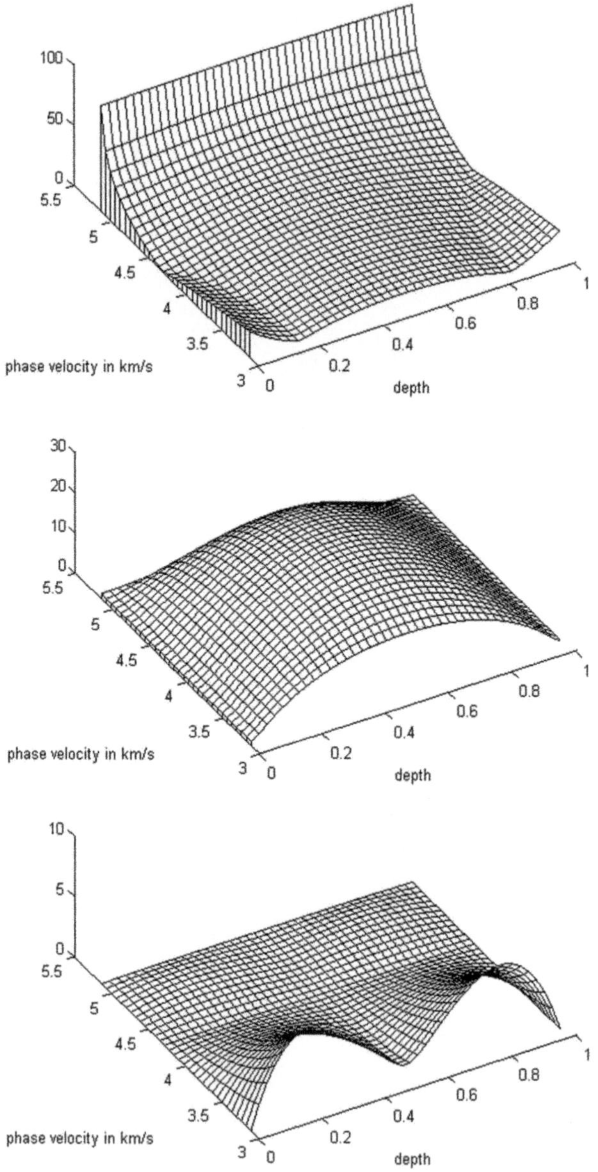

Figure 2.23 Amplitude variations of horizontal normal stress (σ_{11}, top figure), vertical normal stress (σ_{22}, middle figure) and shear stress (σ_{12}, bottom figure) inside the steel plate for S_0 mode, as the phase velocity varies from 3.1 to 5.1 km/s; depth is 0 for the top of the plate and 1 for the bottom of the plate. Dispersion curves for this steel plate are shown in Figure 2.21.

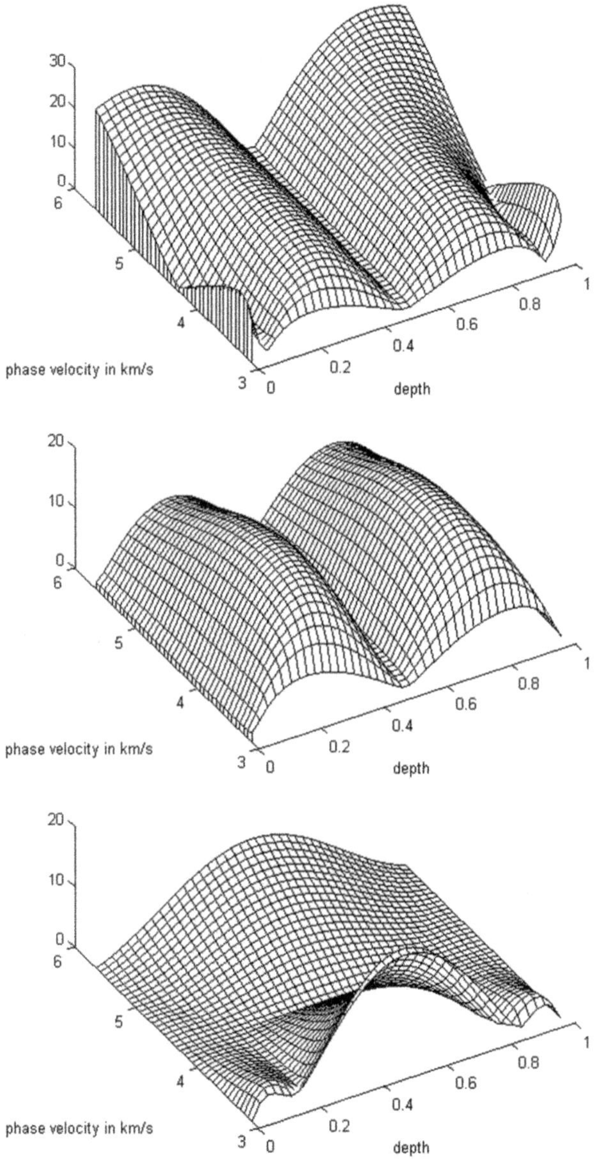

Figure 2.24 Amplitude variations of horizontal normal stress (σ_{11}, top figure), vertical normal stress (σ_{22}, middle figure) and shear stress (σ_{12}, bottom figure) inside the steel plate for A_1 mode, as the phase velocity varies from 3.1 to 5.6 km/s; depth is 0 for the top of the plate and 1 for the bottom of the plate. Dispersion curves for this steel plate are shown in Figure 2.21.

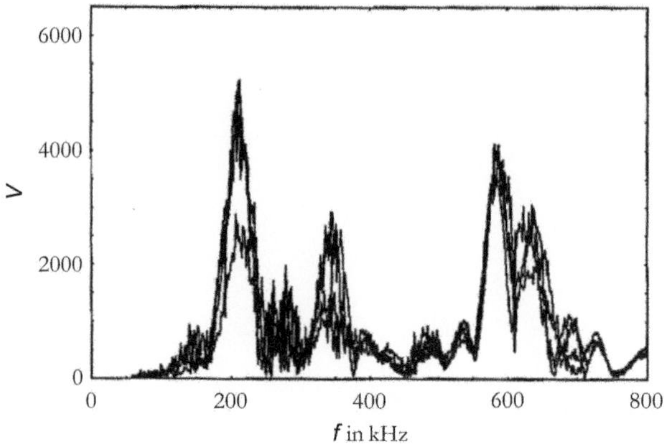

Figure 2.25 *V(f)* curves for the defect-free region of the plate, for 17° angle of incidence [from Ghosh et al. (1998)].

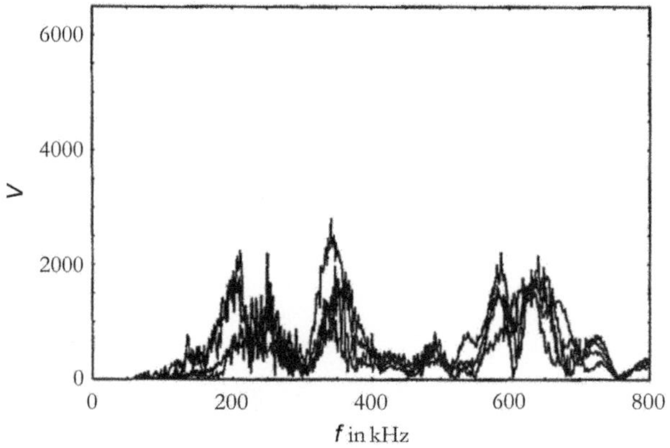

Figure 2.26 *V(f)* curves over the defective region for 17° incident angle [from Ghosh et al. (1998)].

Table 2.2: Lamb Mode Peaks for a 17° Incident Angle (see Figures 2.25 and 2.26)

Mode	S_0		A_1		S_1	
Freq. Range (kHz)	**201–219**		**334–352**		**573–595**	
Amplitude Information	**Non-def.**	**Def.**	**Non-def.**	**Def.**	**Non-def.**	**Def.**
Min.	2523	133	1294	1469	3504	950
Max.	4597	1544	2169	2335	4013	1830
Average amp.	3161	860	1625	1568	3714	1319
Percentage change		–72.79		–3.5		–64.48

Source: Reprinted from Ghosh et al. (1998) with permission from Elsevier.

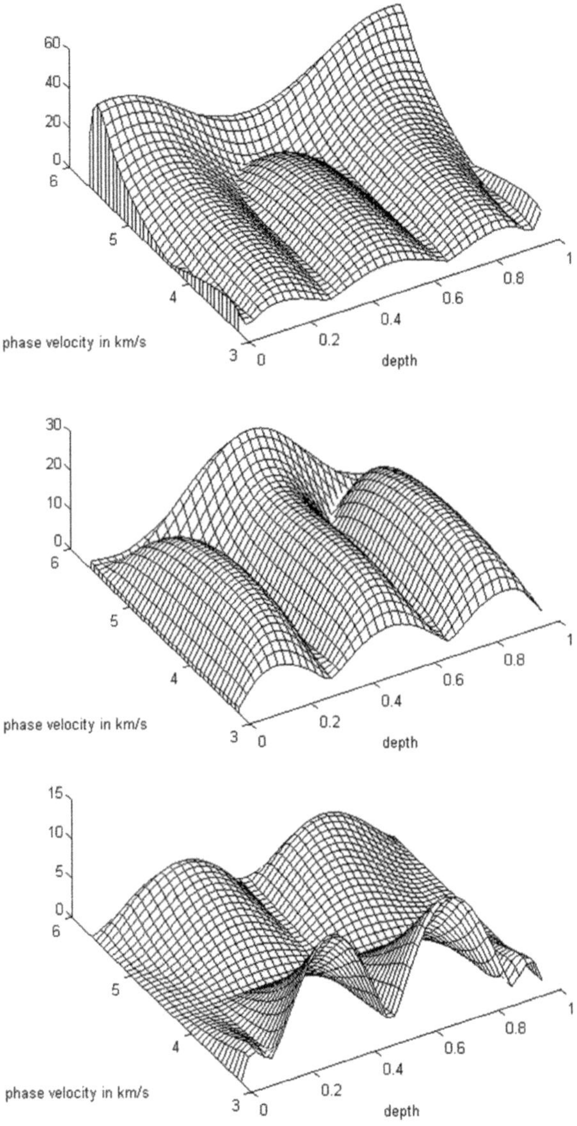

Figure 2.27 Amplitude variations of horizontal normal stress (σ_{11}, top figure), vertical normal stress (σ_{22}, middle figure) and shear stress (σ_{12}, bottom figure) inside the steel plate for S_1 mode, as the phase velocity varies from 3.1 to 5.6 km/s; the depth is 0 for the top of the plate and 1 for the bottom of the plate. Dispersion curves for this steel plate are shown in Figure 2.21.

than the S_0 mode (–72.8%). The A_1 mode, on the other hand, makes the shear stress component σ_{12} nonzero at the central plane, but the normal stress components σ_{11} and σ_{22} are zero there (Figure 2.24); however, although σ_{12} is nonzero, it is small at the central plane for 5.1 km/s phase velocity, and as a result this mode is not significantly affected by the presence of the hole (only –3.5% variation).

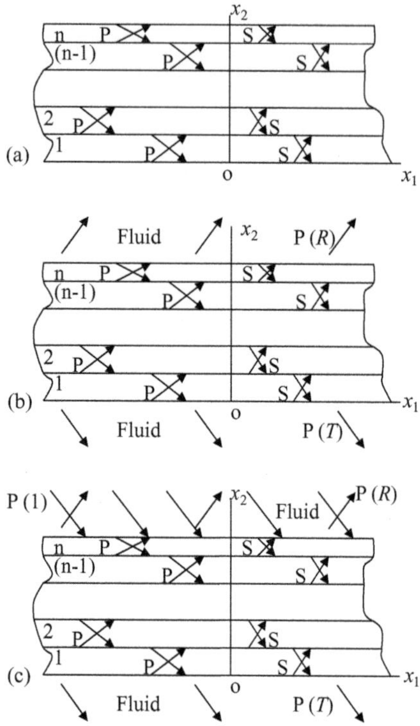

Figure 2.28 Guided wave propagation through the n-layered plate (a) in a vacuum, (b) in a fluid, and (c) the plate immersed in a fluid is struck by a down-going P-wave of amplitude 1.

2.5 GUIDED WAVES IN MULTILAYERED PLATES

So far, we have analyzed Lamb wave propagation in a homogeneous isotropic plate in a vacuum or immersed in a fluid. In this section, the guided wave propagation in a multilayered solid plate is analyzed. Three problems that are considered here are

1. Free vibration of a multilayered solid plate in a vacuum
2. Free vibration of a multilayered solid plate immersed in a fluid
3. Forced vibration of a multilayered solid plate immersed in a fluid – a plane P-wave propagating through the fluid half-space strikes the fluid–solid interface, and thus creates the forcing excitation.

Geometries of these three problems are shown in Figure 2.28.

2.5.1 n-Layered Plates in a Vacuum

The problem geometry is shown in Figure 2.28a. n number of layers of thickness $h_1, h_2, h_3 \ldots h_n$ are perfectly bonded (no slippage condition) along the $(n-1)$ interfaces at $x_2 = y_1, y_2, y_3 \ldots y_{n-1}$ where $y_1 = h_1$, $y_2 = h_1 + h_2$, $y_3 = h_1 + h_2 + h_3$, and $y_{n-1} = h_1 + h_2 + h_3 + \ldots + h_{n-1}$. Therefore, the displacement components (u_1 and u_2), and the normal and shear stress components (σ_{22} and σ_{12}) across every interface must be continuous. In addition, normal and shear stress components (σ_{22} and σ_{12}) on the two boundary surfaces at $x_2 = y_1 = h_1$, and $x_2 = y_n = h_1 + h_2 + h_3 + \ldots + h_n$ must vanish, since the boundaries must be stress-free when the plate is in a vacuum. Thus, it is necessary to solve this boundary value problem, subjected to these boundary conditions and continuity conditions across the interface.

The governing equation inside of a general m-th layer can be satisfied by considering up-going and down-going P- and S-waves in that layer. Note that the subscript m stands for the m-th layer.

$$\begin{aligned}
\phi_m &= \left\{a_m e^{-i\eta_m(x_2-y_{m-1})} + b_m e^{i\eta_m(x_2-y_{m-1})}\right\}e^{ikx_1} \\
\psi_m &= \left\{c_m e^{-i\beta_m(x_2-y_{m-1})} + d_m e^{i\beta_m(x_2-y_{m-1})}\right\}e^{ikx_1} \\
\eta_m &= \sqrt{k_{Pm}^2 - k^2} = i\sqrt{k^2 - k_{Pm}^2} \\
\beta_m &= \sqrt{k_{Sm}^2 - k^2} = i\sqrt{k^2 - k_{Sm}^2}
\end{aligned} \tag{2.32}$$

Where a_m, b_m, c_m and d_m represent wave amplitudes for the down-going and up-going P- and S-waves in the layer; k_{Pm} ($=\omega/c_{Pm}$) and k_{Sm} ($=\omega/c_{Sm}$) are P- and S-wave numbers, respectively. The exponential term is defined such that the phase term becomes zero at the lower interface of the layer. Note that the phase of the exponential terms can be equated to zero at any level y_p simply by replacing y_{m-1} by y_p. Real and imaginary components of a_m, b_m, c_m and d_m are automatically adjusted during the satisfaction of continuity and boundary conditions.

Stress and displacement components at the top ($x_2=y_m$) and bottom ($x_2=y_{m-1}$) of the layer can be defined as

$$\begin{aligned}
\sigma_{12}\big|_{x_2=y_m} &= \mu_m\left\{2\phi_{m,12} + \psi_{m,22} - \psi_{m,11}\right\}_{x_2=y_m} \\
&= \mu_m\left\{2k\eta_m\left(a_m E_m^{-1} - b_m E_m\right) + \left(2k^2 - k_{Sm}^2\right)\left(c_m B_m^{-1} + d_m B_m\right)\right\}e^{ikx_1} \\
\sigma_{12}\big|_{x_2=y_{m-1}} &= \mu_m\left\{2\phi_{m,12} + \psi_{m,22} - \psi_{m,11}\right\}_{x_2=y_{m-1}} \\
&= \mu_m\left\{2k\eta_m\left(a_m - b_m\right) + \left(2k^2 - k_{Sm}^2\right)\left(c_m + d_m\right)\right\}e^{ikx_1} \\
\sigma_{22}\big|_{x_2=y_m} &= -\mu_m\left\{k_{Sm}^2\phi_m + 2\left(\psi_{m,12} + \phi_{m,11}\right)\right\}_{x_2=y_m} \\
&= -\mu_m\left\{\left(k_{Sm}^2 - 2k^2\right)\left(a_m E_m^{-1} + b_m E_m\right)\right. \\
&\quad \left. +2k\beta_m\left(c_m B_m^{-1} - d_m B_m\right)\right\}e^{ikx_1} \\
\sigma_{22}\big|_{x_2=y_{m-1}} &= -\mu_m\left\{k_{Sm}^2\phi_m + 2\left(\psi_{m,12} + \phi_{m,11}\right)\right\}_{x_2=y_{m-1}} \\
&= -\mu_m\left\{\left(k_{Sm}^2 - 2k^2\right)\left(a_m + b_m\right) + 2k\beta_m\left(c_m - d_m\right)\right\}e^{ikx_1} \\
u_2\big|_{x_2=y_m} &= \left\{\phi_{m,2} - \psi_{m,1}\right\}_{x_2=y_m} \\
&= \left\{-i\eta_m\left(a_m E_m^{-1} - b_m E_m\right) - ik\left(c_m B_m^{-1} + d_m B_m\right)\right\}e^{ikx_1} \\
u_2\big|_{x_2=y_{m-1}} &= \left\{\phi_{m,2} - \psi_{m,1}\right\}_{x_2=y_{m-1}} \\
&= \left\{-i\eta_m\left(a_m - b_m\right) - ik\left(c_m + d_m\right)\right\}e^{ikx_1}
\end{aligned} \tag{2.33}$$

$$u_1\big|_{x_2=y_m} = \{\phi_{m,1} + \psi_{m,2}\}_{x_2=y_m}$$

$$= \left\{ik\left(a_m E_m^{-1} + b_m E_m\right) + i\eta_m\left(-c_m B_m^{-1} + d_m B_m\right)\right\}e^{ikx_1}$$

$$u_1\big|_{x_2=y_{m-1}} = \{\phi_{m,1} + \psi_{m,2}\}_{x_2=y_{m-1}}$$

$$= \left\{ik\left(a_m + b_m\right) + i\eta_m\left(-c_m + d_m\right)\right\}e^{ikx_1}$$

where

$$E_m = e^{i\eta_m h_m}$$
$$B_m = e^{i\beta_m h_m} \tag{2.34}$$

From Eq. (2.33) displacement and stress components at the top and bottom of the m-th layer can be written in terms of the wave amplitudes, in the following matrix form:

$$\begin{Bmatrix} u_1 \\ u_2 \\ \sigma_{22} \\ \sigma_{12} \end{Bmatrix}_{x_2=y_m} = \begin{bmatrix} ikE_m^{-1} & ikE_m & -i\eta_m B_m^{-1} & i\eta_m B_m \\ -i\eta_m E_m^{-1} & i\eta_m E_m & -ikB_m^{-1} & -ikB_m \\ \mu_m\left(2k^2 - k_{Sm}^2\right)E_m^{-1} & \mu_m\left(2k^2 - k_{Sm}^2\right)E_m & -2k\mu_m\beta_m B_m^{-1} & 2k\mu_m\beta_m B_m \\ 2k\mu_m\eta_m E_m^{-1} & -2k\mu_m\eta_m E_m & \mu_m\left(2k^2 - k_{Sm}^2\right)B_m^{-1} & \mu_m\left(2k^2 - k_{Sm}^2\right)B_m \end{bmatrix} \begin{Bmatrix} a_m \\ b_m \\ c_m \\ d_m \end{Bmatrix} \tag{2.35a}$$

and

$$\begin{Bmatrix} u_1 \\ u_2 \\ \sigma_{22} \\ \sigma_{12} \end{Bmatrix}_{x_2=y_{m-1}} = \begin{bmatrix} ik & ik & -i\eta_m & i\eta_m \\ -i\eta_m & i\eta_m & -ik & -ik \\ \mu_m\left(2k^2 - k_{Sm}^2\right) & \mu_m\left(2k^2 - k_{Sm}^2\right) & -2k\mu_m\beta_m & 2k\mu_m\beta_m \\ 2k\mu_m\eta_m & -2k\mu_m\eta_m & \mu_m\left(2k^2 - k_{Sm}^2\right) & \mu_m\left(2k^2 - k_{Sm}^2\right) \end{bmatrix} \begin{Bmatrix} a_m \\ b_m \\ c_m \\ d_m \end{Bmatrix} \tag{2.35b}$$

Equations (2.35a) and (2.35b) can be rewritten as

$$\{S_m\} = [G_m]\{C_m\} \tag{2.36a}$$

$$\{S_{m-1}\} = [H_m]\{C_m\} \tag{2.36b}$$

Where $\{S_m\}$ and $\{S_{m-1}\}$ are 4×1 stress–displacement vectors at $x_2 = y_m$ and y_{m-1}, respectively, as given on the left-hand sides of Eq. (2.35), $\{C_m\}$ is the 4×1 coefficient vector, $\begin{bmatrix} a_m & b_m & c_m & d_m \end{bmatrix}^T$, and $[G_m]$, $[H_m]$ are 4×4 square matrices given on the right-hand sides of Eqs. (2.35a) and (2.35b), respectively.

From Eq. (2.36) we can write

$$\{S_m\} = [G_m]\{C_m\} = [G_m][H_m]^{-1}\{S_{m-1}\} = [A_m]\{S_{m-1}\} \tag{2.37}$$

Eq. (2.37) relates the stress–displacement components at the top and bottom of the m-th layer. The matrix $[A_m] = \left([G_m][H_m]^{-1}\right)$ is known as the Layer matrix or Propagator matrix (Thomson, 1950; Haskell, 1953; Kennett, 1983; Kundu and Mal, 1985).

Similarly, for the $m+1$st layer it is possible to write

$$\{S_{m+1}\} = [G_{m+1}]\{C_{m+1}\} \tag{2.38a}$$

$$\{S_m\} = [H_{m+1}]\{C_{m+1}\} \tag{2.38b}$$

$$\{S_{m+1}\} = [A_{m+1}]\{S_m\} \tag{2.38c}$$

Note that from the displacement and stress continuity conditions across the interface at $x_2 = y_m$, $\{S_m\}$ of Eqs. (2.37) and (2.38) must be identical. Therefore,

$$\{S_{m+1}\} = [A_{m+1}]\{S_m\} = [A_{m+1}][A_m]\{S_{m-1}\} \tag{2.39}$$

In this manner, the stress–displacement vector at the top boundary can be related to that at the bottom boundary:

$$\{S_n\} = [A_n][A_{n-1}][A_{n-2}]......[A_2][A_1]\{S_0\} = [J]\{S_0\} \tag{2.40}$$

In Eq. (2.40) $[J]$ is a 4×4 matrix obtained by multiplying n number of layer matrices. Eq. (2.40) is written below in expanded matrix form and the stress-free boundary conditions are enforced. Unknown displacement components at the top and bottom boundaries are denoted as U_n, V_n, U_0 and V_0, as shown below.

$$\begin{Bmatrix} u_1 \\ u_2 \\ \sigma_{22} \\ \sigma_{12} \end{Bmatrix}_{x_2=y_n} = \begin{bmatrix} J_{11} & J_{12} & J_{13} & J_{14} \\ J_{21} & J_{22} & J_{23} & J_{24} \\ J_{31} & J_{32} & J_{33} & J_{34} \\ J_{41} & J_{42} & J_{43} & J_{44} \end{bmatrix} \begin{Bmatrix} u_1 \\ u_2 \\ \sigma_{22} \\ \sigma_{12} \end{Bmatrix}_{x_2=y_0}$$

$$\Rightarrow \begin{Bmatrix} U_n \\ V_n \\ 0 \\ 0 \end{Bmatrix} = \begin{bmatrix} J_{11} & J_{12} & J_{13} & J_{14} \\ J_{21} & J_{22} & J_{23} & J_{24} \\ J_{31} & J_{32} & J_{33} & J_{34} \\ J_{41} & J_{42} & J_{43} & J_{44} \end{bmatrix} \begin{Bmatrix} U_0 \\ V_0 \\ 0 \\ 0 \end{Bmatrix} \tag{2.41}$$

From Eq. (2.41) four algebraic equations with four unknowns U_n, V_n, U_0 and V_0, can be written in the following form:

$$\begin{bmatrix} -1 & 0 & J_{11} & J_{12} \\ 0 & -1 & J_{21} & J_{22} \\ 0 & 0 & J_{31} & J_{32} \\ 0 & 0 & J_{41} & J_{42} \end{bmatrix} \begin{Bmatrix} U_n \\ V_n \\ U_0 \\ V_0 \end{Bmatrix} = \begin{Bmatrix} 0 \\ 0 \\ 0 \\ 0 \end{Bmatrix} \tag{2.42}$$

If the Lamb wave propagates through the plate, then the displacement components at the top and bottom boundaries should not be equal to zero; in other words, the above system of equations must have a nontrivial solution. Therefore the determinant of the coefficient matrix must be zero, as shown below:

$$J_{31}J_{42} - J_{32}J_{41} = 0 \tag{2.43}$$

Equation (2.43) is the dispersion equation for the guided wave propagation through the n-layered plate in a vacuum, as shown in Figure 2.28a.

By assigning a nonzero value to one of the four unknown quantities (U_n, V_n, U_0 and V_0), the remaining three can be obtained from Eq. (2.43). Then the stress–displacement vector $\{S_m\}$ can be computed at any interface $x_2 = y_m$ using the relation $\{S_m\} = [A_m][A_{m-1}]......[A_2][A_1]\{S_0\}$. To obtain the wave amplitudes inside the m-th layer Eq. (2.36) can be used. In this manner a_m, b_m, c_m and d_m are obtained; then the stress and displacement variations (mode shapes) inside the plate are obtained from Eq. (2.33).

2.5.1.1 Numerical Instability

The method, described above, is known as the transfer matrix method, the propagator matrix method, or the Thomson-Haskell matrix method, after Thomson (1950) and Haskell (1953) who first proposed this technique. This technique works well as long as the product of the plate thickness and the wave frequency is not very large.

When this product becomes large then numerical instability occurs. The reason for this numerical instability is that in Eq. (2.43) the difference between two very large but close numbers is computed. As the frequency times thickness increases these two numbers become larger but remain close to each other, and a good number of significant figures are lost when the subtraction operation is carried out. Note that when the number of significant figures lost exceeds 32, then even the double precision computations are not accurate enough to solve the dispersion equation or mode shapes. The numerical precision problem can be removed by means of the delta-matrix operation, a submatrix manipulation technique developed by Dunkin (1965), Dunkin and Corbin (1970), Schwab and Knopoff (1970), and Kundu and Mal (1985).

2.5.1.2 Global Matrix Method

The multilayered plate problem can be solved by an alternate matrix formulation known as the global matrix method (Knopoff, 1964; Mal, 1988). In this approach the numerical precision problem is avoided by following the formulation suggested by Mal (1988).

As mentioned earlier, the source of the numerical problem is the growing exponential terms in Eq. (2.33) and in subsequent equations. From Eq. (2.34) one can see that E_m^{-1} and B_m^{-1} grow exponentially when η_m and β_m become imaginary. It is possible to bypass this problem of growing exponential terms by defining the wave potentials in the m-th layer in the following form:

$$\begin{aligned}
\phi_m &= \left\{a_m e^{-i\eta_m(x_2-y_m)} + b_m e^{i\eta_m(x_2-y_{m-1})}\right\}e^{ikx_1} \\
\psi_m &= \left\{c_m e^{-i\beta_m(x_2-y_m)} + d_m e^{i\beta_m(x_2-y_{m-1})}\right\}e^{ikx_1} \\
\eta_m &= \sqrt{k_{Pm}^2 - k^2} = i\sqrt{k^2 - k_{Pm}^2} \\
\beta_m &= \sqrt{k_{Sm}^2 - k^2} = i\sqrt{k^2 - k_{Sm}^2}
\end{aligned} \tag{2.44}$$

Note that the first terms of the potential expressions of Eq. (2.44) are defined slightly differently from their counter parts in Eq. (2.32). With this definition, for $y_{m-1}<x_2<y_m$ all exponential terms decay instead of growing when η_m and β_m become imaginary.

With these potential expressions the stress and displacement components given in Eq. (2.33) are changed to

$$\begin{aligned}
\sigma_{12}\big|_{x_2=y_m} &= \mu_m\left\{2k\eta_m(a_m - b_mE_m) + (2k^2 - k_{Sm}^2)(c_m + d_mB_m)\right\}e^{ikx_1} \\
\sigma_{12}\big|_{x_2=y_{m-1}} &= \mu_m\left\{2k\eta_m(a_mE_m - b_m) + (2k^2 - k_{Sm}^2)(c_mB_m + d_m)\right\}e^{ikx_1} \\
\sigma_{22}\big|_{x_2=y_m} &= -\mu_m\left\{(k_{Sm}^2 - 2k^2)(a_m + b_mE_m) + 2k\beta_m(c_m - d_mB_m)\right\}e^{ikx_1} \\
\sigma_{22}\big|_{x_2=y_{m-1}} &= -\mu_m\left\{(k_{Sm}^2 - 2k^2)(a_mE_m + b_m) + 2k\beta_m(c_mB_m - d_m)\right\}e^{ikx_1} \\
u_2\big|_{x_2=y_m} &= \left\{-i\eta_m(a_m - b_mE_m) - ik(c_m + d_mB_m)\right\}e^{ikx_1} \\
u_2\big|_{x_2=y_{m-1}} &= \left\{-i\eta_m(a_mE_m - b_m) - ik(c_mB_m + d_m)\right\}e^{ikx_1} \\
u_1\big|_{x_2=y_m} &= \left\{ik(a_m + b_mE_m) + i\eta_m(-c_m + d_mB_m)\right\}e^{ikx_1} \\
u_1\big|_{x_2=y_{m-1}} &= \left\{ik(a_mE_m + b_m) + i\eta_m(-c_mB_m + d_m)\right\}e^{ikx_1}
\end{aligned} \tag{2.45}$$

E_m and B_m are defined in Eq. (2.34).

From Eq. (2.45), displacement and stress components at the top and bottom of the m-th layer can be written in terms of the wave amplitudes, in the following matrix form:

$$\begin{Bmatrix} u_1 \\ u_2 \\ \sigma_{22} \\ \sigma_{12} \end{Bmatrix}_{x_2=y_m} = \begin{bmatrix} ik & ikE_m & -i\eta_m & i\eta_m B_m \\ -i\eta_m & i\eta_m E_m & -ik & -ikB_m \\ \mu_m\left(2k^2-k_{Sm}^2\right) & \mu_m\left(2k^2-k_{Sm}^2\right)E_m & -2k\mu_m\beta_m & 2k\mu_m\beta_m B_m \\ 2k\mu_m\eta_m & -2k\mu_m\eta_m E_m & \mu_m\left(2k^2-k_{Sm}^2\right) & \mu_m\left(2k^2-k_{Sm}^2\right)B_m \end{bmatrix} \begin{Bmatrix} a_m \\ b_m \\ c_m \\ d_m \end{Bmatrix} \quad (2.46a)$$

and

$$\begin{Bmatrix} u_1 \\ u_2 \\ \sigma_{22} \\ \sigma_{12} \end{Bmatrix}_{x_2=y_{m-1}} = \begin{bmatrix} ikE_m & ik & -i\eta_m B_m & i\eta_m \\ -i\eta_m E_m & i\eta_m & -ikB_m & -ik \\ \mu_m\left(2k^2-k_{Sm}^2\right)E_m & \mu_m\left(2k^2-k_{Sm}^2\right) & -2k\mu_m\beta_m B_m & 2k\mu_m\beta_m \\ 2k\mu_m\eta_m E_m & -2k\mu_m\eta_m & \mu_m\left(2k^2-k_{Sm}^2\right)B_m & \mu_m\left(2k^2-k_{Sm}^2\right) \end{bmatrix} \begin{Bmatrix} a_m \\ b_m \\ c_m \\ d_m \end{Bmatrix} \quad (2.46b)$$

Equations (2.46a) and (2.46b) can be written in the short form as given in Eqs. (2.36a) and (2.36b). Similarly, for the m+1st layer, Eqs. (2.38a) and (2.38b) are obtained. The only difference here is that no terms of $[G]$ and $[H]$ matrices contain E_m^{-1} or B_m^{-1}.

From the stress–displacement continuity conditions across the interface

$$\begin{aligned} &\{S_m\} = [G_m]\{C_m\} = [H_{m+1}]\{C_{m+1}\} \\ &\Rightarrow [G_m]\{C_m\} - [H_{m+1}]\{C_{m+1}\} = \{0\} \qquad \text{for } m = 1,2,3,.....(n-1) \end{aligned} \quad (2.47)$$

From the stress-free boundary conditions at $x_2 = y_0$ and y_n we get

$$\begin{aligned} \{S_0\} = [H_1]\{C_1\} = \begin{Bmatrix} U_0 \\ V_0 \\ 0 \\ 0 \end{Bmatrix} \\ \{S_n\} = [G_n]\{C_n\} = \begin{Bmatrix} U_n \\ V_n \\ 0 \\ 0 \end{Bmatrix} \end{aligned} \quad (2.48)$$

In Eq. (2.48) surface displacements U_0, V_0, U_n and V_n are unknowns. Moving all unknowns to the left-hand side of Eq. (2.48), it can be rewritten in the following form:

$$\begin{bmatrix} -1 & 0 & ikE_1 & ik & -i\eta_1 B_1 & i\eta_1 \\ 0 & -1 & -i\eta_1 E_1 & i\eta_1 & -ikB_1 & -ik \\ 0 & 0 & \mu_1\left(2k^2-k_{S1}^2\right)E_1 & \mu_1\left(2k^2-k_{S1}^2\right) & -2k\mu_1\beta_1 B_1 & 2k\mu_1\beta_1 \\ 0 & 0 & 2k\mu_1\eta_1 E_1 & -2k\mu_1\eta_1 & \mu_1\left(2k^2-k_{S1}^2\right)B_1 & \mu_1\left(2k^2-k_{S1}^2\right) \end{bmatrix} \begin{Bmatrix} U_0 \\ V_0 \\ a_1 \\ b_1 \\ c_1 \\ d_1 \end{Bmatrix} = \begin{Bmatrix} 0 \\ 0 \\ 0 \\ 0 \end{Bmatrix} \quad (2.49a)$$

$$\begin{bmatrix} ik & ikE_n & -i\eta_n & i\eta_n B_n & -1 & 0 \\ -i\eta_n & i\eta_n E_n & -ik & -ikB_n & 0 & -1 \\ \mu_n\left(2k^2-k_{Sn}^2\right) & \mu_n\left(2k^2-k_{Sn}^2\right)E_n & -2k\mu_n\beta_n & 2k\mu_n\beta_n B_n & 0 & 0 \\ 2k\mu_n\beta_n & -2k\mu_n\beta_n E_n & \mu_m\left(2k^2-k_{Sm}^2\right) & \mu_n\left(2k^2-k_{Sn}^2\right)B_n & 0 & 0 \end{bmatrix} \begin{Bmatrix} a_n \\ b_n \\ c_n \\ d_n \\ U_n \\ V_n \end{Bmatrix} = \begin{Bmatrix} 0 \\ 0 \\ 0 \\ 0 \end{Bmatrix} \quad (2.49b)$$

Note that there are a total of eight algebraic equations in two matrix equations of (2.49), and a total of $4(n-1)$ algebraic equations in the $(n-1)$ matrix equations of (2.47). The total number of unknown parameters is $(4n+4)$; these are U_0, V_0, U_n, V_n and a_m, b_m, c_m, d_m ($m=1, 2, 3, \dots n$). This system of $4(n+1)$ algebraic equations can be written in the matrix form as given below.

$$\begin{bmatrix} [H_1^*]_{4\times6} & [0]_{4\times4} & [0]_{4\times4} & [0]_{4\times4} & \cdots & [0]_{4\times4} & [0]_{4\times4} & [0]_{4\times6} \\ \left[[0]_{4\times2} \vdots [G_1]_{4\times4}\right] & -[H_2]_{4\times4} & [0]_{4\times4} & [0]_{4\times4} & \cdots & [0]_{4\times4} & [0]_{4\times4} & [0]_{4\times6} \\ [0]_{4\times6} & [G_2]_{4\times4} & -[H_3]_{4\times4} & [0]_{4\times4} & \cdots & [0]_{4\times4} & [0]_{4\times4} & [0]_{4\times6} \\ [0]_{4\times6} & [0]_{4\times4} & [G_3]_{4\times4} & -[H_4]_{4\times4} & \cdots & [0]_{4\times4} & [0]_{4\times4} & [0]_{4\times6} \\ \cdots & \cdots & \cdots & \cdots & \cdots & \cdots & \cdots & \cdots \\ [0]_{4\times6} & [0]_{4\times4} & [0]_{4\times4} & [0]_{4\times4} & \cdots & [G_{n-2}]_{4\times4} & -[H_{n-1}]_{4\times4} & [0]_{4\times6} \\ [0]_{4\times6} & [0]_{4\times4} & [0]_{4\times4} & [0]_{4\times4} & \cdots & [0]_{4\times4} & [G_{n-1}]_{4\times4} & -\left[[H_n]_{4\times4} \vdots [0]_{4\times2}\right] \\ [0]_{4\times6} & [0]_{4\times4} & [0]_{4\times4} & [0]_{4\times4} & \cdots & [0]_{4\times4} & [0]_{4\times4} & [G_n^*]_{4\times6} \end{bmatrix}_{4(n+1)\times4(n+1)} \quad (2.50)$$

$$\times \begin{Bmatrix} \begin{Bmatrix} U_0 \\ V_0 \end{Bmatrix}_{2\times1} \\ \{C_1\}_{4\times1} \\ \{C_2\}_{4\times1} \\ \{C_3\}_{4\times1} \\ \cdots \\ \{C_{n-1}\}_{4\times1} \\ \{C_n\}_{4\times1} \\ \begin{Bmatrix} U_n \\ V_n \end{Bmatrix}_{2\times1} \end{Bmatrix}_{4(n+1)\times1} = \begin{Bmatrix} 0 \\ 0 \\ 0 \\ 0 \\ \cdots \\ 0 \\ 0 \\ 0 \end{Bmatrix}_{4(n+1)\times1}$$

In Eq. (2.50), $\left[H_1^*\right]$ and $\left[G_n^*\right]$ are 4×6 coefficient matrices given in Eqs. (2.49a) and (2.49b), respectively. Expressions of $[G_m]$, $[H_m]$ and $\{C_m\}$ are given on the right-hand side of Eq. (2.46).

For a nontrivial solution of the above system of homogeneous equations (also known as the global system of equations), the determinant of the banded square matrix of Eq. (2.50) must be equal to zero. By equating the determinant to zero the dispersion equation is obtained. Then a unit value is assigned to one of the unknowns (for example, U_0 may be assumed to be 1) and the remaining $(4n+3)$ unknowns can be solved. After solving the wave amplitudes a_m, b_m, c_m, d_m ($m=1, 2, 3, \dots n$), the displacement and stress fields at any point can be computed from the displacement–potential and stress–potential relations.

2.5.2 *n*-Layered Plates in a Fluid

The problem geometry of the n-layered plate in a fluid is shown in Figure 2.28b. The plate dimensions and material properties are identical to those given in Section 2.5.1. The only difference here is that the plate is immersed in a fluid. Fluid properties are denoted by the subscript f.

The potential field in the fluid is given by:

$$\phi_{fL} = Te^{-i\eta_f x_2} e^{ikx_1}$$
$$\phi_{fU} = Re^{i\eta_f (x_2 - y_n)} e^{ikx_1} \tag{2.51}$$
$$\eta_f = \sqrt{k_f^2 - k^2} = i\sqrt{k^2 - k_f^2}$$

Subscripts L and U correspond to the lower and upper fluid half-spaces, respectively. k_f is the wave number in the fluid. Normal stress and displacement components at the fluid–solid boundaries are obtained from the above potential fields:

$$\begin{aligned} \sigma_{22}\big|_{x_2=y_n} &= -\rho_f\omega^2\{\phi_{fU}\}_{x_2=y_n} = -\rho_f\omega^2 Re^{ikx_1} \\ \sigma_{22}\big|_{x_2=y_0} &= -\rho_f\omega^2\{\phi_{fL}\}_{x_2=y_0} = -\rho_f\omega^2 Te^{ikx_1} \\ u_2\big|_{x_2=y_n} &= \{\phi_{fU,2}\}_{x_2=y_n} = i\eta_f Re^{ikx_1} \\ u_2\big|_{x_2=y_0} &= \{\phi_{fL,2}\}_{x_2=y_0} = -i\eta_f Te^{ikx_1} \end{aligned} \tag{2.52}$$

Wave potentials inside the m-th layer are given in Eq. (2.32). Subsequent derivation [up to Eq. (2.40)] remains the same for this problem as well. However, Eq. (2.41) should be different in this case because the normal stress components at the fluid–solid interfaces are not equal to zero. Equating the vertical displacement and normal stress components at the fluid–solid interfaces, computed from the wave potentials in the fluid and solid, the following matrix equation is obtained:

$$\begin{aligned} &\begin{Bmatrix} u_1 \\ u_2 \\ \sigma_{22} \\ \sigma_{12} \end{Bmatrix}_{x_2=y_n} = \begin{bmatrix} J_{11} & J_{12} & J_{13} & J_{14} \\ J_{21} & J_{22} & J_{23} & J_{24} \\ J_{31} & J_{32} & J_{33} & J_{34} \\ J_{41} & J_{42} & J_{43} & J_{44} \end{bmatrix} \begin{Bmatrix} u_1 \\ u_2 \\ \sigma_{22} \\ \sigma_{12} \end{Bmatrix}_{x_2=y_0} \\ \Rightarrow &\begin{Bmatrix} U_n \\ i\eta_f R \\ -\rho_f\omega^2 R \\ 0 \end{Bmatrix} = \begin{bmatrix} J_{11} & J_{12} & J_{13} & J_{14} \\ J_{21} & J_{22} & J_{23} & J_{24} \\ J_{31} & J_{32} & J_{33} & J_{34} \\ J_{41} & J_{42} & J_{43} & J_{44} \end{bmatrix} \begin{Bmatrix} U_0 \\ -i\eta_f T \\ -\rho_f\omega^2 T \\ 0 \end{Bmatrix} \end{aligned} \tag{2.53}$$

Note that the shear stress component is zero at the top and bottom surfaces of the plate because a perfect fluid cannot have any shear stress. Horizontal displacement components, U_0 and U_n, in the solid plate are unknowns; they are not necessarily equal to the horizontal displacement components in the fluid because of the possibility of slippage occurring between the fluid and the solid particles at the fluid–solid interface. Equation (2.53) has four unknowns, R, T, U_0 and U_n, and can be rearranged in the following manner:

$$\begin{bmatrix} 0 & -(i\eta_f J_{12} + \rho_f\omega^2 J_{13}) & J_{11} & -1 \\ -i\eta_f & -(i\eta_f J_{22} + \rho_f\omega^2 J_{23}) & J_{21} & 0 \\ \rho_f\omega^2 & -(i\eta_f J_{32} + \rho_f\omega^2 J_{33}) & J_{31} & 0 \\ 0 & -(i\eta_f J_{42} + \rho_f\omega^2 J_{43}) & J_{41} & 0 \end{bmatrix} \begin{Bmatrix} R \\ T \\ U_0 \\ U_n \end{Bmatrix} = \begin{Bmatrix} 0 \\ 0 \\ 0 \\ 0 \end{Bmatrix} \tag{2.54}$$

For a nontrivial solution of the above system of homogeneous equations, the determinant of the coefficient matrix must be equal to zero. Thus we get

$$\begin{aligned} &i\eta_f\left[J_{31}\left(i\eta_f J_{42}+\rho_f\omega^2 J_{43}\right)-J_{41}\left(i\eta_f J_{32}+\rho_f\omega^2 J_{33}\right)\right] \\ &+\rho_f\omega^2\left[J_{21}\left(i\eta_f J_{42}+\rho_f\omega^2 J_{43}\right)-J_{41}\left(i\eta_f J_{22}+\rho_f\omega^2 J_{23}\right)\right]=0 \end{aligned} \tag{2.55}$$

Equation (2.55) is the dispersion equation for the leaky Lamb wave propagation in an n-layered plate immersed in a fluid.

As in the previous section, after solving the dispersion equation and assuming $U_0=1$, the other three unknown quantities R, T and U_n can be obtained from Eq. (2.54). Then the stress–displacement vector $\{S_m\}$ can be computed at any interface $x_2=y_m$ using the relation $\{S_m\}=[A_m][A_{m-1}]......[A_2][A_1]\{S_0\}$. After evaluating $\{S_m\}$, Eq. (2.36) is used to obtain the wave amplitudes inside the m-th layer. Thus a_m, b_m, c_m and d_m are obtained. The stress and displacement variations (mode shapes) inside the plate can be obtained from the displacement–potential and stress–potential relations, as given in Eq. (2.33).

Example 2.2

Prove that the dispersion equation (2.42) is a special case of the dispersion equation (2.55). In other words, derive Eq. (2.42) from Eq. (2.55).

SOLUTION

Eq. (2.42) is for a plate in a vacuum while Eq. (2.55) is for a plate in a fluid medium. Therefore, as the fluid property approaches the vacuum property, Eq. (2.55) should approach Eq. (2.42). Let us assume that the fluid density (ρ_f) and the acoustic wave speed in the fluid (c_f) are both very small, close to zero. In other words, the fluid is almost like the vacuum. Then the wave number ($k_f=\omega/c_f$) should be very large and we see that

$$\eta_f=\sqrt{k_f^2-k^2}=k_f\sqrt{1-\frac{k^2}{k_f^2}}\neq 0$$

Substituting $\rho_f=0$ into Eq. (2.55) we get

$$\begin{aligned} &i\eta_f\left[J_{31}\left(i\eta_f J_{42}+\rho_f\omega^2 J_{43}\right)-J_{41}\left(i\eta_f J_{32}+\rho_f\omega^2 J_{33}\right)\right] \\ &\qquad+\rho_f\omega^2\left[J_{21}\left(i\eta_f J_{42}+\rho_f\omega^2 J_{43}\right)-J_{41}\left(i\eta_f J_{22}+\rho_f\omega^2 J_{23}\right)\right] \\ &=i\eta_f\left[J_{31}\left(i\eta_f J_{42}\right)-J_{41}\left(i\eta_f J_{32}\right)\right] \\ &=-\eta_f^2\left[J_{31}J_{42}-J_{32}J_{41}\right]=0 \end{aligned}$$

Since η_f is not equal to zero

$$J_{31}J_{42}-J_{32}J_{41}=0$$

2.5.2.1 *Global Matrix Method*

As mentioned in Section 2.5.1, numerical instability occurs for large values of frequency times the plate thickness. This numerical precision problem can be avoided by following the delta-matrix manipulation (Dunkin and Corbin, 1970; Kundu and Mal, 1985), or global matrix method (Knopoff, 1964; Mal, 1988).

For this problem the first few steps of the global matrix formulation are identical to those of the previous problem. Thus, Eqs. (2.44) to (2.47) are valid in this case also. However, since the boundary conditions are different for this problem, Eq. (2.48) should [see Eq. (2.53)] be changed to

$$\{S_0\}=[H_1]\{C_1\}=\begin{Bmatrix} U_0 \\ -i\eta_f T \\ -\rho_f\omega^2 T \\ 0 \end{Bmatrix}$$

$$\{S_n\}=[G_n]\{C_n\}=\begin{Bmatrix} U_n \\ i\eta_f R \\ -\rho_f\omega^2 R \\ 0 \end{Bmatrix} \tag{2.56}$$

In Eq. (2.56) U_0, U_n, R and T are unknowns. The above equations can be rewritten in the following form:

$$\begin{bmatrix} -1 & 0 & ikE_1 & ik & -i\eta_1 B_1 & i\eta_1 \\ 0 & i\eta_f & -i\eta_1 E_1 & i\eta_1 & -ikB_1 & -ik \\ 0 & \rho_f\omega^2 & \mu_1\left(2k^2-k_{S1}^2\right)E_1 & \mu_1\left(2k^2-k_{S1}^2\right) & -2k\mu_1\beta_1 B_1 & 2k\mu_1\beta_1 \\ 0 & 0 & 2k\mu_1\eta_1 E_1 & -2k\mu_1\eta_1 & \mu_1\left(2k^2-k_{S1}^2\right)B_1 & \mu_1\left(2k^2-k_{S1}^2\right) \end{bmatrix}\begin{Bmatrix} U_0 \\ T \\ a_1 \\ b_1 \\ c_1 \\ d_1 \end{Bmatrix}=\begin{Bmatrix} 0 \\ 0 \\ 0 \\ 0 \end{Bmatrix} \tag{2.57a}$$

$$\begin{bmatrix} ik & ikE_n & -i\eta_n & i\eta_n B_n & -1 & 0 \\ -i\eta_n & i\eta_n E_n & -ik & -ikB_n & 0 & -i\eta_f \\ \mu_n\left(2k^2-k_{Sn}^2\right) & \mu_n\left(2k^2-k_{Sn}^2\right)E_n & -2k\mu_n\beta_n & 2k\mu_n\beta_n B_n & 0 & \rho_f\omega^2 \\ 2k\mu_n\beta_n & -2k\mu_n\beta_n E_n & \mu_m\left(2k^2-k_{Sm}^2\right) & \mu_n\left(2k^2-k_{Sn}^2\right)B_n & 0 & 0 \end{bmatrix}\begin{Bmatrix} a_n \\ b_n \\ c_n \\ d_n \\ U_n \\ R \end{Bmatrix}=\begin{Bmatrix} 0 \\ 0 \\ 0 \\ 0 \end{Bmatrix} \tag{2.57b}$$

Note that there are a total of eight algebraic equations in the two matrix equations of (2.57), and a total of $4(n-1)$ algebraic equations in the $(n-1)$ matrix equations of (2.47). The total number of unknown parameters is $(4n+4)$, these are U_0, U_n, R, T and a_m, b_m, c_m, d_m $(m=1, 2, 3, \ldots n)$. This system of $4(n+1)$ algebraic equations can be written in the matrix form as:

$$\begin{bmatrix} \left[H_1^*\right]_{4\times6} & [0]_{4\times4} & \cdots & [0]_{4\times6} \\ \left[[0]_{4\times2}\vdots[G_1]_{4\times4}\right] & -[H_2]_{4\times4} & \cdots & [0]_{4\times6} \\ \cdots & \cdots & \cdots & \cdots \\ [0]_{4\times6} & [0]_{4\times4} & \cdots & \left[G_n^*\right]_{4\times6} \end{bmatrix}_{4(n+1)\times4(n+1)}\begin{Bmatrix} \begin{Bmatrix} U_0 \\ T \end{Bmatrix}_{2\times1} \\ \{C_1\}_{4\times1} \\ \{C_2\}_{4\times1} \\ \{C_3\}_{4\times1} \\ \cdots \\ \{C_{n-1}\}_{4\times1} \\ \{C_n\}_{4\times1} \\ \begin{Bmatrix} U_n \\ R \end{Bmatrix}_{2\times1} \end{Bmatrix}_{4(n+1)\times1}=\begin{Bmatrix} 0 \\ 0 \\ 0 \\ 0 \\ \cdots \\ 0 \\ 0 \\ 0 \end{Bmatrix}_{4(n+1)\times1} \tag{2.58}$$

The banded square matrix expression of Eq. (2.58) is similar to the one given in Eq. (2.50). However, in Eq. (2.58), $\left[H_1^*\right]$ and $\left[G_n^*\right]$ are 4×6 coefficient matrices that are given in Eqs. (2.57a) and (2.57b), respectively. Note that these expressions differ from the ones given in Eq. (2.49). Expressions for $[G_m]$, $[H_m]$ and $\{C_m\}$ are given on the right-hand side of Eq. (2.46). The unknown vectors of Eqs. (2.58) and (2.50) are also slightly different.

For a nontrivial solution of the above system of homogeneous equations, the determinant of the banded square matrix of Eq. (2.58) must be equal to zero. The dispersion equation is obtained by equating this determinant to zero. Then a unit value is assigned to one of the unknowns (for example, U_0 may be assumed to be 1) and the remaining $(4n+3)$ unknowns can be solved. After solving the wave amplitudes a_m, b_m, c_m, d_m ($m=1, 2, 3, \ldots n$), the displacement and stress fields at any point can be computed from the displacement–potential and stress–potential relations.

2.5.3 *n*-Layered Plate Immersed in a Fluid, and Struck by a Plane P-Wave

In the previous two Sections (2.5.1 and 2.5.2) we have solved the free vibration problem – plate in a vacuum or in a fluid, in absence of any external excitation. In this section we consider the forced vibration problem – a multilayered plate is struck by a plane P-wave of amplitude 1, as shown in Figure 2.28c. This external excitation generates a reflected wave of amplitude R, a transmitted wave of amplitude T, and up-going and down-going waves inside the plate layers, as shown in Figure 2.28c. Let the incident angle (measured from the vertical axis) of the incoming P-wave be θ. Then the horizontal wave number should be $k=k_f \sin\theta$.

The potential field in the fluid, in this case, should be similar to Eq. (2.51), but with an additional term for the incoming P-wave in the upper fluid half-space:

$$
\begin{aligned}
\phi_{fL} &= Te^{-i\eta_f x_2}e^{ikx_1} \\
\phi_{fU} &= \left\{e^{-i\eta_f(x_2-y_n)} + Re^{i\eta_f(x_2-y_n)}\right\}e^{ikx_1} \\
\eta_f &= \sqrt{k_f^2-k^2} = k_f\cos\theta; \quad k = k_f\sin\theta
\end{aligned}
\tag{2.59}
$$

Normal stress and displacement components at the fluid–solid boundaries are obtained from the above potential fields:

$$
\begin{aligned}
\sigma_{22}\big|_{x_2=y_n} &= -\rho_f\omega^2\left\{\phi_{fU}\right\}_{x_2=y_n} = -\rho_f\omega^2(1+R)e^{ikx_1} \\
\sigma_{22}\big|_{x_2=y_0} &= -\rho_f\omega^2\left\{\phi_{fL}\right\}_{x_2=y_0} = -\rho_f\omega^2 Te^{ikx_1} \\
u_2\big|_{x_2=y_n} &= \left\{\phi_{fU,2}\right\}_{x_2=y_n} = i\eta_f(-1+R)e^{ikx_1} \\
u_2\big|_{x_2=y_0} &= \left\{\phi_{fL,2}\right\}_{x_2=y_0} = -i\eta_f Te^{ikx_1}
\end{aligned}
\tag{2.60}
$$

Wave potentials inside the *m*-th layer are given in Eq. (2.32). Subsequent derivation [up to Eq. (2.40)] remains identical for this problem as well. However, Eq. (2.41) should be different in this case because the normal stress components at the fluid–solid interfaces are not equal to zero. Equating the vertical displacement and normal stress components at the fluid–solid interfaces computed from the wave potentials in the fluid and solid, the following matrix equation is obtained,

$$\begin{Bmatrix} u_1 \\ u_2 \\ \sigma_{22} \\ \sigma_{12} \end{Bmatrix}_{x_2=y_n} = \begin{bmatrix} J_{11} & J_{12} & J_{13} & J_{14} \\ J_{21} & J_{22} & J_{23} & J_{24} \\ J_{31} & J_{32} & J_{33} & J_{34} \\ J_{41} & J_{42} & J_{43} & J_{44} \end{bmatrix} \begin{Bmatrix} u_1 \\ u_2 \\ \sigma_{22} \\ \sigma_{12} \end{Bmatrix}_{x_2=y_0}$$

$$\Rightarrow \begin{Bmatrix} U_n \\ i\eta_f(-1+R) \\ -\rho_f\omega^2(1+R) \\ 0 \end{Bmatrix} = \begin{bmatrix} J_{11} & J_{12} & J_{13} & J_{14} \\ J_{21} & J_{22} & J_{23} & J_{24} \\ J_{31} & J_{32} & J_{33} & J_{34} \\ J_{41} & J_{42} & J_{43} & J_{44} \end{bmatrix} \begin{Bmatrix} U_0 \\ -i\eta_f T \\ -\rho_f\omega^2 T \\ 0 \end{Bmatrix} \tag{2.61}$$

Horizontal displacement components, U_0 and U_n, in the solid plate, are unknowns and, not necessarily equal to the horizontal displacement components in the fluid because of the possibility of slippage occurring between the fluid and the solid particles at the fluid–solid interface. Equation (2.61) has four unknowns R, T, U_0 and U_n, and can be rearranged in the following manner:

$$\begin{bmatrix} 0 & -(i\eta_f J_{12} + \rho_f\omega^2 J_{13}) & J_{11} & -1 \\ -i\eta_f & -(i\eta_f J_{22} + \rho_f\omega^2 J_{23}) & J_{21} & 0 \\ \rho_f\omega^2 & -(i\eta_f J_{32} + \rho_f\omega^2 J_{33}) & J_{31} & 0 \\ 0 & -(i\eta_f J_{42} + \rho_f\omega^2 J_{43}) & J_{41} & 0 \end{bmatrix} \begin{Bmatrix} R \\ T \\ U_0 \\ U_n \end{Bmatrix} = \begin{Bmatrix} 0 \\ i\eta_f \\ \rho_f\omega^2 \\ 0 \end{Bmatrix} \tag{2.62}$$

The four unknowns R, T, U_0 and U_n can be obtained from the above system of nonhomogeneous equations. Then the stress–displacement vector $\{S_m\}$ can be computed at any interface $x_2=y_m$ using the relation $\{S_m\}=[A_m][A_{m-1}]......[A_2][A_1]\{S_0\}$. After evaluating $\{S_m\}$, Eq. (2.36) is used to obtain the wave amplitudes inside the m-th layer. Thus, a_m, b_m, c_m and d_m are obtained; then the stress and displacement variations inside the plate can be obtained from the displacement–potential and stress–potential relations, as given in Eq. (2.33).

2.5.3.1 Global Matrix Method

As mentioned in Sections 2.5.1 and 2.5.2, the numerical instability occurs for large frequency times the plate thickness values. This numerical precision problem can be avoided by following the delta-matrix manipulation or global matrix method.

For this problem, the first few steps of the global matrix formulation are identical to those of the previous problems. Equations (2.44) to (2.47) apply to this problem as well. However, since the boundary conditions are different here, Eq. (2.48) should be changed to [see Eq. (2.61)]:

$$\{S_0\}=[H_1]\{C_1\}=\begin{Bmatrix} U_0 \\ -i\eta_f T \\ -\rho_f\omega^2 T \\ 0 \end{Bmatrix}$$

$$\{S_n\}=[G_n]\{C_n\}=\begin{Bmatrix} U_n \\ i\eta_f(-1+R) \\ -\rho_f\omega^2(1+R) \\ 0 \end{Bmatrix} \tag{2.63}$$

In Eq. (2.63) U_0, U_n, R and T are unknown quantities. The above equations can be rewritten in the following form:

$$\begin{bmatrix} -1 & 0 & ikE_1 & ik & -i\eta_1 B_1 & i\eta_1 \\ 0 & i\eta_f & -i\eta_1 E_1 & i\eta_1 & -ikB_1 & -ik \\ 0 & \rho_f\omega^2 & \mu_1\left(2k^2 - k_{S1}^2\right)E_1 & \mu_1\left(2k^2 - k_{S1}^2\right) & -2k\mu_1\beta_1 B_1 & 2k\mu_1\beta_1 \\ 0 & 0 & 2k\mu_1\eta_1 E_1 & -2k\mu_1\eta_1 & \mu_1\left(2k^2 - k_{S1}^2\right)B_1 & \mu_1\left(2k^2 - k_{S1}^2\right) \end{bmatrix} \begin{Bmatrix} U_0 \\ T \\ a_1 \\ b_1 \\ c_1 \\ d_1 \end{Bmatrix} = \begin{Bmatrix} 0 \\ 0 \\ 0 \\ 0 \end{Bmatrix} \tag{2.64a}$$

$$\begin{bmatrix} ik & ikE_n & -i\eta_n & i\eta_n B_n & -1 & 0 \\ -i\eta_n & i\eta_n E_n & -ik & -ikB_n & 0 & -i\eta_f \\ \mu_n\left(2k^2 - k_{Sn}^2\right) & \mu_n\left(2k^2 - k_{Sn}^2\right)E_n & -2k\mu_n\beta_n & 2k\mu_n\beta_n B_n & 0 & \rho_f\omega^2 \\ 2k\mu_n\beta_n & -2k\mu_n\beta_n E_n & \mu_m\left(2k^2 - k_{Sm}^2\right) & \mu_n\left(2k^2 - k_{Sn}^2\right)B_n & 0 & 0 \end{bmatrix} \begin{Bmatrix} a_n \\ b_n \\ c_n \\ d_n \\ U_n \\ R \end{Bmatrix} = \begin{Bmatrix} 0 \\ -i\eta_f \\ -\rho_f\omega^2 \\ 0 \end{Bmatrix} \tag{2.64b}$$

Note that Eq. (2.64a) is identical to Eq. (2.57a) but Eq. (2.64b) differs from Eq. (2.57b). There are a total of eight algebraic equations in the two matrix equations of (2.64), and a total of $4(n-1)$ algebraic equations in $(n-1)$ matrix equations of (2.47). The total number of unknown parameters is $(4n+4)$; these are U_0, U_n, R, T and a_m, b_m, c_m, d_m ($m = 1, 2, 3, \ldots n$). This system of $4(n+1)$ algebraic equations can be written in the matrix form as given below:

$$\begin{bmatrix} \left[H_1^*\right]_{4\times6} & [0]_{4\times4} & \cdots & [0]_{4\times6} \\ \left[[0]_{4\times2} \vdots [G_1]_{4\times4}\right] & -[H_2]_{4\times4} & \cdots & [0]_{4\times6} \\ \cdots & \cdots & \cdots & \cdots \\ [0]_{4\times6} & [0]_{4\times4} & \cdots & \left[G_n^*\right]_{4\times6} \end{bmatrix}_{4(n+1)\times4(n+1)} \begin{Bmatrix} \begin{Bmatrix} U_0 \\ T \end{Bmatrix}_{2\times1} \\ \{C_1\}_{4\times1} \\ \{C_2\}_{4\times1} \\ \{C_3\}_{4\times1} \\ \cdots \\ \{C_{n-1}\}_{4\times1} \\ \{C_n\}_{4\times1} \\ \begin{Bmatrix} U_n \\ R \end{Bmatrix}_{2\times1} \end{Bmatrix}_{4(n+1)\times1} = \begin{Bmatrix} 0 \\ 0 \\ 0 \\ 0 \\ 0 \\ \cdots \\ 0 \\ -i\eta_f \\ -\rho_f\omega^2 \\ 0 \end{Bmatrix}_{4(n+1)\times1} \tag{2.65}$$

The banded square matrix expression of Eq. (2.65) is similar to the one given in Eqs. (2.50) and (2.58). In Eq. (2.65), $\left[H_1^*\right]$ and $\left[G_n^*\right]$ are 4×6 coefficient matrices that are given in Eqs. (2.64a) and (2.64b), respectively. Note that these matrices are identical to those given in Eq. (2.57) but different from the ones given in Eq. (2.49). Expressions of $[G_m]$, $[H_m]$ and $\{C_m\}$ are given on the right-hand side of Eq. (2.46). Unknown vectors of Eqs. (2.65) and (2.58) are the same, but the two vectors on the right-hand side of these two equations differ.

The system of non-homogeneous equations [Eq. (2.65)] can be solved to obtain the $(4n+4)$ unknowns. After solving the wave amplitudes a_m, b_m, c_m, d_m ($m = 1, 2, 3, \ldots n$), the displacement and stress fields at any point can be computed from the displacement–potential and stress–potential relations.

2.6 GUIDED WAVES IN SINGLE AND MULTILAYERED COMPOSITE PLATES

Until now, we have analyzed plates made of isotropic elastic layers only. All those analyses excluded fiber-reinforced composite plates and any plate made of anisotropic layers. In this section, guided wave propagation in a unidirectional fiber-reinforced composite plate and multilayered composite plates, made of anisotropic layers, is studied. In the multilayered plate, fiber direction can vary from one layer to the next.

For this plate problem, like before, it is possible to consider three separate problem geometries, as shown in Figure 2.28 – (a) plate in a vacuum, (b) a plate in a fluid, and (c) a plane P-wave striking a plate in a fluid. However, we have already shown in the previous section (Example 2.2) that the plate in a vacuum is a special case of the plate in a fluid; therefore, it is not necessary to consider the plate in a vacuum problem separately. In the dispersion equation for the plate immersed in a fluid, if we set the fluid density equal to zero, and the acoustic wave speed of the fluid to a small number, then we recover the dispersion equation for the plate in a vacuum. In Section 2.5 it is also shown that the dispersion equation for the free vibration problem (plate in a fluid) can be obtained from the forced vibration problem (plane P-wave striking a plate in a fluid) because the only difference between these two problems is that for the free vibration problem we get a system of homogeneous equations, whereas, for the forced vibration problem, it is a system of non-homogeneous equations. However, the coefficient matrix from which the dispersion equation is derived is identical for the two problems [see Eqs. (2.65) and (2.58), or Eqs. (2.62) and (2.54)]. Therefore, if we analyze the forced vibration problem that is a plane P-wave striking a composite plate immersed in a fluid, then we have the solution for the most general case. The two free vibration problems, plate in a vacuum and plate in a fluid, can be derived from this general solution of the forced vibration problem. This problem is solved following the technique outlined by Mal et al. (1991).

Following Mal's notation, the coordinate axis for this problem geometry is slightly changed from our earlier assumptions. The vertical axis is changed from x_2 to x_3 (positive downwards) and the wave propagation direction is inclined at an angle ϕ with respect to the x_1 axis as shown in Figure 2.29.

The stress–strain relation for a transversely isotropic solid is given in Eqs. (1.63) and (1.64). Those relations are given for the x_3 axis being the axis of symmetry. If the x_1 axis now becomes the axis of symmetry, then, following the reasoning given in Chapter 1, it can be shown that Eq. (1.63) is changed to

$$\begin{Bmatrix} \sigma_{11} \\ \sigma_{22} \\ \sigma_{33} \\ \sigma_{23} \\ \sigma_{31} \\ \sigma_{12} \end{Bmatrix} = \begin{bmatrix} C_{11} & C_{12} & C_{12} & 0 & 0 & 0 \\ & C_{22} & C_{23} & 0 & 0 & 0 \\ & & C_{22} & 0 & 0 & 0 \\ & & & C_{44} & 0 & 0 \\ & \text{symm} & & & C_{55} & 0 \\ & & & & & C_{55} \end{bmatrix} \begin{Bmatrix} \varepsilon_{11} \\ \varepsilon_{22} \\ \varepsilon_{33} \\ 2\varepsilon_{23} \\ 2\varepsilon_{31} \\ 2\varepsilon_{12} \end{Bmatrix} \tag{2.66}$$

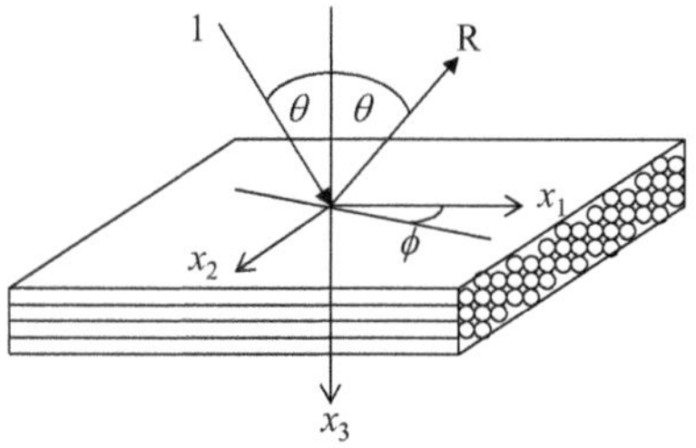

Figure 2.29 Reflection of a plane P-wave by a fiber-reinforced composite plate immersed in a fluid. The fiber direction is the x_1 direction. The plane containing the incident and reflected waves forms an angle ϕ with the x_1 axis.

where

$$C_{44} = \frac{C_{22} - C_{23}}{2} \tag{2.67}$$

This stress–strain relation can be found in any book on mechanics of anisotropic solids (Christensen, 1981). [Also see Eqs. (1.63) and (1.64) of this book for comparison.]

Buchwald (1961) proposed the following displacement–potential relation which is useful for solving this problem:

$$\begin{Bmatrix} u_1 \\ u_2 \\ u_3 \end{Bmatrix} = \begin{bmatrix} \partial/\partial x_1 & 0 & 0 \\ 0 & \partial/\partial x_2 & \partial/\partial x_3 \\ 0 & \partial/\partial x_3 & -\partial/\partial x_2 \end{bmatrix} \begin{Bmatrix} \phi_1 \\ \phi_2 \\ \phi_3 \end{Bmatrix} \tag{2.68}$$

In absence of any body force, the equations of motion (Eq.1.78) become

$$\begin{aligned} \sigma_{11,1} + \sigma_{12,2} + \sigma_{13,3} &= \rho \ddot{u}_1 \\ \sigma_{21,1} + \sigma_{22,2} + \sigma_{23,3} &= \rho \ddot{u}_2 \\ \sigma_{31,1} + \sigma_{32,2} + \sigma_{33,3} &= \rho \ddot{u}_3 \end{aligned} \tag{2.69}$$

Substituting Eq. (2.66) into Eq. (2.69) and specializing the equations for harmonic time dependence $\left(e^{-i\omega t}\right)$ we obtain

$$\begin{aligned} &\left(C_{11}u_{1,11} + C_{12}u_{2,21} + C_{12}u_{3,31}\right) + C_{55}\left(u_{1,22} + u_{2,12}\right) + C_{55}\left(u_{1,33} + u_{3,13}\right) + \rho\omega^2 u_1 = 0 \\ &C_{55}\left(u_{1,21} + u_{2,11}\right) + \left(C_{12}u_{1,12} + C_{22}u_{2,22} + C_{23}u_{3,32}\right) + C_{44}\left(u_{2,33} + u_{3,23}\right) + \rho\omega^2 u_2 = 0 \\ &C_{55}\left(u_{1,31} + u_{3,11}\right) + C_{44}\left(u_{2,32} + u_{3,22}\right) + \left(C_{12}u_{1,13} + C_{23}u_{2,23} + C_{22}u_{3,33}\right) + \rho\omega^2 u_3 = 0 \end{aligned} \tag{2.70}$$

Then substitution of Eq. (2.68) into Eq. (2.70) gives

$$\begin{aligned} &\left(C_{11}\phi_{1,111} + C_{12}\phi_{2,221} + C_{12}\phi_{3,321} + C_{12}\phi_{2,331} - C_{12}\phi_{3,231}\right) \\ &\qquad + C_{55}\left(\phi_{1,122} + \phi_{2,212} + \phi_{3,312}\right) + C_{55}\left(\phi_{1,133} + \phi_{2,313} - \phi_{3,213}\right) + \rho\omega^2\phi_{1,1} = 0 \\ &C_{55}\left(\phi_{1,121} + \phi_{2,211} + \phi_{3,311}\right) + \left(C_{12}\phi_{1,112} + C_{22}\phi_{2,222} + C_{22}\phi_{3,322} + C_{23}\phi_{2,332} - C_{23}\phi_{3,232}\right) \\ &\qquad + C_{44}\left(\phi_{2,233} + \phi_{3,333} + \phi_{2,323} - \phi_{3,223}\right) + \rho\omega^2\phi_{2,2} + \rho\omega^2\phi_{3,3} = 0 \\ &C_{55}\left(\phi_{1,131} + \phi_{2,311} - \phi_{3,211}\right) + C_{44}\left(\phi_{2,232} + \phi_{3,332} + \phi_{2,322} - \phi_{3,222}\right) \\ &\qquad + \left(C_{12}\phi_{1,113} + C_{23}\phi_{2,223} + C_{23}\phi_{3,323} + C_{22}\phi_{2,333} - C_{22}\phi_{3,233}\right) + \rho\omega^2\phi_{2,3} - \rho\omega^2\phi_{3,2} = 0 \end{aligned} \tag{2.71}$$

The first equation of Eq. (2.71) can be written as

$$\begin{aligned} &\Big[\left(C_{11}\phi_{1,11} + C_{12}\phi_{2,22} + C_{12}\phi_{3,32} + C_{12}\phi_{2,33} - C_{12}\phi_{3,23}\right) \\ &\quad + C_{55}\left(\phi_{1,22} + \phi_{2,22} + \phi_{3,32}\right) + C_{55}\left(\phi_{1,33} + \phi_{2,33} - \phi_{3,23}\right) + \rho\omega^2\phi_1\Big]_{,1} = 0 \\ &\Rightarrow \left[\left(C_{11}\phi_{1,11} + C_{12}\phi_{2,22} + C_{12}\phi_{2,33}\right) + C_{55}\left(\phi_{1,22} + \phi_{2,22} + \phi_{1,33} + \phi_{2,33}\right) + \rho\omega^2\phi_1\right]_{,1} = 0 \\ &\Rightarrow \left[\left(C_{11}\phi_{1,11} + C_{12}\nabla_1^2\phi_2\right) + C_{55}\nabla_1^2\left(\phi_1 + \phi_2\right) + \rho\omega^2\phi_1\right]_{,1} = 0 \\ &\Rightarrow \left[\left(C_{11}\phi_{1,11} + C_{55}\nabla_1^2\phi_1 + \rho\omega^2\phi_1\right) + \left(C_{12} + C_{55}\right)\nabla_1^2\phi_2\right]_{,1} = 0 \end{aligned} \tag{2.72}$$

where $\nabla_1^2 = \dfrac{\partial^2}{\partial x_2^2} + \dfrac{\partial^2}{\partial x_3^2}$.

Similarly, the second equation of (2.71) can be rewritten in the following form:

$$\begin{aligned}
&\left[C_{55}\left(\phi_{1,11}+\phi_{2,11}\right)+\left(C_{12}\phi_{1,11}+C_{22}\phi_{2,22}+C_{23}\phi_{2,33}\right)+C_{44}\left(\phi_{2,33}+\phi_{2,33}\right)+\rho\omega^2\phi_2\right]_{,2}\\
&\quad+\left[C_{55}\phi_{3,11}+C_{22}\phi_{3,22}-C_{23}\phi_{3,22}+C_{44}\left(\phi_{3,33}-\phi_{3,22}\right)+\rho\omega^2\phi_3\right]_{,3}=0\\
&\Rightarrow\left[\left(C_{12}+C_{55}\right)\phi_{1,11}+\left(C_{55}\phi_{2,11}\right)+\left(C_{22}\phi_{2,22}+C_{22}\phi_{2,33}-2C_{44}\phi_{2,33}\right)+2C_{44}\phi_{2,33}+\rho\omega^2\phi_2\right]_{,2}\\
&\quad+\left[C_{55}\phi_{3,11}+2C_{44}\phi_{3,22}+C_{44}\left(\phi_{3,33}-\phi_{3,22}\right)+\rho\omega^2\phi_3\right]_{,3}=0\\
&\Rightarrow\left[\left(C_{12}+C_{55}\right)\phi_{1,11}+C_{55}\phi_{2,11}+C_{22}\left(\phi_{2,22}+C_{22}\phi_{2,33}\right)+\rho\omega^2\phi_2\right]_{,2}\\
&\quad+\left[C_{55}\phi_{3,11}+C_{44}\left(\phi_{3,33}+\phi_{3,22}\right)+\rho\omega^2\phi_3\right]_{,3}=0\\
&\Rightarrow\left[\left(C_{12}+C_{55}\right)\phi_{1,11}+C_{55}\phi_{2,11}+C_{22}\nabla_1^2\phi_2+\rho\omega^2\phi_2\right]_{,2}+\left[C_{55}\phi_{3,11}+C_{44}\nabla_1^2\phi_3+\rho\omega^2\phi_3\right]_{,3}=0
\end{aligned}\tag{2.73}$$

and the third equation is rewritten as

$$\begin{aligned}
&C_{55}\left(\phi_{1,131}+\phi_{2,311}-\phi_{3,211}\right)+C_{44}\left(\phi_{2,232}+\phi_{3,332}+\phi_{2,322}-\phi_{3,222}\right)\\
&\quad+\left(C_{12}\phi_{1,113}+C_{23}\phi_{2,223}+C_{23}\phi_{3,323}+C_{22}\phi_{2,333}-C_{22}\phi_{3,233}\right)+\rho\omega^2\phi_{2,3}-\rho\omega^2\phi_{3,2}=0\\
&\left[C_{55}\left(\phi_{1,11}+\phi_{2,11}\right)+C_{44}\left(\phi_{2,22}+\phi_{2,22}\right)+\left(C_{12}\phi_{1,11}+C_{23}\phi_{2,22}+C_{22}\phi_{2,33}\right)+\rho\omega^2\phi_2\right]_{,3}\\
&\quad+\left[-C_{55}\phi_{3,11}+C_{44}\left(\phi_{3,33}-\phi_{3,22}\right)+\left(C_{23}\phi_{3,33}-C_{22}\phi_{3,33}\right)-\rho\omega^2\phi_3\right]_{,2}=0\\
&\left[\left(C_{55}+C_{12}\right)\phi_{1,11}+C_{55}\phi_{2,11}+2C_{44}\phi_{2,22}+\left(C_{22}\phi_{2,22}-2C_{44}\phi_{2,22}+C_{22}\phi_{2,33}\right)+\rho\omega^2\phi_2\right]_{,3}\\
&\quad+\left[-C_{55}\phi_{3,11}+C_{44}\left(\phi_{3,33}-\phi_{3,22}\right)+\left(-2C_{44}\phi_{3,33}\right)-\rho\omega^2\phi_3\right]_{,2}=0\\
&\left[\left(C_{55}+C_{12}\right)\phi_{1,11}+C_{55}\phi_{2,11}+C_{22}\nabla_1^2\phi_2+\rho\omega^2\phi_2\right]_{,3}-\left[C_{55}\phi_{3,11}+C_{44}\nabla_1^2\phi_3+\rho\omega^2\phi_3\right]_{,2}=0
\end{aligned}\tag{2.74}$$

Note that the sufficient conditions for satisfying Eqs. (2.72), (2.73) and (2.74) are

$$\begin{aligned}
&C_{11}\phi_{1,11}+C_{55}\nabla_1^2\phi_1+\rho\omega^2\phi_1+\left(C_{12}+C_{55}\right)\nabla_1^2\phi_2=0\\
&\left(C_{12}+C_{55}\right)\phi_{1,11}+C_{55}\phi_{2,11}+C_{22}\nabla_1^2\phi_2+\rho\omega^2\phi_2=0\\
&C_{55}\phi_{3,11}+C_{44}\nabla_1^2\phi_3+\rho\omega^2\phi_3=0
\end{aligned}\tag{2.75}$$

Eq. (2.75) can be rewritten in the following form:

$$\begin{aligned}
&\left(\frac{C_{55}}{\rho}\nabla_1^2+\frac{C_{11}}{\rho}\frac{\partial^2}{\partial x_1^2}+\omega^2\right)\phi_1+\frac{\left(C_{12}+C_{55}\right)}{\rho}\nabla_1^2\phi_2=0\\
&\frac{\left(C_{12}+C_{55}\right)}{\rho}\frac{\partial^2}{\partial x_1^2}\phi_1+\left(\frac{C_{22}}{\rho}\nabla_1^2+\frac{C_{55}}{\rho}\frac{\partial^2}{\partial x_1^2}+\omega^2\right)\phi_2=0\\
&\left(\frac{C_{44}}{\rho}\nabla_1^2+\frac{C_{55}}{\rho}\frac{\partial^2}{\partial x_1^2}+\omega^2\right)\phi_3=0
\end{aligned}\tag{2.76}$$

Substituting

$$\frac{C_{22}}{\rho}=a_1 \qquad \frac{C_{11}}{\rho}=a_2 \qquad \frac{C_{12}+C_{55}}{\rho}=a_3 \qquad \frac{C_{44}}{\rho}=a_4 \qquad \frac{C_{55}}{\rho}=a_5 \tag{2.77}$$

in Eq. (2.76) we get

$$\begin{gathered}\left(a_5\nabla_1^2+a_2\frac{\partial^2}{\partial x_1^2}+\omega^2\right)\phi_1+a_3\nabla_1^2\phi_2=0\\ a_3\frac{\partial^2}{\partial x_1^2}\phi_1+\left(a_1\nabla_1^2+a_5\frac{\partial^2}{\partial x_1^2}+\omega^2\right)\phi_2=0\\ \left(a_4\nabla_1^2+a_5\frac{\partial^2}{\partial x_1^2}+\omega^2\right)\phi_3=0\end{gathered} \tag{2.78a}$$

Eq. (2.78a) can be written in matrix form:

$$\begin{bmatrix}\left(a_5\nabla_1^2+a_2\frac{\partial^2}{\partial x_1^2}+\omega^2\right) & a_3\nabla_1^2 & 0\\ a_3\frac{\partial^2}{\partial x_1^2} & \left(a_1\nabla_1^2+a_5\frac{\partial^2}{\partial x_1^2}+\omega^2\right) & 0\\ 0 & 0 & \left(a_4\nabla_1^2+a_5\frac{\partial^2}{\partial x_1^2}+\omega^2\right)\end{bmatrix}\begin{Bmatrix}\phi_1\\ \phi_2\\ \phi_3\end{Bmatrix}=\begin{Bmatrix}0\\0\\0\end{Bmatrix} \tag{2.78b}$$

The solution for the above system of differential equations is given by (Mal et al., 1991):

$$\begin{Bmatrix}\phi_1\\ \phi_2\\ \phi_3\end{Bmatrix}=\begin{bmatrix}q_{11} & q_{12} & 0\\ q_{21} & q_{22} & 0\\ 0 & 0 & 1\end{bmatrix}\left(\begin{bmatrix}e^{i\zeta_1x_3} & 0 & 0\\ 0 & e^{i\zeta_2x_3} & 0\\ 0 & 0 & e^{i\zeta_3x_3}\end{bmatrix}\begin{Bmatrix}A_1^+\\ A_2^+\\ A_3^+\end{Bmatrix}\right.$$
$$\left.+\begin{bmatrix}e^{-i\zeta_1x_3} & 0 & 0\\ 0 & e^{-i\zeta_2x_3} & 0\\ 0 & 0 & e^{-i\zeta_3x_3}\end{bmatrix}\begin{Bmatrix}A_1^-\\ A_2^-\\ A_3^-\end{Bmatrix}\right)e^{i(\xi_1x_1+\xi_2x_2)}$$

or,

$$\begin{Bmatrix}\phi_1\\ \phi_2\\ \phi_3\end{Bmatrix}=\begin{bmatrix}q_{11} & q_{12} & 0\\ q_{21} & q_{22} & 0\\ 0 & 0 & 1\end{bmatrix}\begin{Bmatrix}A_1^+e^{i\zeta_1x_3}+A_1^-e^{-i\zeta_1x_3}\\ A_2^+e^{i\zeta_2x_3}+A_2^-e^{-i\zeta_2x_3}\\ A_3^+e^{i\zeta_3x_3}+A_3^-e^{-i\zeta_3x_3}\end{Bmatrix}e^{i(\xi_1x_1+\xi_2x_2)} \tag{2.79}$$

where

$$
\begin{aligned}
q_{11} &= a_3 b_1 \\
q_{12} &= a_3 b_2 \\
q_{21} &= \omega^2 - a_2\xi_1^2 - a_5 b_1 \\
q_{22} &= \omega^2 - a_2\xi_1^2 - a_5 b_2
\end{aligned}
\tag{2.80a}
$$

$$
\begin{aligned}
b_{1,2} &= -\left(\frac{\beta}{2\alpha}\right) \mp \left[\left(\frac{\beta}{2\alpha}\right)^2 - \frac{\gamma}{\alpha}\right]^{1/2} \\
\alpha &= a_1 a_5 \\
\beta &= \left(a_1 a_2 + a_5^2 - a_3^2\right)\xi_1^2 - \omega^2\left(a_1 + a_5\right) \\
\gamma &= \left(a_2\xi_1^2 - \omega^2\right)\left(a_5\xi_1^2 - \omega^2\right)
\end{aligned}
\tag{2.80b}
$$

$$
\begin{aligned}
\zeta_1^2 &= -\xi_2^2 + b_1 \\
\zeta_2^2 &= -\xi_2^2 + b_2 \\
\xi_3^2 &= -\xi_2^2 + \left(\omega^2 - a_5\xi_1^2\right)/a_4
\end{aligned}
\tag{2.80c}
$$

subject to $\operatorname{Im}\left(\zeta_j\right) \geq 0, \quad j = 1, 2, 3.$

From the potential expressions given in Eq. (2.79), the three displacement components and the normal and shear stress components at a surface, $x_3 = \text{constant}$, can be obtained using Eqs. (2.68) and (2.66):

$$
\begin{bmatrix} u_1 & u_2 & u_3 & \sigma_{13} & \sigma_{23} & \sigma_{33} \end{bmatrix}^T = \left\{S(x_3)\right\} e^{i(\xi_1 x_1 + \xi_2 x_2)} = \begin{bmatrix} Q_{11} & Q_{12} \\ Q_{21} & Q_{22} \end{bmatrix} [E] \begin{Bmatrix} A^+ \\ A^- \end{Bmatrix} e^{i(\xi_1 x_1 + \xi_2 x_2)} \tag{2.81}
$$

where

$$
[E] = \begin{bmatrix}
e^{i\zeta_1 x_3} & 0 & 0 & 0 & 0 & 0 \\
0 & e^{i\zeta_2 x_3} & 0 & 0 & 0 & 0 \\
0 & 0 & e^{i\zeta_3 x_3} & 0 & 0 & 0 \\
0 & 0 & 0 & e^{-i\zeta_1 x_3} & 0 & 0 \\
0 & 0 & 0 & 0 & e^{-i\zeta_2 x_3} & 0 \\
0 & 0 & 0 & 0 & 0 & e^{-i\zeta_3 x_3}
\end{bmatrix} \tag{2.82}
$$

$$
\begin{bmatrix} A^+ & A^- \end{bmatrix}^T = \begin{bmatrix} A_1^+ & A_2^+ & A_3^+ & A_1^- & A_2^- & A_3^- \end{bmatrix}^T \tag{2.83}
$$

$$
\begin{aligned}
[Q_{11}] &= \begin{bmatrix}
i\xi_1 q_{11} & i\xi_1 q_{12} & 0 \\
i\xi_2 q_{21} & i\xi_2 q_{22} & i\zeta_3 \\
i\zeta_1 q_{21} & i\zeta_2 q_{22} & -i\xi_2
\end{bmatrix} \\
[Q_{12}] &= \begin{bmatrix}
i\xi_1 q_{11} & i\xi_1 q_{12} & 0 \\
i\xi_2 q_{21} & i\xi_2 q_{22} & -i\zeta_3 \\
-i\zeta_1 q_{21} & -i\zeta_2 q_{22} & -i\xi_2
\end{bmatrix}
\end{aligned}
\tag{2.84}
$$

$$[Q_{21}] = \begin{bmatrix} -\rho a_5\xi_1\zeta_1(q_{11}+q_{21}) & -\rho a_5\xi_1\zeta_2(q_{12}+q_{22}) & \rho a_5\xi_1\xi_2 \\ -2\rho a_4\xi_2\zeta_1 q_{21} & -2\rho a_4\xi_2\zeta_2 q_{22} & \rho a_4(\xi_2^2-\zeta_3^2) \\ \delta_1 & \delta_2 & 2\rho a_4\xi_2\zeta_3 \end{bmatrix}$$

$$[Q_{22}] = \begin{bmatrix} \rho a_5\xi_1\zeta_1(q_{11}+q_{21}) & \rho a_5\xi_1\zeta_2(q_{12}+q_{22}) & \rho a_5\xi_1\xi_2 \\ 2\rho a_4\xi_2\zeta_1 q_{21} & 2\rho a_4\xi_2\zeta_2 q_{22} & \rho a_4(\xi_2^2-\zeta_3^2) \\ \delta_1 & \delta_2 & -2\rho a_4\xi_2\zeta_3 \end{bmatrix}$$

$$\begin{aligned} \delta_1 &= \rho\left[(a_5-a_3)\xi_1^2 q_{11} - (a_1-2a_4)\xi_2^2 q_{21} - a_1\zeta_1^2 q_{21}\right] \\ \delta_2 &= \rho\left[(a_5-a_3)\xi_1^2 q_{12} - (a_1-2a_4)\xi_2^2 q_{22} - a_1\zeta_2^2 q_{22}\right] \end{aligned} \tag{2.84a}$$

The wave field due to the incident wave (see Figure 2.29) is given by

$$e^{i(\xi_1 x_1 + \xi_2 x_2 + \zeta_0 x_3)} \tag{2.85}$$

where

$$\xi_1 = k_0 \sin\theta\cos\phi, \quad \xi_2 = k_0\sin\theta\sin\phi, \quad \zeta_0 = k_0\cos\theta \text{ and } k_0 = \omega/\alpha_0 \tag{2.86}$$

Acoustic wave potentials in the upper and lower fluids are denoted by ϕ_0 and ϕ_b, respectively. Then the displacement and stress components in the fluid are given by

$$\begin{aligned} u_i &= \frac{\partial\phi_\alpha}{\partial x_i}, \qquad i = 1, 2, 3 \\ \sigma_{33} &= -\rho_0\omega^2\phi_\alpha, \qquad \sigma_{13} = \sigma_{23} = 0 \end{aligned} \tag{2.87}$$

where the subscript α is either 0 (for the top fluid) or b (for the bottom fluid).

In terms of the reflection and transmission coefficients, R and T, the wave potentials in the top and the bottom fluid half-spaces are given by

$$\begin{aligned} \phi_0 &= \left(e^{i\zeta_0 x_3} + Re^{-i\zeta_0 x_3}\right)e^{i(\xi_1 x_1 + \xi_2 x_2)} \\ \phi_b &= Te^{i[\zeta_0(x_3-H)+\xi_1 x_1 + \xi_2 x_2]} \end{aligned} \tag{2.88}$$

Normal displacement and stress components should be continuous across the top and bottom interfaces, between the fluid half-space and the plate. Shear stresses should vanish at the interfaces. Fluid and solid can have different horizontal displacement components across an interface. Therefore, the displacement and stress components at the top and bottom surfaces of the plate can be written as

$$\begin{aligned} \{S(x_3)\}^T e^{i(\xi_1 x_1+\xi_2 x_2)} &= \begin{bmatrix} u_1 & u_2 & \phi_{0,3} & 0 & 0 & -\rho_0\omega^2\phi_0 \end{bmatrix}, \qquad x_3 = 0 \\ &= \begin{bmatrix} u_1 & u_2 & \phi_{b,3} & 0 & 0 & -\rho_0\omega^2\phi_b \end{bmatrix}, \qquad x_3 = H \end{aligned} \tag{2.89}$$

Substituting Eq. (2.88) into Eq. (2.89) yields

$$\begin{aligned} \{S(x_3)\}^T e^{i(\xi_1 x_1+\xi_2 x_2)} &= \begin{bmatrix} U_0 & V_0 & i\zeta_0(1-R) & 0 & 0 & -\rho_0\omega^2(1+R) \end{bmatrix} e^{i(\xi_1 x_1+\xi_2 x_2)}, \qquad x_3 = 0 \\ &= \begin{bmatrix} U_1 & V_1 & i\zeta_0 T & 0 & 0 & -\rho_0\omega^2\phi_b T \end{bmatrix} e^{i(\xi_1 x_1+\xi_2 x_2)}, \qquad x_3 = H \end{aligned} \tag{2.90}$$

where ζ_0, ξ_1 and ξ_2 are defined in Eq. (2.86). Propagation term $e^{i(\xi_1 x_1+\xi_2 x_2)}$ is implied in every term. Note, that u_1 and u_2 of Eq. (2.89) are related to U_0, V_0, U_1 and V_1 of Eq. (2.90) in the following manner:

$$u_1\big|_{x_3=0} = U_0 e^{i(\xi_1 x_1+\xi_2 x_2)}, \quad u_2\big|_{x_3=0} = V_0 e^{i(\xi_1 x_1+\xi_2 x_2)},$$

$$u_1\big|_{x_3=H} = U_1 e^{i(\xi_1 x_1+\xi_2 x_2)}, \quad u_2\big|_{x_3=H} = V_1 e^{i(\xi_1 x_1+\xi_2 x_2)}$$

2.6.1 Single Layer Composite Plates Immersed in a Fluid

Equating $\{S(x_3)\}$ of Eqs. (2.90) and (2.81) at $x_3=0$ and H, we get a total of twelve equations, with twelve unknowns $A_1^+, A_2^+, A_3^+, A_1^-, A_2^-, A_3^-, R, T, U_0, V_0, U_1$ and V_1. Since it gives a system of non-homogeneous equations, the twelve unknowns can be uniquely solved using the twelve equations. The dispersion equation is obtained by equating the determinant of the coefficient matrix to zero.

2.6.2 Multilayered Composite Plates Immersed in a Fluid

Symbols and steps given in Mal et al. (1991) are followed in this section as well. A multilayered composite plate is analyzed here; fibers in different layers are oriented in different directions. The global coordinate system xyz is introduced, where the xy-plane is parallel to the surfaces of the plate. For each layer, or lamina, a local coordinate system $x_1x_2x_3$ is also introduced, with the x_1-axis along the fiber direction and the x_3-axis being identical to the z-axis. The fiber direction (x_1-axis) in the m-th lamina makes an angle φ_m to the x-axis; φ_m, in general, varies from one lamina to the next.

The z-dependent parts of the three displacement components in the global coordinate system are denoted by $U(z)$, $V(z)$ and $W(z)$, and the three stress components σ_{13}, σ_{23}, σ_{33} by $X(z)$, $Y(z)$ and $Z(z)$, respectively. The symbols u_i and σ_{i3} ($i=1, 2, 3$) denote the z-dependent parts of the displacement and stress components in each local coordinate system.

The displacement and stress vectors in the global coordinate system are transformed from the local coordinate system by the following relations:

$$\begin{Bmatrix} U_m \\ V_m \\ W_m \end{Bmatrix} = \left[L(m)\right] \begin{Bmatrix} u_1^m \\ u_2^m \\ u_3^m \end{Bmatrix}$$
$$\begin{Bmatrix} X_m \\ Y_m \\ Z_m \end{Bmatrix} = \left[L(m)\right] \begin{Bmatrix} \sigma_{13}^m \\ \sigma_{23}^m \\ \sigma_{33}^m \end{Bmatrix} \quad (2.91)$$

where the subscript and superscript "m" represents the corresponding components in the m-th layer. The transformation matrix $[L(m)]$ is given by

$$\left[L(m)\right] = \begin{bmatrix} \cos(\phi_m) & -\sin(\phi_m) & 0 \\ \sin(\phi_m) & \cos(\phi_m) & 0 \\ 0 & 0 & 1 \end{bmatrix} \quad (2.92)$$

The stress–displacement vector $\{S(z)\}$ must be continuous across all interfaces parallel to the xy-plane. In the m-th lamina [for which $z_{m-1} \le z \le z_m$], the $\{S_m(z)\}$

vector is represented in the following partitioned matrix form, using Eqs. (2.81) and (2.91):

$$\{S_m(z)\}=\begin{bmatrix} L(m) & 0 \\ 0 & L(m) \end{bmatrix}\begin{bmatrix} Q_{11}(m) & Q_{12}(m) \\ Q_{21}(m) & Q_{22}(m) \end{bmatrix}\begin{bmatrix} E^{+}(z,m) & 0 \\ 0 & E^{-}(z,m) \end{bmatrix}\begin{Bmatrix} A^{+}(m) \\ A^{-}(m) \end{Bmatrix} \tag{2.93}$$

where all partitioned submatrices and vectors are of order three. The vectors $\{A^{\pm}(m)\}$ and the matrices $[Q_{ij}(m)]$ have the same definitions as those for the uniform plate [see Eqs. (2.83) and (2.84)]. For computing $[Q_{ij}(m)]$, material properties in Eq. (2.84) should be substituted by those for the m-th lamina. The matrices $\left[E^{\pm}(z,m)\right]$ are given by

$$\begin{aligned}
\left[E^{+}(z,m)\right]&=\begin{bmatrix} e^{i\zeta_1(z-z_{m-1})} & 0 & 0 \\ 0 & e^{i\zeta_2(z-z_{m-1})} & 0 \\ 0 & 0 & e^{i\zeta_3(z-z_{m-1})} \end{bmatrix} \\
\left[E^{-}(z,m)\right]&=\begin{bmatrix} e^{i\zeta_1(z_m-z)} & 0 & 0 \\ 0 & e^{i\zeta_2(z_m-z)} & 0 \\ 0 & 0 & e^{i\zeta_3(z_m-z)} \end{bmatrix}
\end{aligned} \tag{2.94}$$

The continuity conditions across the interface z_m, $\{S_m(z_m)\}=\{S_{m+1}(z_m)\}$ can be written as

$$\left[Q_m^{-}\right]\{A_m\}=\left[Q_{m+1}^{+}\right]\{A_{m+1}\} \tag{2.95}$$

where

$$\begin{aligned}
\{A_m\}&=\begin{Bmatrix} A_m^{+} \\ A_m^{-} \end{Bmatrix} \\
\left[Q_m^{+}\right]&=\begin{bmatrix} -L(m)Q_{11}(m) & -L(m)Q_{12}(m)E_m \\ -L(m)Q_{21}(m) & -L(m)Q_{22}(m)E_m \end{bmatrix} \\
\left[Q_m^{-}\right]&=\begin{bmatrix} L(m)Q_{11}(m)E_m & L(m)Q_{12}(m) \\ L(m)Q_{21}(m)E_m & L(m)Q_{22}(m) \end{bmatrix} \\
\left[E_m\right]&=\begin{bmatrix} e^{i\zeta_1 h_m} & 0 & 0 \\ 0 & e^{i\zeta_2 h_m} & 0 \\ 0 & 0 & e^{i\zeta_3 h_m} \end{bmatrix}
\end{aligned} \tag{2.96}$$

with $h_m=z_m-z_{m-1}$. The superscripts "−" and "+" represent the upper and lower interface of the m-th lamina. Furthermore, the wave numbers ζ_j ($j=1, 2, 3$) are subject to the constraint $\mathrm{Im}(\zeta_j)>0$ so that the diagonal elements of $[E_m]$ are always bounded.

For a multilayered, multiorientation (fibers going in different directions in different lamina) composite plate, having N layers, subjected to a plane acoustic wave striking its top surface, the following equation can be obtained by satisfying the boundary conditions at the fluid–solid interfaces and continuity conditions at the inner interfaces:

$$\begin{bmatrix} Q_0^- & Q_1^+ & 0 & 0 & \cdots & \cdots & \cdots & \cdots & 0 \\ 0 & Q_1^- & Q_2^+ & 0 & \cdots & \cdots & \cdots & \cdots & 0 \\ \cdots & \cdots & \cdots & \cdots & \cdots & \cdots & \cdots & \cdots & \cdots \\ 0 & \cdots & 0 & Q_{m-1}^- & Q_m^+ & 0 & \cdots & \cdots & 0 \\ 0 & \cdots & \cdots & 0 & Q_m^- & Q_{m+1}^+ & 0 & \cdots & 0 \\ \cdots & \cdots & \cdots & \cdots & \cdots & \cdots & \cdots & \cdots & \cdots \\ \cdots & \cdots & \cdots & \cdots & \cdots & \cdots & \cdots & \cdots & \cdots \\ \cdots & \cdots & \cdots & \cdots & \cdots & \cdots & \cdots & \cdots & \cdots \\ 0 & \cdots & \cdots & \cdots & \cdots & \cdots & 0 & Q_N^- & Q_b^+ \end{bmatrix} \begin{Bmatrix} A_0 \\ A_1 \\ \vdots \\ A_m \\ A_{m+1} \\ \vdots \\ \vdots \\ A_N \\ A_{N+1} \end{Bmatrix} = \begin{Bmatrix} P_1 \\ P_2 \\ 0 \\ 0 \\ 0 \\ \vdots \\ \vdots \\ 0 \\ 0 \end{Bmatrix} \tag{2.97}$$

where the matrices $\left[Q_0^-\right], \left[Q_b^+\right]$ and vectors $\{P_1\},\{P_2\}$ are related to the fluid loading and are given by

$$\left[Q_0^-\right] = \begin{bmatrix} -1 & 0 & 0 \\ 0 & -1 & 0 \\ 0 & 0 & i\zeta_0 \\ 0 & 0 & 0 \\ 0 & 0 & 0 \\ 0 & 0 & \rho_0\omega^2 \end{bmatrix}^T, \qquad \left[Q_b^+\right] = \begin{bmatrix} 1 & 0 & 0 \\ 0 & 1 & 0 \\ 0 & 0 & i\zeta_0 \\ 0 & 0 & 0 \\ 0 & 0 & 0 \\ 0 & 0 & -\rho_0\omega^2 \end{bmatrix}^T \tag{2.98}$$

$$\{P_1\} = \{0 \quad 0 \quad i\zeta_0\}, \qquad \{P_2\} = \{0 \quad 0 \quad -\rho_0\omega^2\}$$

The unknown coefficients R and T, and the tangential displacements on the fluid–solid interfaces, are contained in $\{A_0\}$ and $\{A_b\}$ (or $\{A_{N+1}\}$) in the form

$$\{A_0\} = \{U_0 \quad V_0 \quad R\}, \qquad \{A_{N+1}\} = \{U_{N+1} \quad V_{N+1} \quad T\} \tag{2.99}$$

Eq. (2.97) can be solved by standard numerical methods.

2.6.3 Multilayered Composite Plates in a Vacuum (Dispersion Equation)

To obtain the dispersion equation for a multilayered composite plate placed in a vacuum and having different orientations of fibers in different layers, the traction values at the two outer surfaces are set equal to zero. Then Eq. (2.97) gives a system of homogeneous equations. A nontrivial solution for that system of equations exists if

$$\mathrm{Det}\begin{bmatrix} \hat{Q}_1^+ & 0 & 0 & \cdots & \cdots & \cdots & \cdots & \cdots & 0 \\ Q_1^- & Q_2^+ & 0 & 0 & \cdots & \cdots & \cdots & \cdots & 0 \\ 0 & Q_2^- & Q_3^+ & 0 & \cdots & \cdots & \cdots & \cdots & 0 \\ \cdots & \cdots & \cdots & \cdots & \cdots & \cdots & \cdots & \cdots & \cdots \\ 0 & \cdots & 0 & Q_{m-1}^- & Q_m^+ & 0 & \cdots & \cdots & 0 \\ 0 & \cdots & \cdots & 0 & Q_m^- & Q_{m+1}^+ & 0 & \cdots & 0 \\ \cdots & \cdots & \cdots & \cdots & \cdots & \cdots & \cdots & \cdots & \cdots \\ 0 & \cdots & \cdots & \cdots & \cdots & 0 & Q_{N-1}^- & Q_N^+ & 0 \\ 0 & \cdots & \cdots & \cdots & \cdots & \cdots & \cdots & 0 & \hat{Q}_N^- \end{bmatrix} = 0 \tag{2.100a}$$

where

$$\left[\hat{Q}_1^{+}\right]=\left[L(1)Q_{21}(1) \quad L(1)Q_{22}(1)E_1\right]$$
$$\left[\hat{Q}_N^{-}\right]=\left[-L(N)Q_{21}(N)E_N \quad -L(N)Q_{22}(N)\right] \tag{2.100b}$$

For waves propagating in a cross-ply laminate at 0° or 90° orientation to the fibers of the top lamina, the determinant becomes singular since it includes the antiplane (SH) motion. To remove this singularity, it is necessary to remove the elements associated with SH-wave motion and adjust the dimensions of the matrices appropriately (Mal et al., 1991).

2.6.4 Composite Plate Analysis with Attenuation

The solution technique discussed in Sections 2.6.1, 2 and 3 ignores the material attenuation. To incorporate the material attenuation in the formulation, the material constants of Eq. (2.66) are to be made complex. The real parts give the stiffness properties and the imaginary parts are associated with the attenuation properties. Attenuation or dissipation of the waves in fiber-reinforced composite materials is caused by the viscoelastic nature of the matrix and by scattering from fibers and other inhomogenities. Both of these effects can be modeled in the frequency domain by assuming that the material constants, C_{ij}, are complex and frequency dependent. For isotropic viscoelastic solid modeling, P- and S-wave speeds can be made complex and expressed in the following form (Mal et al., 1992):

$$\alpha=\sqrt{\frac{\lambda+2\mu}{\rho}}=\frac{\hat{\alpha}}{1+\dfrac{i}{2Q_\alpha}}$$
$$\beta=\sqrt{\frac{\mu}{\rho}}=\frac{\hat{\beta}}{1+\dfrac{i}{2Q_\beta}} \tag{2.101a}$$

where Q_α and Q_β are called quality factors.

Laboratory measurements on a variety of materials have shown that (Mal et al., 1992).

(a) $\hat{\alpha}$, $\hat{\beta}$, Q_α and Q_β are independent of frequency in a broad frequency range, (b) Q_α and Q_β are proportional to the wave speeds $\hat{\alpha}$ and $\hat{\beta}$, respectively, and (c) numerical values of Q_α and $Q_\beta$for most materials are large. Therefore,

$$\frac{1}{Q_\beta}=\frac{1}{k.\hat{\beta}}=p$$
$$\frac{1}{Q_\alpha}=\frac{1}{k.\hat{\alpha}}=\frac{1}{k.\hat{\beta}}\frac{\hat{\beta}}{\hat{\alpha}}=p\frac{\hat{\beta}}{\hat{\alpha}} \tag{2.101b}$$

From Eqs. (2.101a) and (2.101b)

$$\alpha=\frac{\hat{\alpha}}{1+\dfrac{i}{2Q_\alpha}}=\frac{\hat{\alpha}}{1+\dfrac{1}{2}ip\left(\dfrac{\hat{\beta}}{\hat{\alpha}}\right)}=\frac{\hat{\alpha}}{\sqrt{1+ip\left(\dfrac{\hat{\beta}}{\hat{\alpha}}\right)}}$$
$$\beta=\frac{\hat{\beta}}{1+\dfrac{i}{2Q_\beta}}=\frac{\hat{\beta}}{1+\dfrac{1}{2}ip}=\frac{\hat{\beta}}{\sqrt{1+ip}} \tag{2.101c}$$

or,

$$\alpha^2 = \frac{(\hat{\alpha})^2}{1 + ip\left(\frac{\hat{\beta}}{\hat{\alpha}}\right)}$$

$$\beta^2 = \frac{(\hat{\beta})^2}{1 + ip} \tag{2.101d}$$

Note that the material attenuation is expressed in terms of only one material parameter p. This is possible due to the experimental fact that the quality factor, which is proportional to the inverse of the damping factor ($p/2$), is approximately proportional to the wave speed.

For a transversely isotropic solid, it can be shown that the five bulk wave speeds in the material are proportional to $\sqrt{a_i}$, for $i=1$ to 5; see Eq. (2.77) for the definition of a_i (Mal et al., 1992). Thus, for such anisotropic solids with attenuation, a_i of Eq. (2.77) can be made complex in the same manner as in Eq. (2.101d) [Mal et al., (1992)]:

$$\frac{C_{22}}{\rho} = a_1 = \hat{a}_1\left(1 + ip\sqrt{\frac{\hat{a}_5}{\hat{a}_1}}\right)^{-1}$$

$$\frac{C_{11}}{\rho} = a_2 = \hat{a}_2\left(1 + ip\sqrt{\frac{\hat{a}_5}{\hat{a}_2}}\right)^{-1}$$

$$\frac{C_{12} + C_{55}}{\rho} = a_3 = \hat{a}_3\left(1 + ip\sqrt{\frac{\hat{a}_5}{\hat{a}_3}}\right)^{-1} \tag{2.102}$$

$$\frac{C_{44}}{\rho} = a_4 = \hat{a}_4\left(1 + ip\sqrt{\frac{\hat{a}_5}{\hat{a}_4}}\right)^{-1}$$

$$\frac{C_{55}}{\rho} = a_5 = \hat{a}_5(1 + ip)^{-1}$$

Note that in the above definition there is only one independent parameter p that is associated with the material attenuation, and p_2 is called the damping factor or damping ratio. The damping factor is frequency independent in the low frequency range, when the wavelengths are long compared to the internal microstructure dimensions (grain size, fiber diameter etc.). At higher frequencies the damping factor increases with the frequency because of the wave scattering by the internal microstructure. Mal et al. (1992) proposed the following frequency dependence of the damping factor:

$$p = p_0\left[1 + a_0\left(\frac{f}{f_0} - 1\right)^n H(f - f_0)\right] \tag{2.103}$$

where:

$H(f)$ is the Heaviside step function

p_0 and a_0 are material constants

n =2 for two-dimensional models, and 3 for three-dimensional models

Eq. (2.103) implies that wave attenuation is constant for frequencies less than f_0, and it increases for frequencies greater than f_0. The material parameter a_0 determines the rate of increase of attenuation with frequency.

The "quality factor" and 'damping factor', described above, are only two of many definitions associated with the material damping and attenuation. Different terminologies and symbols used for modeling material damping are listed below:

Ψ = Specific Damping Capacity

η = Loss Factor

δ = Logarithmic Decrement

ϕ = Phase Angle by which stress leads strain

E'' = Loss Modulus

$\xi = p/2$ = Damping Ratio or Damping Factor

ΔW = Energy Loss per cycle

α = Attenuation

Relations between these various definitions of material damping are given below for small values of material damping ($\tan\phi < 0.1$) (Kinra and Wolfenden, 1992):

$$\frac{1}{Q} = \frac{\Psi}{2\pi} = \eta = \frac{\delta}{\pi} = \tan\phi = \phi = \frac{E''}{E'} = 2\zeta = \frac{\Delta W}{2\pi W} = \frac{\lambda\alpha}{\pi} \tag{2.104}$$

where:

Q = Quality factor
E' = Storage modulus
W = Maximum Elastic Stored Energy
λ = Wavelength of Elastic Wave

2.7 DEFECT DETECTION IN MULTILAYERED COMPOSITE PLATES – EXPERIMENTAL INVESTIGATION

Following the theory described in Section 2.6, Mal et al. (1991) calculated the reflected spectra of defect-free and damaged (delaminated) composite plates. They have shown, theoretically as well as experimentally, that delamination has a strong effect on the reflected signal spectra. Detecting internal defects in a plate by scanning it with propagating Lamb waves has been proposed by Nagy et al. (1989), Chimenti and Martin (1991), Ditri et al. (1992), Kundu et al. (1996), Maslov and Kundu (1997), Kundu and Maslov (1997) and Yang and Kundu (1998) among others. Yang and Kundu (1998) used the theory of guided wave propagation in multilayered anisotropic composite plates to determine which Lamb mode should be used to detect defects in a specific layer of a 12-ply composite plate. Later, Kundu et al. (2001) have shown that by analyzing the stress profiles in a multilayered composite plate generated by an ultrasonic beam striking the plate at an incident angle close to the Lamb critical angle but not exactly at the critical angle, it is possible to detect different types of defect (delamination, broken fibers, missing fibers etc.) in the composite plate and predict in which ply the defect is located. A five-layer metal matrix (Ti-6Al-4V) composite plate reinforced by SCS-6 fibers was studied by Kundu et al. (2001). The numerical and experimental results of this study are presented below, after the specimen description.

2.7.1 Specimen Description

The specimen is a five-layered metal matrix composite plate of dimension $80 \times 33 \times 1.97$ mm^3. Five layers or plies of SCS-6 fibers in Ti-6Al-4V matrix are oriented in 90° and 0° directions in alternate layers. SCS is a copyrighted/registered name by the fiber manufacturer, Textron Inc. This fiber has a carbon core of about 25 μm diameter, two concentric layers of silicon carbide surrounding the carbon core, and two very thin (a few microns thick) layers of carbon coating on the outside. The overall fiber diameter is about 152 μm. The fibers in the top, middle and bottom layers are oriented in the x_2 direction or along the length of the plate; the other two plies are in the x_1 direction or along the width of the plate (see Figure 2.30). The composite was made by the foil-fiber-foil technique. The internal flaws, shown in Figure 2.30, were intentionally introduced in the plate during the fabrication process. The first (top) and the fifth (bottom) layers of fiber did not have any flaws. The left part of the second layer fibers was coated with boron nitride to impede the formation of good bonding between the fibers and the matrix as schematically shown in Figure 2.30. The fibers in the third layer were intentionally broken near the middle. The fourth layer had two areas of missing fibers; five fibers on the left side and ten on the right were removed. Photographs of the third and fourth layers are shown in Figure 2.31. These photographs were taken before fabricating the specimen. Broken and missing fiber zones can be clearly seen in these photographs.

Figure 2.30 shows how the specimen was scanned by propagating Lamb waves in the direction normal to the fiber direction of layers 1, 3 and 5 and parallel to the fibers of layers 2 and 4.

Before investigating the images generated by Lamb waves, the specimen was first scanned by the conventional C-scan technique where a P-wave strikes the plate at normal incidence. The C-scan images are shown in Figure 2.32. Three

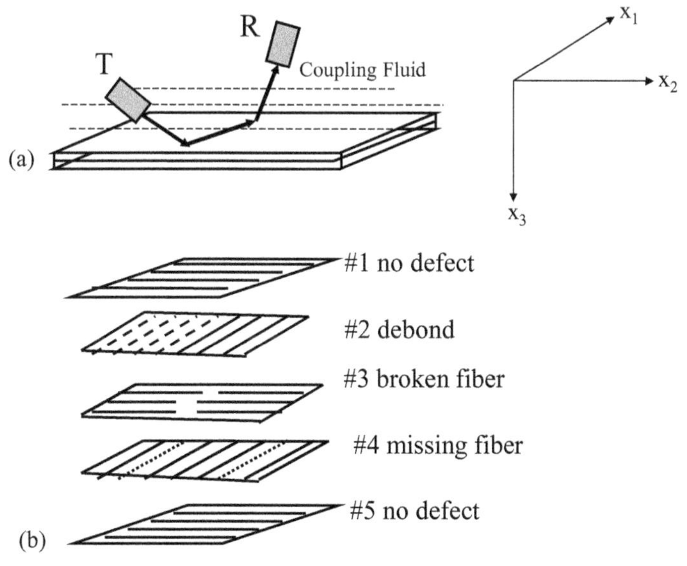

Figure 2.30 (a) Relative orientations of the transmitter, receiver and the plate specimen and (b) schematic of the internal defects in the five layers of the composite plate specimen [from Kundu et al. (2001)].

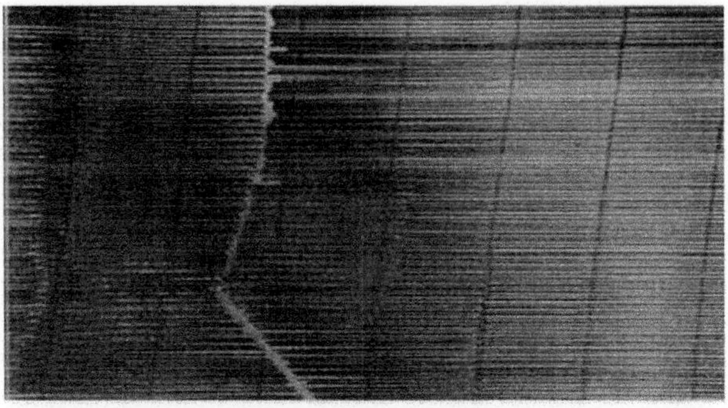

(a)

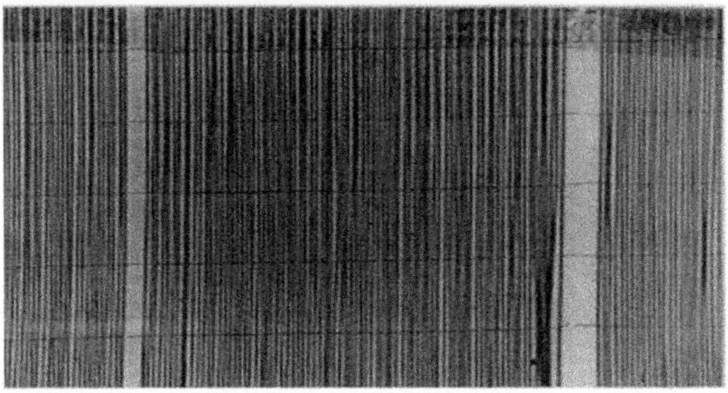

(b)

Figure 2.31 Photograph of (a) the broken fibers of the third layer and (b) missing fibers of the fourth layer, taken before fabricating the specimen [from Kundu et al. (1996)].

images of Figure 2.32 were generated by 10 MHz (top and middle) and 75 MHz (bottom) focused transducers used in the pulse-echo mode. The transducer axis is positioned normal to the plate specimen. For the top and bottom images, the gate position is such that the reflected signals from the middle of the layer are received and the back surface echo is omitted; therefore, the internal defects should clearly be seen in these two images; for the middle image the back surface echo is also recorded. In all of these three images the debond can be clearly seen. The missing and broken fibers can be faintly seen in some images.

To understand and analyze the Lamb wave generated images, it is necessary to compute the internal stress and displacement profiles when Lamb waves propagate in the direction normal to the fiber direction of layers 1, 3 and 5. In other words, the fiber orientation relative to the Lamb wave propagation direction is 90° for layers 1, 3 and 5, and 0° for layers 2 and 4. During the experiment the specimen was immersed in water.

2.7.2 Numerical and Experimental Results

To compute internal stresses and displacements in a multilayered plate, one needs to know all elastic constants of individual layers. However, the five independent elastic constants of the individual layers were not known and could not be measured easily. Only the density (5.1 gm/cc) of the plate could be measured without any difficulty. The P-wave speed (1.49 km/s) and the density (1 gm/cc) of the coupling fluid (water) are also known quantities.

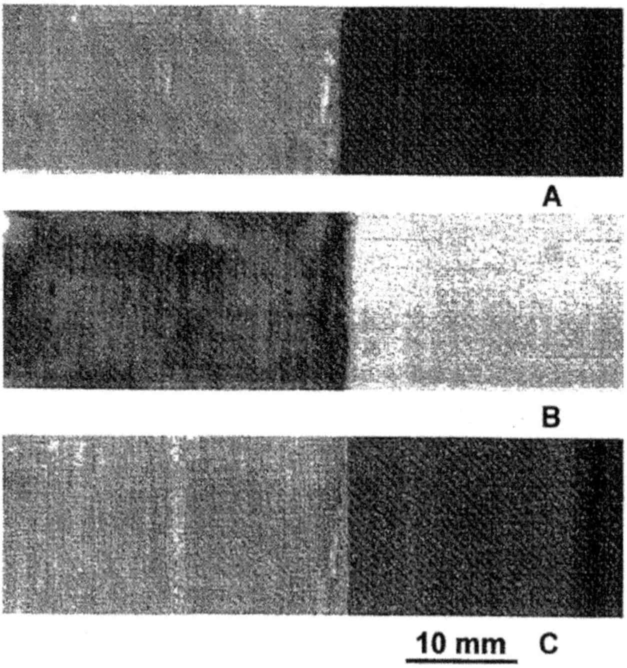

Figure 2.32 Conventional C-scan images generated by 10 MHz (top and middle) and 75 MHz (bottom) focused transducers, used in the pulse-echo mode. The back surface echo is omitted for constructing the top and bottom images but it is considered for the middle image [from Kundu et al. (1996)].

Huang et al. (1997), and Yang and Mal (1996) gave elastic properties of titanium (Ti) and silicon carbide (SiC) for SCS-6 fiber-reinforced titanium matrix. These properties are listed in Table 2.3.

From the above table the stress–strain relation for Ti and SiC can be written in the following form. In the constitutive matrix (or [C] matrix) a range is given for each element. This range is obtained from the two sets of values of the elastic constants given in Table 2.3.

Stress–strain relation of Ti:

$$\begin{Bmatrix} \sigma_{11} \\ \sigma_{22} \\ \sigma_{33} \\ \sigma_{23} \\ \sigma_{31} \\ \sigma_{12} \end{Bmatrix} = \begin{bmatrix} 130.1-193.5 & 55.9-103.3 & 55.9-103.3 & 0 & 0 & 0 \\ & 130.1-193.3 & 55.9-103.3 & 0 & 0 & 0 \\ & & 130.1-193.3 & 0 & 0 & 0 \\ & & & 37.1-45.1 & 0 & 0 \\ & & & & 37.1-45.1 & 0 \\ & & & & & 37.1-45.1 \end{bmatrix} \begin{Bmatrix} \varepsilon_{11} \\ \varepsilon_{22} \\ \varepsilon_{33} \\ 2\varepsilon_{23} \\ 2\varepsilon_{31} \\ 2\varepsilon_{12} \end{Bmatrix} \quad (2.105a)$$

and for SiC:

$$\begin{Bmatrix} \sigma_{11} \\ \sigma_{22} \\ \sigma_{33} \\ \sigma_{23} \\ \sigma_{31} \\ \sigma_{12} \end{Bmatrix} = \begin{bmatrix} 446-520 & 91.4-176 & 91.4-176 & 0 & 0 & 0 \\ & 446-520 & 91.4-176 & 0 & 0 & 0 \\ & & 446-520 & 0 & 0 & 0 \\ & & & 172-177.4 & 0 & 0 \\ & & & & 172-177.4 & 0 \\ & & & & & 172-177.4 \end{bmatrix} \begin{Bmatrix} \varepsilon_{11} \\ \varepsilon_{22} \\ \varepsilon_{33} \\ 2\varepsilon_{23} \\ 2\varepsilon_{31} \\ 2\varepsilon_{12} \end{Bmatrix} \quad (2.105b)$$

Table 2.3: Elastic Properties of Ti and SiC [after Huang et al. (1997)[1], Yang and Mal (1996)[2]]

Material	Young's Modulus (E, GPa)	Poisson's ratio (ν)	Lame's First constant (λ, GPa)	Shear Modulus (G, GPa)	Density (ρ, gm/cc)
Titanium[1] (Ti)	121.6	0.35	103.3	45.1	5.4
Titanium[2] (Ti)	96.5		55.9	37.1	4.5
Silicon-Carbide[1] (SiC)	415.0	0.17	91.4	177.4	3.2
Silicon-Carbide[2] (SiC)	431.0		176	172.0	3.2

Values Given in these Two References are marked by superscripts 1 and 2.
Source: Reprinted from Kundu et al. (2001) with permission from Elsevier.

Constitutive matrices of both Ti and SiC, in Eqs. (2.105a) and (2.105b), are isotropic and have two independent elastic constants. However, the SiC fiber-reinforced Ti matrix composite has hexagonal symmetry. Therefore, the [C] matrix for the composite should be anisotropic and have five independent elastic constants. Experimental values of phase velocity for different Lamb modes at various frequencies, as obtained by Kundu et al. (1996), are shown by open triangles in Figure 2.33. A total of twenty triangles are shown in this figure.

The [C] matrix for the individual layers of the five-layer composite plate was obtained by trial and error method by matching the theoretical dispersion curves with the experimental points. After a number of trials, the following stress–strain relation gave the best fit between the theoretical curves and the experimental points,

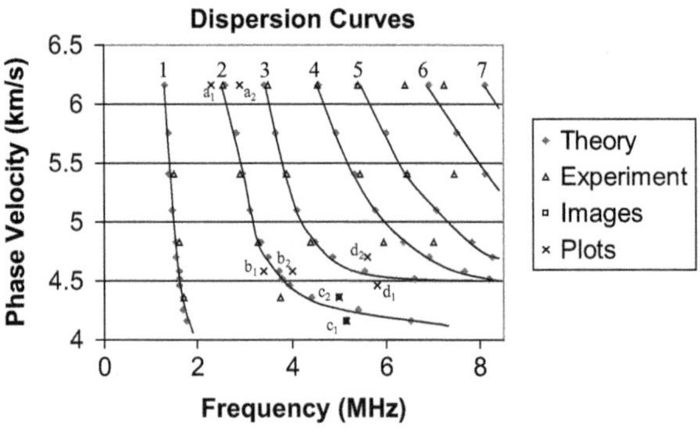

Figure 2.33 Numerically computed dispersion curves (diamond symbols connected by continuous lines). Twenty experimental points are shown by triangular symbols. Stress plots of Figures 2.34, 2.35 and 2.37 are generated for eight different frequency–phase velocity combinations (a_j, b_j, c_j and d_j, $j=1$ and 2); those points are shown by cross markers. Square markers (points c_1 and c_2) show the frequency–phase velocity combinations used for generating the two images of Figure 2.36. Seven modes, shown here, are numbered from 1 to 7 [from Kundu et al. (2001)].

$$\begin{Bmatrix} \sigma_{11} \\ \sigma_{22} \\ \sigma_{33} \\ \sigma_{23} \\ \sigma_{31} \\ \sigma_{12} \end{Bmatrix} = \begin{bmatrix} 325 & 103 & 103 & 0 & 0 & 0 \\ & 194 & 92 & 0 & 0 & 0 \\ & & 194 & 0 & 0 & 0 \\ & & & 51 & 0 & 0 \\ & & & & 100 & 0 \\ & & & & & 100 \end{bmatrix} \begin{Bmatrix} \varepsilon_{11} \\ \varepsilon_{22} \\ \varepsilon_{33} \\ 2\varepsilon_{23} \\ 2\varepsilon_{31} \\ 2\varepsilon_{12} \end{Bmatrix} \quad (2.106)$$

where x_1 is the fiber direction, elastic constants are given in GPa. Note that $C_{44}=(C_{22}-C_{23})/2$.

Theoretical leaky Lamb wave dispersion curves for the five-layer composite plate of 1.97 mm total thickness are shown in Figure 2.33 by black diamond symbols connected by continuous lines. These are computed for the individual layer properties given in Eq. (2.106), for Lamb waves propagating normal to the fiber direction of the top layer, see Figure 2.30. Lamb modes are numbered from 1 to 7 from left (low frequency) to right (high frequency). It should be noted here that the matching between the experimental values and the theoretical dispersion curves is reasonably good for the five modes. Fourteen out of the seventeen experimental values for these five modes almost coincide with the theoretical curves but the sixth and seventh modes did not match very well with the theoretical curves. This matching can be further improved by adjusting the elastic properties of the layers by more iterations or by implementing sophisticated optimization schemes such as the simplex algorithm [Nelder and Mead (1965), Karim, Mal and Bar-Cohen (1990), Kundu (1992), Kinra and Iyer (1995)].

After obtaining the elastic properties, stress profiles are computed for different frequency–phase velocity combinations – on and around the second and third Lamb modes. These frequency–phase velocity combinations, for which stress profiles have been computed, are denoted by a_j, b_j, c_j and d_j ($j=1$ and 2) and shown by eight crosses in Figure 2.33. The plots are obtained for the plane longitudinal wave of a given frequency striking the composite plate at a specified angle. The corresponding phase velocity is obtained from the incident angle using Snell's law (Eq. 2.31). If the phase velocity–frequency combination is such that it is near a leaky Lamb mode but does not coincide exactly with the dispersion curve, then the stress and displacement components would differ from those for the leaky Lamb wave propagation. It should be mentioned here that, for the stress field computations near Lamb modes, the contributions of incident, reflected and transmitted waves in the upper and lower fluid half-spaces are considered. The incident angle is set such that the phase velocity, computed from Snell's law, becomes close to a Lamb mode phase velocity. Because of the presence of the incident, reflected and transmitted signals the normal stress components, at the top and bottom surfaces of the plate, do not have the same values.

Images of the composite plate have been generated for two different frequency–phase velocity combinations (c_1 and c_2); those two points are marked by two squares in Figure 2.33.

Figures 2.34 and 2.35 show the computed stress profiles along the thickness or depth of the plate for six frequency–phase velocity combinations: two pairs (a_j and b_j) near the second mode and one pair (d_j) near the third mode. Figure 2.34 shows the shear stress (σ_{13}) variations along the depth of the plate. Figure 2.35 shows the normal stress variations (σ_{33} in the left column and σ_{11} in the right column). The phase velocity (c_L) and the incident angle (θ) are related by Snell's law [Eq. (2.31)].

The horizontal axes of Figures 2.34 and 2.35 show the depth along the plate thickness [in the x_3 direction (see Figure 2.30a)] varying from zero (top of the

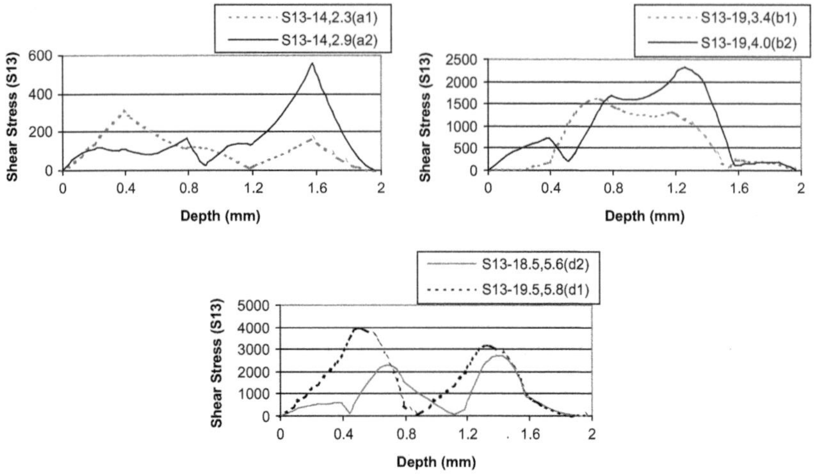

Figure 2.34 Shear stress variations inside the composite plate near second (top figures) and third (bottom figure) Lamb modes. Dotted lines have been generated for the frequency–phase velocity combinations denoted by points a_1, b_1 and d_1 in Figure 2.33. Points a_2, b_2 and d_2 of Figure 2.33 generate the continuous curves [from Kundu et al. (2001)].

plate) to 1.97 mm (bottom of the plate). Since the plate has five layers of identical thickness, the layer interfaces are located at 0.394 mm, 0.788 mm, 1.182 mm and 1.576 mm. In Figures 2.34 and 2.35 the horizontal axis is marked at 0.4, 0.8, 1.2 and 1.6 mm, very close to the interface positions. In each plot of Figures 2.34 and 2.35 two curves are shown. Dotted lines correspond to the frequency–phase velocity combinations that are located slightly below or left of the Lamb modes and the continuous lines correspond to the points slightly above or right of the Lamb modes. The curves in Figures 2.34 and 2.35 are marked as SIJ-θ,*f*(αj) where θ is the angle of incidence in degree, *f* is the signal frequency in MHz, α*j* identifies the point (a_j, b_j, c_j or d_j) of Figure 2.33, and SIJ stands for S13 for shear stress (σ_{13}), and S33 or S11 for normal stresses (σ_{33} or σ_{11}). Note that the curves are not symmetric about the central plane of the plate. Therefore, these near Lamb modes, which are generated as the total effect of the incident, reflected and transmitted waves near the Lamb critical angle of incidence, should be able to distinguish identical defects in two layers of mirror symmetry about the central plane of the plate.

It should be noted here that the continuous curves give relatively higher values in the lower half of the plate. Note that, for the frequency–phase velocity combinations a_2 and b_2 (of Figure 2.33), σ_{13} and σ_{11} in the fourth layer are much greater than those in the second layer. On the other hand, for a_1 and b_1 points, the first and second layer responses are greater than the fourth and fifth layer responses. This difference is less prominent for σ_{33}. This general trend is true for points d_1 and d_2 also. However, for the point d_2 the differences in the stress amplitude between the upper and lower halves of the plate are not as strong as those for points a_2 and b_2.

Results in Figures 2.34 and 2.35 are summarized below:

1. Stress fields in the neighborhood of a Lamb mode are not symmetric with respect to the central plane of symmetry of the plate.

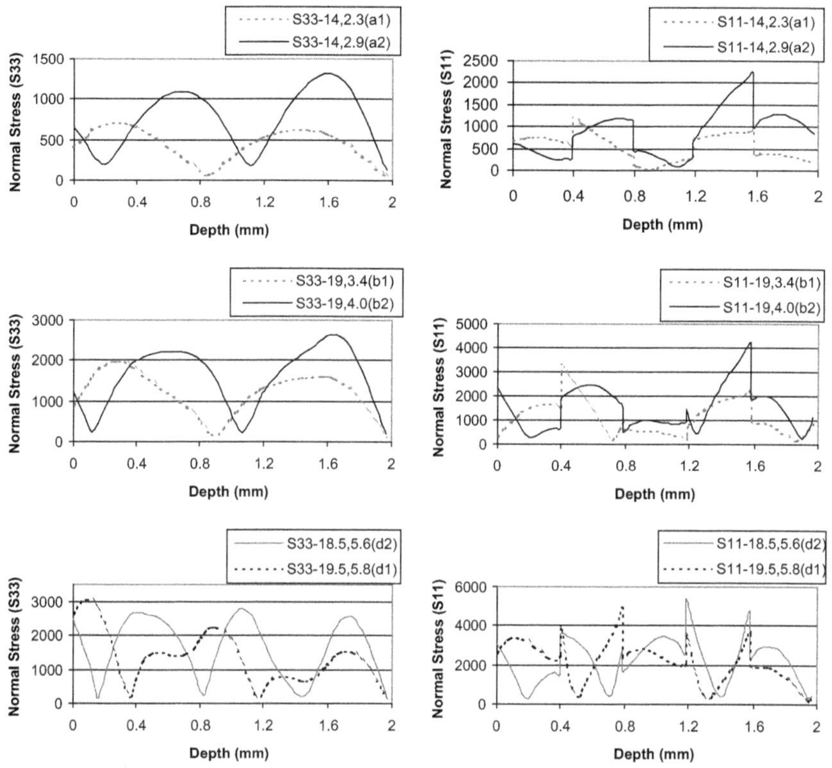

Figure 2.35 Normal stress variations inside the composite plate near second and third Lamb modes. Dotted lines have been generated for the frequency–phase velocity combinations denoted by points a_1, b_1 and d_1 in Figure 2.33. Points a_2, b_2 and d_2 of Figure 2.33 generate the continuous curves. σ_{33} and σ_{11} are shown in the left and right columns, respectively [from Kundu et al. (2001)].

2. If moving in one direction relative to the Lamb mode causes the stresses to grow in the upper half of the plate, then an opposite direction movement causes the stresses to grow in the lower half of the plate.
3. The percentage difference in the stress values between two layers of mirror symmetry in the lower and upper halves of the plate is not same for all stress components.
4. The percentage difference in the stress values between two layers of mirror symmetry in the neighborhood of a Lamb mode varies from one Lamb mode to another.

Is this difference in stress amplitudes sufficient to distinguish between the defects in the upper and lower halves of the plate? To investigate it, two images of the specimen were generated with the frequency–phase velocity combinations corresponding to points c_1 and c_2 of Figure 2.33. It should be mentioned here that point c_1 corresponds to a 21° incident angle and 5.15 MHz signal frequency, and for point c_2 the incident angle is 20° and the signal frequency is 5 MHz. A laboratory-made ultrasonic scanner was used to generate the ultrasonic images. A broad band Panametrics transducer (0.5 inch diameter) was excited using a

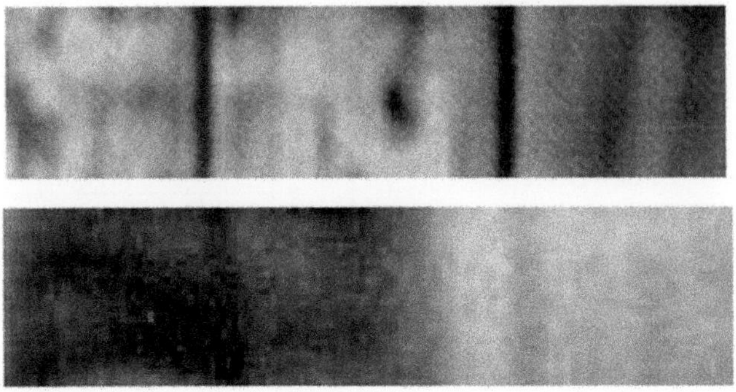

Figure 2.36 Two images of the five-layer composite plate specimen generated by two different frequency–phase velocity combinations, shown by points c_1 and c_2 in Figure 2.33. The top image has been generated by 5.0 MHz signal incident at 20° (point c_2) and the bottom image has been produced by 5.15 MHz signal incident at 21° (point c_1) [from Kundu et al. (2001)].

Matec 310 gated amplifier with tone-burst signals from the Wavetek function generator. The reflected signal was received by a Matec receiver and was digitized by a GAGE 40 MHz data acquisition board, and then the received signal was analyzed. The image generating software computed either the peak-to-peak or the average amplitude of the signal in a given time window and then plotted it as a gray scale image against the horizontal (x_1,x_2) position of the transducers. The window was set near the first arrival time of the signal, thus reflections from the plate boundary were avoided.

The generated images are shown in Figure 2.36. The 5 MHz signal, incident at 20°, clearly shows the missing fiber defects of the fourth layer and the 5.15 MHz signal, incident at 21°, shows the delamination defect (darker region) of the second layer. It also faintly shows the missing fibers of the fourth layer.

Delamination and missing fibers reduce the shear stress carrying capacity at the defect position. Note that the compressive normal stress σ_{33} can be present at the defect position from the non-vanishing contact pressure. Therefore, a study of the σ_{13} profile is critical for predicting the sensitivity of the propagating waves to delamination and missing fiber type defects. σ_{13} profiles for points c_1 and c_2 of Figure 2.33 are shown in Figure 2.37. It should be noted here that, for

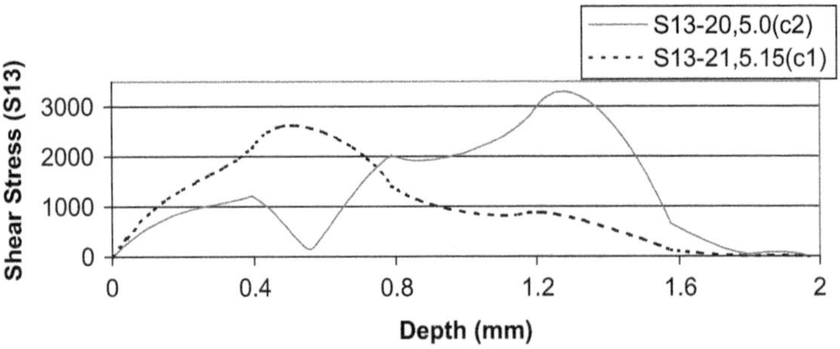

Figure 2.37 Shear stress variations inside the composite plate for frequency–phase velocity combinations shown in Figure 2.33 by points c_1 (for the dotted curve) and c_2 (for the continuous curve). Corresponding ultrasonic images are shown in Figure 2.36 [from Kundu et al. (2001)].

the 5 MHz signal, σ_{13} is very small in the second layer and it is maximum in the fourth layer. For this reason, in Figure 2.36 we see that the image generated by the 5 MHz signal clearly shows the missing fiber defects of the fourth layer and completely ignores the delamination defect of the second layer. On the other hand, for the 5.15 MHz signal (dotted line of Figure 2.37), the shear stress is maximum in the second layer and very small in the fourth layer. This explains why the image generated by the 5.15 MHz signal shows the delamination defect (darker region) of the second layer while the missing fiber defect of the fourth layer is not as clear.

The advantage of the near Lamb mode imaging is clearly demonstrated here. In the conventional C-scan image (Figure 2.32) the delamination defect guards the missing fiber defects, but in the near Lamb wave image (Figure 2.36) the delamination defect does not have much effect on the detection of missing fiber defects, when the appropriate combination of the striking angle and the signal frequency is selected.

2.8 GUIDED WAVE PROPAGATION IN THE CIRCUMFERENTIAL DIRECTION OF A PIPE

Elastic waves can propagate in axial and circumferential directions of a pipe, or at an angle other than 90° relative to the axial direction. In this section the elastic wave propagation in the circumferential direction is discussed. Longitudinal stress corrosion cracks in large diameter pipes can be detected more efficiently by propagating guided waves in the circumferential direction. The difference between the guided wave propagation in a flat plate (discussed above) and in the circumferential direction of a pipe (discussed in this section) is that the curvature of a flat plate is zero but for the pipe it is not. Therefore, if we study the guided wave propagation in a curved plate having a nonzero curvature in the direction of the wave propagation and a zero curvature in the perpendicular direction, then that study is essentially same as the guided wave propagation in the circumferential direction of a pipe. Towfighi et al. (2002) first solved this problem for a general anisotropic material. Since the solution for the isotropic curved plate is a special case of the anisotropic curved plate problem solved by Towfighi et al. (2002), their solution technique is discussed here in detail and then specialized to obtain the solution for an isotropic curved plate. Unlike isotropic materials for which the Stokes-Helmholtz decomposition technique simplifies the governing equation, in anisotropic cases no such general decomposition technique works. Therefore, coupled differential equations must be solved.

Viktorov (1958) analyzed the guided wave propagation in a curved surface. He introduced the angular wave number concept and then derived, decomposed and solved the governing differential equations. He considered only one curved surface and found the solutions for convex and concave cylindrical surfaces. In order to analyze curved plates, Qu et al. (1996) added the boundary conditions for the second surface and solved the problem of guided wave propagation in isotropic curved plates. Different aspects of the circumferential direction wave propagation along one or multiple curved surfaces have been analyzed by Grace and Goodman (1966), Brekhovskikh (1968), Cerv (1988), Liu and Qu (1998a,b) and Valle et al. (1999). In all these works the material has been modeled as an isotropic elastic material. Wave propagation in the circumferential direction of an anisotropic cylinder was first given by Towfighi et al. (2002). They provided a systematic solution method, which is capable of solving a set of coupled differential equations, and thus can be utilized to solve a variety of wave propagation problems.

2.8.1 Fundamental Equations

The formulation given by Towfighi et al. (2002) for the wave propagation in a cylinder in the circumferential direction is presented here. The wave propagation direction is shown in Figure 2.38. In this section the wave carrier is interchangeably called a "curved plate", "cylinder", "pipe segment", or simply a "pipe", all meaning the same thing. What we are interested in is computing the dispersion curves in the curved plate for waves propagating from section T to R (see Figure 2.38). This analysis only considers the convex and concave surfaces of the curved plate but does not include the reflected guided waves from the edge or boundary of the plate. The problem geometry can be a segment of a cylinder or a complete cylinder.

The wave propagation in the circumferential direction in pipes with isotropic material properties is usually modeled as a plane strain problem, i.e. the displacement component along the longitudinal axis of the pipe is set equal to zero. For a few other types of anisotropy this situation remains valid. However, for general anisotropy the longitudinal component of displacement must be considered in the mathematical modeling. The symmetry of both geometry and material properties is required for plane strain idealization. In absence of such symmetry a three-dimensional mathematical modeling is necessary.

In cylindrical coordinates, the strain components in terms of displacements can be written as (see Chapter 1, Table 1.2)

$$
\begin{aligned}
\varepsilon_{rr} &= \frac{\partial u_r}{\partial r} \\
\varepsilon_{\theta\theta} &= \frac{1}{r}\frac{\partial u_\theta}{\partial \theta} + \frac{u_r}{r} \\
\varepsilon_{zz} &= \frac{\partial u_z}{\partial z} \\
\varepsilon_{rz} &= \frac{1}{2}\left(\frac{\partial u_r}{\partial z} + \frac{\partial u_z}{\partial r}\right) \\
\varepsilon_{r\theta} &= \frac{1}{2}\left(\frac{1}{r}\frac{\partial u_r}{\partial \theta} - \frac{u_\theta}{r} + \frac{\partial u_\theta}{\partial r}\right) \\
\varepsilon_{z\theta} &= \frac{1}{2}\left(\frac{1}{r}\frac{\partial u_z}{\partial \theta} + \frac{\partial u_\theta}{\partial z}\right)
\end{aligned}
\tag{2.107}
$$

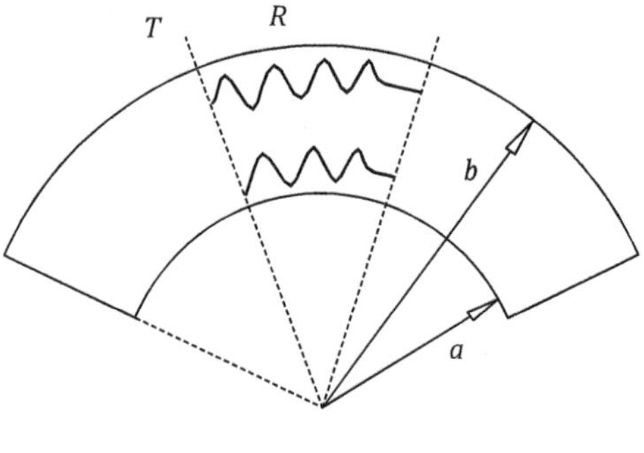

Figure 2.38 Circumferential direction wave propagation in a pipe segment or a curved plate from section T to section R. Wave speed is assumed to be proportional to the radius of curvature [from Towfighi et al. (2002)].

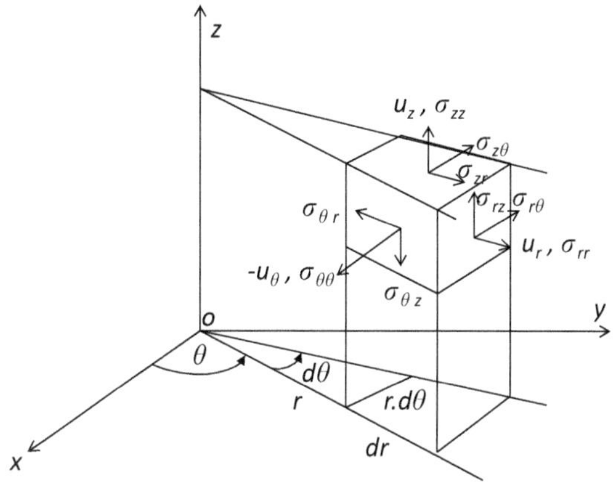

Figure 2.39 Stress and displacement components in cylindrical coordinate system [from Towfighi et al. (2002)].

The stress and displacement components are shown in Figure 2.39. Constitutive matrix for general anisotropy containing 21 independent elastic constants can be written in cylindrical coordinate system from Eq. (1.59):

$$\begin{Bmatrix} \sigma_{\theta\theta} \\ \sigma_{zz} \\ \sigma_{rr} \\ \sigma_{\theta z} \\ \sigma_{r\theta} \\ \sigma_{rz} \end{Bmatrix} = \begin{bmatrix} C_{11} & C_{12} & C_{13} & C_{14} & C_{15} & C_{16} \\ & C_{22} & C_{23} & C_{24} & C_{25} & C_{26} \\ & & C_{33} & C_{34} & C_{35} & C_{36} \\ & & & C_{44} & C_{45} & C_{46} \\ & \text{symm} & & & C_{55} & C_{56} \\ & & & & & C_{66} \end{bmatrix} \begin{Bmatrix} \varepsilon_{\theta\theta} \\ \varepsilon_{zz} \\ \varepsilon_{rr} \\ 2\varepsilon_{\theta z} \\ 2\varepsilon_{r\theta} \\ 2\varepsilon_{rz} \end{Bmatrix} \quad (2.108)$$

Equations of motion for three components of displacement in cylindrical coordinates can be obtained by combining the equilibrium equations in cylindrical coordinates given in Table 1.2 of Chapter 1 and equation 1.78.

$$\begin{aligned} &\frac{\partial \sigma_{rr}}{\partial r} + \frac{\partial \sigma_{rz}}{\partial z} + \frac{1}{r}\frac{\partial \sigma_{r\vartheta}}{\partial \theta} + \frac{1}{r}(\sigma_{rr} - \sigma_{\theta\theta}) - \rho \frac{\partial^2 u_r(r,\theta,t)}{\partial t^2} = 0 \\ &\frac{\partial \sigma_{r\theta}}{\partial r} + \frac{\partial \sigma_{z\theta}}{\partial z} + \frac{1}{r}\frac{\partial \sigma_{\theta\vartheta}}{\partial \theta} + \frac{2}{r}\sigma_{r\theta} - \rho \frac{\partial^2 u_\theta(r,\theta,t)}{\partial t^2} = 0 \\ &\frac{\partial \sigma_{rz}}{\partial r} + \frac{\partial \sigma_{zz}}{\partial z} + \frac{1}{r}\frac{\partial \sigma_{z\vartheta}}{\partial \theta} + \frac{1}{r}\sigma_{rz} - \rho \frac{\partial^2 u_z(r,\theta,t)}{\partial t^2} = 0 \end{aligned} \quad (2.109)$$

Stress components in the above equations can be expressed in terms of displacement components. The displacement components then can be expressed in propagating wave forms as discussed in the following section.

2.8.2 Wave Form

In cylindrical geometry the generation of surface waves in the circumferential direction having a plane wave front requires the circumferential wave speed to be a function of the radial distance. Viktorov (1958) introduced this concept and called it the angular wave number, as shown below.

$$\begin{aligned} u_r(r,\theta,t) &= U_r(r)\exp(ip\theta - i\omega t) \\ u_\theta(r,\theta,t) &= U_t(r)\exp(ip\theta - i\omega t) \\ u_z(r,\theta,t) &= U_z(r)\exp(ip\theta - i\omega t) \end{aligned} \tag{2.110}$$

where $U_r(r)$, $U_t(r)$ and $U_z(r)$ represent the amplitudes of vibration in the radial, tangential and axial directions, respectively. 'i' is the imaginary number $i = \sqrt{-1}$. It should be noted here that the phase velocity is not a constant and changes with radius. As shown in Figure 2.38 the phase velocity has to be proportional to the radius to have a plane wave front. Therefore, if c_b is assumed to be the phase velocity at the outer surface with radius b, for other positions having a radius r the phase velocity would be

$$v_{ph}(r) = \frac{c_b r}{b} \tag{2.111}$$

For a flat plate the wave number k is defined as ω / v_{ph} because the curvature does not change. However, for a curved plate the same definition should be r dependent. Thus, the angular wave number p, which is independent of r, is defined as

$$p = \frac{\omega}{v_{ph}(r)/r} = \frac{\omega b}{c_b} \tag{2.112}$$

2.8.3 Governing Differential Equations

Subsequent substitution of equations (2.107–108) and (2.110–112) into equation (2.109) yields the following governing differential equations [Towfighi et al. (2002)]:

$$\begin{aligned} &-2C_{55}U_r(r)p^2 - 2C_{15}U_t(r)p^2 - 2C_{45}U_z(r)p^2 - 2iC_{11}U_t(r)p \\ &-2iC_{55}U_t(r)p - 2iC_{14}U_z(r)p + 4irC_{35}U_r'(r)p \\ &+2irC_{13}U_t'(r)p + 2irC_{55}U_t'(r)p + 2irC_{34}U_z'(r)p \\ &+2irC_{56}U_z'(r)p + 2r^2\rho\omega^2U_r(r) - 2C_{11}U_r(r) \\ &+2C_{15}U_t(r) + 2rC_{33}U_r'(r) - 2rC_{15}U_t'(r) - 2rC_{16}U_z'(r) \\ &+2rC_{36}U_z'(r) + 2r^2C_{33}U_r''(r) + 2r^2C_{35}U_t''(r) + 2r^2C_{36}U_z''(r) = 0 \end{aligned}$$

$$\begin{aligned} &-2C_{15}U_r(r)p^2 - 2C_{11}U_t(r)p^2 - 2C_{14}U_z(r)p^2 + 2iC_{11}U_r(r)p \\ &+2iC_{55}U_r(r)p + 2iC_{45}U_z(r)p + 2irC_{13}U_r'(r)p \\ &+2irC_{55}U_r'(r)p + 4irC_{15}U_t'(r)p + 2irC_{16}U_z'(r)p \\ &+2irC_{45}U_z'(r)p + 2r^2\rho\omega^2U_t(r) + 2C_{15}U_r(r) \\ &-2C_{55}U_t(r) + 2rC_{15}U_r'(r) + 4rC_{35}U_r'(r) + 2rC_{55}U_t'(r) \\ &+4rC_{56}U_z'(r) + 2r^2C_{35}U_r''(r) + 2r^2C_{55}U_t''(r) + 2r^2C_{56}U_z''(r) = 0 \end{aligned} \tag{2.113}$$

$$-2C_{45}U_r(r)p^2 - 2C_{14}U_t(r)p^2 - 2C_{44}U_z(r)p^2 + 2iC_{14}U_r(r)p$$
$$-2iC_{45}U_t(r)p + 2irC_{34}U_r'(r)p + 2irC_{56}U_r'(r)p$$
$$+2irC_{16}U_t'(r)p + 2irC_{45}U_t'(r)p + 4irC_{46}U_z'(r)p$$
$$+2rC_{16}U_r'(r) + 2r^2\rho\omega^2U_z(r) + 2rC_{36}U_r'(r)$$
$$+2rC_{66}U_z'(r) + 2r^2C_{36}U_r''(r) + 2r^2C_{56}U_t''(r) + 2r^2C_{66}U_z''(r) = 0$$

2.8.4 Boundary Conditions

In order to obtain the dispersion curves, the traction-free boundary conditions (zero stress values on the inner and outer surfaces of the pipe) must be satisfied.

Therefore, at $r=a$ and $r=b$

$$\begin{aligned}
&C_{13}U_r(r) + ipC_{35}U_r(r) + ipC_{13}U_t(r) - C_{35}U_t(r) + ipC_{34}U_z(r)\\
&\quad + rC_{33}U_r'(r) + rC_{35}U_t'(r) + rC_{36}U_z'(r) = 0\\
&C_{15}U_r(r) + ipC_{55}U_r(r) + ipC_{15}U_t(r) - C_{55}U_t(r) - ipC_{45}U_z(r)\\
&\quad + rC_{35}U_r'(r) + rC_{55}U_t'(r) + rC_{56}U_z'(r) = 0\\
&C_{16}U_r(r) + ipC_{56}U_r(r) + ipC_{16}U_t(r) - C_{56}U_t(r) + ipC_{46}U_z(r)\\
&\quad + rC_{36}U_r'(r) + rC_{56}U_t'(r) + rC_{66}U_z'(r) = 0
\end{aligned} \tag{2.114}$$

2.8.5 Solution

It can be seen that all differential equations are functions of three displacement components, $U_r(r)$, $U_t(r)$ and $U_z(r)$, and their derivatives. It should also be noted that $U_r(r)$, $U_t(r)$ and $U_z(r)$ are functions of the radius only and they appear in all equations. Therefore, there are three coupled differential equations and six boundary conditions that must be satisfied simultaneously.

To solve this system of coupled differential equations, the unknown functions are expanded in Fourier Series (FS). Substitution of the FS expansions into the differential equations provides three algebraic equations that must be satisfied for the entire problem domain. To satisfy the equations for a given number of FS terms, weighted residual integration with a linear weight function is adopted [Towfighi et al. (2002)]:

$$R = \int_a^b wf(r, x_i)dr = 0 \tag{2.115}$$

The radius corresponding to the peak value of the linear weight function can take any value between the inner and the outer radii, each resulting in one independent equation. Thus, from every differential equation any number of algebraic equations can be obtained.

It should be also noted that the general solution is a linear combination of all solution functions that can be obtained. Therefore, the general solution should contain combinatorial parameters. The number of combinatorial parameters is the same as the number of individual solutions. These combinatorial parameters are necessary to satisfy the boundary conditions. Satisfaction of six boundary conditions requires six parameters and six equations. Therefore, the necessary and sufficient number of combinatorial parameters is six and it indicates the existence of six independent solutions.

Substitution of solution functions into the differential equations leads to three equations, each containing all of the FS parameters. In other words, all FS parameters for the three amplitude functions appear in every equation. Because of this coupling, the values of parameters obtained for the FS expansion of $U_r(r)$, $U_t(r)$ and $U_z(r)$ are not independent and a solution must yield all parameters as one set of results. Since the equations are linear and the results must be combined using combinatorial parameters, only their relative values must be found. Therefore, one of the FS parameters can be assumed equal to one. Then the relative values for other FS parameters can be calculated in terms of this unit value. Each set of the parameter values defines a set of dependent shapes for the above amplitude functions; these are called basic shapes. Since the number of equations must be equal to the number of unknowns, a specific number of weight functions are required.

The FS expansion for $U_r(r)$ is given by

$$U_r(r) = x_0 + \sum_{n=1}^{m} \left\{ \cos\left(\frac{n\pi r}{L}\right) x_n + \sin\left(\frac{n\pi r}{L}\right) y_n \right\} \tag{2.116}$$

which contains (2m + 1) parameters or coefficients x_n and y_n. With two other expressions for $U_t(r)$ and $U_z(r)$ the total number of unknowns becomes (6m + 3). Applying weighted residuals method as discussed above, the following set of linear equations is obtained:

$$\begin{pmatrix} a_{11}x_1 & a_{12}x_2 & \cdots & a_{1s}x_s & a_{1(s+1)}x_{(s+1)} & \cdots & a_{1(s+6)}x_{(s+6)} \\ a_{21}x_1 & a_{22}x_2 & \cdots & a_{2s}x_s & a_{2(s+1)}x_{(s+1)} & \cdots & a_{2(s+6)}x_{(s+6)} \\ \cdots & \cdots & \cdots & \cdots & \cdots & \cdots & \cdots \\ a_{s1}x_1 & a_{s2}x_2 & \cdots & a_{ss}x_s & a_{s(s+1)}x_{(s+1)} & \cdots & a_{s(s+6)}x_{(s+6)} \end{pmatrix} = \begin{pmatrix} 0 \\ 0 \\ \cdots \\ 0 \end{pmatrix} \tag{2.117}$$

where $x_{s+1}, x_{s+2}, \ldots x_{s+6}$ represent the last sine and cosine terms of FS expansions. Assigning six independent unit vectors to the last six parameters as shown in equation (2.118)

$$\begin{pmatrix} x_{s+1}^1 & x_{s+1}^2 & x_{s+1}^3 & x_{s+1}^4 & x_{s+1}^5 & x_{s+1}^6 \\ x_{s+2}^1 & x_{s+2}^2 & x_{s+2}^3 & x_{s+2}^4 & x_{s+2}^5 & x_{s+2}^6 \\ x_{s+3}^1 & x_{s+3}^2 & x_{s+3}^3 & x_{s+3}^4 & x_{s+3}^5 & x_{s+3}^6 \\ x_{s+4}^1 & x_{s+4}^2 & x_{s+4}^3 & x_{s+4}^4 & x_{s+4}^5 & x_{s+4}^6 \\ x_{s+5}^1 & x_{s+5}^2 & x_{s+5}^3 & x_{s+5}^4 & x_{s+5}^5 & x_{s+5}^6 \\ x_{s+6}^1 & x_{s+6}^2 & x_{s+6}^3 & x_{s+6}^4 & x_{s+6}^5 & x_{s+6}^6 \end{pmatrix} = \begin{pmatrix} 1 & 0 & 0 & 0 & 0 & 0 \\ 0 & 1 & 0 & 0 & 0 & 0 \\ 0 & 0 & 1 & 0 & 0 & 0 \\ 0 & 0 & 0 & 1 & 0 & 0 \\ 0 & 0 & 0 & 0 & 1 & 0 \\ 0 & 0 & 0 & 0 & 0 & 1 \end{pmatrix} \tag{2.118}$$

six independent solutions are obtained. Therefore, the number of equations has to be s = 6m-3. Consequently, the general solution can be obtained as a linear combination of the above solutions as shown below.

$$A_1 \begin{pmatrix} x_1^1 \\ x_2^1 \\ x_3^1 \\ \cdots \\ \cdots \\ x_s^1 \end{pmatrix} + A_2 \begin{pmatrix} x_1^2 \\ x_2^2 \\ x_3^2 \\ \cdots \\ \cdots \\ x_s^2 \end{pmatrix} + A_3 \begin{pmatrix} x_1^3 \\ x_2^3 \\ x_3^3 \\ \cdots \\ \cdots \\ x_s^3 \end{pmatrix} + A_4 \begin{pmatrix} x_1^4 \\ x_2^4 \\ x_3^4 \\ \cdots \\ \cdots \\ x_s^4 \end{pmatrix} + A_5 \begin{pmatrix} x_1^5 \\ x_2^5 \\ x_3^5 \\ \cdots \\ \cdots \\ x_s^5 \end{pmatrix} + A_6 \begin{pmatrix} x_1^6 \\ x_2^6 \\ x_3^6 \\ \cdots \\ \cdots \\ x_s^6 \end{pmatrix} \tag{2.119}$$

The superscript for the FS parameters shows the solution set number. Substitution of the obtained FS parameters into stress components on the inner and outer surfaces of the pipe leads to an eigenvalue problem. The determinant of the coefficients of A_i should be zero for any point located on the dispersion curves.

2.8.6 Numerical Results

Following the above mathematical modeling steps, dispersion curves have been computed by Towfighi et al. (2002) using Mathematica programming. Results computed in this manner have been compared with the dispersion curves for anisotropic flat plates obtained by different techniques, as discussed in the earlier sections of this chapter and other publications [Rose (1999)]. It should be noted that for small thickness to radius of curvature ratios the curved plates can be approximated as flat plates. Additionally, the results computed by the above formulation have been compared with the published results for isotropic pipes [Qu et al. (1996)]. After these comparisons Towfighi et al. (2002) computed the dispersion curves for anisotropic pipes that are presented below.

2.8.6.1 Comparison with Isotropic Flat Plate Results

Dispersion curves for a flat plate are given in Mal and Singh (1991) (see Figure 2.40). This figure is very similar to Figure 2.06 although the material properties used for generating Figures 2.06 and 2.40 are slightly different. Dispersion curves generated by the FS expansion method discussed above for a curved plate having an outer radius of 1 m with thickness and material properties same as those used to generate Figure 2.40 are shown in Figure 2.41. A comparison of Figures 2.40 and 2.41 shows a very good match between the two when only 20 terms are used in the FS expansion.

2.8.6.2 Comparison with Anisotropic Flat Plate Results

Mathematical steps for computing the dispersion curves of anisotropic flat plates have been discussed in the earlier sections of this chapter and also available in the literature [Karim et al. (1990), Yang and Kundu (1998), Rose (1999)]. In this

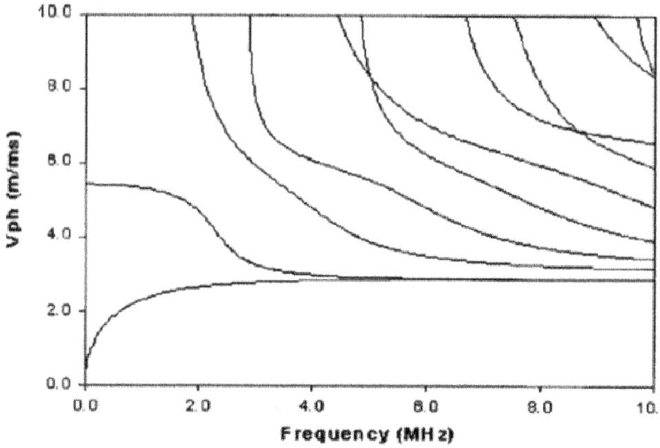

Figure 2.40 Dispersion curves for an isotropic flat plate [an aluminum plate, see Mal and Singh (1991)]. Plate thickness = 1 mm, density = 2.8 gm/cc, P-wave speed = 6.4 km/s, S-wave speed = 3.1 km/s.

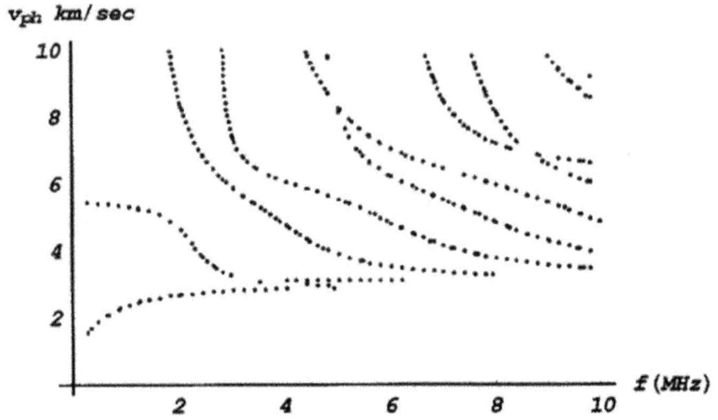

Figure 2.41 Dispersion curves generated by the method described in Section 2.8. Plate thickness = 1 mm. Pipe outer radius = 1.0 m, density = 2.8 gm/cc, P-wave speed = 6.4 km/s, S-wave speed = 3.1 km/s [from Towfighi et al. (2002)].

section the results computed by the above technique are compared with those given in Rose (1999).

For the unidirectional composite plate or pipe with 0° angle between the wave propagation direction and the fiber direction (see Figure 2.42), the material and the geometric symmetry conditions are maintained; therefore, the plain strain formulation remains valid.

Consequently the constitutive matrix reduces to the following form:

$$\begin{Bmatrix} \sigma_{\theta\theta} \\ \sigma_{zz} \\ \sigma_{rr} \\ \sigma_{r\theta} \end{Bmatrix} = \begin{bmatrix} 128.2 & 6.9 & 6.9 & 0 \\ 6.9 & 14.95 & 7.33 & 0 \\ 6.9 & 7.33 & 14.95 & 0 \\ 0 & 0 & 0 & 6.73 \end{bmatrix} \begin{Bmatrix} \varepsilon_{\theta\theta} \\ 0 \\ \varepsilon_{rr} \\ 2\varepsilon_{r\theta} \end{Bmatrix} \tag{2.120}$$

In Eq. (2.120) the stiffness values are given in GPa. Flat plate results are shown in Figure 2.43. Results for the curved plate are shown in Figures 2.44 and 2.45.

The result of Figure 2.44 is obtained using 30 terms ($m = 30$) in the Fourier Series Expansion. To show the effect of the number of terms (m) on the computed

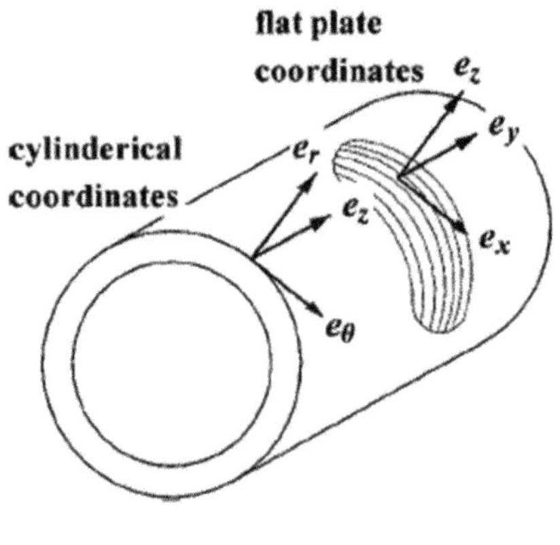

Figure 2.42 Cartesian and cylindrical coordinate systems used for flat plate and pipe analyses [from Towfighi et al. (2002)].

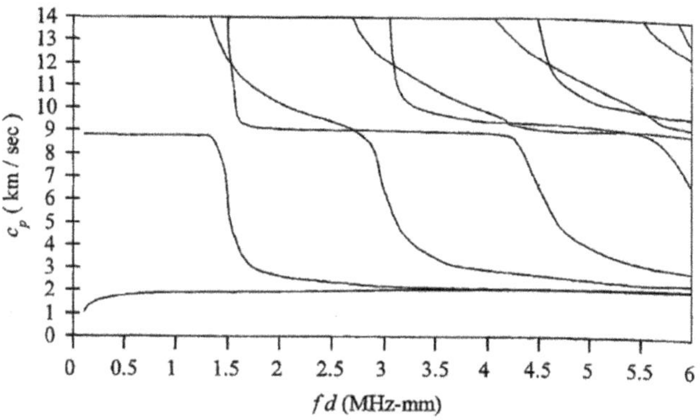

Figure 2.43 Dispersion curves of a unidirectional composite plate for waves propagating in the fiber direction. Material properties are given in Eq. (2.120), ρ=1580 kg/m³ [Rose (1999)].

results, the same dispersion curves are computed with $m=20$ and shown in Figure 2.45. It is interesting to note that a smaller value of m produces broken lines. Therefore, the user can easily realize the need for a greater number of terms in the FS expansion when the lines in the dispersion curve plot are found broken. Figure 2.44 also has some missing parts, which can be improved further by increasing m. However, for $m=30$ we get enough points on the dispersion curve plot for a good comparison with the results given in Rose (1999).

For the same material with fibers going in the longitudinal direction of the pipe, the constitutive matrix changes, as shown in the following equation:

$$\begin{Bmatrix} \sigma_{\theta\theta} \\ \sigma_{zz} \\ \sigma_{rr} \\ \sigma_{r\theta} \end{Bmatrix} = \begin{bmatrix} 14.95 & 6.9 & 7.33 & 0 \\ 6.9 & 128.2 & 6.9 & 0 \\ 7.33 & 6,9 & 14.95 & 0 \\ 0 & 0 & 0 & 3.81 \end{bmatrix} \begin{Bmatrix} \varepsilon_{\theta\theta} \\ 0 \\ \varepsilon_{rr} \\ 2\varepsilon_{r\theta} \end{Bmatrix} \tag{2.121}$$

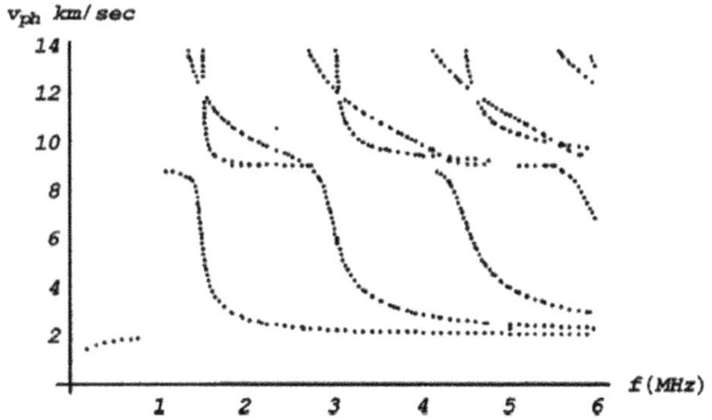

Figure 2.44 Dispersion curves for the circumferential direction wave propagation in a large diameter pipe made of a unidirectional fiber-reinforced composite material – fibers are oriented in the circumferential direction as well. Material properties are given in Eq. (2.120). Pipe wall thickness = 1 mm. Pipe outer radius = 1000 mm, $m=30$ [from Towfighi et al. (2002)].

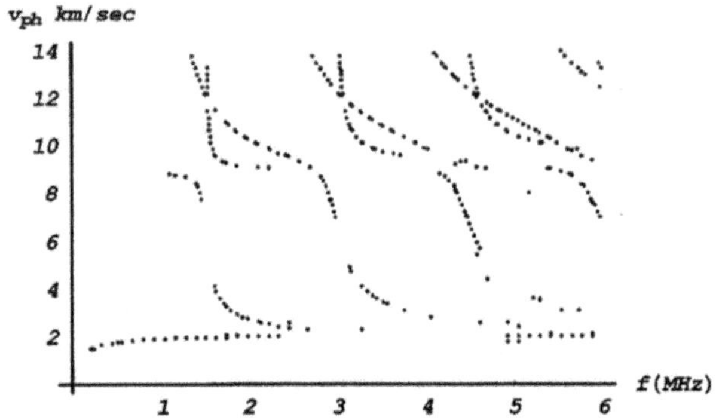

Figure 2.45 Dispersion curves for the anisotropic pipe with $m=20$. Pipe dimensions and material properties are same as those in Figure 2.44, the only difference is in the m value [from Towfighi et al. (2002)].

Obtained results for this case also match with the corresponding dispersion curves given in Rose (1999) (see Figures 2.46 and 2.47.)

For the case where fibers are oriented at 45 degree relative to the pipe axis, plane strain assumptions are no longer valid. The constitutive matrix for this case is obtained by transforming the coordinate system as shown in Eq. (2.122). Figures 2.48 and 2.49 show dispersion curves for this case; these curves also show good matching.

$$\begin{Bmatrix} \sigma_{\theta\theta} \\ \sigma_{zz} \\ \sigma_{rr} \\ \sigma_{zr} \\ \sigma_{r\theta} \\ \sigma_{\theta z} \end{Bmatrix} = \begin{bmatrix} 45.9675 & 32.5075 & 7.115 & 0 & 0 & -28.3125 \\ 32.5075 & 45.9675 & 7.115 & 0 & 0 & -28.3125 \\ 7.115 & 7.115 & 14.95 & 0 & 0 & 0.215 \\ 0 & 0 & 0 & 5.27 & -1.46 & 0 \\ 0 & 0 & 0 & -1.46 & 5.27 & 0 \\ -28.3125 & -28.3125 & 0.215 & 0 & 0 & 32.3375 \end{bmatrix} \begin{Bmatrix} \varepsilon_{\theta\theta} \\ \varepsilon_{zz} \\ \varepsilon_{rr} \\ 2\varepsilon_{zr} \\ 2\varepsilon_{r\theta} \\ 2\varepsilon_{\theta z} \end{Bmatrix} \tag{2.122}$$

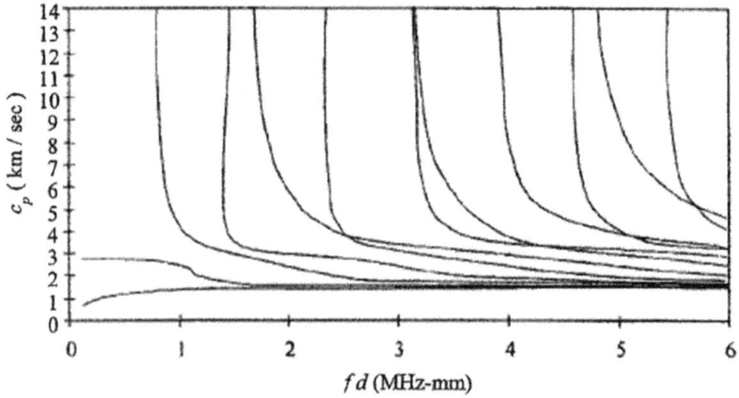

Figure 2.46 Dispersion curves of unidirectional composite plate for waves propagating perpendicular to the fiber direction (fibers are oriented in the y-direction while waves propagate in the x-direction). Material properties are given in Eq. (2.121). Plate thickness = 1 mm, ρ=1580 kg/m^3 [Rose (1999)].

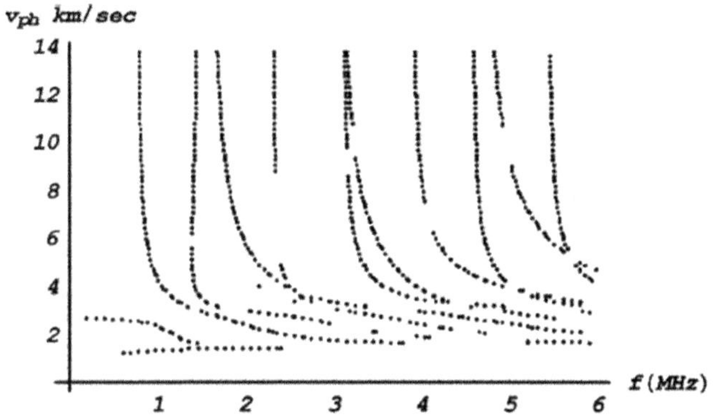

Figure 2.47 Computed dispersion curves for an anisotropic large diameter pipe, when fiber and wave propagation directions are perpendicular to each other (fibers are oriented in the longitudinal direction while waves propagate in the circumferential direction). Material properties are given in Eq. (2.121). Pipe wall thickness = 1 mm. Pipe outer radius = 1000 mm [from Towfighi et al. (2002)].

Unlike the flat plate case for the curved plate, the mid-plane is not the plane of symmetry. Therefore, the dispersion curves cannot be grouped as symmetric and anti-symmetric modes in Figure 2.49. That is why all modes are shown together in Figure 2.49 for a large diameter anisotropic pipe.

2.8.6.3 Comparison of Results for Isotropic Pipes

Qu et al. (1996) have derived dispersion curves for isotropic aluminum pipes. Their results match well with Figure 2.50 obtained by this method with non-dimensional $\bar{k}$ and $\bar{\omega}$ where $\bar{k} = k(b-a)$ and $\bar{\omega} = \omega(b-a)\sqrt{\frac{\rho}{\mu}}$.

2.8.6.4 Anisotropic Pipe of Smaller Radius

To show the effect of the radius of curvature on the dispersion curves, the pipe radius is varied from 1000 mm to 2.5 mm, keeping the wall thickness and material properties same as those for Figures 2.44 and 2.46. Dispersion curves obtained by the 30 terms FS expansion for $r = 1000$, 10, 5 and 2.5 mm are shown in Figures 2.51 and 2.52. Figure 2.51 shows dispersion curves for fibers going in the circumferential direction and Figure 2.52 is for fibers going in the axial direction while the waves propagate in the circumferential directions in both cases.

From Figure 2.51 one can see that for fibers oriented in the circumferential direction the dispersion curves do not change significantly as the outer radius (r) is reduced from 1000 mm to 10 mm. However, when r is reduced further, the deviation of the dispersion curves from the large radius case is no longer negligible. For fibers oriented in the axial direction (Figure 2.52) the dispersion curves remain almost unchanged for $r = 1000$ mm down to 2.5 mm. For r = 2.5 mm the dispersion curves are obtained with $m = 45$ in the FS expansion of amplitude functions. For this computation m was increased to 45 because the computation with $m = 30$ gave too many broken lines in the dispersion curve plot for $r = 2.5$ mm.

In summary, a comparison between Figures 2.51 and 2.52 shows that the effect of the curvature is stronger when the fibers are oriented along the circumferential direction. It should be noted that the fibers also have a curvature if they are oriented in the circumferential direction. When the fibers are oriented in the

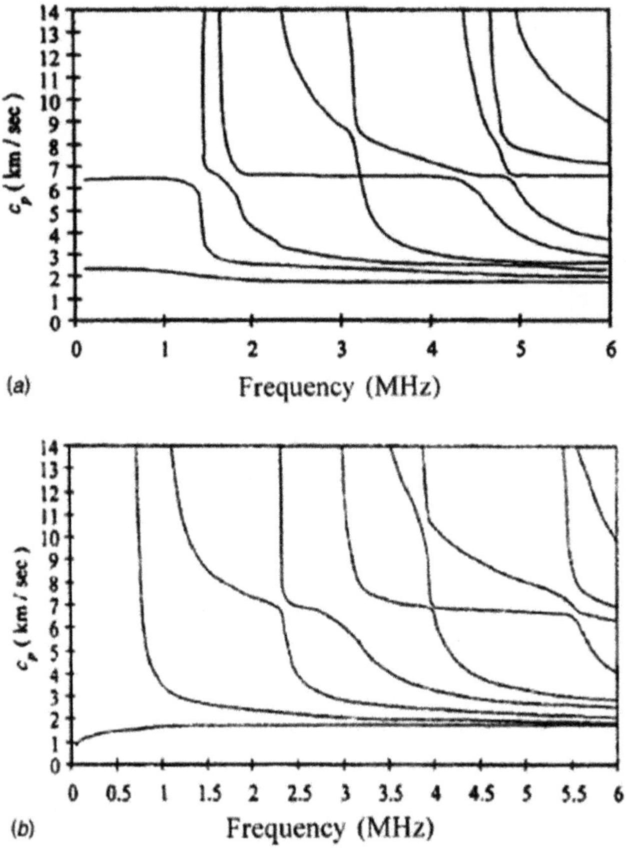

Figure 2.48 Dispersion curves for (a) symmetric and (b) anti-symmetric modes in a unidirectional composite plate when waves propagate at 45° angle relative to the fiber orientation direction. Plate thickness = 1 mm and ρ=1580 kg/m^3 [Rose (1999)].

axial direction and therefore have no curvature, the flat plate approximation can be extended to pipes of much lower radius.

Dispersion curves for the 5 mm outer radius pipe with fibers oriented along the 45° direction are shown in Figure 2.53. These curves are obtained for the material properties given in Eq. (2.122). In the right plot of this figure, although the curves look fine for frequencies greater than 1 MHz, in the frequency range 0 to 1 MHz a number of vertical lines appear due to the numerical errors when the number (m) of FS terms is 25. When m is increasrd to 35 those lines disappear. The results for $m = 35$ in the frequency range 0 to 1 MHz are shown on the left plot of Figure 2.53.

2.9 GUIDED WAVE PROPAGATION IN THE AXIAL DIRECTION OF A PIPE

In spite of having nonzero curvatures, the wave propagation in rods and hollow cylinders is analogous to that in plates. The SH and symmetric modes of plates are analogous to torsional and longitudinal modes in rods. The anti-symmetric

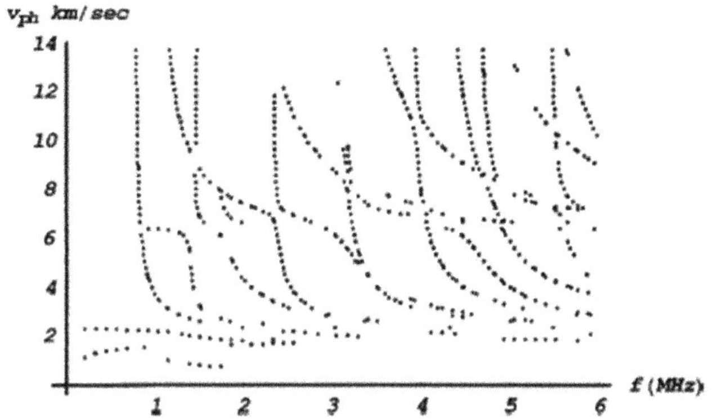

Figure 2.49 Dispersion curves for the circumferential direction wave propagation in a large diameter pipe made of unidirectional fiber-reinforced composite material – fibers are oriented at an angle 45° relative to the wave propagation direction. Material properties are given in Eq. (2.122). Pipe wall thickness = 1 mm. Pipe outer radius = 1000 mm, $m = 25$ [from Towfighi et al. (2002)].

plate modes are analogous to the non-axisymmetric flexural rod modes. The propagation of time-harmonic waves in an infinitely long solid cylinder was first formulated by Pochhammer (1876) and Chree (1886). Similar waves in a hollow circular cylinder for axisymmetric cases were investigated by McFadden (1954) and Herrmann and Mirsky (1956). Gazis (1959a,b) first solved the non-axisymmetric harmonic wave propagation in the axial direction of an infinitely long elastic hollow cylinder. Greenspon (1960a,b) studied dispersion curves and displacement fields for an elastic cylindrical shell. Zemanek (1972) presented numerical analyses of the frequency equation in an elastic cylinder. In the following Gazis' solution is given for cylindrical guided wave propagation in an isotropic pipe.

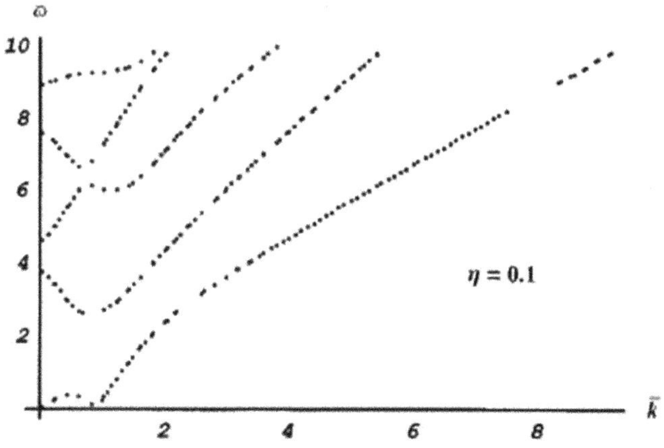

Figure 2.50 Dispersion curves for an aluminum pipe obtained by the method discussed in Section 5.8. η (ratio of inner to outer radius) = 0.1, density = 2.7 gm/cc, P-wave speed = 6.42 km/s, S-wave speed = 3.02 km/s [from Towfighi et al. (2002)].

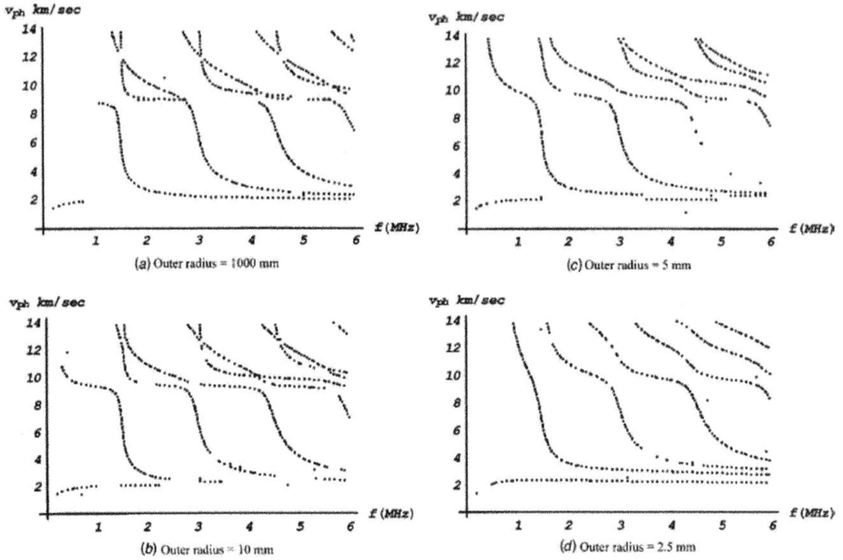

Figure 2.51 Dispersion curves for circumferential direction wave propagation in fiber-reinforced composite cylinders when fibers are oriented in the circumferential direction as well; outer radius of the pipe is (a) 1000 mm, (b) 10 mm, (c) 5 mm, and (d) 2.5 mm. Pipe wall thickness = 1 mm, material properties are given in Eq. (2.120) [from Towfighi et al. (2002)].

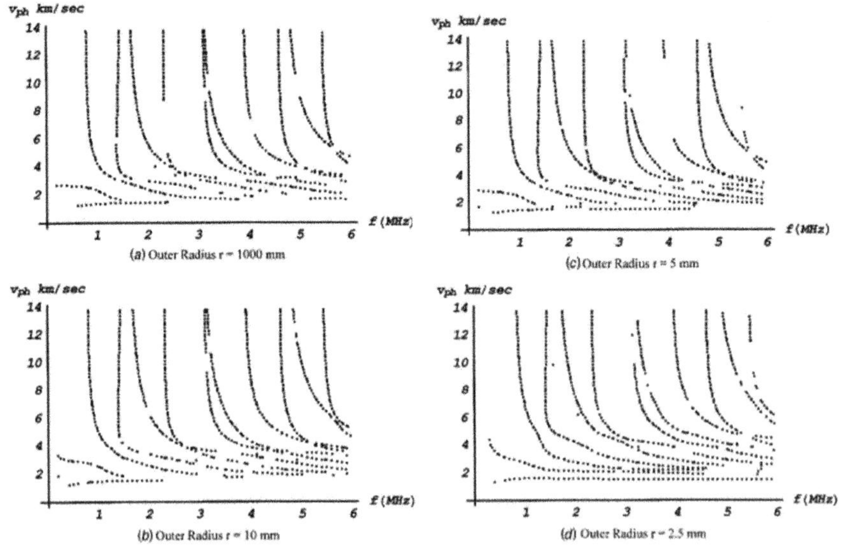

Figure 2.52 Dispersion curves for circumferential direction wave propagation in a fiber-reinforced composite pipe when fibers are oriented in the axial direction, outer radius of the pipe is (a) 1000 mm, (b) 10 mm, (c) 5 mm, and (d) 2.5 mm. Pipe wall thickness = 1 mm, material properties are given in Eq. (2.121) [from Towfighi et al. (2002)].

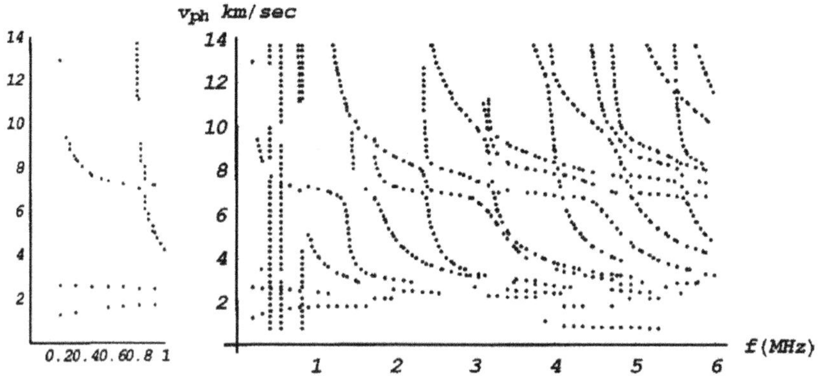

Figure 2.53 Dispersion curves for circumferential direction wave propagation in a fiber-reinforced composite pipe when fibers are oriented in the 45° direction relative to the wave propagation direction. Material properties are given in Eq. (2.122). The outer radius and thickness of the pipe are 5 mm and 1 mm, respectively. Right figure is for $m = 25$, and the left figure is for $m = 35$. Frequency range for the left figure is 0 to 1 MHz while for the right figure it is 0 to 6 MHz [from Towfighi et al. (2002)].

2.9.1 Formulation

A homogeneous cylinder of infinite length as shown in Figure 2.54 is the problem geometry considered by Gazis (1959a). The fundamental governing equation that he solved is the Navier's equation that was introduced in Chapter 1.

$$\mu \nabla^2 \mathbf{u} + (\lambda + \mu) \underline{\nabla}\, \underline{\nabla}.\mathbf{u} = \rho \left(\frac{\partial^2 \mathbf{u}}{\partial t^2} \right) \tag{2.123}$$

In Eq. (2.123) $\mathbf{u}$ is the displacement vector, ρ is the density, λ and μ are Lamé's constants and ∇^2 is the three-dimensional Laplace operator. Following Helmholtz decomposition as discussed in Chapter 1 the vector $\mathbf{u}$ can be expressed in terms of a dilatational scalar potential φ and an equivoluminal vector potential $\mathbf{H}$.

$$\mathbf{u} = \underline{\nabla} \phi + \underline{\nabla} \times \mathbf{H}$$

where (2.124)

$$\underline{\nabla} \cdot \mathbf{H} = F(\mathbf{r}, t)$$

In the above equation, F is a function of coordinate vector $\mathbf{r}$ and time t. The equation of motion is satisfied if the potential functions ϕ and $\mathbf{H}$ satisfy the following wave equations:

$$c_P^2 \nabla^2 \phi = \frac{\partial^2 \phi}{\partial t^2} \tag{2.125}$$

$$c_S^2 \nabla^2 \mathbf{H} = \frac{\partial^2 \mathbf{H}}{\partial t^2} \tag{2.126}$$

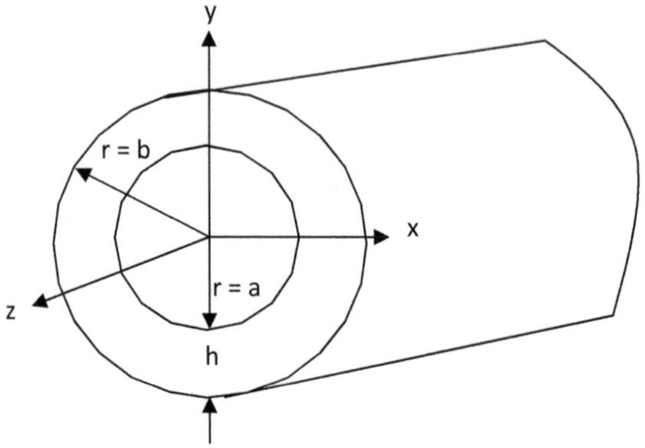

Figure 2.54 Coordinate directions used for the wave propagation analysis in an axisymmetric cylinder in the longitudinal or axial direction.

where c_P and c_S are longitudinal and shear wave speeds, respectively, defined in Chapter 1.

Gazis (1959a) suggested that the above equations can be solved by assuming

$$\begin{aligned}
\phi &= f(r)\cos n\theta\cos(\omega t+\xi z)\\
H_r &= g_r(r)\sin n\theta\sin(\omega t+\xi z)\\
H_\theta &= g_\theta(r)\cos n\theta\sin(\omega t+\xi z)\\
H_z &= g_3(r)\sin n\theta\cos(\omega t+\xi z)
\end{aligned} \tag{2.127}$$

Substituting the above expressions of ϕ and $\mathbf{H}$ into Eqs. 2.125 and 2.126, respectively, one obtains

$$\begin{aligned}
&\left(\nabla^2+\omega^2/c_P^2\right)\phi=0\\
&\left(\nabla^2+\omega^2/c_S^2\right)H_z=0\\
&\left(\nabla^2-1/r^2+\omega^2/c_S^2\right)H_r-\left(2/r^2\right)\left(\partial H_\theta/\partial\theta\right)=0\\
&\left(\nabla^2-1/r^2+\omega^2/c_S^2\right)H_\theta+\left(2/r^2\right)\left(\partial H_r/\partial\theta\right)=0
\end{aligned} \tag{2.128}$$

The differential operator $\mathrm{B}_{n,x}$ is then defined in the following form:

$$\mathrm{B}_{n,x}=\left[\frac{\partial^2}{\partial x^2}+\frac{1}{x}\frac{\partial}{\partial x}-\left(\frac{n^2}{x^2}-1\right)\right] \tag{2.129}$$

Substitution of Eq. (2.127) into Eq. (2.128) yields the following equations after some manipulations:

$$\begin{aligned}
&B_{n,\alpha r}\left[f\right]=0\\
&B_{n,\beta r}\left[g_3\right]=0\\
&B_{n+1,\beta r}\left[g_r-g_\theta\right]=0\\
&B_{n-1,\beta r}\left[g_r+g_\theta\right]=0
\end{aligned} \tag{2.130}$$

where

$$\alpha^2=\omega^2/c_P^2-\xi^2, \qquad \beta^2=\omega^2/c_S^2-\xi^2, \tag{2.131}$$

The general solution of Eq. (2.130) can be given in terms of the n-th order Bessel functions J_n and Y_n, or the modified Bessel functions I_n and K_n of arguments $\alpha_1 r=|\alpha r|$ and $\beta_1 r=|\beta r|$, depending on whether α and β obtained from Eq. (2.131) are real or imaginary. The general solution of Eq. (2.130) is

$$\begin{aligned}
&f=AZ_n(\alpha_1 r)+BW_n(\alpha_1 r)\\
&g_3=A_3Z_n(\beta_1 r)+B_3W_n(\beta_1 r)\\
&2g_1=(g_r-g_\theta)=2A_1Z_{n+1}(\beta_1 r)+2B_1W_{n+1}(\beta_1 r)\\
&2g_2=(g_r+g_\theta)=2A_2Z_{n-1}(\beta_1 r)+2B_2W_{n-1}(\beta_1 r)
\end{aligned} \tag{2.132}$$

where:
- Z is used to denote J or I function
- W denotes Y or K function

The proper selection of Bessel functions in different intervals of frequency ω is shown in Table 2.4.

The property of Gauge Invariance [Morse and Feshback (1954)] can now be utilized in order to eliminate two of the integration constants. Any one of the three potentials g_i, ($i=1$, 2 or 3) of Eq. (2.123) can be set equal to zero, without loss of the generality of solution. Physically, it implies that the displacement field corresponding to an equivoluminal potential g_i of Eq. (2.132) can also be derived by a combination of two other equivoluminal potentials, setting one of them equal to zero. If we take $g_2=0$, then we obtain

$$g_r=-g_\theta=g_1 \tag{2.133}$$

Table 2.4: Types of Bessel Functions used at Different Intervals of Frequency ω

$\omega>c_P\xi$	$c_P\xi>\omega>c_S\xi$	$\omega<c_S\xi$
$Z_n(\alpha_1 r)=J_n(\alpha_1 r)$	$Z_n(\alpha_1 r)=I_n(\alpha_1 r)$	$Z_n(\alpha_1 r)=I_n(\alpha_1 r)$
$W_n(\alpha_1 r)=Y_n(\alpha_1 r)$	$W_n(\alpha_1 r)=K_n(\alpha_1 r)$	$W_n(\alpha_1 r)=K_n(\alpha_1 r)$
$Z_n(\beta_1 r)=J_n(\beta_1 r)$	$Z_n(\beta_1 r)=J_n(\beta_1 r)$	$Z_n(\beta_1 r)=I_n(\beta_1 r)$
$W_n(\beta_1 r)=Y_n(\beta_1 r)$	$W_n(\beta_1 r)=Y_n(\beta_1 r)$	$W_n(\beta_1 r)=K_n(\beta_1 r)$

The displacement field in cylindrical coordinates can be obtained from the following equations [Auld (1973)]:

$$
\begin{aligned}
u_r &= \frac{\partial \phi}{\partial r} + \frac{1}{r}\frac{\partial H_z}{\partial \theta} - \frac{\partial H_\theta}{\partial z} \\
u_\theta &= \frac{1}{r}\frac{\partial \phi}{\partial \theta} + \frac{\partial H_r}{\partial z} - \frac{\partial H_z}{\partial r} \\
u_z &= \frac{\partial \phi}{\partial z} + \frac{1}{r}\frac{\partial}{\partial z}\left(r H_\theta\right) - \frac{1}{r}\frac{\partial H_r}{\partial \theta}
\end{aligned}
\tag{2.134}
$$

Applying Eq. (2.134) into Eq. (2.132) and utilizing the relation (2.133), the displacement field is obtained in the following form:

$$
\begin{aligned}
u_r &= \left(f' + \xi g_1 + \frac{n}{r} g_3\right)\cos(n\theta)\cos(\omega t + \xi z) \\
u_\theta &= \left(-\frac{n}{r} f - + \xi g_1 - g_3{}'\right)\sin(n\theta)\cos(\omega t + \xi z) \\
u_z &= -\left(\xi f + \frac{(n+1)}{r} g_1 + g_1{}'\right)\cos\,(n\theta)\sin(\omega t + \xi z)
\end{aligned}
\tag{2.135}
$$

Where the prime notation indicates the derivatives with respect to *r*, *n* denotes the circumferential order of the mode, which represents the integer number of wavelengths around the circumference of the pipe.

The strain–displacement relations in cylindrical coordinate system has been given in Chapter 1:

$$
\begin{aligned}
\varepsilon_{rr} &= \frac{\partial u_r}{\partial r} \\
\varepsilon_{rz} &= (1/2)\left[\frac{\partial u_r}{\partial z} + \frac{\partial u_z}{\partial r}\right] \\
\varepsilon_{r\theta} &= (1/2)\left[r\frac{\partial}{\partial r}\left(\frac{u_\theta}{r}\right) + \frac{1}{r}\frac{\partial u_r}{\partial \theta}\right]
\end{aligned}
\tag{2.136}
$$

and the stress–strain relations are

$$
\begin{aligned}
\sigma_{rr} &= \lambda\left(\varepsilon_{rr} + \varepsilon_{\theta\theta} + \varepsilon_{zz}\right) + 2\mu\,\varepsilon_{rr} \\
\sigma_{rz} &= 2\mu\,\varepsilon_{rz} \\
\sigma_{r\theta} &= 2\mu\,\varepsilon_{r\theta}
\end{aligned}
\tag{2.137}
$$

Note that

$$
\left(\varepsilon_{rr} + \varepsilon_{\theta\theta} + \varepsilon_{zz}\right) = \nabla^2\phi = -\left(\alpha^2 + \xi^2\right) f \cos(n\theta)\cos(\omega t + \xi z) \tag{2.138}
$$

Using the stress–strain and the strain–displacement relations, the stresses in terms of the displacement potentials are obtained:

$$\sigma_{rr}=\left\{-\lambda\left(\alpha^2+\xi^2\right)f+2\mu\left[f''+\frac{n}{r}\left(g_3'-\frac{g_3}{r}\right)+\xi g_1'\right]\right\}\cos(n\theta)\cos(\omega t+\xi z)$$

$$\sigma_{r\theta}=\mu\left\{-\frac{2n}{r}\left(f'-\frac{f}{r}\right)-\left(2g_3''-\beta^2 g_3\right)-\xi\left(\frac{n+1}{r}g_1-g_1'\right)\right\}\sin(n\theta)\cos(\omega t+\xi z) \quad (2.139)$$

$$\sigma_{rz}=\mu\left\{-2\xi f'-\frac{n}{r}\left[g_1'+\left(\frac{n+1}{r}-\beta^2+\xi^2\right)g_1\right]-\frac{n\xi}{r}g_3\right\}\cos(n\theta)\sin(\omega t+\xi z)$$

Traction-free boundary conditions on outer and inner surfaces of the pipe implies that at $r=a$ and b

$$\sigma_{rr}=\sigma_{r\theta}=\sigma_{rz}=0 \quad (2.140)$$

Substituting Eq. (2.139) into Eq. (2.140) we obtain a system of six homogeneous equations in the following form:

$$[D]\{A\}=\{0\} \quad (2.141)$$

where:

[D] is a 6×6 square matrix

{A} is a 6×1 vector whose elements are A, B, A_1, B_1, A_3 and B_3 of Eq. (2.132)

For a nontrivial solution of {A} the determinant of matrix [D] must vanish. This condition, Det[D] = 0, gives the characteristic equation or the dispersion equation for the guided wave propagation in the axial direction of an isotropic pipe. Elements of this [D] matrix have been given by Gazis (1959a) as shown below:

$$D_{11}=\left[2n(n-1)\left(\beta^2-\xi^2\right)a^2\right]Z_n(\alpha_1 a)+2\lambda_1\alpha_1 a Z_{n+1}(\alpha_1 a)$$

$$D_{12}=2\xi\beta_1 a^2 Z_n(\beta_1 a)\,(\beta_1 a)-2\xi a(n+1)Z_{n+1}(\beta_1 a)$$

$$D_{13}=-2n(n-1)Z_n(\beta_1 a)+2\lambda_2 n\beta_1 a Z_{n+1}(\beta_1 \alpha)$$

$$D_{14}=\left[2n(n-1)-\left(\beta^2-\xi^2\right)a^2\right]W_n(\alpha_1 a)+2\alpha_1 a W_{n+1}(\alpha_1 a)$$

$$D_{15}=2\lambda_2\xi\beta_1 a^2 W_n(\beta_1 a)-2(n+1)\xi a W_{n+1}(\beta_1 a)$$

$$D_{16}=-2n(n-1)W_n(\beta_1 a)+2n\beta_1 a W_{n+1}(\beta_1 a)$$

$$D_{21}=2n(n-1)Z_n(\alpha_1 a)-2\lambda_1 n\alpha_1 a Z_{n+1}(\alpha_1 a) \quad (2.142)$$

$$D_{22}=\xi\beta_1 a^2 Z_n(\beta_1 a)+2\xi a(n+1)Z_{n+1}(\beta_1 a)$$

$$D_{23}=-\left[2n(n-1)-\beta^2 a^2\right]Z_n(\beta_1 a)-2\lambda_2\beta_1 a Z_{n+1}(\beta_1 a)$$

$$D_{24}=2n(n-1)W_n(\alpha_1 a)-2n\alpha_1 a W_{n+1}(\alpha_1 a)$$

$$D_{25}=-\lambda_2\xi\beta_1 a^2 W_n(\beta_1 a)+2(n+1)\xi a W_{n+1}(\beta_1 a)$$

$$D_{26}=-\left[2n(n-1)-\beta^2 a^2\right]W_n(\beta_1 a)-2\beta_1 a W_{n+1}(\beta_1 a)$$

$$D_{31}=2n\xi\alpha_1 Z_n(\alpha_1 a)-2\lambda_1\xi\alpha_1 a^2 Z_{n+1}(\alpha_1 a)$$

$$D_{32} = n\beta_1 a Z_n(\beta_1 a) - (\beta^2 - \xi^2)a^2 Z_{n+1}(\beta_1 a)$$

$$D_{33} = -n\xi a Z_n(\beta_1 a)$$

$$D_{34} = 2n\xi a W_n(\alpha_1 a) - 2\xi\alpha_1 a^2 W_{n+1}(\alpha_1 a)$$

$$D_{35} = \lambda_2 n\beta_1 a W_n(\beta_1 a) - (\beta^2 - \xi^2)a^2 W_{n+1}(\beta_1 a)$$

$$D_{36} = -n\xi a W_n(\beta_1 a)$$

The remaining three rows are obtained from the first three rows simply by substituting a by b. In Eq. (2.142) λ_1 and λ_2 are 1 when J and Y functions are used and –1 when I and K functions are used. Refer to Table 2.4 to see when these parameter values should be equated to 1 or –1.

2.9.2 Use of Cylindrical Guided Waves for Damage Detection in Pipe wall

In the past, guided wave modes were used to detect wall thinning in cylindrical pipes [Silk and Bainton (1979)]. More recently, Lowe et al. (1998a,b), Guo and Kundu (2000, 2001), Cawley et al. (2003), Alleyne and Cawley (2003), Demma et al. (2004), Rose et al. (1994a,b, 2003), Hay and Rose (2002) and Barshinger et al. (2002) among others successfully used cylindrical guided waves for detecting cracks, holes and corrosion damages in pipes. Hay and Rose (2002) have developed comb transducers for guided wave mode generation in a pipe using flexible PVDF films. Guo and Kundu (2000, 2001) designed an innovative transducer holder mechanism for generating cylindrical guided waves for pipe inspection. Using these transducer holders Na and Kundu (2002a,b) Na et al. (2002) and Ahmad et al. (2009) generated cylindrical guided waves for detecting pipe wall damage in underwater and underground empty, water carrying, and concrete filled pipes. Vasiljevic et al. (2008) detected pipe wall anomalies by generating guided waves using noncontact EMAT transducers.

The steps necessary for defect detection in pipes using cylindrical guided waves are similar to the steps used for defect detection in plates by Lamb waves, discussed earlier in this chapter. These steps are briefly discussed below.

Step 1 – Generate Dispersion Curves: Following the formulation outlined in Section 2.9.1, generate the dispersion curves for the given pipe geometry and material properties. Dispersion curves for an aluminum pipe (22.23 mm outer diameter and 1.59 mm wall thickness) are shown in Figure 2.55 [Na and Kundu (2002a)]. In this figure different wave modes are marked following the convention discussed by Silk and Bainton (1979). This convention tracks the modes by their type, their circumferential order, and their consecutive order. This labeling system assigns each mode to one of three types –

a. Longitudinal axially symmetric modes – L(0,m) (m = 1, 2, 3, 4 …),

b. Torsional axially symmetric modes – T(0,m) (m = 1, 2, 3, 4 …) and

c. Non-axially symmetric flexural modes – F(n,m) (n = 1, 2, 3, 4 … m = 1, 2, 3, 4 …)

In this convention the first integer n reflects modes of flexing of the pipe as a whole while the second integer m reflects the modes of vibration within the wall of the pipe. For the second index m, the fundamental modes are given the value 1 and the higher order modes are numbered consecutively. Using this convention,

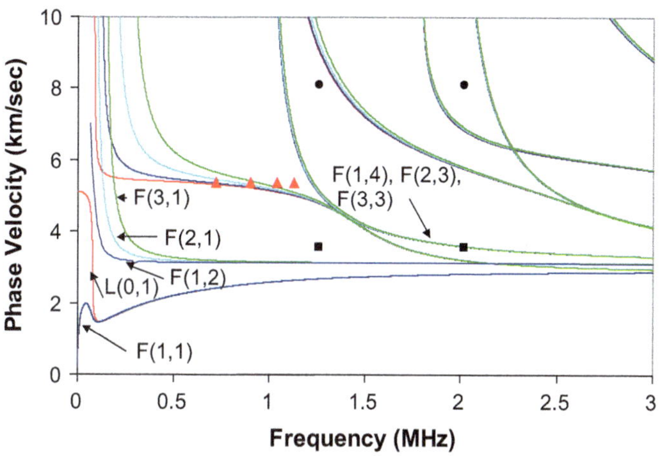

Figure 2.55 Phase velocity dispersion curves for the axial direction wave propagation in the aluminum pipe in vacuum – longitudinal, 1st, 2nd and 3rd flexural modes are shown. Solid squares, triangles and circles correspond to the experimentally obtained peak positions of the *V(f)* curves for 51°, 31° and 20° transmitter angles, respectively [from Na and Kundu (2002a)].

for example, the second longitudinal mode is L(0,2) and the fundamental torsional mode is T(0,1). In the limiting case as the radius of curvature of the pipe approaches infinity the class of torsional modes T(0,*m*) correspond to the antiplane guided wave modes in the plate (see Figure 1.27) and the class of longitudinal modes L(0,*m*) approach the Lamb wave modes in the plate when L(0,1) and L(0,2) correspond to A_0 and S_0 modes, respectively [Silk and Bainton (1979)]. L(0,*m*) and T(0,*m*) modes are sensitive to changes in the wall thickness. The same can be said for the F(*n*,*m*) modes, but these will also be sensitive to anything that affects the pipe flexural properties. This will include such factors as the presence of mountings and constraints which are not of direct interest for pipe wall damage detection and it is thus prudent to confine testing as far as possible to the L(0,*m*) or T(0,*m*) classes.

Step 2 – Generate Cylindrical Guided Waves in the Pipe: Guided waves can be generated in the pipe by placing the transmitter in direct contact with the pipe as shown in Figure 2.56a. Many investigators use this simple method for generating guided waves. However, then multiple guided wave modes may be generated. Guo and Kundu (2001) suggested to place the transmitter at an inclined position relative to the pipe axis and use a coupling fluid (typically water) in a small container between the pipe and the transducer as shown in Figure 2.56b, c to generate a specific guided wave mode using Snell's law [see Eq. (2.31)]. Note that the acoustic wave speed in the coupling fluid c_f is known and the phase velocity c_L of the desired guided wave mode in the pipe is obtained from the dispersion curves generated in step 1. A cylindrical pool and a conical pool are shown in Figure 2.56b, c respectively. Between these two arrangements the conical pool was found to produce better (more consistent and less noisy) wave modes. When pipes are not placed horizontally, Guo and Kundu (2001) used a solid coupler made of Plexiglas with a conical inner surface but spherical outer surface on which the transmitters can easily slide (Figure 2.56d). Photographs of the conical water container and solid coupler are shown in Figure 2.57. Hay and Rose (2002) used comb transducers to

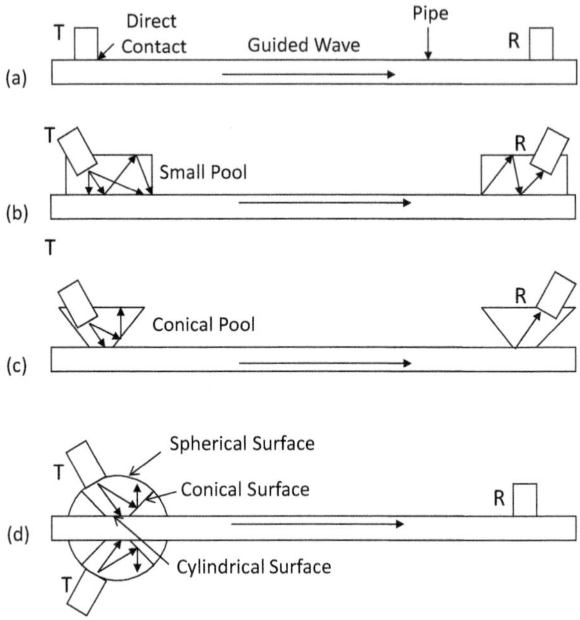

Figure 2.56 Different possible coupling mechanisms between the transmitter and the pipe: (a) direct contact, (b) a small pool containing coupling fluid, (c) a conical pool containing coupling fluid, and (d) an annular plexiglas having spherical outer surface, conical inner surface and a cylindrical surface tightly fitted on the outer surface of the pipe [from Guo and Kundu (2001)].

generate specific guided wave modes in pipes instead of the coupling mechanisms discussed here.

Step 3 – Inspect the Sensitivity of Guided Wave Modes to Defects: Guided waves are generated by a transmitting transducer as discussed in step 2 and received by a receiving transducer placed at another location of the pipe as shown in Figure 2.56. Although the transmitter should be inclined at an angle (equal to the critical angle determined from Snell's law) the receiving transducer can be placed in direct contact with the pipe as shown in Figure 2.56d. The received signal is expected to be altered by any defect located on the path of the propagating guided wave. Sensitivity of different guided wave modes to various types of pipe wall defect at different frequencies are expected to be different. Therefore, the sensitivity of different guided wave modes to various types of pipe wall damage requires a thorough investigation to design an efficient technique of pipe inspection using guided waves. One interesting observation made and reported by Guo and Kundu (2001) and Na and Kundu (2002a) is that when the guided wave modes are generated near the horizontal asymptotic lines of the dispersion curves (where the square and triangular markers are shown in Figure 2.55) the signals are more sensitive to defects than when the modes are generated at other locations (for example the locations marked by the circular markers in Figure 2.55). Horizontal curves imply phase velocity being independent of the frequency, hence no or very little dispersion. Therefore, modes having small dispersion are more sensitive to defects. Figure 2.58 shows strengths of the received signal as a function of

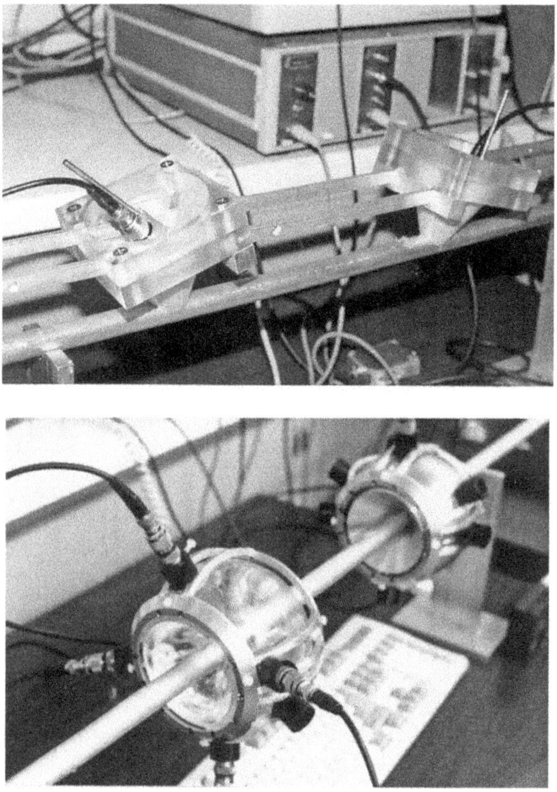

Figure 2.57 The top and bottom photographs correspond to two different coupling mechanisms between the transmitter and the pipe – corresponding schematic diagrams are shown in Figure 2.56c,d, respectively.

frequency when the transmitter is excited by a tone-burst (multiple cycles of single frequency) signal while sweeping the frequency between 300 and 2600 kHz. Five curves are plotted in each figure – one corresponding to the defect-free pipe and the remaining four corresponding to four different types of defect on the pipe wall [Na and Kundu (2002a)]. Two distinct peaks in each figure correspond to different guided wave modes. The frequency–phase velocity combinations corresponding to these two peaks are denoted by two solid circles and two solid squares in Figure 2.55. Circles (showing phase velocity greater than 8 km/s) correspond to the two peaks at the top of Figure 2.58 while squares (phase velocity below 4 km/s) represent the two peaks of the bottom of Figure 2.58. The phase velocity can be easily changed by simply varying the inclination angle of the transmitter, as stated earlier. Since dispersion curves for several modes are located near these points, any of these modes can contribute to these two peaks. The fact that these experimental points (denoted by circles and squares in Figure 2.55) don't perfectly coincide with the theoretical dispersion curves simply means the aluminum material properties used for generating the dispersion curves are slightly different from the properties of the aluminum pipe used in the experiment. It is interesting to note that the curves in Figure 2.58 (in particular the peak near 1300 kHz frequency) are more sensitive to defects. This is because the frequency–phase

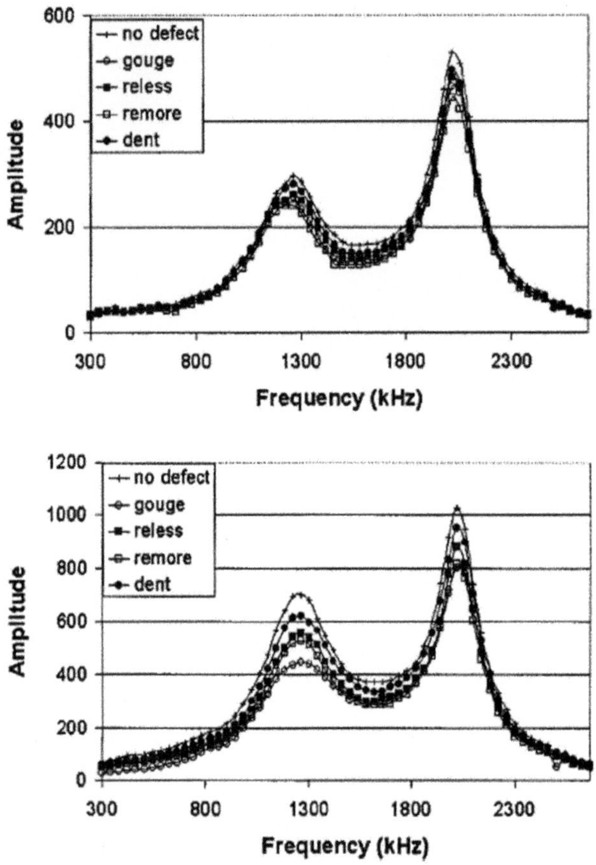

Figure 2.58 *V(f)* curves or voltage versus frequency curves recorded by the receiver for axial direction wave propagation in an aluminum pipe (whose dispersion curves are shown in Figure 2.55) for 20° (top figure) and 51° (bottom figure) transmitter angles. Five curves in each figure correspond to four defective pipes and one defect-free pipe. The transmitter–receiver arrangements used for generating these curves are shown in Figure 2.56d [from Na and Kundu (2002a)].

velocity combinations used for generating those plots are closer to the horizontal asymptotic line of the dispersion curves (square markers in Figure 2.55) than those for the upper figure (circular markers in Figure 2.55).

Step 4 – Use the Most Sensitive Mode for Defect Detection: After identifying the guided wave mode that is most sensitive to the pipe wall defects, use that mode for pipe wall damage detection if that mode can propagate sufficient distance along the pipe. If it attenuates too fast, then select another mode that can propagate sufficient distance along the pipe and is also reasonably sensitive to the defect. Attenuation of a specific mode can be studied experimentally by generating that specific mode at one location of the pipe and then investigating how far that mode can propagate before losing its strength. It can also be determined analytically from the attenuation dispersion curves of the pipe [Rose (1999)].

Although the curves of Figure 2.58 are generated for the pipe wall damages located between the transmitter and the receiver, it is not necessary for the defect to always be between the transducers to be detected. After being reflected at the pipe ends or at the defect, guided waves propagate several times through the entire length of the pipe before losing their strength completely. Therefore, if the defect is not located in the pipe segment between the transmitter and the receiver it can still be detected. Readers are referred to Vasiljevic et al. (2008) and Shelke et al. (2012) to study more on detecting pipe wall defects that are not located between the transmitter and the receiver.

2.10 CONCLUDING REMARKS

The first seven sections of this chapter are devoted to a detailed analysis of guided wave propagation in plates. Cases considered include isotropic and anisotropic plates, homogeneous and multilayered plates, plates with and without material attenuation, and plates vibrating naturally (free vibration) and under external excitations (forced vibration). These seven sections present the fundamental theory and its applications in designing ultrasonic nondestructive evaluation (NDE) experiments for damage detection in plates. Some exercise problems from these sections are also provided at the end of this chapter.

In Sections 2.8 and 2.9 theory and NDE applications of guided wave propagations in pipes are presented. Section 2.8 deals with the wave propagation theory in the circumferential direction while Section 2.9 analyzes wave propagation in the axial direction of a pipe. The solution technique presented in Section 2.8 is a more general technique capable of handling anisotropic pipes, and therefore more powerful than the classical methods based on Stokes-Helmhotz decomposition used for studying wave propagations in isotropic pipes. Presented results show the effect of the curvature change on the dispersion curves. The solution technique used in this section is a general solution technique proposed by Towfighi (2001) for solving coupled partial differential equation sets. Applicability of this technique to other problems such as the wave propagation in spherical shell structures has been demonstrated by Towfighi and Kundu (2003). The technique presented here for the homogeneous anisotropic pipe can easily be extended to multilayered geometries [Vasudeva et al. (2008)] and is not discussed in this chapter. Although the wave propagation in the circumferential direction of an isotropic pipe is a special case of the general solution presented in Section 2.8 for the anisotropic pipe, the technique for solving wave propagation in isotropic pipes using Stokes-Helmholtz decomposition [Qu and Jacobs (2004)] is computationally much simpler and should be followed for the isotropic case. Finally, in Section 2.9 the wave propagation in the axial direction of a pipe is presented, limiting the solution technique and the NDE applications to the isotropic case only.

EXERCISE PROBLEMS

PROBLEM 2.1

Give the expression of σ_{11} for symmetric and anti-symmetric modes for the Lamb wave propagating in the x_1 direction in a plate of thickness 2h as shown in Figure 1.29.

PROBLEM 2.2

For a solid plate in a vacuum the dispersion equations are given in Eq. (2.1), and when it is immersed in a fluid the dispersion equations are given in Eqs. (2.21) and (2.24).

1. If the P-wave speed in the fluid is a finite nonzero value α_f, and the fluid density is close to zero, then should Eq. (2.1) approximately give the dispersion equations for the plate immersed in that fluid? Justify your 'yes' or 'no' answer.
2. If the P-wave speed in the fluid is close to 0 and the fluid density is a finite nonzero value ρ_f, then should Eq. (2.1) approximately give the dispersion equations for the plate immersed in that fluid? Justify your 'yes' or 'no' answer.

PROBLEM 2.3

Stonely-Scholte waves propagate along a fluid–solid interface (at $x_2 = 0$) with wave amplitudes decaying exponentially in both fluid ($x_2 > 0$) and solid ($x_2 < 0$) media.

1. Obtain an equation from which the Stonely-Scholte wave speed can be computed.
2. Is this Stonely-Scholte wave dispersive?
3. Specialize the dispersion equation in the limiting case as ρ_f approaches zero.

PROBLEM 2.4

Stonely waves propagate along the interface (at $x_2 = 0$) of two solids with wave amplitudes decaying exponentially in both solids (at $x_2 > 0$ and $x_2 < 0$). Obtain the dispersion equation from which the Stonely wave speed can be computed. [Note that when one solid half-space is replaced by a fluid half-space then the interface wave generated at the fluid–solid interface is a special case of Stonely wave and is known as Stonely-Scholte wave or Scholte wave].

PROBLEM 2.5

1. Phase velocity of the guided antiplane wave propagating through a plate of thickness 2 h that is fixed at the bottom surface and free (traction-free) at the top surface (see Figure 2.59) is given by

$$c_m = \frac{c_S}{\sqrt{1 - f(c_S, m)}}$$

   Derive the expression of the function $f(c_S, m)$.

2. Plot the variation of the phase velocity c_0 as a function of frequency ω for the mode that has the lowest cut-off frequency. This mode is denoted as the fundamental mode. In your plot show the cut-off frequency (if any), and the asymptotic value of the curve. Is this mode dispersive?

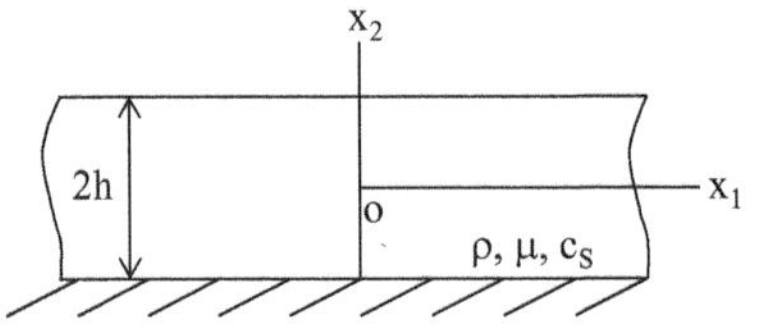

Figure 2.59 Problem Geometry for Problem 2.5.

3. Obtain and plot the mode shape for the fundamental mode.
4. Show how the mode shape that you have obtained in part c is satisfying the boundary conditions.

PROBLEM 2.6

By expanding the Lamb wave dispersion equation for anti-symmetric (or flexural) modes at low frequencies, prove that the wave speed (c_L) for the fundamental flexural mode at low frequency is approximately given by

$$c_L = \left(\frac{4\mu(\lambda+\mu)h^2}{3\rho(\lambda+2\mu)} \right)^{1/4} \omega^{1/2}$$

where:

- 2h is the plate thickness
- ρ the density
- ω the wave frequency in radian/second
- λ and μ are Lame's first and second constants, respectively

PROBLEM 2.7

ϕ_{fL} and ϕ_{fU} of Eqs. (2.17) and (2.18) are defined differently while keeping φ and Ψ definitions unchanged, as shown below:

Symmetric Motion

$$\phi = B\cosh(\eta x_2)e^{ikx_1}$$

$$\psi = C\sinh(\beta x_2)e^{ikx_1}$$

$$\phi_{fL} = \phi_{fU} = M\cosh(\eta_f x_2)e^{ikx_1}$$

Anti-symmetric Motion

$$\phi = A\sinh(\eta x_2)e^{ikx_1}$$

$$\psi = D\cosh(\beta x_2)e^{ikx_1}$$

$$\phi_{fL} = \phi_{fU} = N\sinh(\eta_f x_2)e^{ikx_1}$$

Prove that for the above definitions the dispersion equations take the following form:

For Symmetric Motion

$$\left(2k^2 - k_S^2\right)^2 \cosh(\eta h)\sinh(\beta h) - 4k^2\eta\beta\sinh(\eta h)\cosh(\beta h) = \frac{\rho_f \eta k_S^4}{\rho\eta_f} \frac{\sinh(\eta h)\sinh(\beta h)}{\tanh(\eta_f h)}$$

For Anti-symmetric Motion

$$\left(2k^2 - k_S^2\right)^2 \sinh(\eta h)\cosh(\beta h) - 4k^2\eta\beta\cosh(\eta h)\sinh(\beta h) = \frac{\rho_f \eta k_S^4}{\rho\eta_f} \tanh(\eta_f h)\cosh(\eta h)\cosh(\beta h)$$

Note that the above dispersion equations are different from those given in Eqs. (2.21) and (2.24).

Explain why this difference appears. In other words, what are the shortcomings of the above definitions of ϕ_{fL} and ϕ_{fU}?

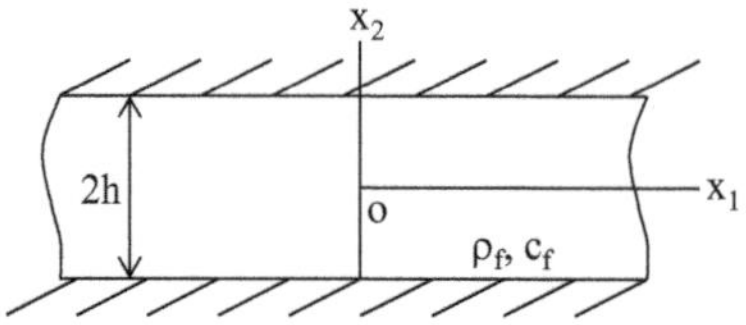

Figure 2.60 Fluid layer of thickness 2h, trapped between two rock half-spaces, for Problem 2.9.

PROBLEM 2.8

Calculate the inclination angles (measured from the axis normal to the plate surface) of two narrowband ultrasonic transducers of central frequency 1 MHz and 5 MHz to generate A_0, S_0, A_1 and S_1 modes in a 1 mm thick aluminum plate when the plate is immersed in water (wave speed = 1.48 km/s) and acetone (wave speed = 1.17 km/s). You can assume that the dispersion curves for aluminum, shown in Figure 2.6, are not significantly affected when the plate is immersed in water or acetone.

PROBLEM 2.9

1. A fluid layer of thickness 2 h is trapped between two huge rocks as shown in Figure 2.60. The fluid density is ρ_f and P-wave speed in the fluid is c_f. Obtain the dispersion equation for the guided wave propagation through the trapped fluid layer. In your derivation model the rock as rigid.
2. Solve the dispersion equation and plot the variation of the phase velocity as a function of frequency ω, for the first three modes. In your plot show the cut-off frequency (if any), and the asymptotic value of each curve.
3. Obtain and plot the mode shapes (displacement and stress variations) for all three modes.

REFERENCES

Ahmad, R., S. Banerjee, and T. Kundu, "Pipe wall damage detection in buried pipes using guided waves", *ASME Journal of Pressure Vessel Technology*, Vol. 131(1), Article 011501, pp. 011501–8. (2009).

Alleyne, D. N. and P. Cawley, "The interaction of lamb waves with defects", *IEEE Transactions on Ultrasonics, Ferroelectrics, and Frequency Control*, Vol. 39(3), pp. 381–397 (1992).

Alleyne, D. N. and P. Cawley, "Long range propagation of lamb waves in chemical plant pipe work", *Materials Evaluation*, Vol. 55, pp. 504–508 (2003).

Auld, B. A., *Acoustic Fields and Waves in Solids: Volume I*, John Wiley & Sons, Inc, New York, NY, (1973).

Barshinger, J., J. L. Rose, and M. J. Avioli, "Guided wave resonance tuning for pipe inspection", *Journal of Pressure Vessel Technology*, Vol. 124(3), pp. 303–310 (2002).

Brekhovskikh, L. M., "Surface waves confined to the curvature of the boundary in solid", *Soviet Physics – Acoustics*, Vol. 13, pp. 462–472 (1968).

Buchwald, V. T., "Rayleigh waves in transversely isotropic media", *The Quarterly Journal of Mechanics and Applied Mathematics,* Vol. 14(3), pp. 293–318 (1961).

Cawley, P., M. J. S. Lowe, D. N. Alleyne, B. Pavlakovic, and P. Wilcox, "Practical long range guided wave testing: Application to pipes and rail", *Materials Evaluation,* Vol. 61, pp. 66–74 (2003).

Cerv, J., "Dispersion of elastic waves and rayleigh-type waves in a thin disc", *Acta Technica CSAV,* Vol. 89, pp. 89–99 (1988).

Chimenti, D. E. and R. W. Martin, "Nondestructive evaluation of composite laminates by leaky lamb waves", *Ultrasonics,* Vol. 29(1), pp. 13–21 (1991).

Chree, C., "Longitudinal vibrations of a circular bar", *Quarterly Journal of Mathematics,* Vol. 21, pp. 287–298 (1886).

Christesen, R. M., *Mechanics of Composite Materials,* John Wiley, New York (1981), Ch.4.

Demma, A., P. Cawley, M. J. S. Lowe, A. G. Roosenbrand, and B. Pavlakovic, "The reflection of guided waves from notches in pipes: A guide for interpreting corrosion measurements", *NDT and E International,* Vol. 37(3), pp. 167–180 (2004).

Ditri, J. J., J. L. Rose, and G. Chen, "Mode selection criteria for defect detection optimization using lamb waves', In *Review of Progress in Quantitative Nondestructive Evaluation",* eds. D. O. Thompson and D. E. Chimenti, *Pub,* Plenum Press, N.Y., Vol. 11, pp. 2109–2115 (1992).

Dunkin, G.W., "Computation of modal solutions in layered elastic media at high frequencies", *Bulletin of the Seismological Society of America,* Vol. 55, pp. 335–358 (1965).

Dunkin, G. W. and D. G. Corbin, "Deformation of a layered elastic half space by uniformly moving loads", *Bulletin of the Seismological Society of America,* Vol. 60, pp. 167–191, (1970).

Gazis, D. C., "Three dimensional investigation of propagation of waves in hollow circular cylinders. I", *Journal of the Acoustical Society of America,* Analytical Foundation, Vol. 31, pp. 568–573 (1959a).

Gazis, D. C., "Three dimensional investigation of propagation of waves in hollow circular cylinders. II. numerical results", *Journal of the Acoustical Society of America,* Analytical Foundation, Vol. 31, pp. 573–578 (1959b).

Ghosh, T. and T. Kundu, "A new transducer holder mechanism for efficient generation and reception of lamb modes in large plates", *The Journal of the Acoustical Society of America,* Vol. 104(3), pp. 1498–1502 (1998).

Ghosh, T., T. Kundu, and P. Karpur, "Efficient use of lamb modes for detecting defects in large plates", *Ultrasonics,* Vol. 36(7), pp. 791–801 (1998).

Grace, O. D. and R. R. Goodman, "Circumferential waves on solid cyliners", *The Journal of the Acoustical Society of America,* Vol. 39(1), pp. 173–174 (1966).

Greenspon, J. E., "Vibration of a thick-walled cylindrical shell – comparison of the exact theory with approximate theories", *The Journal of the Acoustical Society of America*, Vol. 32(5), pp. 571–578 (1960a).

Greenspon, J. E., "Axially symmetric vibrations of a thick cylindrical shell in an acoustic medium", *The Journal of the Acoustical Society of America*, Vol. 32(8), pp. 1017–1025 (1960b).

Guo, D. and T. Kundu, "A new sensor for pipe inspection by lamb waves", *Materials Evaluation*, Vol. 58(8), pp. 991–994 (2000).

Guo, D. and T. Kundu, "A new transducer holder mechanism for pipe inspection", *The Journal of the Acoustical Society of America*, Vol. 110(1), pp. 303–309 (2001).

Haskell, N. A., "The dispersion of surface waves on multilayered media", *Bulletin of the Seismological Society of America*, Vol. 43, pp. 17–34 (1953).

Hay, T. R. and J. L. Rose, "Flexible PVDF comb transducers for excitation of axisymmetric guided waves in pipe", *Sensors and Actuators A: Physical*, Vol. 100(1), pp. 18–23 (2002).

Herrmann, G. and I. Mirsky, "Three-dimensional and shell-theory analysis of axially symmetric motions of cylinders", *Transactions of the ASME Journal of Applied Mechanics*, Vol. 78, pp. 563–568 (1956).

Huang, W., S. I. Rokhlin, and Y. J. Wang, "Analysis of different boundary condition models for study of wave scattering from fiber-matrix interphases", *The Journal of the Acoustical Society of America*, Vol. 101(4), pp. 2031–2042 (1997).

Karim, M. R., A. K. Mal, and Y. Bar-Cohen, "Inversion of leaky lamb wave data by simplex algorithm", *The Journal of the Acoustical Society of America*, Vol. 88(1), pp. 482–491 (1990).

Kennett, B. L. N., *Seismic Wave Propagation in Stratified Media*, Cambridge University Press, London (1983).

Kinra, V. K. and V. R. Iyer, "Ultrasonic measurement of the thickness, phase velocity, density or attenuation of a thin viscoelastic plate. Part II: The inverse problem", *Ultrasonics*, Vol. 33(2), pp. 111–122 (1995).

Kinra, V. K. and A. Wolfenden, eds., "*M^3D: Mechanics and Mechanisms of Material Damping*", American Society for Testing and Materials, Philadelphia, "Relationship Amongst Various Measures of Damping", p. 3 (1992).

Knopoff, L., "A matrix method for elastic wave problems", *Bulletin of the Seismological Society of America*, Vol. 54, pp. 431–438 (1964).

Kundu, T., "Inversion of acoustic material signature of layered solids", *The Journal of the Acoustical Society of America*, Vol. 91(2), pp. 591–600 (1992).

Kundu, T. and A. K. Mal, "Elastic waves in a multilayered solid due to a dislocation source", *Wave Motion*, Vol. 7(5), pp. 459–471 (1985).

Kundu, T. and K. I. Maslov, "Material interface inspection by lamb waves", *International Journal of Solids and Structures*, Vol. 34(29), pp. 3885–3901 (1997).

Kundu, T., K. I. Maslov, P. Karpur, T. E. Matikas, and P. D. Nicolaou, "A lamb wave scanning approach for mapping defects in [0/90] titanium matrix composites", *Ultrasonics*, Vol. 34(1), pp. 43–49 (1996).

Kundu, T., C. Potel, and J. F. de Belleval, "Importance of the near lamb mode imaging of multilayered composite plates", *Ultrasonics*, Vol. 39(4), pp. 283–290 (2001).

Lamb, H., "On waves in an elastic plate", *Proceedings of the Royal Society A: Mathematical, Physical and Engineering Sciences*, Vol. 93(648), p. 114–128 (1917).

Liu, G. and J. Qu, "Guided circumferential waves in a circular annulus", *ASME Journal of Applied Mechanics*, Vol. 65(2), pp. 424–430 (1998a).

Liu, G. and J. Qu, "Transient wave propagation in a circular annulus subjected to impulse excitation on its outer surface", *Journal of the Acoustical Society of America*, Vol. 103, pp. 1210–1220 (1998b).

Lowe, M. J. S., D. N. Alleyne, and P. Cawley, "Defect detection in pipes using guided waves", *Ultrasonics*, Vol. 36(1–5), pp. 147–154 (1998a).

Lowe, M. J. S., D. N. Alleyne, and P. Cawley, "Mode conversion of a guided wave by a part-circumferential notch in a pipe", *Journal of Applied Mechanics, Transactions of the ASME*, Vol. 65, pp. 649–656 (1998b).

Mal, A. K., "Wave propagation in layered composite laminates under periodic surface loads", *Wave Motion*, Vol. 10(3), pp. 257–266 (1988).

Mal, A. K. and S. J. Singh, *Deformation of Elastic Solids*, Prentice-Hall Inc., Englewood Cliffs, New Jersey, p. 313 (1991).

Mal, A. K., C.-C. Yin, and Y. Bar-Cohen, "Ultrasonic nondestructive evaluation of cracked composite laminates", *Composites Engineering*, Vol. 1(2), pp. 85–101 (1991).

Mal, A. K., Y. Bar-Cohen, and S. -S. Lih, "Wave attenuation in fiber-reinforced composites", In *M^{3}D: Mechanics and Mechanisms of Material Damping, ASTM STP 1169*, eds. V. K. Kinra and A. Wolfenden, American Society for Testing and Materials, Philadelphia, pp. 245–261 (1992).

Maslov, K. I. and T. Kundu, "Selection of lamb modes for detecting internal defects in composite laminates", *Ultrasonics*, Vol. 35(2), pp. 141–150 (1997).

McFadden, J. A., "Radial vibrations of thick-walled hollow cylinders", *The Journal of the Acoustical Society of America*, Vol. 26(5), pp. 714–715 (1954).

Morse, P. M., H. Feshback, and E. L. Hill, "Methods of theoretical physics", *American Journal of Physics*, McGraw Hill Book Company, Inc., New York, Vol. 22(6), pp. 410–413, (1954).

Na, W. B. and T. Kundu, "Underwater pipeline inspection using guided waves", *ASME Journal of Pressure Vessel Technology*, Vol. 124(2), pp. 196–200 (2002a).

Na, W. B. and T. Kundu, "EMAT-based inspection of concrete filled steel pipes for internal voids and inclusions", *ASME Journal of Pressure Vessel Technology*, Vol. 124, pp. 265–272 (2002b).

Na, W. B., T. Kundu, and M.R. Ehsani, "Ultrasonic guided waves for steel bar-concrete interface inspection", *Materials Evaluation*, Vol. 60(3), pp. 437–444 (2002).

Nagy, P. B., L. Adler, D. Mih, and W. Sheppard, "'Single mode lamb wave inspection of composite laminates'", In *Review of Progress in Quantitative NDE*, eds. D. O. Thompson and D. E. Chimenti, *Pub*, Plenum Press, New York, Vol. 8B, pp. 1535–1542 (1989).

Nelder, J. A. and R. Mead, "A simplex method for function minimization", *The Computer Journal*, Vol. 7(4), pp. 308–313 (1965).

Pochhammer, L., "Ueber die Fortpflanzungsgeschwindigkeiten Kleiner Schwingungen in einem unbegrenzten isotropen Kreiscylinder", *Zeitschrift für Mathematik*, Vol. 81, pp. 324–336 (1876).

Qu, J. and L. Jacobs, "Cylindrical waveguides and their applications in ultrasonic evaluation", In *Ultrasonic Nondestructive Evaluation: Engineering and Biological Material Characterization*, ed. T. Kundu, CRC Press, Boca Raton, FL, pp. 311–363 (2004).

Qu, J., Y. Berthelot, and Z. Li, "Dispersion of guided circumferential waves in a circular annulus", In *Review of Progress in Quantitative Nondestructive Evaluation*, eds. D. O. Thompson and D. E. Chimenti, Plenum Press Publishing, New York, Vol. 15, pp. 169–176 (1996).

Rayleigh, Lord, "On waves propagated along the plane surfaces of an elastic solid", *Proceedings of the London Mathematical Society*, Vol. 17, pp. 4–11 (1885).

Rose, J. L. *Ultrasonic waves in solid media*, Cambridge University Press, Cambridge, U.K. (1999).

Rose, J. L., J. J. Ditri, A. Pilarski, K. Rajana, and F. Carr, "A guided wave inspection technique for nuclear steam generator tubing", *NDT & E International*, Vol. 27(6), pp. 307–310 (1994a).

Rose, J. L., K. M. Rajana, and F. T. Carr, "Ultrasonic guided wave inspection concepts for steam generator tubing", *Materials Evaluation*, Vol. 52(2), pp. 134–139 (1994b).

Rose, J. L., Z. Sun, P. J. Mudge, and M. J. Avioli, "Guided wave flexural mode tuning and focusing for pipe inspection", *Materials Evaluation*, Vol. 61, pp. 162–167 (2003).

Scholte, J. G., "On the stonely wave equation", *Proceedings of the Koninklijke Nederlandse Akademie van Wetenschappen*, Vol. 45, pp. 159–164 (1942).

Schwab, F. and L. Knopoff "Surface wave dispersion computations", *Bulletin of the Seismological Society of America*, Vol. 60, pp. 321–344 (1970).

Shelke, A., M. Vasiljevic, T. Kundu, U. Amjad, and W. Grill, "Extracting quantitative information on pipe wall damage in absence of clear signals from defect", *ASME Journal of Pressure Vessel Technology*, Vol. 134(5) (2012).

Silk, M. G. and K. F. Bainton, "The propagation in metal tubing of ultrasonic wave modes equivalent to lamb waves", *Ultrasonics*, Vol. 17(1), pp. 11–19 (1979).

Stonely, R., "Elastic waves at the surface of separation of two solids", *Proceedings of the Royal Society A: Mathematical, Physical and Engineering Sciences*, Vol. 106(738), pp. 416–428 (1924).

Thomson, W. T., "Transmission of elastic waves through a stratified solid medium", *Journal of Applied Physics*, Vol. 21(2), pp. 89–93 (1950).

Towfighi, S., "Elastic wave propagation in circumferential direction in anisotropic pipes" Ph.D. Dissertation, University of Arizona, Tucson, AZ (2001).

Towfighi, S. and T. Kundu, "Elastic wave propagation in anisotropic spherical curved plates", *International Journal of Solids and Structures*, Vol. 40(20), pp. 5495–5510 (2003).

Towfighi, S., T. Kundu, and M. Ehsani, "Elastic wave propagation in circumferential direction in anisotropic cylindrical curved plates", *ASME Journal of Applied Mechanics*, Vol. 69(3), pp. 283–291 (2002).

Valle, C., J. Qu, and L. J. Jacobs, "Guided circumferential waves in layered cylinders", *International Journal of Engineering Science*, Vol. 37(11), pp. 1369–1387 (1999).

Vasiljevic, M., T. Kundu, W. Grill, and E. Twerdowski, "Pipe wall damage detection by electromagnetic acoustic transducer generated guided waves in absence of defect signals", *The Journal of the Acoustical Society of America*, Vol. 123(5), pp. 2591–2597 (2008).

Vasudeva, R. Y., G., Sudheer, and A. R.Verma, "Dispersion of circumferential waves in cylindrically anisotropic layered pipes in plane strain", *The Journal of the Acoustical Society of America*, Vol. 123(6), pp. 4147–4151 (2008).

Viktorov, I. A., "Rayleigh-type waves on a cylindrical surface", *Soviet Physics – Acoustics*, Vol. 4, pp. 131–136 (1958).

Viktorov, I. A., *Rayleigh and Lamb Waves – Physical Theories and Applications*, Plenum Press, New York, NY (1967).

Yang, R.-B. and A. K. Mal, "Elastic waves in a composite containing inhomogeneous fibers", *International Journal of Engineering Science*, Vol. 34(1), pp. 67–79 (1996).

Yang, W. and T. Kundu, "Guided waves in multilayered anisotropic plates for internal defect detection", *ASCE Journal of Engineering Mechanics*, Vol. 124(3), pp. 311–318 (1998).

Zemanek, J., Jr., "An experimental and theoretical investigation of elastic wave propagation in a cylinder", *The Journal of the Acoustical Society of America*, Vol. 51(1B), pp. 265–283 (1972).

3 Modeling Elastic Waves by Distributed Point Source Method (DPSM)

In Chapter 1 we learned how plane waves and spherical waves propagate through a solid or fluid medium. Spherical waves are generated by a point source in an infinite medium, the cylindrical waves are generated by a line source, while an infinite plane source can generate plane waves, as shown in Figure 3.1.

Harmonic waves are generated from harmonic sources that have time dependence $e^{-i\omega t}$. The equation of the propagating spherical wave generated by a point source in a fluid space is given by Eq. (1.223) and equations of a propagating plane waves in a fluid are given in Eq. (1.198) in terms of pressure and in Eq. (1.208) in terms of the wave potential [specializing Eq. (1.208) for x_1 direction wave propagation and dropping harmonic time dependence in Eq. (1.198)]:

$$G(r) = \frac{e^{ik_f r}}{4\pi r} \tag{1.223}$$

$$p(x_1) = e^{ik_f x_1} \tag{1.198}$$

$$\phi(x_1) = e^{ik_f x_1} \tag{1.208}$$

G in Eq. (1.223) can be either pressure or wave potential. The potential–pressure relation is given in Eq. (1.209). k_f is the wave number of the fluid and has been defined in Chapter 1, after Eq. (1.197). In Eq. (1.223), r is the radial distance of the spherical wave front from the point source at the origin. In Eqs. (1.198) and (1.208) x_1 is the propagation direction of the plane wave front.

If the wave sources of Figure 3.1 are located in a homogeneous solid instead of the fluid medium, then only compressional waves are generated in the solid and their expressions can be obtained by simply substituting k_f by k_P where k_P is the P-wave number of the solid. In the absence of any interface or boundary, the mode conversion does not occur and shear waves are not generated from the compressional waves.

In many NDE (nondestructive evaluation) applications elastic waves are generated by a source of finite dimension and the wave fronts are not spherical, cylindrical or plane. Typical dimensions of the commercially available ultrasonic transducers that are most commonly used for ultrasonic wave generation vary from a quarter of an inch to one inch in diameter. Of course, in special applications the ultrasonic source can be much smaller (in the order of microns for high frequency, 1 GHz, acoustic microscopy applications) or much larger (several inches for large structure inspection). To correctly compute the wave field (displacement, stress or pressure), generated by a finite source, it is necessary to follow some numerical or semi-analytical techniques, as discussed in this chapter.

3.1 MODELING A FINITE PLANE SOURCE BY A DISTRIBUTION OF POINT SOURCES

The pressure field due to a finite plane source can be assumed to be the summation of pressure fields generated by a number of point sources distributed over the finite source, as shown in Figure 3.2. The finite source can be, for example, the front face of a transducer, as shown in this figure.

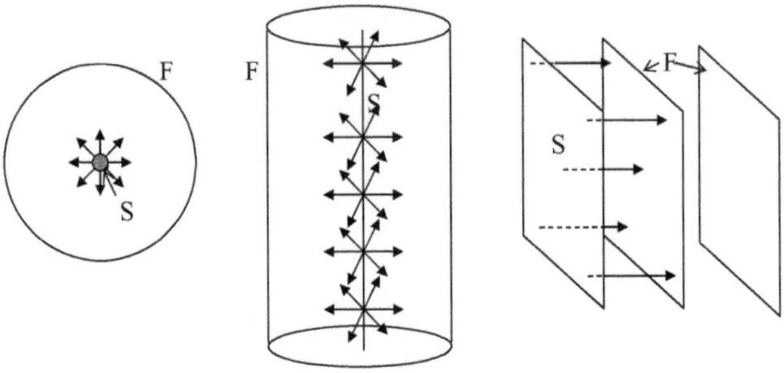

Figure 3.1 Point source (left), line source (middle) and infinite plane source (right) generating spherical, cylindrical and plane wave fronts, respectively. Sources are denoted by S and the wave fronts by F.

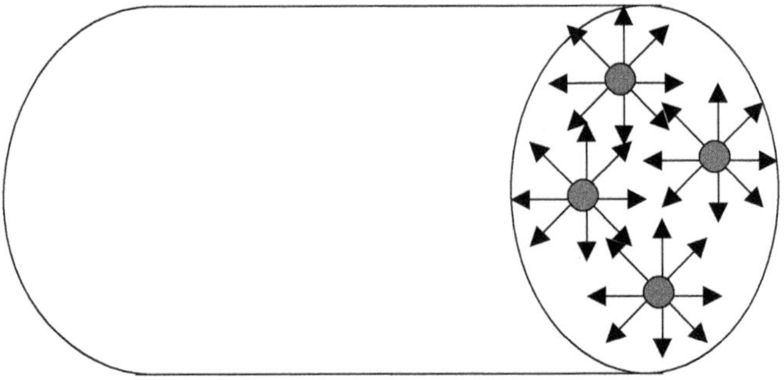

Figure 3.2 Four point sources distributed over a finite source.

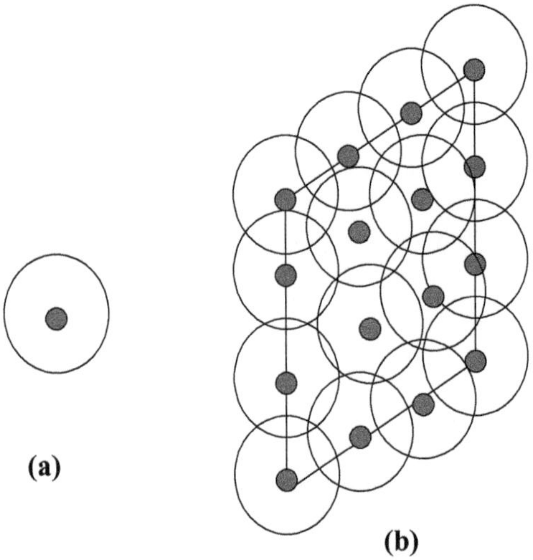

Figure 3.3 (a) Single point source and (b) distribution of point sources placed on a plane.

This assumption can be justified in the following manner:

A harmonic point source, which alternately expands and contracts, can be represented by a point and a sphere as shown in Figure 3.3a. The point represents the contracted position and the sphere (circle in a two-dimensional figure) represents the expanded position. When a large number of these point sources are placed side-by-side on a plane surface, then the simultaneous contraction and expansion of the point sources approximately model the contraction and expansion of the entire plane on which the point sources are distributed, as shown in Figure 3.3b.

The combined effect of a large number of point sources distributed on a plane surface is the vibration of the particles in the direction normal to the plane surface because non-normal components of motion of neighboring source points cancel one another. However, non-normal components do not vanish along the edge of the surface. As a result, the particles on the edge not only vibrate normal to the surface but also expand to a hemisphere and contract back to the point. If this edge effect does not have a strong contribution on the total motion, then the normal vibration of a finite plane surface can be approximately modeled by replacing the finite surface by a large number of point sources distributed over the surface.

3.2 PLANAR PISTON TRANSDUCER IN A FLUID

Let us compute the pressure field in a fluid medium for the planar piston transducer of finite diameter, as shown in Figure 3.4. This problem can be solved analytically, numerically and semi-analytically [see Kundu et al. (2010)].

3.2.1 Analytical Solution

The pressure generated by an acoustic transducer is given by the Rayleigh integral, also known as the Rayleigh-Sommerfeld integral (see Chapter 1)

$$p(\mathbf{x},\omega)=\frac{-i\omega\rho}{2\pi}\int_S v_0(\mathbf{y},\omega)\frac{e^{ik_f r}}{r}dS(\mathbf{y}) \tag{3.1}$$

In equation (3.1), p is the fluid pressure at a general position $\mathbf{x}(x,y,z)$ (see Figure 3.4), for the transducer surface area S vibrating at frequency ω (rad/s) in a perfect fluid of density ρ. The normal velocity component (v_z) at the transducer surface at position $\mathbf{y}$ is v_0. The distance between points x and y is denoted by r, and the wave number k_f is equal to ω/c_f, where c_f is the acoustic wave speed in the fluid.

For a circular transducer of radius a vibrating uniformly with velocity v_0 throughout the transducer surface, the pressure field on the z-axis at point $\mathbf{x}(0, 0, z)$ (see Figure 3.4) can be analytically computed. In this case Eq. (3.1) can be analytically evaluated as shown below [Kundu et al. (2010)]:

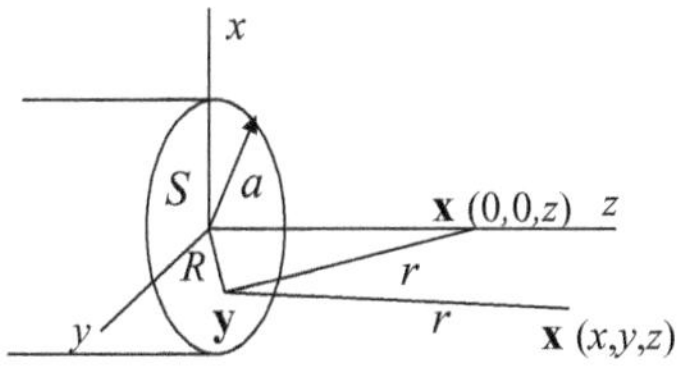

Figure 3.4 Point **y** is on the planar piston transducer face and point **x** is in the fluid medium in front of the transducer. Point **x** is where the ultrasonic field is computed – it can be on the central axis or away from it. The distance between points **x** and **y** is denoted by r [from Kundu et al. (2010)].

$$p(\mathbf{x},\omega)=\frac{-i\omega\rho}{2\pi}\int_S v_0(\mathbf{y},\omega)\frac{e^{ik_f r}}{r}dS(\mathbf{y})$$

$$\therefore p(z,\omega)=\frac{-i\omega\rho}{2\pi}\int_0^a\int_0^{2\pi} v_0\frac{e^{ik_f r}}{r}Rd\theta dR$$

$$=\frac{-i\omega\rho 2\pi v_0}{2\pi}\int_0^a\frac{e^{ik_f r}}{r}RdR$$

$$=-i\omega\rho v_0\int_0^a\frac{e^{\left(ik_f\sqrt{z^2+R^2}\right)}}{\sqrt{z^2+R^2}}RdR \tag{3.2}$$

Substituting $\sqrt{z^2+R^2}=u$, or $du=\dfrac{RdR}{\sqrt{z^2+R^2}}$ in Eq. (3.2) one obtains

$$\begin{aligned}p(z,\omega)&=-i\omega\rho v_0\int_0^a\frac{\exp\left(ik_f\sqrt{z^2+R^2}\right)}{\sqrt{z^2+R^2}}RdR\\&=\rho v_0 c_f\left\{\exp(ik_f z)-\exp\left(ik_f\sqrt{z^2+a^2}\right)\right\}\end{aligned} \tag{3.3}$$

However, if the transducer surface is rectangular, or has some other non axi-symmetric geometry, then Eq. (3.1) cannot be computed analytically even for the case when the observation point **x** is located on the z-axis; then the integral must be computed numerically. When the observation point **x** is not located on the z-axis, then no closed-form analytical expression of the Rayleigh integral exists, and the integral must be computed numerically for all transducer geometries.

3.2.2 Numerical Solution

Finite element method (FEM) is the most popular numerical technique in engineering and science. However, in ultrasonic applications the advancement of FEM has been relatively slow for two reasons. At high frequencies the wavelength of the ultrasonic waves is very small. To obtain reliable results the dimensions of individual finite elements must only be a fraction of the wave length. Thus, the number of elements needed to solve an ultrasonic problem becomes very large. However, with the advancement of computation power and development of more sophisticated finite element codes, such as PZFLEX (2001) and COMSOL (2008), geared toward handling ultrasonic problems more efficiently, the FEM is becoming more popular for solving ultrasonic problems [Hosten and Blateau (2008), Hosten and Castaings (2005, 2006), Moreau et al. (2006)].

A second difficulty in FE analysis arises while solving ultrasonic problems in an infinite or semi-infinite medium. Finite element analysis requires a finite geometry. Therefore, artificial boundaries must be introduced in the FE model of an unbounded medium. For static problems the artificial boundaries of the discretized volumes are placed at large distances from the points of application of the load so that the stress, strain and displacements at the artificial boundary position become negligibly small. However, for ultrasonic wave propagation problems, since the wave can propagate a long distance it is difficult to find a boundary location where the displacement almost vanishes. To avoid reflections from the artificial boundaries dampers are placed at the boundary locations. However, even with the damped boundary conditions some wave energy can still be reflected by the artificial boundary. Investigators finally could overcome this spurious reflection problem by introducing an artificial layer (instead of a simple boundary) surrounding the problem geometry. The damping coefficient in this artificial layer is varied gradually, starting from a low value (same as the damping coefficient of the problem geometry) at the inner boundary and then gradually increasing to a much higher value at the outer boundary of the layer [Hosten and Blateau (2008), Hosten and Castaings (2005, 2006), Moreau et al. (2006)]. Introduction of such energy absorbing layers with gradually varying attenuation avoids any artificial interface having a sharp mismatch in material properties that reflects the wave. Therefore, the only drawback that still remains in the FE analysis is the requirement of a large number of finite elements, especially for three-dimensional (3D) analyses which is still beyond the capacity of many computers. For this reason FEM-based ultrasonic wave propagation analyses available in the literature today mostly solved 2D problems [Hosten and Blateau (2008), Hosten and Castaings (2005, 2006)]. For plane-stress, plane-strain and axi-symmetric problems FEM works very well without much difficulty. However, for a true 3D problem such as a square transducer or when the scatterer is not located on the axis of symmetry of a circular transducer, it is difficult to solve the ultrasonic problem by FEM even today.

3.2.3 Semi-Analytical DPSM Solution

As briefly mentioned in section 3.1, a finite sized transducer can be modeled by a number of point sources. To avoid the singularity in the point source solution at $r = 0$ [see Eq. (1.223)] the point sources are placed inside the solid transducer slightly behind the transducer face as shown in Figure 3.5. Therefore, for any point in front of the transducer surface or even on the transducer face, r does

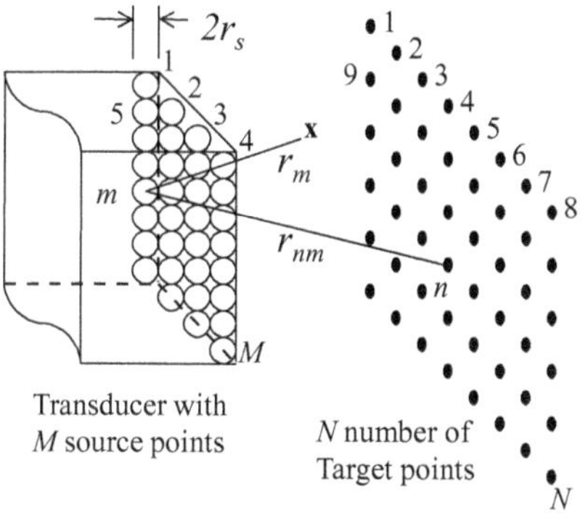

Figure 3.5 Point source allocation for DPSM analysis – M number of point sources are placed slightly behind the transducer face at the centers of small spheres of diameter $2r_s$. Spheres touch the transducer surface; therefore, the point sources are placed at distance r_s behind the transducer front face. The distance from the m-th point source to any general point **x** is shown as r_m. The distance between the m-th point source and n-th target point is denoted by r_{nm}. Target points are a set of N points where the ultrasonic field is to be computed. n varies from 1 to N, as shown in the figure [from Kundu et al. (2010)].

not become zero and the singularity is completely avoided. Figure 3.5 shows M number of source points placed at the centers of M spheres of radius r_s. Spheres touch the front face of the transducer. Therefore the point sources are located at a distance of r_s behind the transducer face. The reason for such placement of point sources is described in detail below.

From Eq. (1.223) the pressure field in the fluid at point **x** at a distance r_m from the m-th point source of strength A_m is given by

$$p_m(\mathbf{x}) = p_m(r_m) = A_m \frac{\exp(ik_f r_m)}{r_m} \tag{3.4}$$

If there are M point sources distributed over the transducer surface, as shown in Figure 3.5, then the total pressure at point **x** is given by

$$p(\mathbf{x}) = \sum_{m=1}^{M} p_m(r_m) = \sum_{m=1}^{M} A_m \frac{\exp(ik_f r_m)}{r_m} \tag{3.5}$$

It should be noted that Eqs. (3.1) and (3.5) are analogous. Eq. (3.5) is the discretized form of Eq. (3.1) with $A_m = \dfrac{-i\omega\rho v_0 dS_m}{2\pi}$.

The pressure–velocity relation in a fluid is given in Chapter 1:

$$-\frac{\partial p}{\partial n} = \rho \frac{\partial v_n}{\partial t} \tag{3.6}$$

For harmonic waves ($e^{-i\omega t}$ time dependence), the derivative of the velocity is obtained by simply multiplying v_n by negative $i\omega$:

$$-\frac{\partial p}{\partial n} = \rho \frac{\partial v_n}{\partial t} = -i\omega\rho v_n \;\Rightarrow v_n = \frac{1}{i\omega\rho}\frac{\partial p}{\partial n} \tag{3.7}$$

Therefore, the velocity in the radial direction, at a distance r from the m-th point source, is given by

$$\begin{aligned} v_m(r) &= \frac{A_m}{i\omega\rho}\frac{\partial}{\partial r}\left(\frac{\exp(ik_f r)}{r}\right) = \frac{A_m}{i\omega\rho}\left(\frac{ik_f \exp(ik_f r)}{r} - \frac{\exp(ik_f r)}{r^2}\right) \\ &= \frac{A_m}{i\omega\rho}\frac{\exp(ik_f r)}{r}\left(ik_f - \frac{1}{r}\right) \end{aligned} \tag{3.8}$$

and the z-component of the velocity is

$$v_{zm}(r) = \frac{A_m}{i\omega\rho}\frac{\partial}{\partial z}\left(\frac{\exp(ik_f r)}{r}\right) = \frac{A_m}{i\omega\rho}\frac{z\exp(ik_f r)}{r^2}\left(ik_f - \frac{1}{r}\right) \tag{3.9}$$

When contributions of all M sources are added then the total velocity in the z direction at point **x** is obtained:

$$v_z = \sum_{m=1}^{M} v_{zm}(r_m) = \sum_{m=1}^{M} \frac{A_m}{i\omega\rho}\frac{z_m \exp(ik_f r_m)}{r_m^2}\left(ik_f - \frac{1}{r_m}\right) \tag{3.10}$$

where z_m is the distance of point **x** in the z direction (perpendicular to the transducer face) measured from the m-th source.

If the transducer surface velocity in the z direction is given by v_0, then for all **x** values on the transducer surface the velocity should be equal to v_0. Therefore,

$$v_z(x) = \sum_{m=1}^{M} v_{zm}(r_m) = \sum_{m=1}^{M} \frac{A_m}{i\omega\rho} \frac{z_m \exp(ik_f r_m)}{r_m^2}\left(ik_f - \frac{1}{r_m}\right) = v_0 \tag{3.11}$$

By taking M points (or M different **x** coordinates, see Figure 3.5) on the transducer surface where the spheres touch the transducer face it is possible to obtain a system of M linear equations with M unknowns (A_1, A_2, A_3, ... A_M). However, the difficulty arises when the point source location and the point of interest **x** (referred to as the target point or observation point from here on) coincide, because then r_m is zero and v_{zm} in Eq. (3.11) becomes infinity. If both point sources and the target points are located on the transducer surface then only r_m can be zero. To avoid this possibility, as stated earlier, the point sources are located slightly behind the transducer surface, at a distance of r_S, as shown in Figure 3.5. Note that in this arrangement the smallest value that r_m can have is r_S.

If point **x** in Figure 3.5 coincides with the n-th target point and the velocity at the n-th target point or the observation point is to be computed then from Equation (3.11) one can write

$$v_z(x_n) = \sum_{m=1}^{M} \frac{A_m}{i\omega\rho} \frac{z_{nm} \exp(ik_f r_{nm})}{r_{nm}^2}\left(ik_f - \frac{1}{r_{nm}}\right) \tag{3.12}$$

In equation (3.12) the first subscript n (of z and r) corresponds to the n-th target point and the second subscript m refers to the m-th source point; therefore, r_{nm} is the distance of the n-th target point from the m-th source point while z_{nm} is the distance along the z direction between the n-th target point and m-th source point. If N target points are placed on the transducer surface and M source points are located slightly behind the transducer surface, then Eq. (3.12) can be written in matrix form

$$\mathbf{V}_S = \mathbf{M}_{SS}\mathbf{A}_S \tag{3.13}$$

where:

- $\mathbf{V}_S$ is the $(N \times 1)$ vector of the velocity components at N number of target points on the transducer surface
- $\mathbf{A}_S$ is the $(M \times 1)$ vector containing the strengths of M number of point sources
- $\mathbf{M}_{SS}$ is the $(N \times M)$ matrix relating the two vectors $\mathbf{V}_S$ and $\mathbf{A}_S$

$$\{V_S\}^T = \begin{bmatrix} v_z^1 & v_z^2 & \ldots & v_z^N \end{bmatrix} \tag{3.14}$$

The transpose of the column vector $\mathbf{V}_S$ is a row vector of dimension $(1 \times N)$. The superscript n $(1 \le n \le N)$ of the element v_z^n corresponds to the n-th target point. If the transducer surface vibrates with a constant velocity amplitude v_0 then all elements of Eq. (3.14) should have the same value v_0.

Vector $\mathbf{A}_S$ of the source strengths is given by

$$\{A_S\}^T = \begin{bmatrix} A_1 & A_2 & A_3 & \ldots & A_{M-2} & A_{M-1} & A_M \end{bmatrix} \tag{3.15}$$

The matrix $\mathbf{M}_{SS}$ is obtained from Equation (3.12)

$$\mathbf{M}_{SS} = \begin{bmatrix} f(z_{11}, r_{11}) & f(z_{12}, r_{12}) & \cdots & f(z_{1M}, r_{1M}) \\ f(z_{21}, r_{21}) & f(z_{22}, r_{22}) & \cdots & f(z_{2M}, r_{2M}) \\ \vdots & \vdots & \ddots & \vdots \\ f(z_{N1}, r_{N1}) & f(z_{N2}, r_{N2}) & \cdots & f(z_{NM}, r_{NM}) \end{bmatrix}_{N \times M} \tag{3.16}$$

where

$$f(z_{nm}, r_{nm}) = \frac{z_{nm} \exp(ik_f r_{nm})}{i\omega\rho \cdot r_{nm}^2} \left(ik_f - \frac{1}{r_{nm}} \right) \tag{3.17}$$

$\mathbf{M}_{SS}$ matrix can be made a square matrix by taking the number of target points equal to the number of source points, both equal to M or N. For this case Eq. (3.13) can be inverted to obtain the point source strengths

$$\mathbf{A}_S = [\mathbf{M}_{SS}]^{-1} \mathbf{V}_S \tag{3.18}$$

Instead of velocity, if the pressure is specified on the transducer surface, then pressure at N target points on the transducer face due to M source points is to be computed. Applying Eq. (3.5) at N target points one obtains a matrix relation similar to Eq. (3.13),

$$\mathbf{P}_S = \mathbf{Q}_{SS} \mathbf{A}_S \tag{3.19}$$

where $\mathbf{A}_S$ has been defined in Eq. (3.15), $\mathbf{P}_S$ is the pressure values at N target points on the transducer surface,

$$\{\mathbf{P}_S\}^T = \begin{bmatrix} P^1 & P^2 & \ldots & P^N \end{bmatrix} \tag{3.20}$$

and $\mathbf{Q}_{SS}$ can be obtained from Equation (3.5),

$$\mathbf{Q}_{SS} = \begin{bmatrix} g(r_{11}) & g(r_{12}) & \cdots & g(r_{1M}) \\ g(r_{21}) & g(r_{22}) & \cdots & g(r_{2M}) \\ \vdots & \vdots & \ddots & \vdots \\ g(r_{N1}) & g(r_{N2}) & \cdots & g(r_{NM}) \end{bmatrix}_{N \times M} \tag{3.21}$$

where

$$g(r_{nm}) = \frac{\exp(ik_f r_{nm})}{r_{nm}} \tag{3.22}$$

r_{nm} is shown in Figure 3.5.

The $\mathbf{Q}_{SS}$ matrix can be made a square matrix by taking the number of target points equal to the number of source points. The source strength vector is then obtained by inverting Eq. (3.19)

$$\mathbf{A}_S = [\mathbf{Q}_{SS}]^{-1} \mathbf{P}_S \tag{3.23}$$

After evaluating the source strength vector $\mathbf{A}_S$ from Eq. (3.18) or (3.23), the pressure $p(\mathbf{x})$ or velocity vector $\mathbf{V}(\mathbf{x})$ at any point (on the transducer surface or away) can be obtained from the following equations:

$$\begin{aligned}\mathbf{P}_T &= \mathbf{Q}_{TS}\mathbf{A}_S \\ \mathbf{V}_T &= \mathbf{M}_{TS}\mathbf{A}_S \\ \mathbf{V}_T^* &= \mathbf{M}_{TS}^*\mathbf{A}_S\end{aligned} \tag{3.24}$$

where:

- $\mathbf{P}_T$ is a $(N\times 1)$ vector containing pressure values at N target points
- $\mathbf{V}_T$ is a $(N\times 1)$ vector containing only z-components of velocity at N target points
- $\mathbf{V}_T^*$ is a $(3N\times 1)$ vector containing all three components of velocity at N target points

$\mathbf{P}_T$ and $\mathbf{V}_T$ of Eq. (3.24) are similar to $\mathbf{P}_S$ [Eq. (3.19)] and $\mathbf{V}_S$ [Eq. (3.13)], respectively; the only difference is that now the target points are moved away from the transducer surface. To compute $\mathbf{P}_T$ and $\mathbf{V}_T$ from equation (3.24) no matrix inversion is needed, and therefore $\mathbf{Q}_{TS}$ and $\mathbf{M}_{TS}$ matrices need not be square matrices. Therefore, for this calculation the number of target points (N) can be different from the number of source points (M). Their expressions are same as those given in Eqs. (3.21) and (3.16).

As mentioned above, the vector $\mathbf{V}_T^*$ contains all three components of the velocity vector at target points and is given by

$$\{\mathbf{V}_T^*\}^T = \begin{bmatrix} v_x^1 & v_y^1 & v_z^1 & v_x^2 & v_y^2 & v_z^2 & \cdots & v_x^N & v_y^N & v_z^N \end{bmatrix} \tag{3.25}$$

and the matrix $\mathbf{M}_{TS}^*$ is given by

$$\mathbf{M}_{TS}^* = \begin{vmatrix} f(x_{11},r_{11}) & f(x_{12},r_{12}) & f(x_{13},r_{13}) & \cdots & f(x_{1M},r_{1M}) \\ f(y_{11},r_{11}) & f(y_{12},r_{12}) & f(y_{13},r_{13}) & \cdots & f(y_{1M},r_{1M}) \\ f(z_{11},r_{11}) & f(z_{12},r_{12}) & f(z_{13},r_{13}) & \cdots & f(z_{1M},r_{1M}) \\ f(x_{21},r_{21}) & f(x_{22},r_{22}) & f(x_{23},r_{23}) & \cdots & f(x_{2M},r_{2M}) \\ f(y_{21},r_{21}) & f(y_{22},r_{22}) & f(y_{23},r_{23}) & \cdots & f(y_{2M},r_{2M}) \\ f(z_{21},r_{21}) & f(z_{22},r_{22}) & f(z_{23},r_{23}) & \cdots & f(z_{2M},r_{2M}) \\ \vdots & \vdots & \vdots & \ddots & \vdots \\ f(x_{N1},r_{N1}) & f(x_{N2},r_{N2}) & f(x_{N3},r_{N3}) & \cdots & f(x_{NM},r_{NM}) \\ f(y_{N1},r_{N1}) & f(y_{N2},r_{N2}) & f(y_{N3},r_{N3}) & \cdots & f(y_{NM},r_{NM}) \\ f(z_{N1},r_{N1}) & f(z_{N2},r_{N2}) & f(z_{N3},r_{N3}) & \cdots & f(z_{NM},r_{NM}) \end{vmatrix}_{3N\times M} \tag{3.26}$$

As the z-component of the velocity [given in Eq. (3.9)] is obtained from Eq. (3.8), the x and y components of velocity can be also obtained from Eq. (3.8) [also see Placko and Kundu (2004)]

$$f(w_{nm},r_{nm}) = \frac{w_{nm}\exp(ik_f r_{nm})}{i\omega\rho\cdot r_{nm}^2}\left(ik_f - \frac{1}{r_{nm}}\right) \tag{3.27}$$

Note that w in Eq. (3.27) stands for x, y or z in Eq. (3.26).

3.2.4 Computed Results

The ultrasonic pressure field in front of a circular transducer along the central axis (z-axis in Figure 3.4) is first computed using three techniques – analytical, semi-analytical DPSM and numerical finite element method (FEM) – and compared. Then the field is computed for a point focused concave transducer and a square transducer, for which no closed-form analytical solution is available.

The spacing between two neighboring point sources should not exceed one-third wavelength. See the convergence criterion derived in section 3.2.5. Satisfying this convergence criterion, the pressure field in front of a circular transducer is first computed for the following transducer dimensions and material parameters:

Transducer diameter = 2.54 mm, acoustic wave speed in water = 1.49 km/s, water density = 1 gm/cc, and signal frequency = 5 MHz.

The pressure variation along the central axis (z-axis in Figure 3.4) of the circular transducer is shown in Figure 3.6. The dashed line is the theoretical curve [Eq. (3.3)] and the continuous line is obtained from the DPSM formulation [Eq. (3.24)] with 500 (or 499 to be exact) point sources modeling the transducer face. Although the peaks and dips of the two curves match very well in Figure 3.6 their magnitudes do not. To investigate the reason for this mismatch the velocity field on the transducer face is computed at N number of points where N is much greater than the number of point sources M modeling the transducer face (in this case $M = 500$). The velocity variation on the transducer face is shown in Figure 3.7. Note that the velocity is not uniform on the transducer face. It reaches a peak value of v_0 at points where the small spheres of Figure 3.5 touch the transducer face but in the region between these points they become significantly smaller.

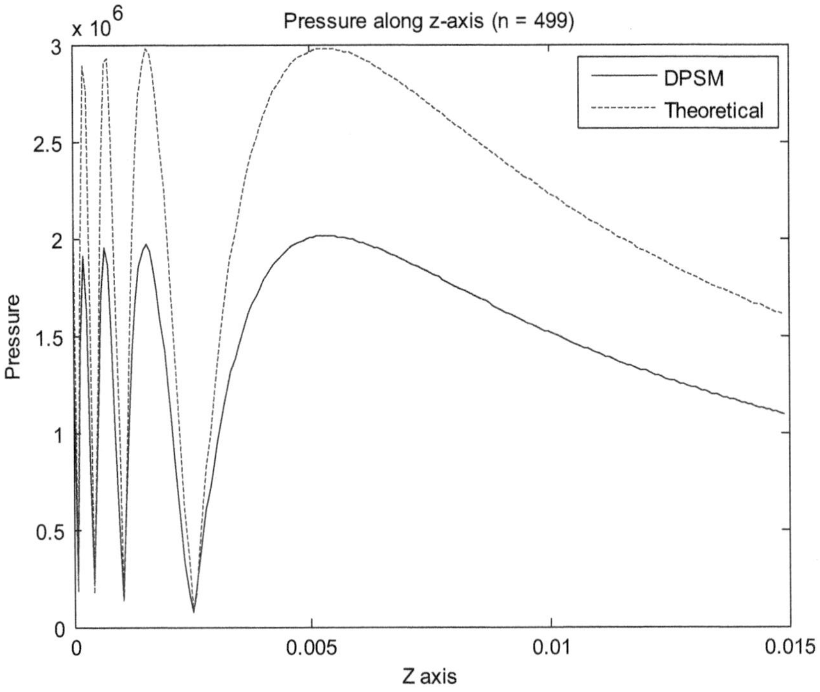

Figure 3.6 Pressure field variation along the central axis of a circular transducer obtained by DPSM analysis [using Eq. (3.18), continuous line] and analytical expression [Eq. (3.3), dashed line) – the transducer diameter is 2.54 mm and the exciting frequency is 5 MHz [from Kundu et al. (2010)].

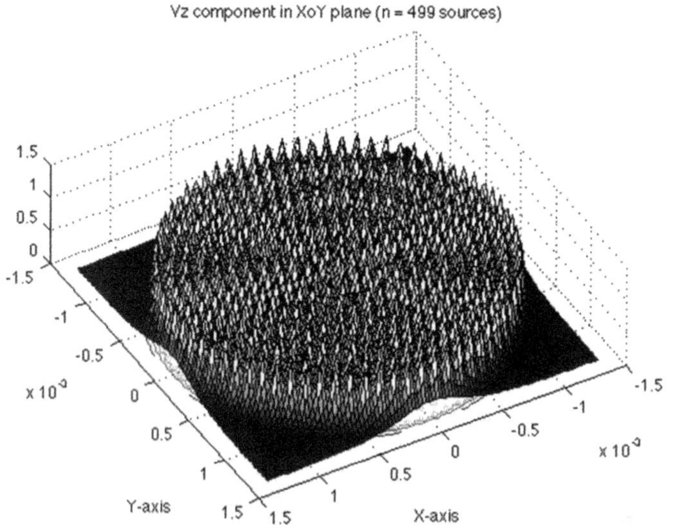

Figure 3.7 Normal velocity variation at the transducer face obtained by assigning a velocity value of 1 at 499 points on the transducer face where the small spheres shown in Figure 3.5 touch the transducer surface. Point source strengths are obtained from Eq. (3.18). 499 peaks correspond to velocity values equal to 1 but the valleys in between the peaks make the average velocity of the transducer surface less than 1. This is the reason for the difference between the two curves shown in Figure 3.6 [from Kundu et al. (2010)].

Therefore, the average velocity of the transducer face becomes less than v_0 giving rise to smaller pressure values from the DPSM analysis as shown in Figure 3.6.

Since the theoretical curve of Figure 3.6 is generated assuming a constant transducer surface velocity of 1, while the DPSM results of Figure 3.6 are for an average transducer surface velocity smaller than 1, the DPSM results must be normalized by dividing the computed values by the average transducer surface velocity. The theoretical curve and the normalized DPSM results obtained in this manner match much better in Figure 3.8, although there is still some small mismatch. The matching can be improved further by considering uniform strength distribution for the 499 point sources that model the transducer face, as one can see in Figure 3.9. Assumption of uniform source strength distribution avoids the need of computing source strengths from Eq. (3.18) by inverting the $[\mathbf{M}_{SS}]$ matrix. When the source strength is assumed uniform then the pressure field variation predicted by the DPSM is obtained from the formulation given in Eq. (3.24) while the analytical solution is obtained from Eq. (3.3). Figure 3.9 shows very good matching between these two results. Matching between the theoretical analysis and DPSM prediction can be improved even further by increasing the number of point sources from 499 to, for example, 1200. Results generated by 1200 point sources are not shown here. Since the DPSM results for the uniform source strengths (Figure 3.9) match very well with the theoretical results in comparison to the DPSM solutions that do not assume uniform source strength (Figure 3.8), we can consider the DPSM solutions with uniform source strengths as the true or ideal solutions for transducer geometries when the closed-form analytical expressions of the Rayleigh-Sommerfeld integral is not available, such as for the case of a rectangular transducer.

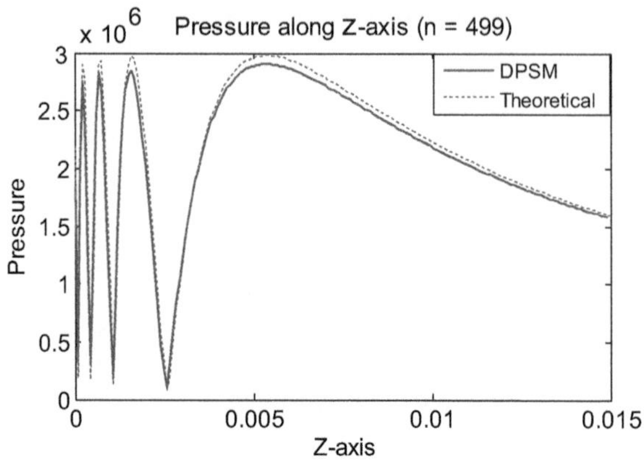

Figure 3.8 Pressure field variation along the central axis of a circular transducer of diameter 2.54 mm vibrating at 5 MHz frequency modeled by 499 point sources. Point source strengths are obtained using Eq. (3.18). This figure is same as Figure 3.6, the only difference is that here the DPSM results (continuous line) are normalized with respect to the average transducer surface velocity [from Kundu et al. (2010)].

Next, the semi-analytical results obtained by the DPSM technique and the numerical results obtained from the finite element analysis using COMSOL (2008) are compared with the theoretical results. Figure 3.10 is similar to Figures 3.8 and 3.9, except that in this figure the pressure along the central axis of a circular transducer is plotted from all three analyses – analytical, semi-analytical and numerical. Results generated by the DPSM analysis with 600 point sources is shown by the continuous line, while the dotted line shows the theoretical curve and the finite element results are shown by the dashed line. Since this problem is axi-symmetric, the finite element result could be obtained relatively easily from the two-dimensional axi-symmetric analysis with 174,927 triangular elements [see Kundu et al. (2010)]. In Figure 3.10 one can see that the error

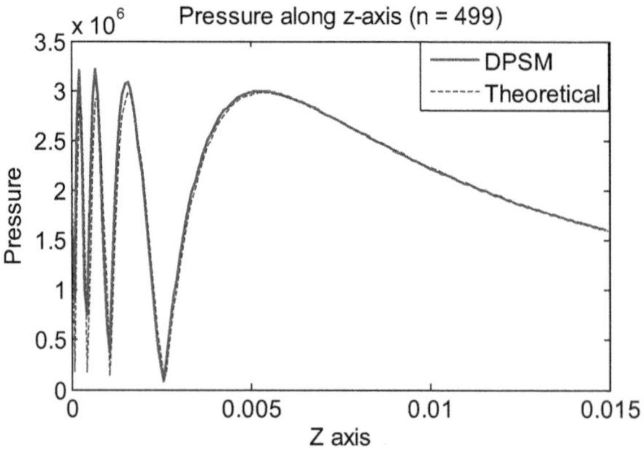

Figure 3.9 Same as Figure 3.8 but here the point sources used in the DPSM formulation are assumed to have uniform strengths and obtained from Eq. (3.24) [from Kundu et al. (2010)].

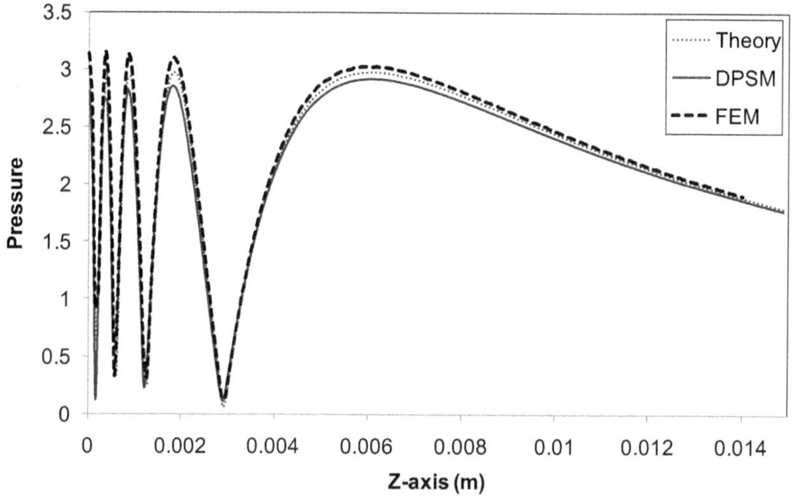

Figure 3.10 Pressure field variation along the central axis of a flat circular transducer – comparison between the theoretical curve (dotted line), finite element solution (dashed line) and DPSM solution (continuous line) [from Kundu et al. (2010)].

in DPSM and FEM results are of the same order that can be reduced further by increasing the number of point sources in the DPSM analysis or the number of elements in the FEM analysis.

Pressure fields in front of square transducers are then computed. Figure 3.11 shows the field for a 0.5 mm × 0.5 mm square transducer. No closed-form

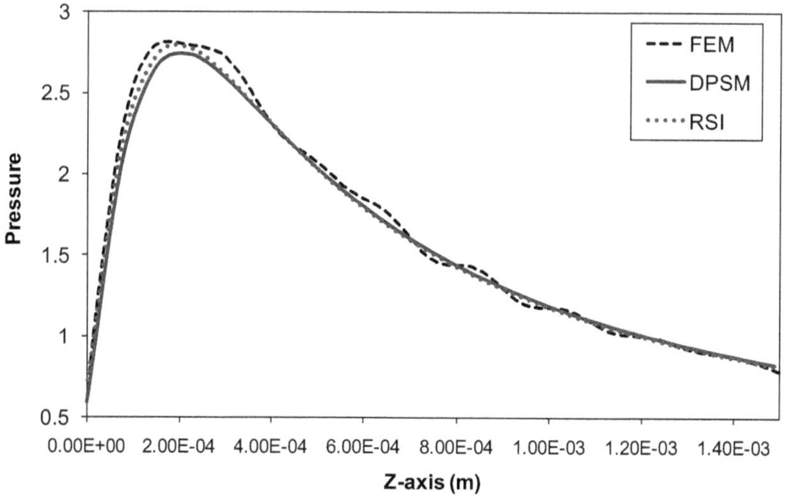

Figure 3.11 Pressure field variation along the central axis of a flat square transducer – Comparison between the RSI solution (dotted line), finite element solution (dashed line) and DPSM solution (continuous line). Transducer dimension is 0.5 mm x 0.5 mm, signal frequency is 5 MHz. Note that the DPSM matches better with the RSI solution [from Kundu et al. (2010)].

analytical solution exists for this transducer geometry. However, in Figure 3.9 we have seen that the DPSM solution with uniform source strength gives a solution that is very close to the RSI solution. Since the DPSM technique can be applied to any transducer geometry, the DPSM solution with uniform source strength is treated as the ideal solution or the RSI solution. This result is shown by the dotted line in Figure 3.11. The continuous line shows the DPSM results when the source strengths are obtained by matching the prescribed velocity condition on the transducer surface. This result is obtained with 1225 point sources modeling the transducer face. The dashed line is obtained from the finite element analysis. Since this problem geometry is not axi-symmetric, the two-dimensional finite element model could not be used here. A three-dimensional finite element mesh with 146,098 3D tetrahedral elements was used to generate the numerical results. Note that in this case DPSM gives better results than FEM.

As the square transducer size was increased from 0.5 mm side length to 1.55 mm side length the finite element method encountered some difficulty but the DPSM solution did not. Results for the 1.55 mm × 1.55 mm square transducer are shown in Figure 3.12a. The DPSM results, shown as a continuous line, matched very well with the RSI solution, the dotted line. However, the three-dimensional finite element mesh with 752,681 3D tetrahedral elements had difficulty in converging as shown by the dashed line. It can be also investigated how well a two-dimensional finite element analysis can model this transducer. With this goal in mind, a two-dimensional plane-strain finite element model with 234,785 triangular elements was analyzed. It generated the dashed-dotted line of Figure 3.12a. Clearly, the 2D approximation is not adequate for modeling this 3D problem. By discretizing a smaller volume in front of the transducer it was possible to generate the pressure field using the 3D finite element analysis, as shown

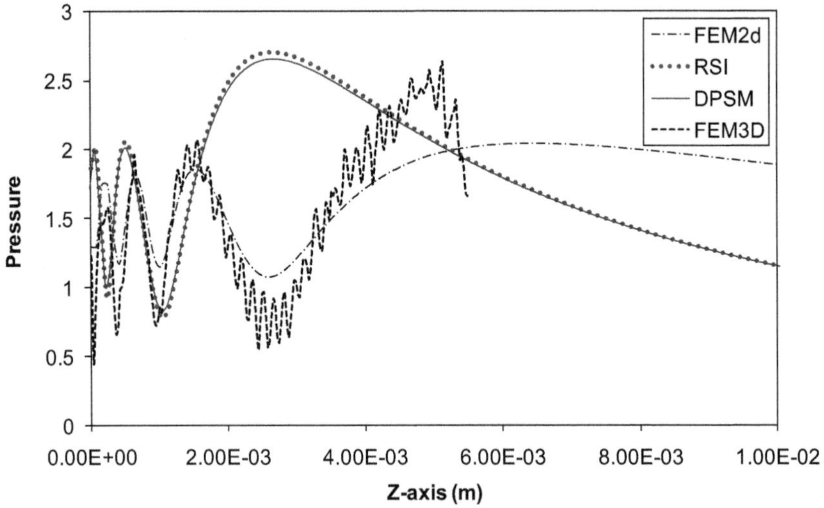

Figure 3.12a Pressure field variation along the central axis of a flat square transducer – Comparison between the RSI solution (dotted line), finite element solution (dashed line) and DPSM solution (continuous line). Transducer dimension is 1.55 mm x 1.55 mm, signal frequency is 5 MHz. A fourth curve (dashed-dotted line) is generated by 2D finite element model for a 1.5 mm wide transducer. Neither 2D nor 3D finite element model produces the RSI solution [from Kundu et al. (2010)].

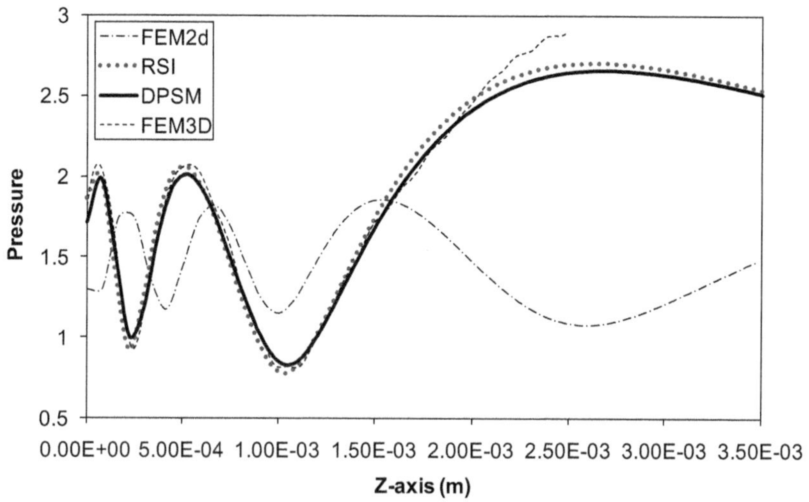

Figure 3.12b Pressure field variation along the central axis of a flat square (1.55 mm x 1.55 mm) transducer – Same as Figure 3.12a, the only difference is that in the axial direction the computation is carried out only up to 3.5 mm instead of 10 mm to have the finite elements of much smaller size. 3D finite element analysis now shows better matching with the RSI and DPSM solutions compared to that in Figure 3.12a [from Kundu et al. (2010)].

in Figure 3.12b. The finite element result was obtained by discretizing a fluid volume of dimension $3\times3\times3$ mm^3 in front of the transducer face into 739,590 tetrahedral finite elements. It took 35 hours of computational time in a CPU Intel Xeon 2x2.66 GHz dual Core, RAM=6 GB computer. In spite of this massive computational effort, the finite element solution still started to deviate from the true solution at a distance greater than 2 mm from the transducer face.

Results presented so far are generated with the assumption that the transducer surface is vibrating with uniform velocity amplitude. Instead of uniform velocity, if it is assumed that the pressure is uniform on the transducer surface, then the point source strengths should be obtained from Eq. (3.23) instead of Eq. (3.18). The pressure field on the central axis is plotted in Figure 3.13 for a flat circular transducer of 2.4 mm^2 surface area vibrating with uniform pressure on the transducer face at 5 MHz frequency. It should be noted that uniform pressure on the transducer face does not imply uniform velocity field or uniform point source strengths on the transducer face. Type I RSI solution corresponds to the uniform pressure condition, which is also known as the resilient disk source in the acoustic literature. For this problem Mellow (2008) provided a simplified analytical solution for the central axis pressure variation. This solution is

$$p = p_0\left[e^{ik_f z} - \frac{z}{\sqrt{z^2+a^2}}e^{ik_f\sqrt{z^2+a^2}}\right] \tag{3.28}$$

where:

- p_0 is the pressure on the transducer face
- z is the distance of the point of interest from the transducer face
- a is the transducer radius
- k_f is the wave number in the fluid

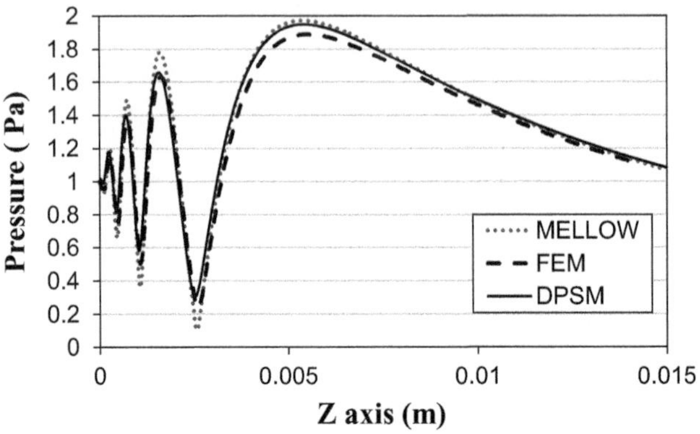

Figure 3.13 Pressure field variation along the central axis of a flat circular transducer of surface area 2.4 mm² vibrating at 5 MHz frequency with uniform pressure distribution at the transducer surface – Comparison between the Mellow's solution (dotted line), finite element solution (dashed line) and DPSM solution (continuous line) [from Kundu et al. (2010)].

This analytical solution is plotted in Figure 3.13 as the dotted line along with the DPSM solution (continuous line) and the FEM solution (dashed line). The DPSM solution is generated by 600 point sources while the FEM solution is obtained with 174,927 triangular axi-symmetric elements.

Note that in Figure 3.13 the amplitude of the oscillating pressure along the central axis gradually increases in the near-field region as the observation point moves away from the transducer face. It should be noted here that when the normal velocity is assumed to be uniform on the transducer face then the amplitude of the oscillating pressure remains almost constant in the near-field region as the observation point moves away from the transducer face (see Figures 3.8 and 3.9). In Figure 3.13 in the far field the DPSM and FEM solutions match well but the Mellow's solution is slightly off. Since Mellow's solution is only an approximate solution (it assumes dipole source) the slight mismatch between the DPSM (or FEM) solution and Mellow's solution in the near field does not imply presence of any computational error in the DPSM (or FEM) solution in the near field. In fact good matching between the DPSM and FEM solutions in the near field reinforces the reliability of both these solutions.

After computing the ultrasonic field along the central axis of the transducer for different transducer geometries (Figures 3.8 to 3.13) the complete field in front of circular and square transducers is generated by DPSM. For complete field computation DPSM is the obvious choice because FEM encounters difficulty in 3D modeling as illustrated in Figures 3.12(a) and 3.12(b) and the closed-form analytical solution exists only for circular transducers along the central axis. A number of recent studies [Kelly and McGough (2007), Mast and Yu (2005), Mellow (2006, 2008)] provided solutions of Rayleigh Integrals of types I and II with elegant mathematical steps, but none of these solutions can produce closed-form expressions for the complete ultrasonic field, and therefore dependence on semi-analytical techniques such as DPSM is unavoidable. Figures 3.14a, b and c show the total fields in front of a circular (3.14a) and two square transducers (3.14b and c) when vibrating with uniform velocity amplitude. Figure 3.14a clearly shows four dips (dark spots) along the central axis. Figures 3.14b and 3.14c show one and two dips, respectively, along the central axis. It should

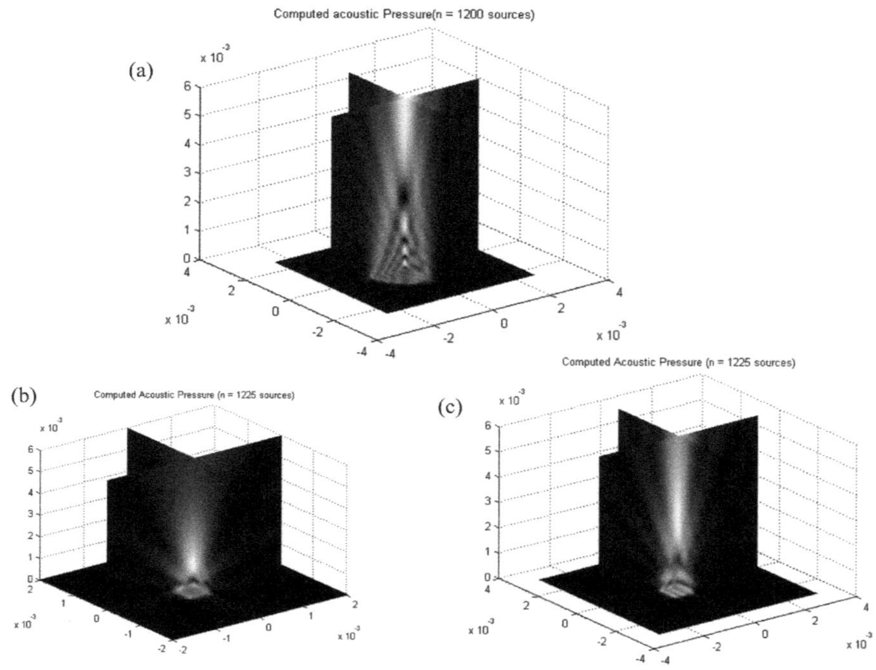

Figure 3.14 Pressure field variations in front of flat transducers vibrating at 5 MHz frequency with uniform velocity at the transducer surface (a) circular (2.54 mm diameter) transducer face, (b) square (0.5 mm x 0.5 mm) transducer face, (c) square (1.55 mm × 1.55 mm) transducer face [from Kundu et al. (2010)].

also be noted that the generated beam becomes more collimated in Figure 3.14c compared to 3.14b as the transducer surface area is increased from 0.25 mm² for Figure 3.14b to 2.4 mm² for Figure 3.14c.

3.2.5 Required Spacing Between Neighboring Point Sources

If the transducer is modeled by M number of point sources and the total surface area S of the transducer face is discretized into M hemispheres of radius r_s, associated with the M point sources, then M and S can be related in the following manner:

$$S = 2\pi r_s^2 M \tag{3.29}$$

$$\rightarrow r_s^2 = \frac{S}{2\pi M} \tag{3.30}$$

It is evident from Figure 3.5 that, as the number of point sources to model the transducer surface is increased, r_S should decrease. It is expected that with a larger number of point sources both the computation time and the accuracy should increase. However, the following question arises: what is the optimum number of point sources to produce reliable results? To answer this question the following analysis is carried out.

For a very small transducer of surface area dS vibrating with a velocity of amplitude v_0 in the x_3 direction the pressure at point $\mathbf{x}$ (at a distance r from the source at point $\mathbf{y}$) can be computed from Eq. (3.1)

$$p(\mathbf{x}) = -\frac{i\omega\rho v_0}{2\pi}\frac{\exp(ik_f r)}{r}dS \tag{3.31}$$

Using Eq. (3.7), the particle velocity in the radial direction can be computed from the above pressure field

$$\begin{aligned} v_r &= \frac{1}{i\omega\rho}\frac{\partial p}{\partial r} = \frac{1}{i\omega\rho}\left(\frac{-i\omega\rho v_0}{2\pi}\right)\left(\frac{ik_f \exp(ik_f r)}{r} - \frac{\exp(ik_f r)}{r^2}\right)dS \\ &= -\frac{v_0(ik_f r - 1)}{2\pi r^2}\exp(ik_f r)dS \end{aligned} \tag{3.32}$$

and the velocity in the x_3 direction

$$v_3 = \frac{1}{i\omega\rho}\frac{\partial p}{\partial x_3} = \frac{1}{i\omega\rho}\frac{\partial p}{\partial r}\frac{\partial r}{\partial x_3} = -\frac{v_0(ik_f r - 1)}{2\pi r^2}\exp(ik_f r)dS\frac{x_3 - y_3}{r} \tag{3.33}$$

where x_3 and y_3 are the x_3 coordinate values of points $\mathbf{x}$ and $\mathbf{y}$, respectively.

If the point $\mathbf{x}$ is taken on the surface of sphere of radius r_S then $r = r_S = x_3 - y_3$, and v_3 of Eq. (3.33) is simplified to

$$\begin{aligned} v_3 &= -\frac{v_0(ik_f r_S - 1)}{2\pi r_S^2}\exp(ik_f r_S)dS = v_0(1 - ik_f r_S)\left(1 + ik_f r_S + O(k_f^2 r_S^2)\right)\frac{dS}{2\pi r_S^2} \\ &\approx v_0\left(1 + k_f^2 r_S^2\right)\frac{dS}{2\pi r_S^2} \end{aligned} \tag{3.34}$$

The right-hand side of Eq. (3.34) should be equal to v_0, since the pressure computed in Eq. (3.31) is obtained from the transducer surface velocity v_0 in the x_3

direction. Therefore, the velocity at $\mathbf{x}$ should be equal to v_0 when $\mathbf{x}$ is taken on the transducer surface. The right-hand side of Eq. (3.34) is v_0 when $dS = 2\pi r_S^2$ and $k_f^2 r_S^2 << 1$. Therefore, dS should be the surface area of a hemisphere of radius r_S, and the second condition implies the following:

$$k_f^2 r_S^2 = \left(\frac{2\pi f}{c_f} r_S \right)^2 << 1.$$

$$\Rightarrow r_S << \frac{c_f}{2\pi f} \tag{3.35}$$

$$\Rightarrow r_S << \frac{\lambda_f}{2\pi}$$

where λ_f is the wavelength in the fluid. Eq. (3.35) is used to compute the number of point sources in the following manner: take a value of r_S satisfying the condition specified in Eq. (3.35), then compute the number of point sources M from the transducer surface area S from the relation

$$M = \frac{S}{2\pi r_S^2} \tag{3.36}$$

Note that the spacing between two neighboring point sources is different from r_S. If the point sources are arranged uniformly at the four corners of squares of side length a, then each point source should be associated with an area of a^2 of the flat transducer face. This area is then equated to the hemispherical surface area of every point source to obtain

$$a^2 = 2\pi r_S^2$$
$$\Rightarrow a = r_S\sqrt{2\pi} \tag{3.37a}$$

Substituting Eq. (3.35) in the above equation we get

$$a = r_S\sqrt{2\pi} << \sqrt{2\pi}\frac{\lambda_f}{2\pi}$$
$$\Rightarrow a << \frac{\lambda_f}{\sqrt{2\pi}} \tag{3.37b}$$

This convergence criterion can be also derived by satisfying the condition that the distance between the point source and the field point must be greater than the Rayleigh distance. At such a distance the field generated by a flat surface area a^2 of the transducer face starts to resemble the field generated by a point source. Since the closest distance between a point source and a field point is r_s and the Rayleigh distance for a flat transducer of surface area a^2 can be approximately given by $\frac{a^2}{\lambda_f}$ [see Cobbold (2007)], r_s should satisfy the following condition:

$$r_s \gg \frac{a^2}{\lambda_f} \tag{3.38a}$$

Substituting $r_S = \frac{a}{\sqrt{2\pi}}$ from Eq. (3.37a) into Eq. (3.38a) we get

$$\frac{a}{2\pi} \gg \frac{a^2}{\lambda_f}$$

$$\therefore a \ll \frac{\lambda_f}{\sqrt{2\pi}} \tag{3.38b}$$

Note that Eqs. (3.37b) and (3.38b) are identical but obtained differently. In summary, the spacing a between two neighboring point sources should be less than $\frac{\lambda_f}{\sqrt{2\pi}}$ or in simple words should be less than one-third of the wavelength. If it is much less than this value, then the computed velocity at the points where the small spheres touch the transducer surface (see Figure 3.5) should be v_0. Otherwise, it will be different from v_0. However, since the DPSM results can be always normalized (see discussions on Figures 3.6 to 3.9) it is not necessary to make this spacing significantly less than one-third wavelength.

The relation between r_s (the distance between the point sources and the transducer face) and 'a' (the spacing between the point sources) given in Eq. (3.37a) is not necessarily unique. It is possible to place the point sources further away from the transducer face and set the distance between the point sources and the transducer face at any value between r_s and $2r_s$. For example, Cheng et al. (2011) changed the distance between the point sources and the transducer face from $\frac{a}{\sqrt{2\pi}}$ to $a/2$. Increasing this distance has the advantage of requiring fewer point sources in the modeling. Cheng et al. (2011) in their modeling satisfied the convergence criterion by taking $a << \frac{\lambda_f}{2}$. It is recommended, however, not to increase this distance beyond $2r_s$ or $\frac{a\sqrt{2}}{\sqrt{\pi}}$.

3.3 FOCUSED TRANSDUCER IN A HOMOGENEOUS FLUID

For a focused transducer, as shown in Figure 3.15, the ultrasonic field in the fluid can be modeled by distributing the point sources along the curved transducer face. O'Neil (1949) argued that for transducers with small curvature Rayleigh-Sommerfield integral representation [Eq. (3.1)] holds if the surface integral is carried out over the curved surface. The DPSM technique can be used to solve this problem as well. The only difference is that in this case the point sources are distributed over a curved surface, instead of a flat surface.

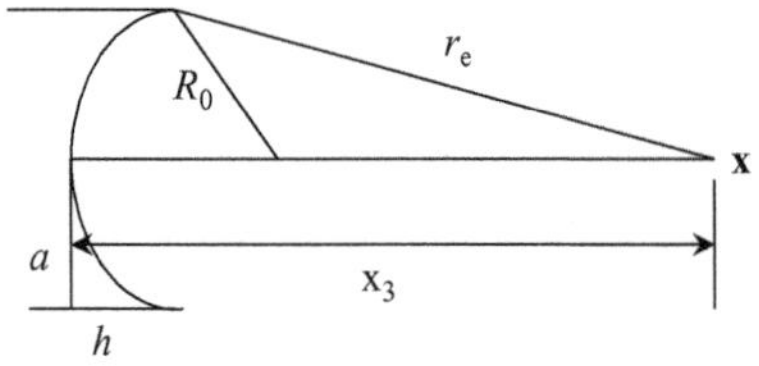

Figure 3.15 A concave transducer of radius a and radius of curvature R_0.

The integral representation of the pressure field in the fluid for a focused transducer should be same as Eq. (3.1). This integral can be evaluated in closed-form, for computing the pressure variation on the central axis of the transducer. The on-axis pressure field is given by [see Schmerr (1998)]

$$p(x_3) = \frac{\rho c v_0}{q_0}\left[\exp(ik_f x_3) - \exp\left(ik_f\sqrt{x_3^2 + a^2}\right)\right] = \frac{\rho c v_0}{q_0}\left[\exp(ik_f x_3) - \exp\left(ik_f r_e\right)\right] \tag{3.39}$$

where

$$q_0 = 1 - \frac{x_3}{R_0} \tag{3.40}$$

R_0 is the radius of curvature of the transducer face, and r_e is the distance of the point of interest from the transducer edge.

At the geometric focus point $x_3 = R_0$, the pressure is given by [Schmerr (1998)]

$$p(R_0) = -i\rho c v_0 k_f h \exp(ik_f R_0) \tag{3.41}$$

If, at $x_3 = z$, the on-axis pressure is maximum then z should satisfy the following equation [Schmerr (1998)]:

$$\cos\left(\frac{k_f\delta}{2}\right) = \frac{2(\delta + z)\sin\left(\frac{k_f\delta}{2}\right)}{(\delta + h)q_0 k_f R_0} \tag{3.42}$$

where

$$\delta = r_e - z = \left[(z-h)^2 + a^2\right]^{\frac{1}{2}} - z \tag{3.43}$$

3.3.1 Computed Results for a Focused Transducer

Figure 3.16 shows the pressure variation along the central axis of a circular concave lens with 4 mm radius of curvature and 40° lens angle. Because of the axi-symmetric nature of the problem the two-dimensional axi-symmetric finite elements could be used to model the ultrasonic field generated by this transducer. The DPSM curve was generated with 600 point sources while 75,893 triangular axi-symmetric finite elements were used to generate the numerical curve. Note that both DPSM and FEM results match equally well with the theoretical curve.

3.4 ULTRASONIC FIELD IN A NON-HOMOGENEOUS FLUID IN PRESENCE OF AN INTERFACE

If the fluid, in front of the transducer, is not homogeneous but is made of two fluids (fluids 1 and 2) with an interface between the two, then the ultrasonic signal generated by the transducer placed in fluid 1 will go through reflection and transmission at the interface. In the DPSM technique, the interface is replaced by two layers of equivalent point sources to model the reflected field in fluid 1 and transmitted field in fluid 2 as discussed in the following.

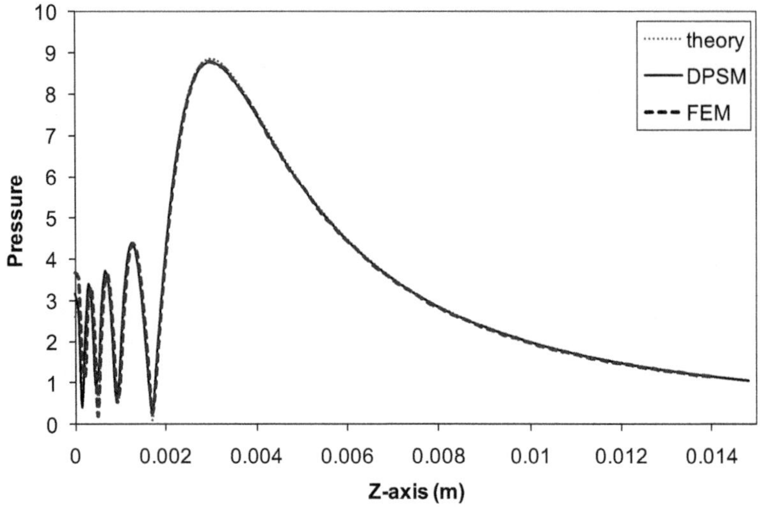

Figure 3.16 Pressure field variation along the central axis of a concave point focused circular transducer – Comparison between the theoretical curve (dotted line), finite element solution (dashed line) and DPSM solution (continuous line). Radius of curvature of the transducer is 4 mm and it forms 40° angle at the center of curvature. All three curves coincide.

3.4.1 Field Computation in Fluid 1

Field in fluid 1 is computed by superimposing the contributions of two layers of point sources distributed over the transducer face and the interface, respectively, as shown in Figure 3.17. Two layers of sources are located at a distance of r_S away from the transducer face and the interface, respectively, such that the apex of every sphere (of radius r_S) touches the transducer face or interface, as shown in the figure. Let M be the number of point sources modeling the transducer face and N the number of point sources being placed along the interface to model the reflected field, as shown in Figure 3.17. $\mathbf{A}_S$ and $\mathbf{A}_I^*$ are two vectors containing the source strengths along the transducer face and the interface, respectively.

Let us take a point P at $\mathbf{x}_n$ on the interface. In Figure 3.18, the point P is shown very close to the interface. It receives two types of ultrasonic signal – direct signal from the transducer, shown as ray 1 in the figure, and reflected signal from the interface, shown as Ray 2. Let us now move this point to the interface. There are M point sources ($\mathbf{y}_m$, m = 1, 2, … M) on the transducer surface and N point sources on the interface. At N target points or observation points ($\mathbf{x}_n$, n = 1, 2, … N) where the small spheres touch the interface (see Figure 3.18) the continuity conditions should be satisfied. If the continuity conditions are to be satisfied for three components of velocity at all N points of the interface, then there are a total of 3 N velocity continuity conditions. On the other hand, if the continuity condition for only the normal velocity component at the interface is to be satisfied, as for a fluid–fluid interface, then there are a total of N velocity continuity conditions at the interface.

Normal velocity components at N target points due to these two layers of point sources can be written as

$$\mathbf{V}_T^1 = \mathbf{M}_{TS}\mathbf{A}_S + \mathbf{M}_{TI}^*\mathbf{A}_I^* \tag{3.44}$$

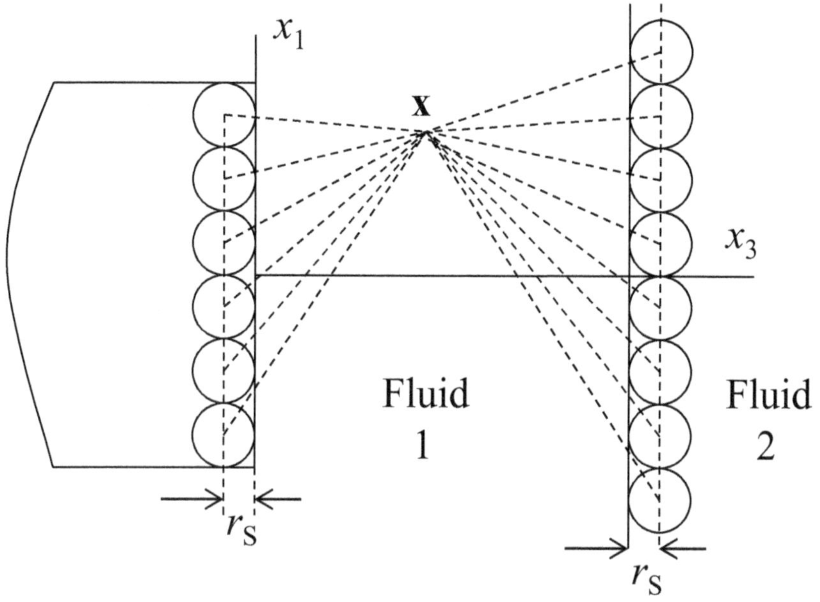

Figure 3.17 Two layers of point sources (at the centers of small circles) contribute to the ultrasonic field in fluid 1. One layer models the incident field from the transducer while the second layer represents the reflected field from the interface. The transducer surface is not necessarily parallel to the fluid–solid interface.

where $\mathbf{V}_T^1$ is the $(N \times 1)$ vector of the normal velocity components in fluid 1 at the target points $(\mathbf{x}_n)$ placed on the interface. $\mathbf{A}_S$ is the $(M \times 1)$ vector of the point source strengths on the transducer surface. Since the normal velocity component (v_0) at the transducer surface is known, $\mathbf{A}_S$ can be obtained from Eq. (3.18). $\mathbf{M}_{TS}$ is the $(N \times M)$ matrix that gives the velocity vector at the target points $\mathbf{x}_n$ due to the transducer sources $\mathbf{A}_S$, and $\mathbf{M}_{TI}^*$ is the $(N \times N)$ matrix that gives the velocity

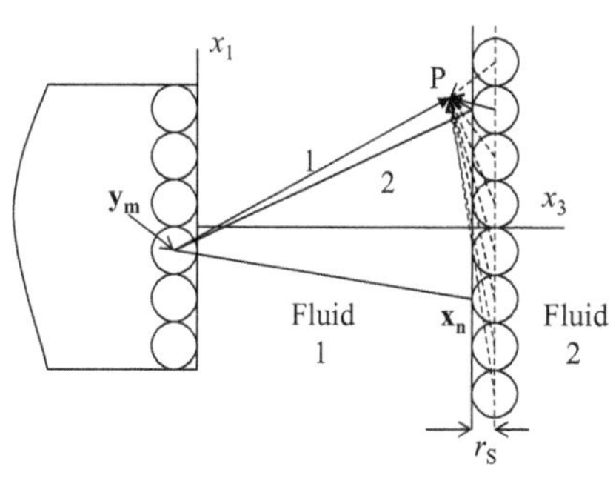

Figure 3.18 Point P can receive two rays from a single point source $\mathbf{y}_m$ near the transducer face – direct ray 1 from the source and ray 2 after being reflected by the interface. The transducer surface is not necessarily parallel to the fluid–solid interface.

vector at the target points due to the interface sources $\mathbf{A}_I^*$. These two matrices have the following forms:

$$\mathbf{M}_{TS} = \begin{bmatrix} f(z_{11}, r_{11}) & \cdots & f(z_{1M}, r_{1M}) \\ \vdots & \ddots & \vdots \\ f(z_{N1}, r_{N1}) & \cdots & f(z_{NM}, r_{NM}) \end{bmatrix}_{N\times M} \tag{3.45}$$

$$\mathbf{M}_{TI}^* = \begin{bmatrix} f(z_{11}^*, r_{11}^*) & \cdots & f(z_{1N}^*, r_{1N}^*) \\ \vdots & \ddots & \vdots \\ f(z_{N1}^*, r_{N1}^*) & \cdots & f(z_{NN}^*, r_{NN}^*) \end{bmatrix}_{N\times N} \tag{3.46}$$

Note that the components of $\mathbf{M}_{TS}$ and $\mathbf{M}_{TI}^*$ are similar to the elements of $\mathbf{M}_{TS}^*$ that are given in Eq. (3.27). It should be mentioned here that in Eq. (3.46) r_{ij}^* and z_{ij}^* are measured from the *j*-th source on the interface to the *i*-th target point.

In Eq. (3.44), and in subsequent equations, the subscripts have the following meanings:

S – ultrasonic source or transducer points

I – interface points

T – target points or observation points (these points can be placed anywhere – inside fluid 1, fluid 2, on the transducer surface, or on the interface).

3.4.2 Field Computation in Fluid 2

Fluid 2 only gets the transmitted ultrasonic energy. Therefore, for ultrasonic field computation in fluid 2, only one layer of point sources, adjacent to the interface is needed, as shown in Figure 3.19. Total field at **x** should be the superposition of fields generated by all these point sources, located at various distances from **x**, as shown by the dotted lines in Figure 3.19. Let $\mathbf{A}_I$ denote the vector of the point source strengths placed along the interface to model the transmitted field.

Normal velocity components at *N* target points placed on the interface in fluid 2 due to the layer of point sources placed along the interface can be written as

$$\mathbf{V}_T^2 = \mathbf{M}_{TI}\mathbf{A}_I \tag{3.47}$$

In Eq. (3.47),

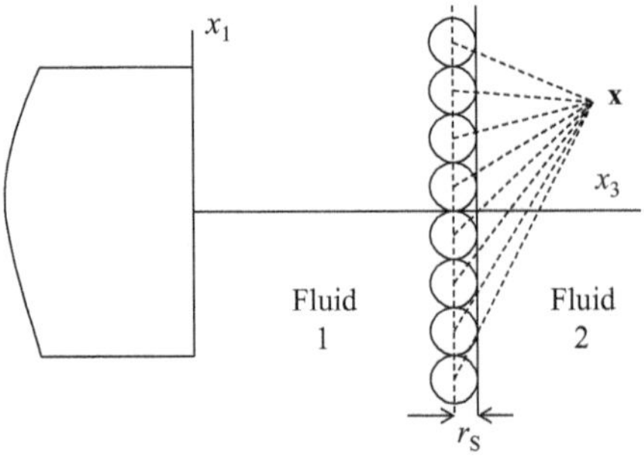

Figure 3.19 One layer of point sources (at the centers of small circles) placed near the interface of two fluids model the transmitted ultrasonic field in fluid 2.

$$M_{TI} = \begin{bmatrix} f_2\left(z^*_{11}, r^*_{11}\right) & \cdots & f_2\left(z^*_{1N}, r^*_{1N}\right) \\ \vdots & \ddots & \vdots \\ f_2\left(z^*_{N1}, r^*_{N1}\right) & \cdots & f_2\left(z^*_{NN}, r^*_{NN}\right) \end{bmatrix}_{N\times N} \tag{3.48}$$

Note that $\mathbf{M}_{TI}$ is similar to $\mathbf{M}^*_{TI}$ where r^*_{ij} and z^*_{ij} are measured from the j-th point source on the interface to the i-th target point. Its elements are defined in the following manner:

$$f_2\left(z^*_{ij}, r^*_{ij}\right) = \frac{z^*_{ij} \cdot \exp\left(ik_{f2} r^*_{ij}\right)}{i\omega\rho_2 \cdot r^{*2}_{ij}}\left(ik_{f2} - \frac{1}{r^*_{ij}}\right) \tag{3.49}$$

where, ρ_2 and k_{f2} correspond to the density and wave number of fluid 2, respectively.

3.4.3 Satisfaction of Continuity Conditions and Evaluation of Unknowns

Equating $\mathbf{V}^1_T$ of Eq. (3.44) and $\mathbf{V}^2_T$ of Eq. (3.47) one gets the following matrix equation:

$$\mathbf{M}_{TS}\mathbf{A}_S + \mathbf{M}^*_{TI}\mathbf{A}^*_I = \mathbf{M}_{TI}\mathbf{A}_I \tag{3.50}$$

Note that the matrix Eq. (3.50) contains N scalar equations and $2N$ unknowns – elements of vectors $\mathbf{A}^*_I$ and $\mathbf{A}_I$. The source strength vector $\mathbf{A}_S$ is known from the known velocity values on the transducer surface. Clearly, we need another set of N equations to solve for the unknowns. The second set of equations is obtained by satisfying the pressure continuity condition across the interface in the following manner:

$$\mathbf{P}^1 = \mathbf{Q}_{TS}\mathbf{A}_S + \mathbf{Q}^*_{TI}\mathbf{A}^*_I = \mathbf{Q}_{TI}\mathbf{A}_I = \mathbf{P}^2 \tag{3.51}$$

The above equation shows that the pressure values at the target points along the interface in fluid 1, denoted as $\mathbf{P}^1$, must be equal to the pressure values at the same target points in fluid 2, denoted as $\mathbf{P}^2$. Matrices $\mathbf{Q}_{TS}$, $\mathbf{Q}_{TI}$ and $\mathbf{Q}^*_{TI}$ give pressure values at the target points generated by point sources placed on the transducer surface and on two sides of the interface. These matrices have the following form:

$$\mathbf{Q}_{TS} = \begin{bmatrix} g\left(r_{11}\right) & g\left(r_{12}\right) & \cdots & g\left(r_{1M}\right) \\ g\left(r_{21}\right) & g\left(r_{22}\right) & \cdots & g\left(r_{2M}\right) \\ \vdots & \vdots & \ddots & \vdots \\ g\left(r_{N1}\right) & g\left(,r_{N2}\right) & \cdots & g\left(r_{NM}\right) \end{bmatrix}_{N\times M} \tag{3.52}$$

$$\mathbf{Q}^*_{TI} = \begin{bmatrix} g\left(r^*_{11}\right) & g\left(r^*_{12}\right) & \cdots & g\left(r^*_{1N}\right) \\ g\left(r^*_{21}\right) & g\left(r^*_{22}\right) & \cdots & g\left(r^*_{2M}\right) \\ \vdots & \vdots & \ddots & \vdots \\ g\left(r^*_{N1}\right) & g\left(r^*_{N2}\right) & \cdots & g\left(r^*_{NN}\right) \end{bmatrix}_{N\times N} \tag{3.53}$$

$$\mathbf{Q}_{TI} = \begin{bmatrix} g_2\left(r_{11}^*\right) & g_2\left(r_{12}^*\right) & \cdots & g_2\left(r_{1N}^*\right) \\ g_2\left(r_{21}^*\right) & g_2\left(r_{22}^*\right) & \cdots & g_2\left(r_{2M}^*\right) \\ \vdots & \vdots & \ddots & \vdots \\ g_2\left(r_{N1}^*\right) & g_2\left(r_{N2}^*\right) & \cdots & g_2\left(r_{NN}^*\right) \end{bmatrix}_{N\times N} \tag{3.54}$$

where

$$g\left(r_{ij}\right) = \frac{\exp\left(ik_f r_{ij}\right)}{r_{ij}} \text{ and } g_2\left(r_{ij}^*\right) = \frac{\exp\left(ik_{f2} r_{ij}^*\right)}{r_{ij}^*} \tag{3.55}$$

As before, r_{ij} denotes the radial distance to the i-th target point from the j-th point source on the transducer surface and r_{ij}^* denotes the radial distance to the i-th target point from the j-th point source on the interface; k_f and k_{f2} are wave numbers in fluids 1 and 2, respectively.

3.5 ULTRASONIC FIELD IN PRESENCE OF A SCATTERER

Interaction between a scatterer such as a cavity or air bubble in a liquid and a converging ultrasonic beam generated by a point focused transducer is investigated in this section. This problem is selected because its analytical solutions are available in the literature [Atalar (1978), Lobkis and Zinin (1990)]. DPSM predicted results are compared with these analytical results and the finite element solutions that are obtained using COMSOL (2008). Detailed discussion on the solution method of this problem can be found in Placko et al. (2010) and Kundu et al. (2010).

3.5.1 DPSM Modeling

Figure 3.20 shows a spherical cavity or an air bubble in a fluid located in front of a point focused acoustic transducer (also known as the acoustic lens). To model this problem geometry by the DPSM technique, several point sources are placed along the lens–fluid interface and the bubble surface. These point sources have

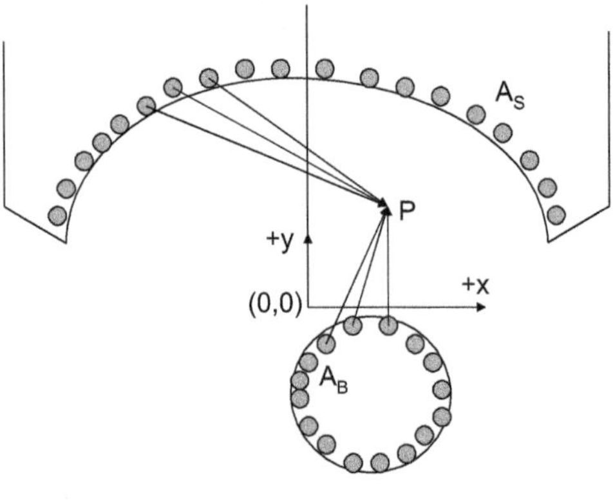

Figure 3.20 Problem geometry showing the distribution of point sources (small circles) along the point focused acoustic lens and the surface of the air bubble or cavity [from Placko et al. (2010)].

unknown source strengths. $\mathbf{A}_S$ and $\mathbf{A}_B$ denote the unknown source strength vectors along the ultrasonic transducer and the bubble, respectively. The pressure and displacement expressions at a general point P can be expressed in terms of these unknown source strengths. A matrix equation in the following form is obtained by simultaneously satisfying the specified velocity boundary conditions on the transducer surface and the pressure boundary conditions on the bubble surface:

$$\begin{bmatrix} \mathbf{M}_{SS} & \mathbf{M}_{SB} \\ \mathbf{Q}_{BS} & \mathbf{Q}_{BB} \end{bmatrix} \begin{Bmatrix} \mathbf{A}_S \\ \mathbf{A}_B \end{Bmatrix} = \begin{Bmatrix} \mathbf{V}_S \\ \mathbf{P}_B \end{Bmatrix} \tag{3.56}$$

Let M and N be the numbers of point sources used to model the transducer surface and the cavity surface, respectively. Then $\mathbf{A}_S$ and $\mathbf{A}_B$ in Eq. (3.56) should have M and N elements, respectively. Vectors $\mathbf{V}_S$ and $\mathbf{P}_B$ on the right-hand side are obtained from the prescribed boundary conditions. For example, we can assume that the pressure $\mathbf{P}_B$ on the cavity surface is zero and the transducer surface velocity V_S is uniform. $\mathbf{V}_S$ vector contains the prescribed velocity values on the lens surface at M points where small spheres in Figure 3.20 touch the transducer surface. Similarly, vector $\mathbf{P}_B$ contains the pressure values at N points on the cavity surface where the small spheres touch the cavity surface. As in the previous sections, matrices $\mathbf{M}$ and $\mathbf{Q}$ relate the source strengths to the particle velocity and pressure, respectively. Note that $\mathbf{M}_{SS}$, $\mathbf{M}_{SB}$, $\mathbf{Q}_{BS}$ and $\mathbf{Q}_{BB}$ matrices have dimensions $M\times M$, $M\times N$, $N\times N$ and $N\times M$, respectively. To obtain the source strengths, the system of equations given in Eq. (3.56) are to be solved. For a large number of point sources the solution of the simultaneous equations requires significant computation time and causes a memory problem in the 32-bit MATLAB. In the 64-bit MATLAB the memory problem is avoided but the computation time is still a concern for solving a large system of equations. The following modification in the formulation can be followed for analyzing a small cavity without requiring one to solve a large system of linear equations.

For a small cavity the number of point sources (N) needed to model the stress-free boundary of the cavity surface is small. However, the point sources (M) needed for modeling the transducer surface is high. Thus the dimension $(M+N)\times(M+N)$ of the square matrix on the left-hand side of Eq. (3.56) is still very high and one needs to solve a large number $(M+N)$ of simultaneous equations. To avoid this difficulty, the pressure field in the fluid can be first computed in absence of the cavity. Since the transducer surface velocity is known, it is not necessary to solve a system of linear equations for obtaining the point source strengths on the transducer surface. In this manner Kundu et al. (2006, 2009) computed the ultrasonic field in a fluid in front of an acoustic microscope lens. The pressure field generated by the incident field in absence of the cavity at N number of points on the cavity surface is first computed. Then N point sources are placed inside the cavity. Their strengths are adjusted by satisfying the condition that the pressure field obtained from these N sources of layer A_B should be equal to the negative of the pressure field obtained from the incident field since the cavity surface must be stress free. In matrix notation this condition can be written as

$$[\mathbf{Q}_{BB}]\{\mathbf{A}_B\} = \{-\mathbf{P}_{B1}\} \tag{3.57}$$

where

$$\{\mathbf{P}_{B1}\} = [\mathbf{Q}_{BS}]\{\mathbf{A}_S\} \tag{3.58}$$

Note that $\mathbf{P}_{B1}$ is the vector containing the pressure values at N number of points on the cavity surface obtained from the incident field. The matrix $\mathbf{Q}_{BB}$ has a dimension of $N \times N$ and therefore, Eq. (3.58) can be quickly solved. The source strength $\mathbf{A}_B$ is obtained from Eqs. (3.57) and (3.58),

$$\{\mathbf{A}_B\} = -[\mathbf{Q}_{BB}]^{-1}[\mathbf{Q}_{BS}]\{\mathbf{A}_S\} \tag{3.59}$$

After obtaining $\mathbf{A}_S$ and $\mathbf{A}_B$, the ultrasonic field in the fluid is obtained by adding the contributions of the two layers of sources.

The above simplified solution [Eqs. (3.57) to (3.59)] is valid for analyzing a small cavity. Neither the above solution nor the analytical solution is applicable when the cavity radius is comparable to the focal distance. In that case Eq. (3.56) is to be solved.

In the DPSM formulation it is possible to introduce various levels of sophistication in the solution process for different dimensions of the cavity. Smaller to bigger size cavities can be analyzed by moving from the simplified to more rigorous solutions, as discussed below.

3.5.1.1 Very Small Cavity Modeled by a Single Point Source

For a very small cavity only one point source can be used to model the cavity. After computing the pressure $\mathbf{P}_{B1}$ on the cavity surface the point source strength $\mathbf{A}_B$ modeling the cavity is obtained from the following scalar equation:

$$\mathbf{A}_B = -\mathbf{Q}_{BB}^{-1}\,\mathbf{P}_{B1} \tag{3.60}$$

3.5.1.2 Small Cavity Modeled with Multiple Point Sources

If the cavity is small, but too big to model by a single point source, then the matrix Eq. (3.59) is used to obtain strengths of multiple point sources modeling the cavity.

3.5.1.3 Complete Solution for Large Cavity

For large cavities where the cavity dimension is comparable to the focal distance of the lens the multiple reflections of waves between the cavity and the microscope lens cannot be ignored. The boundary condition on the transducer surface is influenced by the wave reflected from the cavity. The zero pressure condition on the cavity surface must be satisfied in spite of multiple wave reflections between the transducer and the cavity. In this case, if the resultant velocity on the transducer surface is specified as $\{\mathbf{V}_S\}$ and the pressure on the cavity surface is zero, then the source strengths are obtained from the following system of equations:

$$\begin{bmatrix} \mathbf{M}_{SS} & \mathbf{M}_{SB} \\ \mathbf{Q}_{BS} & \mathbf{Q}_{BB} \end{bmatrix} \begin{Bmatrix} \mathbf{A}_S \\ \mathbf{A}_B \end{Bmatrix} = \begin{Bmatrix} \mathbf{V}_S \\ 0 \end{Bmatrix} \tag{3.61}$$

Note that Eq. (3.61) is very similar to Eq. (3.56), the only difference is that in Eq. (3.61) the cavity surface pressure is set to zero. Therefore, for an air bubble that has non-zero air pressure Eq. (3.56) should be used, while for a cavity with zero internal pressure Eq. (3.61) should be used.

If the pressure $\{\mathbf{P}_S\}$ on the transducer surface is specified instead of the velocity, then the source strengths are obtained from the following equations:

$$\begin{bmatrix} \mathbf{Q}_{SS} & \mathbf{Q}_{SB} \\ \mathbf{Q}_{BS} & \mathbf{Q}_{BB} \end{bmatrix} \begin{Bmatrix} \mathbf{A}_S \\ \mathbf{A}_B \end{Bmatrix} = \begin{Bmatrix} \mathbf{P}_S \\ 0 \end{Bmatrix} \tag{3.62}$$

Both equations (3.61) and (3.62) require the solution of a large system of linear equations. This solution is valid for any size of cavity. It should be noted here that even for a large cavity, if one is interested in computing the ultrasonic field for only the first echo, ignoring the effects of subsequent echoes, then equation (3.59) can also be used.

3.5.2 Analytical Solution

The closed-form expressions suggested by Lobkis and Zinin (1990) and Atalar (1978) give the output of the reflection acoustic microscopy for spherical particles. The analytical solution is based on the Fraunhofer approximation which restricts the location of the spherical object in a specific range near the focal point ($kZ^2/f<<1$, $kR^2/f<<1$; k = wave number; Z = vertical eccentricity of the particle from the focal point; R = horizontal eccentricity; f = focal length or transducer radius). The analytical expression is not valid for problem geometries violating these assumptions. For example, for the transducer with frequency = 1 MHz and focal length = 20 mm, the eccentricity Z must be much less than $\sqrt{\frac{f}{k}} = 0.0022\,\text{m}$. Therefore, 2.5 mm eccentricity does not satisfy the Fraunhofer assumptions. The other relation presented by Lobkis and Zinin (1990) and Atalar (1978) is based on the angular spectrum domain and using the conventional theorem of Fourier optics analysis. All these works consider the reciprocity principle proved in Atalar (1980), in which the derivation neglects the multiple reflections between the reflecting object and transducer. In other words, it uses the pulse-echo technique and looks at the first reflection only.

The reflected signal strength from a spherical cavity with radius 'a' moving along the axis of the focused transducer is given by Lobkis and Zinin (1990):

$$V(e) = \frac{2V_0}{1-\cos\alpha}\sum_{n=0}^{\infty}(-1)^n A_n I_n^2(e) \tag{3.63}$$

where

$$I_n(e) = \sqrt{(2n+1)/2}\int_0^{\alpha}\exp(-ike\cos\theta)P_n(\cos\theta)\sin\theta d\theta$$

$$V_0 = 2\pi f^2(1-\cos\alpha)\left(\frac{v_0}{k}\right)\exp\left(i(2kf-\pi/2)\right)$$

P_n = Legendre polynomials of degree n
e = cavity eccentricity from the focal point
α = half transducer angle or lens angle
f = transducer radius (focal length)
v_0 = oscillating velocity on the transducer surface
k = wave number in the fluid medium
a = cavity radius

$$A_n = -\frac{\rho(k_i a)j_n(ka)j_n'(k_i a) - \rho_i(ka)j_n'(ka)j_n(k_i a)}{\rho(k_i a)h_n(ka)j_n'(k_i a) - \rho_i(ka)h_n'(ka)j_n(k_i a)}$$

j_n = spherical Bessel functions
h_n = spherical Hankel functions
ρ_i = cavity density (air density)
k_i = cavity wave number
ρ = fluid medium density

3.5.3 Numerical Results for the Cavity Problem

Following the formulation given in the above section, the ultrasonic field in front of the acoustic lens is computed in presence of a cavity. The computed field is compared with the field generated in absence of the cavity to see the effect of the cavity.

The opening angle of the transducer (also known as the lens angle) and the lens radius are 100° and 20 mm, respectively. The transducer is excited at 1 MHz frequency. Note that modeling this large lens at 1 MHz frequency is equivalent to modeling a much smaller lens of, for example, 50 μm radius with 100° lens opening angle, operating at 400 MHz frequency because in these two cases the lens radius/wavelength ratio is the same.

The lens surface is modeled by 4003 point sources, distributed just behind the lens face as shown in Figure 3.20. The first set of results presented in Figures 3.21 and 3.22 are obtained for a very small cavity that is modeled by a single point source as discussed in section 3.5.1.1.

The pressure field computed in the fluid in the absence and presence of the cavity are shown in Figures 3.21a and b, respectively. The cavity is placed at the focal point. Note that the high pressure values (represented by the white color in the gray scale images) at the focal point are strongly affected in presence of the cavity. The scattered field away from the focal point is also affected.

The cavity location is then moved toward the lens (positive y direction) and away from the lens (negative y direction). Results in Figure 3.22 are presented for the cavity locations of +/–2.5 mm. The left Figure 3.22a shows the pressure field when the cavity moves closer to the lens by 2.5 mm while the right figure shows the pressure field when the cavity moves 2.5 mm away from the lens. One can clearly see from these figures that the pressure field close to the cavity is strongly affected by its location but the field away from the cavity is not significantly altered.

However, such changes in pressure values very close to the cavity do not help us much in detecting the cavity because the sensing element is not necessarily placed at the cavity location. For cavity detection the acoustic lens is used for transmitting the ultrasonic beam and also receiving the reflected signal from the cavity. The pressure field generated by the reflected ultrasonic energy from the cavity is integrated over the lens surface to compute the acoustic force sensed by the lens due to the reflected signal in comparison to the transmitted force. This ratio is found to be 21.29% when the cavity is located at the focal point but is reduced to 4.85% and 5.43% as the cavity is moved from the focal point by an amount 2.5 mm toward the lens and away from the lens, respectively.

The cavity is then moved around the focal point in the black square region, shown in Figure 3.23a. The variation of the acoustic force on the transducer lens generated by the reflected energy is shown in Figure 3.23b. As expected, the strongest reflection is obtained when the cavity is at the focal point because in this position the cavity surface reflects the maximum amount of energy back to the lens. For any other position of the cavity it receives less energy and as a consequence the reflected energy strength decays as the cavity moves away from the focal point in any direction. It is also interesting to note that the acoustic force (or the integration of the pressure on the lens surface) is detectable (white or light gray color in Figure 3.23b) as the cavity is moved between –3 and +3 mm in the vertical direction (y-axis in Figure 3.20). Figure 3.23b also shows that the acoustic force generated by the reflected signal from the cavity is detectable only when the cavity is moved between –1 and +1 mm in the horizontal direction (x-axis in Figure 3.20). The effect of the cavity size on the reflected energy strength and the scattering pattern is presented below.

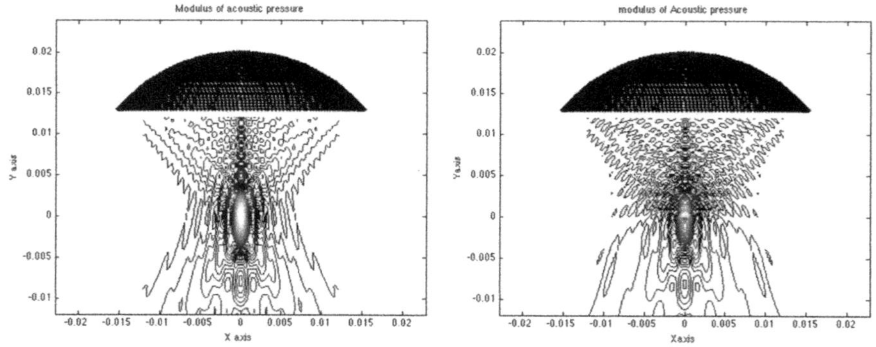

Figure 3.21 Computed pressure field in front of the acoustic lens – (a) in absence of the cavity, (b) in presence of the cavity located at the focal point; note the change in the pressure field near the focal point due to the cavity's presence [from Placko et al. (2010)].

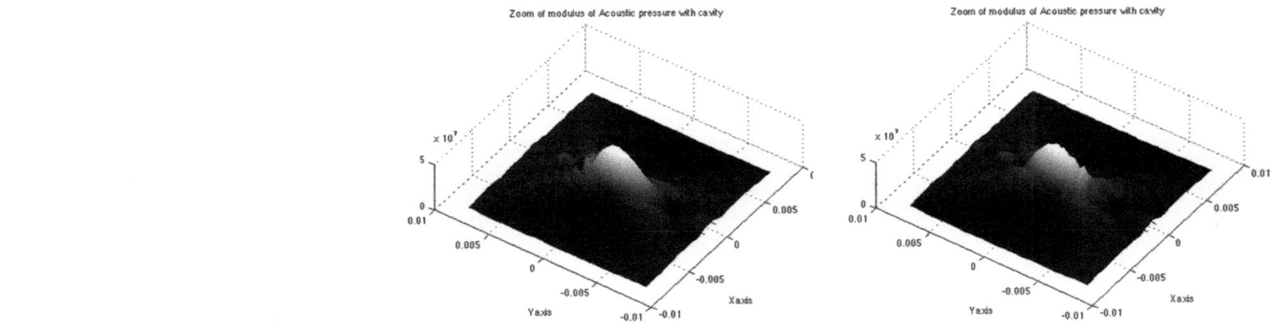

Figure 3.22 Acoustic pressure field variation in the fluid in presence of the cavity for two different cavity locations – (a) y = 2.5 mm (left) and (b) y = –2.5 mm (right); note that in both cases the cavity is placed 2.5 mm away from the focal point – in one case it moves closer to the lens (left), and in the other case it moves away from the lens (right); single point source modeling for the cavity has been used for this computation [from Placko et al. (2010)].

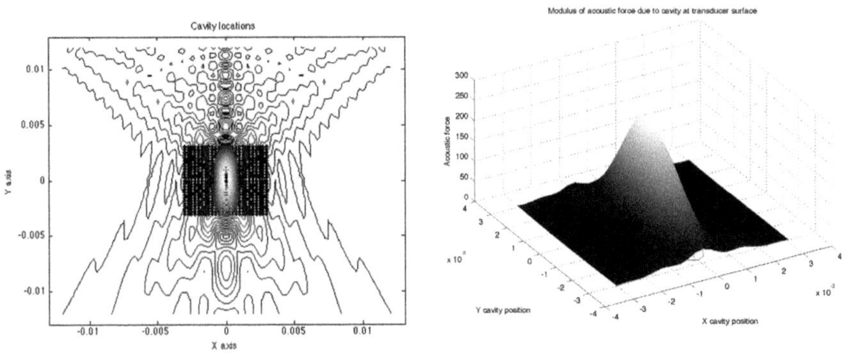

Figure 3.23 (a) Cavity position is moved in the rectangular region marked by the black square near the focal point of the lens, (b) variation of the acoustic force felt by the lens due to the reflected energy from the cavity as it moves within the rectangular region, shown in Figure 3.23(a). Note how the force magnitude decays differently in the x and y directions [from Placko et al. (2010)].

To investigate how accurately a single point source can model a cavity the results generated by the single point source model (section 3.5.1.1) are compared with the results obtained from the multiple point source model (section 3.5.1.2) and that obtained from the complete DPSM analysis (section 3.5.1.3). Figures 3.24 and 3.25 show computed results for 0.2 mm and 1 mm diameter cavities, respectively. In the multiple point source model 100 point sources are used for the cavity modeling. For the complete DPSM analysis the transducer was modeled by 2676 point sources and the cavity by 370 point sources. Normalized reflection forces at the transducer face produced by a cavity as the cavity center moves along the lens axis (y-axis) between $y = -3$ mm to $+3$ mm (the focal point is at $y = 0$) are plotted in Figures 3.24 and 3.25. Three curves are obtained as the cavity is modeled by a single point source (dotted line), 100 point sources (solid line) and 370 point sources in full DPSM analysis (dashed line).

In Figure 3.24 the three curves almost coincide, however in Figure 3.25 the full analysis curve (see section 3.5.1.3) shows some oscillations about the curve generated by the simplified multiple point source model for the small cavity (section 3.5.1.2). These oscillations are the effect of multiple reflections of the signal between the transducer and the cavity; this multiple reflection effect is ignored in the simplified analysis as well as in all theoretical solutions available today.

A clear horizontal shift of about 0.5 mm between the single and multiple point source model generated results can be noticed in Figure 3.25. A much smaller shift (about 0.1 mm) between the same two curves is seen in Figure 3.24. This offset can be explained in the following manner. For the single point source model of the cavity at zero eccentricity, the zero pressure condition is satisfied at the focal point, but for the multiple point source model it is satisfied at +/−0.5 mm for the 1 mm cavity and at +/−0.1 mm for the 0.2 mm cavity when the eccentricity is zero. This slight difference in the problem geometry causes the observed offset.

Next, the DPSM predictions are compared with the finite element results and analytical solutions. Total forces sensed by the acoustic lens as a function of the cavity eccentricity generated by different methods are shown in Figures 3.26 and 3.27. As in Figures 3.24 and 3.25, here also the force is generated by the wave reflected from the cavity and striking the lens. The cavity diameter is 0.2 mm for

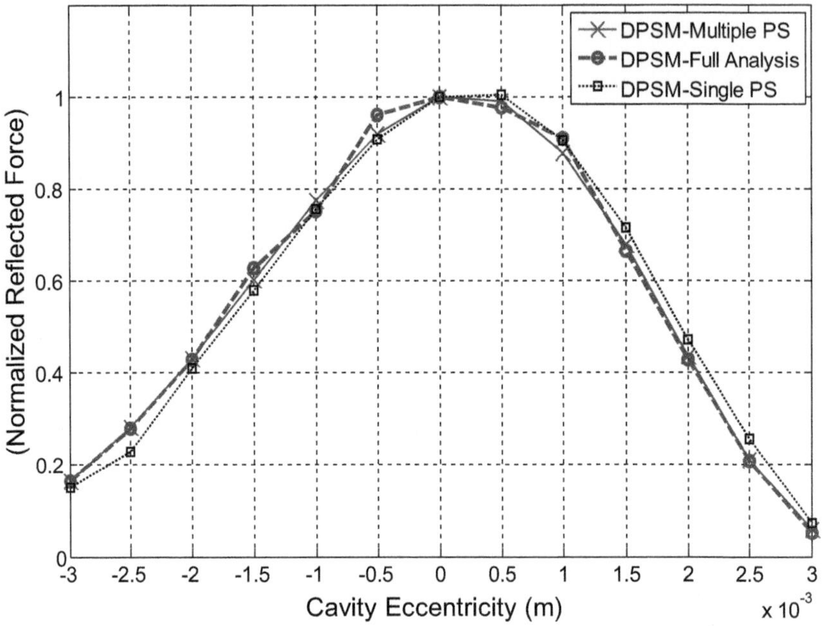

Figure 3.24 Normalized reflection force at the transducer face obtained from 0.2 mm diameter cavity as the cavity center moves along the lens axis around the focal point between −3 mm (measured from the focal point in the direction opposite to the lens) to +3 mm (toward the lens). Three curves are obtained as the cavity is modeled by a single point source (dotted line), 100 point sources (solid line) and full DPSM analysis (dashed line) [from Placko et al. (2010)].

Figure 3.26 and 1 mm for Figure 3.27. Three curves of Figure 3.26 are generated by the finite element analysis (dotted curve with square markers) and the DPSM analysis. 77,891 axi-symmetric triangular elements were used in the finite element mesh while in the DPSM modeling 4003 point sources modeled the acoustic lens and 370 point sources modeled the cavity. DPSM results are generated in two different ways. The complete or full DPSM analysis (section 3.5.1.3) and the simple DPSM analysis (section 3.5.1.2). Note that in the simple DPSM analysis only the incident and the scattered fields are considered, ignoring the effect of the multiple reflections of the waves between the cavity and the lens. All three curves show good matching in Figure 3.26. However, for small eccentricity (between −1 and +1 mm) the curves slightly deviate from one another. This is because the beam strength striking the cavity is high when the cavity eccentricity is low; therefore, the strength of the multiple reflected waves is stronger at low eccentricity. At higher eccentricity, since the multiple reflection effect is negligible, the results obtained from the simple DPSM analysis that ignores multiple reflection and the complete DPSM as well as the FE analysis that considers multiple reflections almost coincide. If the cavity diameter is increased to 1 mm the multiple reflection effect becomes stronger. As a result, for the bigger cavity the difference between the simple and complete DPSM analysis increases as shown in Figure 3.27. The theoretical curve [Eq. (3.63)] is shown by the solid line in Figure 3.27. Note that the theoretical curve is closer to the simple DPSM analysis since both these techniques ignore the multiple reflection effect. Here

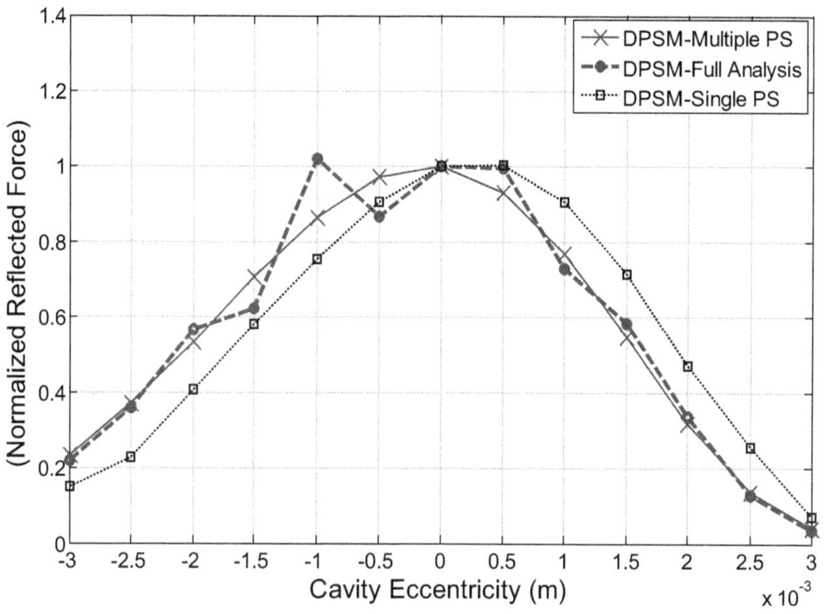

Figure 3.25 Normalized reflection force at the transducer face obtained from 1 mm diameter cavity as the cavity center moves along the lens axis around the focal point between –3 mm (measured from the focal point in the direction opposite to the lens) to +3 mm (toward the lens). Three curves are obtained as the cavity is modeled by a single point source (dotted line), 100 point sources (solid line) and full DPSM analysis (dashed line) [from Placko et al. (2010)].

also the three curves match very well at higher eccentricity because the multiple reflection effect becomes negligible.

If the cavity is moved horizontally violating the axi-symmetric condition then the finite element solution becomes too expensive because a full 3D analysis is required. Then the analytical solution ignoring the multiple reflection effect or the DPSM solution is to be followed. If the cavity is moved further away from the central axis violating the simplifying assumptions associated with the analytical solution then DPSM is the only technique that can solve this problem.

3.6 ULTRASONIC FIELD IN MULTILAYERED FLUID MEDIUM

The multilayered fluid medium shown in Figure 3.28 is modeled in this section by DPSM. Let the complete medium be composed of n different materials separated by (n–1) interfaces. If one wants to model the ultrasonic field generated by two ultrasonic transducers S and T in this inhomogeneous medium, then two layers of point sources must be placed above and below every interface and one layer of point sources should be placed to model each transducer. In this problem geometry, although there are $2n$ different layers of point source ($\mathbf{A}_S$, $\mathbf{A}_T$, A_1, $\mathbf{A}_1^*$, A_2, $\mathbf{A}_2^*$, ... $\mathbf{A}_{n-1}$, $\mathbf{A}^*_{n-1}$), only two layers of source contribute to the field at any point. For example, the total field at a point inside the m-th layer is the superposition of the contributions of all point sources of layers A_m and $\mathbf{A}_{m-1}^*$). For the points in medium 1 it should be the total contribution of layers A_1 and $\mathbf{A}_S$ while points in medium n receive ultrasonic energy from layers A_T and $\mathbf{A}_{n-1}^*$. If in medium 1 or n the transducer is not present, then the field in that medium is

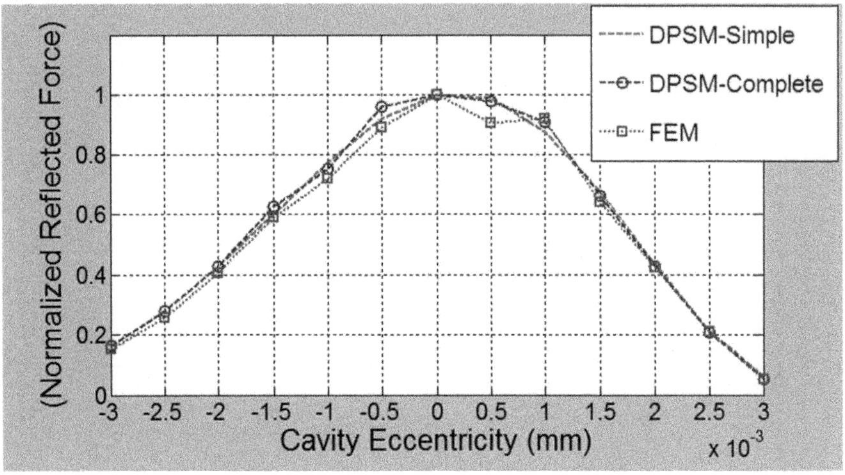

Figure 3.26 Normalized reflection force at the acoustic lens (or transducer, shown in Figure 3.20) obtained from 0.2 mm diameter cavity as the cavity center moves along the lens axis around the focal point between –3 mm (measured from the focal point in the direction opposite to the lens) to +3 mm (toward the lens). Three curves are obtained from three types of modeling – the finite element analysis (dotted line with square markers), complete DPSM analysis that considers multiple reflection effect (dashed line with circular markers), and simple DPSM analysis that ignores the multiple reflection effect (dashed line without any marker) [from Kundu et al. (2010)].

obtained from only one layer of source placed at the interface. For more discussion on this formulation the readers are referred to Kundu and Placko (2007) and Banerjee et al. (2006).

Results for a three-layered medium subjected to ultrasonic excitation by one or two transducers are shown in Figure 3.29. Two different fluids (water and glycerin) constitute the three-layered medium. Glycerin is placed between two water half-spaces. Different orientations of the two transducers have been considered. The density and the P-wave speed of the fluids considered in this study are given in Table 3.1.

Figure 3.29 shows the ultrasonic fields generated in the water–glycerin–water structure for different transducer orientations. The transducers have 4 mm diameter. Ultrasonic fields generated by 1 MHz transducers have been presented in the left column of Figure 3.29 and 2.2 MHz transducers are presented in the right column. The thickness of the glycerin layer is 20 mm for both 1 MHz and 2.2 MHz excitations. 1 MHz transducers are kept at 10 mm distance from the water–glycerin interfaces but 2.2 MHz transducers are placed at 20 mm distance from the same interfaces. Additional distance for the 2.2 MHz transducers is necessary to make sure that the interface is not placed within the near-field region of the transducers. Therefore, the total width of the WGW structure is 40 mm for the 1 MHz transducers and 60 mm for the 2.2 MHz transducers. It is well known that the pressure field generated at a point in front of a transducer depends on the frequency of excitation and the distance of the point from the transducer face. As the frequency increases the isobars gradually shift away from the transducer face. To generate approximately the same pressure value at

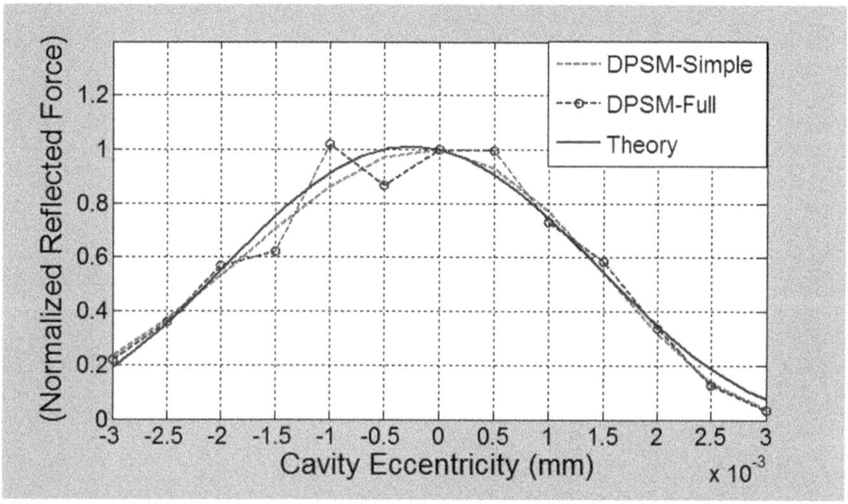

Figure 3.27 Normalized reflection force at the acoustic lens (or transducer, shown in Figure 3.20) obtained from a 1 mm diameter cavity as the cavity center moves along the lens axis around the focal point between −3 mm (measured from the focal point in the direction opposite to the lens) to +3 mm (toward the lens). Three curves are obtained from three types of modeling – analytical solution ignoring the multiple reflection effect (continuous line), complete DPSM analysis that considers multiple reflection effect (dashed line with circular markers), and simple DPSM analysis that ignores multiple reflection effect (dashed line without any marker) [from Kundu et al. (2010)].

the interface position the distance D between the transducers needs to be varied when the transducer frequency is changed from 1 to 2.2 MHz.

In Figures 3.29 the transducer generated signals either strike the interface at normal incidence or at an inclination angle $\theta = 30°$, measured from the vertical axis. The top row of Figure 3.29 is generated when only transducer T is turned on. The middle row is generated when both transducers S and T are turned on and generate ultrasonic beams that strike the interfaces at angle $\theta = 30°$. In the middle row, the ultrasonic fields in glycerin form nice symmetric patterns due to interaction between the transmitted and reflected fields. Note that this pattern is dependent on the signal frequency. The bottom transducer position is then changed to form a beam striking the bottom interface vertically while the other beam strikes the top interface at an angle. Ultrasonic fields produced for this configuration are shown in the bottom row for 1 and 2.2 MHz frequencies.

3.7 ULTRASONIC FIELD COMPUTATION IN PRESENCE OF A FLUID–SOLID INTERFACE

Up to this point we have computed the ultrasonic fields in fluid media only. Let us now investigate how to compute the ultrasonic field in presence of a fluid–solid interface or solid–solid interface.

3.7.1 Fluid–Solid Interface

Across the fluid–solid interface (say at $z=0$) the following four quantities must be continuous – (1) normal displacement u_z (or velocity v_z) in fluid and solid,

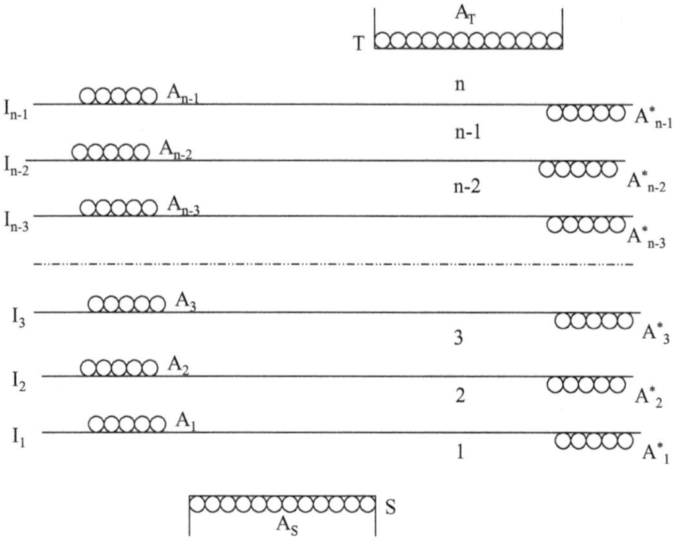

Figure 3.28 An inhomogeneous medium made of n different materials and (n–1) interfaces is excited by two transducers S and T placed in medium #1 and #n. Each transducer is modeled by one layer of point sources while every interface is modeled by two layers of point source. Point sources are located at the centers of the small circles shown in the figure.

(2) normal pressure p (or stress σ_{zz}), (3 and 4) two shear stress components (σ_{xz} and σ_{yz}) must be zero at the interface. Therefore, at every point on the fluid–solid interface the following four conditions must be satisfied:

$$u_z^f = u_z^s,\ \sigma_{zz}^f = \sigma_{zz}^s,\ \sigma_{xz}^s = 0,\ \sigma_{yz}^s = 0 \tag{3.64}$$

In Eq. (3.64) superscripts f and s indicate fluid and solid, respectively. The fluid–solid interface can be modeled by $2N$ point sources (N on the fluid side and N on the solid side) as shown in Figure 3.30. In this figure the top half-space is made of fluid and the bottom half-space is solid. Two layers of small spheres associated with the point sources touch the interface at N points. By satisfying all four continuity conditions given in Eq. (3.64) at these N points, $4N$ equations are obtained to solve for $4N$ unknowns.

For the fluid–fluid interface $2N$ equations were obtained by satisfying the continuity of pressure and normal velocity components at N points on the interface. Those $2N$ equations are just enough to solve for $2N$ unknown source strengths. However, for the fluid–solid interface, since there are $4N$ equations, there must be $4N$ unknowns associated with the $2N$ point sources. While the point source for the fluid uniformly expands and contracts and has only one unknown value for its strength, the point source for the solid has three unknowns. For the solid point source inside every small sphere there are three point forces vibrating in x, y and z directions with three independent strengths as shown in Figure 3.30. Therefore, every point source for solid modeling is composed of three point forces vibrating at the same frequency but with different amplitudes in x, y and z directions. The strengths of these three point forces are the three unknowns associated with every point source modeling the ultrasonic field in the solid.

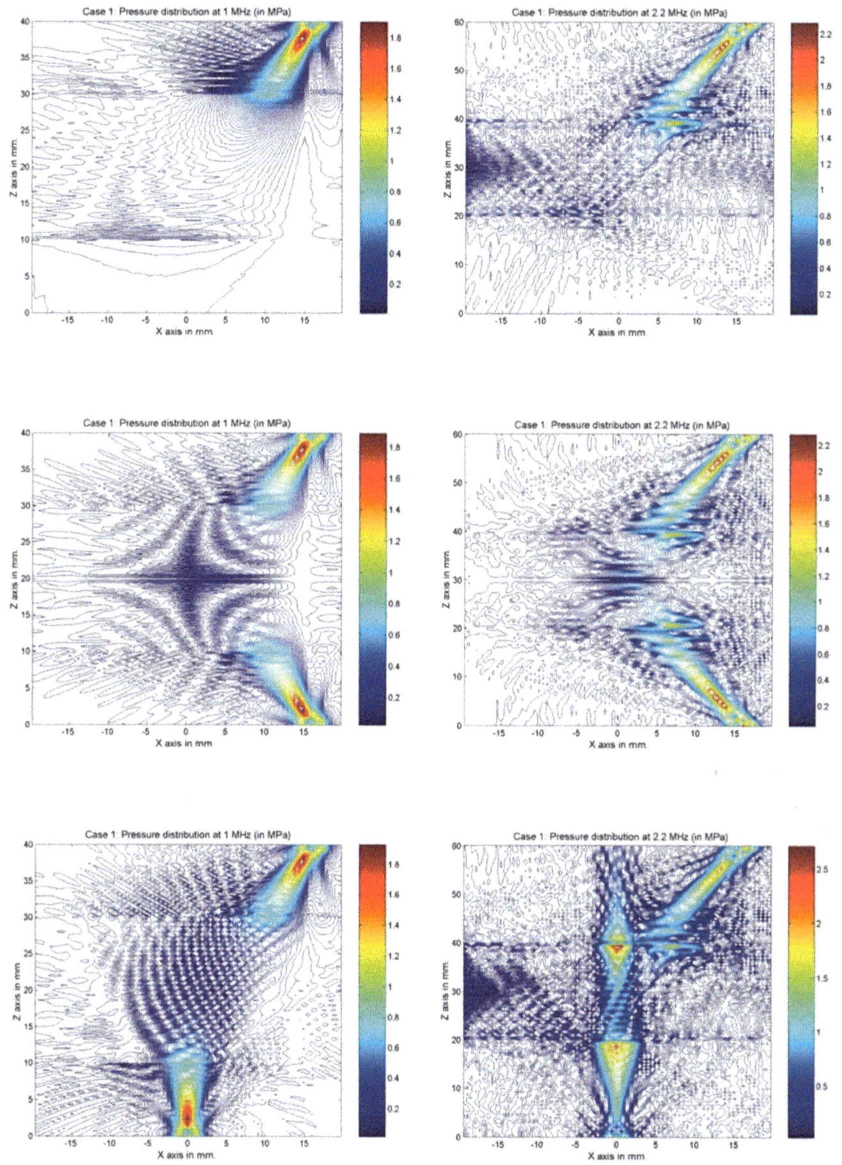

Figure 3.29 Water–Glycerin–Water medium subjected to 1 MHz (left column) and 2.2 MHz (right column) ultrasonic excitations when only one transducer is transmitting ultrasonic energy (top row) and both transducers are active (bottom two rows) [from Banerjee et al. (2006)].

3.7.2 A Fluid Wedge Over a Solid Half-Space – DPSM Formulation

The solution for a fluid wedge problem with a free fluid surface and a fluid–solid interface is given in this section following the work of Dao (2007) and Dao et al. (2009). The problem geometry is shown in Figure 3.31. A homogenous solid half-space is partially immersed in a homogenous fluid at an angle forming a fluid

Table 3.1: Fluid Properties

Fluids	P-wave Speed (km/sec)	Density (gm/cc)
Water	1.48	1.00
Glycerine	1.92	1.26

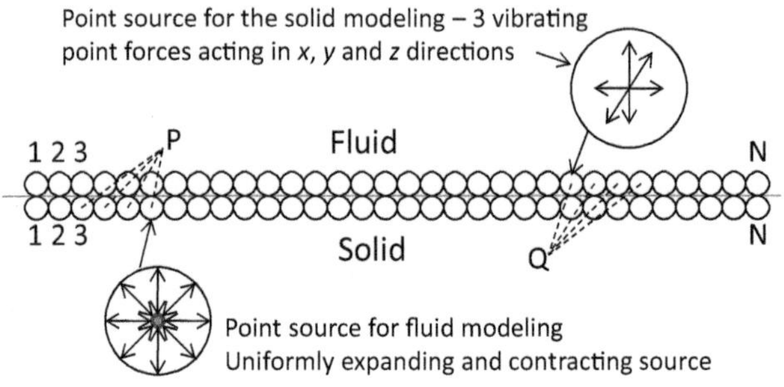

Figure 3.30 2N number of point sources, located at the centers of the small circles, model the fluid–solid interface. The total field at a general point P in the fluid is obtained by superimposing the contributions of all fluid point sources placed on the solid side of the interface. In the same manner the total field at a general point Q in the solid is obtained by superimposing the contributions of all solid point sources placed on the fluid side of the interface [from Placko and Kundu (2007)].

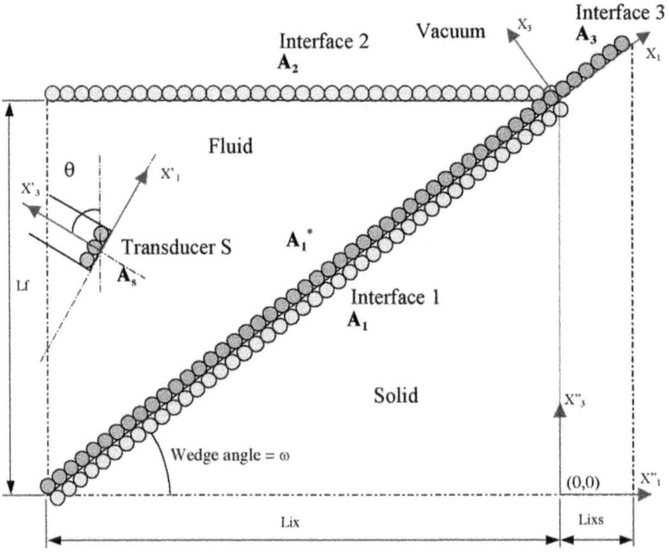

Figure 3.31 Geometry of the wedge problem with point sources shown by small circles, distributed along the transducer face, the fluid–solid interface, the fluid boundary and the solid boundary [from Dao et al. (2009)].

wedge. A finite size transducer immersed in the fluid acts as the ultrasonic energy source. The ultrasonic beam strikes the inclined fluid–solid interface. The ultrasonic fields inside the fluid wedge and the solid half-space are to be computed.

Following the DPSM technique discussed above, one layer of point sources is distributed near the transducer face and another four layers of point source are distributed on two boundaries and one interface. Point sources are placed at the centers of small circles shown in Figure 3.31. For simplicity, both boundaries and the interface are denoted as the interface. Interface 1 is the inclined fluid–solid interface. Interface 2 is the traction-free fluid surface that can be also denoted as the fluid–vacuum interface, and interface 3 is the inclined solid–vacuum interface or the traction-free boundary of the solid. Strengths of the five layers of point sources are denoted as $\mathbf{A}_s$, A_1, $\mathbf{A}_1^*$, A_2 and A_3 in Figure 3.31. $\mathbf{A}_s$, A_1, $\mathbf{A}_1^*$, A_2 and A_3 are five vectors whose elements denote the strengths of individual point sources in these five layers of source.

The ultrasonic fields in fluid and solid media can be written in matrix form. If T is a set of target points in the fluid medium, then the velocity at the target points can be expressed in terms of transducer sources $\mathbf{A}_s$ and interface sources A_1, A_2 in the following manner:

$$\mathbf{V}_T = \mathbf{M}_{TS}\mathbf{A}_S + \mathbf{M}_{T1}\mathbf{A}_1 + \mathbf{M}_{T2}\mathbf{A}_2 \tag{3.65}$$

Velocity at the transducer's surface is given as follows:

$$\mathbf{V}_S = \mathbf{M}_{SS}\mathbf{A}_S + \mathbf{M}_{S1}\mathbf{A}_1 + \mathbf{M}_{S2}\mathbf{A}_2 \tag{3.66}$$

In equations (3.65) and (3.66) vectors $\mathbf{V}_T$ and $\mathbf{V}_S$ are particle velocities at target points and transducer surface points, respectively. Matrix $\mathbf{M}_{IJ}$ when multiplied by the source strength vector $\mathbf{A}_J$ ($J = S$, 1 or 2) gives velocity at points I ($I = T$ for the target points and S for the transducer surface points). Matrix $\mathbf{M}$ has been defined earlier. In the above equation and in all subsequent equations the only unknowns are the source strength vectors $\mathbf{A}_s$, A_1, $\mathbf{A}_1^*$, A_2, and A_3. In terms of these unknown source strength vectors the pressure fields at the target points in the fluid medium are given by

$$\mathbf{P}_T = \mathbf{Q}_{TS}\mathbf{A}_S + \mathbf{Q}_{T1}\mathbf{A}_1 + \mathbf{Q}_{T2}\mathbf{A}_2 \tag{3.67}$$

$\mathbf{Q}_{TS}$, $\mathbf{Q}_{T1}$, and $\mathbf{Q}_{T2}$ are matrices that relate pressure values at the target points to the point source strength vectors $\mathbf{A}_s$, A_1 and A_2, respectively.

At the target points, the displacement field along the x_3 direction in the fluid is written as

$$\mathbf{U3}_T = \mathbf{DF3}_{TS}\mathbf{A}_S + \mathbf{DF3}_{T1}\mathbf{A}_1 + \mathbf{DF3}_{T2}\mathbf{A}_2 \tag{3.68}$$

To model the ultrasonic field in a solid, every point source should contain three different point forces in three mutually perpendicular directions, as discussed above and shown in Figure 3.30. Normal stresses in the x_3 direction on interface 1 generated by the point sources are given by

$$\mathbf{S33}_{11^*}\mathbf{A}_1^* + \mathbf{S33}_{13}\mathbf{A}_3 \tag{3.69}$$

Similarly, the shear stresses developed at interface 1 are

$$\begin{gathered}\mathbf{S31}_{11^*}\mathbf{A}_1^* + \mathbf{S31}_{13}\mathbf{A}_3 \\ \mathbf{S32}_{11^*}\mathbf{A}_1^* + \mathbf{S32}_{13}\mathbf{A}_3\end{gathered} \tag{3.70}$$

Normal stresses generated in the x_3 direction on interface 3 can be written as

$$\mathbf{S33}_{31^*}\mathbf{A}_1^* + \mathbf{S33}_{33}\mathbf{A}_3 \tag{3.71}$$

and shear stresses developed on interface 3 are

$$\begin{aligned} &\mathbf{S31}_{31^*}\mathbf{A}_1^* + \mathbf{S31}_{33}\mathbf{A}_3 \\ &\mathbf{S32}_{31^*}\mathbf{A}_1^* + \mathbf{S32}_{33}\mathbf{A}_3 \end{aligned} \tag{3.72}$$

Displacement along the x_3 direction in the solid on interfaces 1 and 3 are

$$\begin{aligned} &\mathbf{DS3}_{11^*}\mathbf{A}_1^* + \mathbf{DS3}_{13}\mathbf{A}_3 \\ &\mathbf{DS3}_{31^*}\mathbf{A}_1^* + \mathbf{DS3}_{33}\mathbf{A}_3 \end{aligned} \tag{3.73}$$

The transducer surface velocity V_{SO} is known. Therefore, the boundary condition at the transducer face can be written as

$$\mathbf{M}_{SS}\mathbf{A}_S + \mathbf{M}_{S1}\mathbf{A}_1 + \mathbf{M}_{S2}\mathbf{A}_2 = \mathbf{V}_{SO} \tag{3.74}$$

Across the fluid–solid interface (interface 1 in Figure 3.31) negative of the normal stress ($-S_{33}$) in the solid and the pressure in the fluid should be continuous. Also, across interface 1, the displacement component normal to the interface (x_3 direction in Figure 3.31) should be continuous. Shear stresses in the solid at the interface should vanish. Satisfying these continuity conditions the following equations are obtained

$$\begin{aligned} &\mathbf{Q}_{1S}\mathbf{A}_S + \mathbf{Q}_{11}\mathbf{A}_1 + \mathbf{Q}_{12}\mathbf{A}_2 = -\mathbf{S33}_{11^*}\mathbf{A}_1^* - \mathbf{S33}_{13}\mathbf{A}_3 \\ &\mathbf{DF3}_{1S}\mathbf{A}_S + \mathbf{DF3}_{11}\mathbf{A}_1 + \mathbf{DF3}_{12}\mathbf{A}_2 = \mathbf{DS3}_{11^*}\mathbf{A}_1^* + \mathbf{DS3}_{13}\mathbf{A}_3 \\ &\mathbf{S31}_{11^*}\mathbf{A}_1^* + \mathbf{S31}_{13}\mathbf{A}_3 = 0 \\ &\mathbf{S32}_{11^*}\mathbf{A}_1^* + \mathbf{S32}_{13}\mathbf{A}_3 = 0 \end{aligned} \tag{3.75}$$

Since the pressure on interface 2 must be zero (traction-free fluid surface) the boundary condition at this interface is given by

$$\mathbf{Q}_{2S}\mathbf{A}_S + \mathbf{Q}_{21}\mathbf{A}_1 + \mathbf{Q}_{22}\mathbf{A}_2 = 0 \tag{3.76}$$

Similarly, the traction-free boundary condition on interface 3 can be written as

$$\begin{aligned} &\mathbf{S33}_{31^*}\mathbf{A}_1^* + \mathbf{S33}_{33}\mathbf{A}_3 = 0 \\ &\mathbf{S31}_{31^*}\mathbf{A}_1^* + \mathbf{S31}_{33}\mathbf{A}_3 = 0 \\ &\mathbf{S32}_{31^*}\mathbf{A}_1^* + \mathbf{S32}_{33}\mathbf{A}_3 = 0 \end{aligned} \tag{3.77}$$

Equations (3.74) to (3.77) are written in the following matrix form:

$$\begin{bmatrix} \mathbf{M}_{SS} & \mathbf{M}_{S1} & \mathbf{M}_{S2} & 0 & 0 \\ \mathbf{Q}_{1S} & \mathbf{Q}_{11} & \mathbf{Q}_{12} & \mathbf{S33}_{11^*} & \mathbf{S33}_{13} \\ \mathbf{DF3}_{1S} & \mathbf{DF3}_{11} & \mathbf{DF3}_{12} & -\mathbf{DS3}_{11^*} & -\mathbf{DS3}_{13} \\ 0 & 0 & 0 & \mathbf{S31}_{11^*} & \mathbf{S31}_{13} \\ 0 & 0 & 0 & \mathbf{S32}_{11^*} & \mathbf{S32}_{13} \\ \mathbf{Q}_{2S} & \mathbf{Q}_{21} & \mathbf{Q}_{22} & 0 & 0 \\ 0 & 0 & 0 & \mathbf{S33}_{31^*} & \mathbf{S33}_{33} \\ 0 & 0 & 0 & \mathbf{S31}_{31^*} & \mathbf{S31}_{33} \\ 0 & 0 & 0 & \mathbf{S32}_{31^*} & \mathbf{S32}_{33} \end{bmatrix} \begin{Bmatrix} \mathbf{A}_S \\ \mathbf{A}_1 \\ \mathbf{A}_2 \\ \mathbf{A}_1^* \\ \mathbf{A}_3 \end{Bmatrix} = \begin{Bmatrix} \mathbf{V}_{SO} \\ 0 \\ 0 \\ 0 \\ 0 \end{Bmatrix} \tag{3.78}$$

In the above equation matrices $\mathbf{M}_{IJ}$ and $\mathbf{Q}_{IJ}$ have been defined earlier. However, $\mathbf{DF3}_{IJ}$, $\mathbf{S31}_{IJ}$, $\mathbf{S32}_{IJ}$, $\mathbf{S33}_{IJ}$ and $\mathbf{DS3}_{IJ}$ are introduced for the first time here. They compute the following field parameters:

DF3 – u_3 component of displacement inside the fluid

DS3 – u_3 component of displacement inside the solid

S31 – Shear stress component (σ_{31}) inside the solid

S32 – Shear stress component (σ_{32}) inside the solid

S33 – Normal stress component (σ_{33}) inside the solid

From equation (3.78) the unknown source strengths can be obtained. Step-by-step derivation of the above equations and elements of different matrices can be found in Dao (2007) and Banerjee and Kundu (2007).

Numerical results are obtained from the above formulation. The results are presented for a 2.25 MHz transducer with a diameter of 6.35 mm (0.25 inch) generating the ultrasonic beam. The material properties for the aluminum and water used in this computation are as follows: aluminum – P-wave speed (c_P) = 6.35 km/s, S-wave speed (c_S) = 3.04 km/s, density (ρ_s) = 2.7 gm/cc; water – P-wave speed (c_f) = 1.48 km/s, density (ρ_f) = 1 gm/cc.

350 point sources are distributed behind the circular transducer face to model the transducer and additional point sources are placed along the fluid–solid interface (interface 1), the free surfaces of the liquid wedge (interface 2), and the solid (interface 3), as shown in Figure 3.31. At the plane of symmetry (or the central plane which is the plane of the paper) four lines of point sources near the interfaces 1, 2 and 3 are placed, as shown in Figure 3.31. The ultrasonic field is first computed on the central plane with this one plane of point sources consisting of four lines along interfaces 1, 2 and 3. Then two more planes of point sources are added on two sides of the central plane and the field is computed again at the central plane with these three planes of point sources. This process of adding two planes of point sources on two sides of the central plane is continued until the computed field at the central plane is converged. Note that additional planes of point sources on two sides of the central plane increase the total number of point sources along interfaces 1, 2 and 3. For the transducer modeling 350 point sources are placed near the transducer face from the very beginning and not changed.

On each side of the fluid–solid interface (interface 1) 135 point sources are distributed on the central plane. Sources are placed in the illuminated region and also well beyond the illuminated region of the interface. Therefore, to model the problem geometry with three planes of point source, a total of 405 point sources are necessary on each side of the fluid–solid interface. Increasing the number of point sources to five planes of source did not significantly change the computed ultrasonic field at the central plane. Total number of point sources on the free surface of the liquid wedge and solid half-space (interfaces 2 and 3) are 405 and 51, respectively.

Results presented in Figure 3.32 show ultrasonic pressures inside the liquid wedge and normal stress (S11) in the solid half-space for three different angles of strike. Note that the x_1 axis is parallel to the interface. Ultrasonic fields are plotted for 15.42°, 30.42°, and 45.42° angles of strike measured from the normal to the fluid–solid interface. Plots are generated for the projected length of 60 mm along the x_1 direction. In Figure 3.32 (a,b,c,d,e,f) side scale bars are provided to give an idea of the strengths of the ultrasonic fields in different plots. Note that the scale bars are not the same in all figures. The signal strengths in both fluid and solid

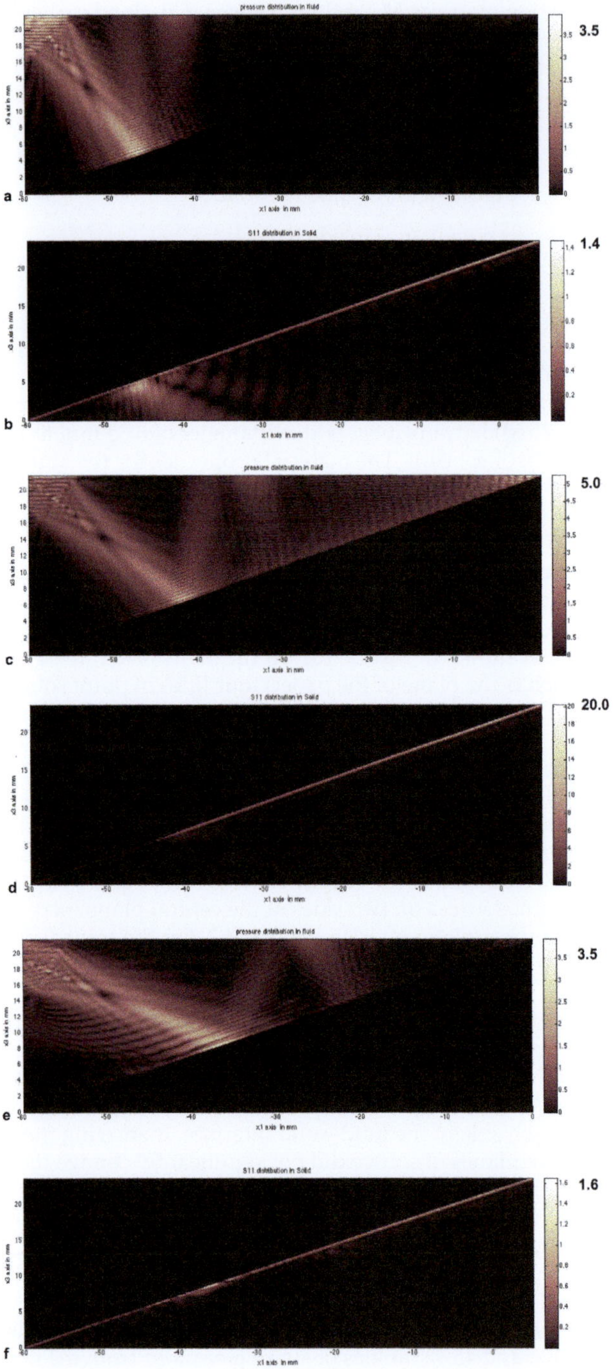

Figure 3.32 Pressure in the 20° fluid wedge (a, c, e) and normal stress (S_{11}) in the solid half-space (b, d, f) for 15.42° (a, b), 30.42° (c, d) and 45.42° (e, f) angles of strike. 30.42° is the Rayleigh critical angle [from Dao et al. (2009)].

are significantly higher for 30.42° angle of incidence. This is expected since this angle corresponds to the Rayleigh critical angle.

Note that for 15.42° incident angle part of the ultrasonic energy is reflected back into the fluid wedge (Figure 3.32a) and part is transmitted into the solid (Figure 3.32b). In Figure 3.32b two transmitted beams look like the P-wave and S-wave beams. However, both of those are S-wave beams generated by the main striking beam and one of its side lobes. The main beam and the side lobes strike the interface at different angles resulting in two transmitted S-wave beams with different inclinations. Figure 3.32b also shows a weak beam propagating along the interface. This is the surface skimming P-wave generated by a part of the diverging beam striking the interface at P-critical angle, $\sin^{-1}\left(\frac{1.48}{6.35}\right) = 13.5°$. For the Rayleigh critical angle of incidence (30.42°) it is interesting to note that the entire liquid wedge, between the transducer and the wedge corner, is illuminated (Figure 3.32c); a propagating Rayleigh wave in the solid can also be seen clearly in Figure 3.32d. For 45.42° angle of strike (Figure 3.32e) strong reflections from the fluid–solid interface as well as from the free liquid surface are observed. As expected, relatively weak ultrasonic energy inside the solid is observed in Figure 3.32f.

It should be mentioned here that although the ultrasonic energy inside the solid, near the fluid–solid interface beyond the point of strike is observed for all three angles of incidence, it is strongest (one order of magnitude higher) for the critical angle (30.42°) of strike. The maximum value of the bar scale of Figure 3.32d is 20 while those for Figures 3.32b and 3.32f are 1.4 and 1.6, respectively. Amplification of the ultrasonic energy near the liquid wedge corner should also be noted, especially in Figures 3.32c and e.

3.7.3 Solid–Solid Interface

Across the solid–solid interface all three components of displacement and three (one normal and two shear) components of stress must be continuous. If the normal to the interface is in the z direction, then three displacement components u_x, u_y and u_z along with the three stress components σ_{zz}, σ_{xz} and σ_{yz} must be continuous across the interface. Therefore, at every point on the solid–solid interface six conditions must be satisfied.

The solid–solid interface can be modeled by 2*N* point sources (*N* on each side of the interface) as shown in Figure 3.30. Since both half-spaces are solid, every point source will have three unknowns – amplitudes of the three point forces vibrating in three mutually perpendicular directions. Thus, in this case, for one interface with 2*N* point sources we will have 6*N* continuity conditions giving rise to 6*N* equations to solve for 6*N* unknowns.

Detailed description and step-by-step derivation for developing the DPSM models for problems with solid–solid interface can be found in Banerjee and Kundu (2007, 2008).

3.8 DPSM MODELING FOR TRANSIENT PROBLEMS

The DPSM formulation discussed so far is based on the steady-state Green's functions. Therefore, all ultrasonic problems discussed above are steady-state problems. Although the steady-state solution has many advantages and is sometimes desirable it does not provide any information about the time of arrival of different ultrasonic waves. In the steady-state solution it may be difficult to identify a weak defect signal reflected from a crack or inclusion because those weak signals do not significantly affect the steady-state response. However, in the transient response, since the times of arrival are different for different ultrasonic signals, it is possible to identify weak defect signals if they are separated from the strong signals in the time scale. Therefore, it is important to have the transient solution as well. Since DPSM can efficiently give steady-state solutions for complex geometries it can be a very strong method if it can solve transient problems as well. In the following the DPSM is extended to solve the transient ultrasonic problems and solutions to some sample problems are provided.

3.8.1 Fluid–Solid Interface Excited by a Bounded Beam – DPSM Formulation

The schematic diagram of the problem geometry for a solid half-space excited by a finite size transducer placed in the fluid half-space, as shown in Figure 3.33, is solved first. The transducer is modeled by one layer of point source. The fluid–solid interface is modeled by two layers of point source. $\mathbf{A}_S$ is the source strength vector of the point sources modeling the transducer. These sources contribute to the ultrasonic field in the fluid. A_1 is the source strength vector of the point sources that are placed above the fluid–solid interface and generates the reflected ultrasonic field in the fluid. $\mathbf{A}_1^*$ is the source strength vector of the point sources that are distributed below the fluid–solid interface and model the transmitted field in the solid. In Figure 3.33 two points D and E are shown in the fluid and solid media, respectively. The ultrasonic field at point D is the summation of the contributions of all point sources having source strengths A_1 and $\mathbf{A}_S$. Similarly, the ultrasonic field at point E is the summation of the contributions of all point sources having the source strength $\mathbf{A}_1^*$.

At a given set of target points the particle velocity and pressure inside the fluid or the displacement and stress inside the solid can be expressed in matrix form. The strength of the point sources can be obtained by satisfying the boundary conditions and the interface continuity conditions. If the normal velocity of the transducer face is assumed to be $\mathbf{V}_{S0}$, then following the steps discussed above the boundary condition on the transducer surface can be written as

$$\mathbf{M}_{SS}\mathbf{A}_S + \mathbf{M}_{S1}\mathbf{A}_1 = \mathbf{V}_{S0} \tag{3.79}$$

where $\mathbf{A}_S$ is the $(M\times 1)$ vector containing the strength of M number of point sources, modeling the transducer face. If N number of point sources are distributed on each side of the fluid–solid interface then A_1 has $(N\times 1)$ elements. $\mathbf{M}_{SS}$ and $\mathbf{M}_{S1}$ are two matrices of dimensions $M\times M$ and $M\times N$, respectively.

As discussed in section 3.7.1, across the fluid–solid interface the normal displacement (u_3) should be continuous. Also at the interface the pressure in the

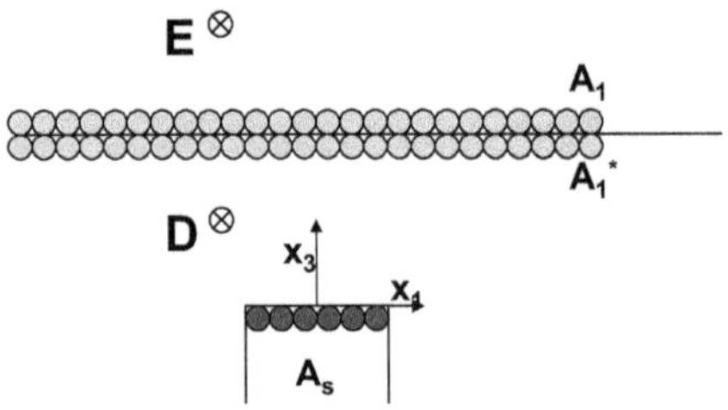

Figure 3.33 DPSM model for the solid half-space adjacent to a fluid half-space containing an ultrasonic transducer. The fluid–solid interface is modeled by two layers of point sources while the transducer is modeled by one layer of point sources. Points D and E are located in the fluid and solid, respectively. The transducer face and the interface are not necessarily parallel [from Das et al. (2010)].

fluid and the negative of the normal stress ($-\sigma_{33}$) in the solid should be continuous and the shear stress in the solid at the interface must vanish. Therefore, on the interface

$$\mathbf{Q}_{IS}\mathbf{A}_S + \mathbf{Q}_{I1}\mathbf{A}_1 = -\mathbf{S33}_{I1^*}\mathbf{A}_1^* \tag{3.80}$$

$$\mathbf{DF3}_{IS}\mathbf{A}_S + \mathbf{DF3}_{I1}\mathbf{A}_1 = \mathbf{DS3}_{I1^*}\mathbf{A}_1^* \tag{3.81}$$

$$\mathbf{S31}_{I1^*}\mathbf{A}_1^* = 0 \tag{3.82}$$

$$\mathbf{S32}_{I1^*}\mathbf{A}_1^* = 0 \tag{3.83}$$

M and **Q** used in the above equations (3.79 to 3.83) have been defined earlier. What other matrices **S33**, **DF3**, **DS3**, **S31** and **S32** compute have been stated after Eq. (3.78). The first subscript (S or I) of these matrices indicate if the target points are placed on the transducer surface or the interface. The second subscript is used to denote the set of point sources used to compute the stress, pressure, displacement or velocity fields. The second subscripts S, 1 and 1^* correspond to point sources $\mathbf{A}_S$, A_1 and $\mathbf{A}_1^*$, respectively.

It should also be kept in mind that every point source that calculates the transmitted field in the solid has three different point forces in three mutually perpendicular directions as unknowns. Thus $\mathbf{A}_1^*$ is a $(3N\times 1)$ vector, while A_1 is a $(N\times 1)$ vector and $\mathbf{A}_S$ is a $(M\times 1)$ vector. The elements of **S33, DF3, DS3, S31** and **S32** matrices are functions of different Green's functions and are given by Banerjee and Kundu (2007).

In the matrix form Eqs. (3.79) to (3.83) can be written as

$$\begin{bmatrix} \mathbf{M}_{SS} & \mathbf{M}_{S1} & 0 \\ \mathbf{Q}_{1S} & \mathbf{Q}_{11} & \mathbf{S33}_{11^*} \\ \mathbf{DF3}_{1S} & \mathbf{DF3}_{11} & -\mathbf{DS3}_{11^*} \\ 0 & 0 & \mathbf{S31}_{11^*} \\ 0 & 0 & \mathbf{S32}_{11^*} \end{bmatrix}_{(M+4N)x(M+4N)} \begin{Bmatrix} \mathbf{A}_S \\ \mathbf{A}_1 \\ \mathbf{A}_1^* \end{Bmatrix}_{(M+4N)} = \begin{Bmatrix} \mathbf{V}_{S0} \\ 0 \\ 0 \end{Bmatrix}_{M+4N} \tag{3.84}$$

or,

$$[\underline{\mathbf{MT}}]\{\mathbf{A}\} = \{\mathbf{V}\} \tag{3.85}$$

Note that Eq. (3.85) is obtained by satisfying simultaneously all boundary and interface continuity conditions of the problem where M and N are the numbers of point sources used to model the transducer and the interface, as shown in Figure 3.33. The point sources are located at a distance of r_S from the transducer face and the interface. The boundary and interface conditions are satisfied at the apex points where the small spheres touch the transducer face and the fluid–solid interface. The total number of boundary and interface conditions is equal to the total number of unknowns; thus, the system of equations given in Eq. (3.84) or (3.85) has a unique solution. The number of point sources needed for the ultrasonic field computation is obtained by satisfying the convergence criterion (see section 3.2.5) that the spacing between two neighboring point sources should be less than $\frac{\lambda}{\sqrt{2\pi}}$. All results here are presented for the spacing between two neighboring sources less than or equal to λ / π.

Solving Eq. (3.85), the source strength vector for the complete problem geometry is obtained:

$$\{\mathbf{A}\} = [\mathbf{MT}]^{-1}\{\mathbf{V}\} \tag{3.86}$$

Matrix [MT] is a well-conditioned matrix and its inversion can be carried out without any difficulty.

3.8.1.1 Transient Analysis

As the solution of the problem with the steady-state boundary condition is known and the problem is linear, the Fourier transform technique can be followed to obtain the transient solution as was done by Das et al. (2010). The harmonic time dependence for the prescribed steady-state external excitation is given by $e^{-i\omega t}$ and the steady-state solution for this problem is given by $r(\mathbf{x},\omega)e^{-i\omega t}$. Note that $r(\mathbf{x},\omega)$ is a function of the position vector ($\mathbf{x}$) of the observation point and the frequency (ω) of excitation, but not of time. Instead of harmonic time dependence, if $f(t)$ is the prescribed time dependent excitation and $F(\omega)$ is its Fourier transform, then for a given value of ω, $F(\omega)r(\mathbf{x},\omega)e^{-i\omega t}$ will be the steady-state solution for the prescribed excitation $F(\omega)e^{-i\omega t}$. Therefore, for the external excitation of $f(t) = \frac{1}{2\pi}\int_{-\infty}^{\infty} F(\omega)e^{-i\omega t}\,d\omega$ the transient solution is given by

$$R(\mathbf{x},t) = \frac{1}{2\pi}\int_{-\infty}^{\infty} F(\omega)r(\mathbf{x},\omega)e^{-i\omega t}\,d\omega \tag{3.87}$$

In Eq. (3.87) $R(\mathbf{x},t)$ represents any field response – displacement, velocity, pressure or stress. DPSM can be used to obtain the steady-state solution ($r(\mathbf{x},\omega)e^{-i\omega t}$) discussed above. For a given $f(t)$ one can obtain $F(\omega)$ analytically, or numerically using the Fast Fourier Transform (FFT) technique [Cooley and Tukey (1965)]. Then the DPSM solution is multiplied by $F(\omega)$ and the inverse fast Fourier transform (IFFT) of the product is taken to obtain the transient solution.

If there are N sampled values used in the FFT and Δt is the sampling time interval, then the total time period is $T = N\Delta t$. The frequency sampling interval is $\Delta f = \frac{1}{T}$ and the sampling frequency is given by $f_s = \frac{1}{\Delta t}$. From Shannon's Sampling theorem, the sampling frequency should be at least twice the maximum significant frequency contained in the signal being sampled. Thus, only the frequency values less than or equal to the Nyquist frequency (half the sampling frequency) are used for the transient analysis.

3.8.1.2 Computed Results

Transient results are generated based on the above formulation. A finite size transducer (4 mm diameter) is placed inside an unbounded fluid medium (water with wave speed 1.48 km/sec) near a fluid–solid interface. For the homogeneous solid half-space aluminum is used as the solid material (see Figure 3.34). For aluminum density, longitudinal and shear wave speeds are 2.7 gm/cc, 6.25 km/sec and 3.04 km/sec, respectively. For this problem the loading function $f(t)$ is given by

$$f(t) = e^{-2(t-t_0)^2/2}\sin(2\pi f_c t) \tag{3.88}$$

where:

f_c is the central frequency of the signal

t_0 is the time delay

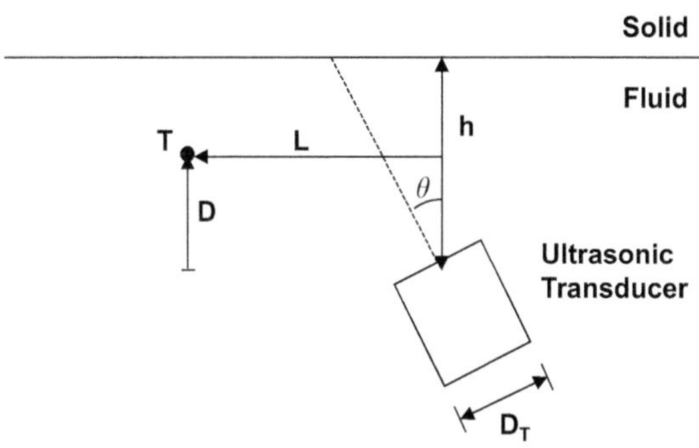

Figure 3.34 Problem Geometry showing the fluid–solid interface, the diameter and position of the transducer in the fluid half-space, and the target point (**T**) where the pressure is computed [from Das et al. (2010)].

We take $f_c = 1\text{MHz}$ and $t_0 = 3\mu\text{sec}$. The number of data points taken in the time history curve for constructing the signal is 1024 with the sampling time interval (Δt) equal to 0.25 μsec. T in Figure 3.34 is the target point where the ultrasonic field (in this case, pressure in the fluid) is computed and L is the horizontal distance of the target point from the vertical line passing through the center of the transducer face, and D is the vertical distance of the target point from the center of the transducer face. θ is the angle of inclination of the transducer with respect to the vertical axis and h is the distance of the center of the transducer face from the fluid–solid interface.

It should be mentioned that a three-dimensional problem is solved here. The interface lengths in the in-plane and the out-of-plane directions are much greater than the transducer diameter. At the plane of symmetry (central vertical plane) point sources are placed along two straight lines near the fluid–solid interface, as shown in Figure 3.33. The following results are presented with three or five planes of point source in the out-of-plane direction. Therefore, there is a total of six to ten straight lines of sources – half of those above and half below the interface are used to model the fluid–solid interface. The transducer is modeled by 100 to 150 point sources. On the central plane along one line 95 to 130 point sources are distributed on each side of the 30 mm long interface. The lower values (100 point sources) modeling the transducer and 570 (= $95 \times 2 \times 3$) point sources modeling the interface produced acceptable results; however, for better accuracy some plots are generated with higher values of point sources.

Results are presented for normal and inclined incidence of the ultrasonic beam on a fluid–solid interface as shown in Figure 3.35. For the normal incidence the signal from the transducer can reach point T multiple times after going through several reflections between the transducer and the fluid–solid interface, as shown in Figure 3.35a. In this figure path 1 corresponds to the signal traveling directly from the transducer to the point of observation or the target point T whose position is shown by a dashed line in the figure. Paths 2, 3 and 4 correspond to the signal arriving at the same point after going through one or more reflections at the interface and the transducer face. How many such reflections can occur before the signal dies down depends on the size of the transducer, strength of the signal etc. To answer this question one must obtain the transient solution.

For the inclined incidence, as shown in Figure 3.35b, if the angle of strike matches with the Rayleigh critical angle, then the Rayleigh wave is generated

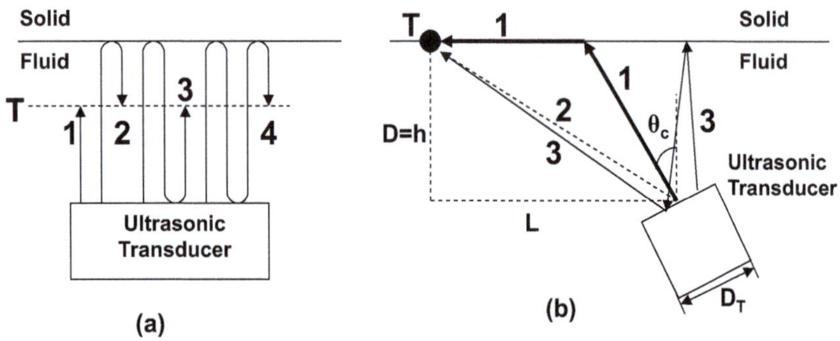

Figure 3.35 (a) Problem geometry for normal incidence showing multiple reflections of the ultrasonic beam between the fluid–solid interface and the transducer face (b) problem geometry for inclined incidence showing different rays emanating from the ultrasonic transducer [from Das et al. (2010)].

at the fluid–solid interface and point T is struck by the leaky Rayleigh wave as shown by ray 1 in the figure. However, the entire ultrasonic energy generated by the transducer does not propagate at the same angle; some ultrasonic energy has a higher inclination angle and reaches point T directly from the transducer as shown by ray 2. Some part of the signal propagating at a lower inclination angle than the Rayleigh angle can also reach point T after being reflected by both the interface and the transducer face, as shown by ray 3 in Figure 3.35b.

Figure 3.36 shows the steady-state solution of the ultrasonic pressure field in water and the normal stress (σ_{33}) in aluminum for normal incidence of the ultrasonic beam. Signal frequency is 1 MHz. This figure is generated with five layers

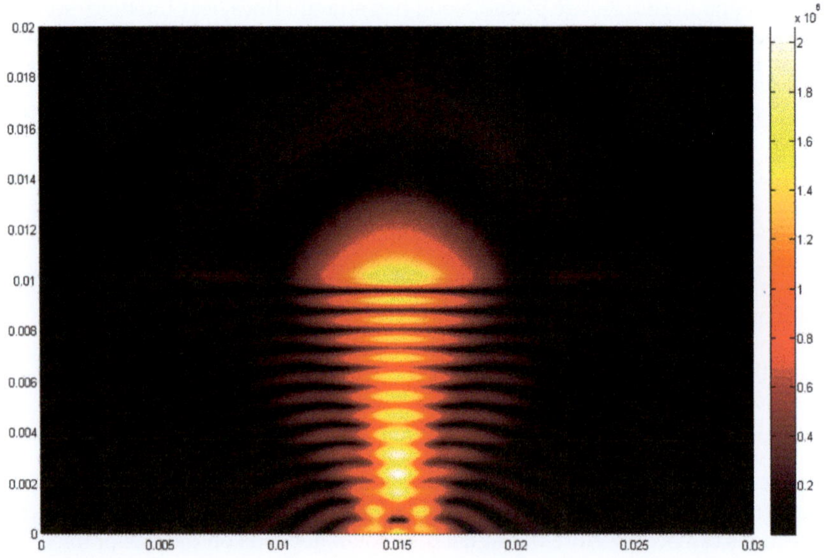

Figure 3.36 Steady-state ultrasonic pressure field in water (lower half-space in the image) and normal stress (σ_{33}) in aluminum (upper half-space in the image) for normal incidence of the ultrasonic beam.

of point sources in the out-of-plane direction modeling the interface. Although this figure clearly shows the interference pattern between the transducer and the interface it does not provide any information on the time of arrival of signals 1, 2, 3 and 4 shown in Figure 3.35a. It also does not provide any information on the number of reflections after which the signal dies down. Figure 3.37 shows the pressure value in time domain at the target point T (for L = 0 mm, D = 6 mm, h = 10 mm and $\theta = 0°$; see Figure 3.34) when an ultrasonic beam with 1 MHz central frequency [see Eq. (3.88)] strikes the interface. The same problem geometry is used to obtain the steady-state solution given in Figure 3.36. Four plots correspond to four different models of the same problem. For the top left plot (Figure 3.37a) 570 point sources model the interface, 285 sources are placed on each side of the interface in three planes in the out-of-plane direction, and 100 point sources model the transducer. The bottom left plot (Figure 3.37b) is generated by 1100 point sources modeling the interface and 100 point sources modeling the

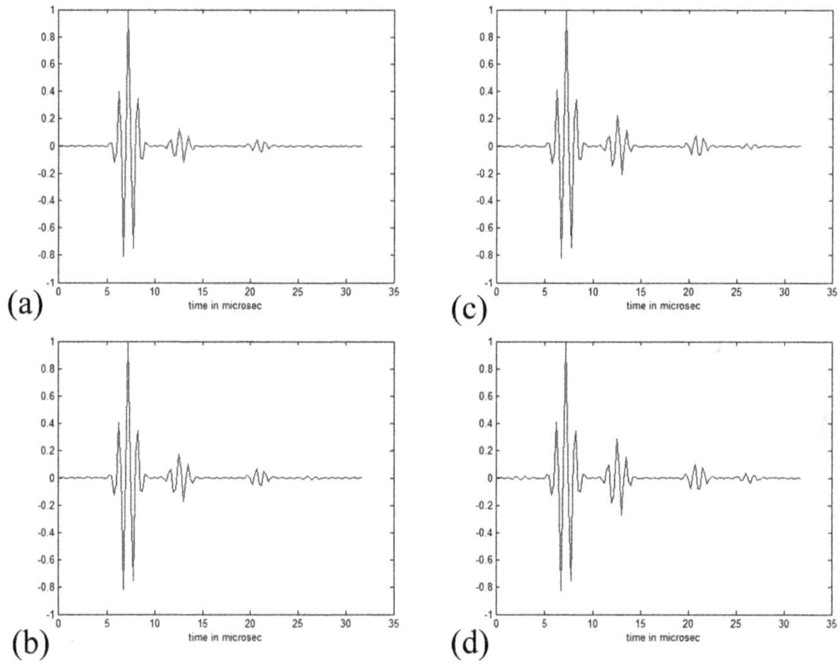

Figure 3.37 Ultrasonic pressure in time domain at the observation point **T** (in Figure 3.34, L = 0 mm, D = 6 mm, h = 10 mm and $\theta = 0°$). For the same problem geometry the steady-state solution is given in Figure 3.36. Four plots correspond to four different models of the same problem. (a) For the top left plot 570 point sources model the interface and 100 point sources model the transducer. (b) The bottom left plot is generated by 1100 point sources modeling the interface and 100 point sources modeling the transducer. (c) Number of point sources for the top right plot is same as that for Figure 3.37b. However, the distance between the point sources and the transducer face (and the interface) is $2r_s$ for Figure 3.37c while it is r_s [see Eq. (3.30)] for Figure 3.37b. (d) For the bottom right plot 1300 point sources model the interface and 150 point sources model the transducer. For this plot point sources are also placed at a distance of $2r_s$ from the transducer face and the interface.

transducer; for this plot 550 point sources are placed on each side of the interface in five parallel planes. The top right plot (Figure 3.37c) is generated by the same number of point sources modeling the interface and the transducer, as shown in Figure 3.37b. The only difference between Figures 3.37b and c is that the distance between the point sources and the transducer face (and the interface) is r_s [see Eq. (3.30)] for Figure 3.37b and it is $2r_s$ for Figure 3.37c. For the bottom right plot (Figure 3.37d) 1300 point sources model the interface, 650 sources are placed on each side of the interface in five planes and 150 point sources model the transducer. For this plot the point sources are also placed at a distance of $2r_s$ from the transducer face and the interface. Note that although the strongest signal is almost identical in these four figures the subsequent signals become slightly stronger as the number of point sources increases. However, arrival times of the four pulses do not show any variation as the point source distributions are changed.

The first strong signal arriving before 10 μsec is the direct signal traveling from the transducer to point T, ray 1 of Figure 3.35a. The second, third and fourth signals of Figure 3.37 correspond to rays 2, 3 and 4, respectively in Figure 3.35a. Arrival times of these four signals are then compared with the theoretically expected values. Since the wave speed in water is 1.48 km/sec the travel time from point T to the interface and then returning from the interface to point T is given by $(4\times 2)/1.48=5.41$ μsec. From Figure 3.37 the time difference between the first two major signals [the first one corresponding to the direct signal (ray 1 in Figure 3.35a) and the second one corresponding to the first reflected signal from the interface (ray 2 in Figure 3.35a)] is found to be 5.30 μsec. The difference between these two values is only 2%. Similarly, other travel times can be checked with the theoretical values and in all cases the difference is found to be less than 3%.

Then the inclined beam incidence case is modeled. The angle of incidence is set at the Rayleigh critical angle to generate Rayleigh waves in the solid. The Rayleigh wave speed in the solid is given by (see Chapter 1)

$$c_R = \frac{0.862 + 1.14\upsilon}{1+\upsilon} c_S \tag{3.89}$$

where:

υ is the Poisson's ratio of the solid material

c_S is the shear wave speed in the solid

For the wave speed and density of aluminum (given earlier) the Poisson's ratio is found to be 0.35 and the Rayleigh wave speed is 2.84 km/sec. Thus, to generate the Rayleigh wave at the fluid–solid interface, the angle of inclination of the transducer should be equal to $\sin^{-1}\left(\frac{1.48}{2.84}\right) = 31.41°$. With $\theta = 31.41°$, $L=24$ mm, $D=h=4$ mm (see Figure 3.35b or 3.34), and the interface length = 50 mm, the pressure value in water at point T is computed for $f_c=1$ MHz. The steady-state result is shown in Figure 3.38 and the transient result is given in Figure 3.39.

In Figure 3.38a x_1 varies from −25 to +25 mm while x_3 varies from 0 to 8 mm. Note that the scales in the horizontal and vertical directions are different and for this reason the striking angle appears to be much smaller than 31.4°. Figure 3.38a clearly shows some energy penetrating into the solid and some generating leaky Rayleigh waves along the fluid–solid interface, as shown by ray path 1 in Figure 3.35b. Figure 3.38a also shows some energy propagating at a much

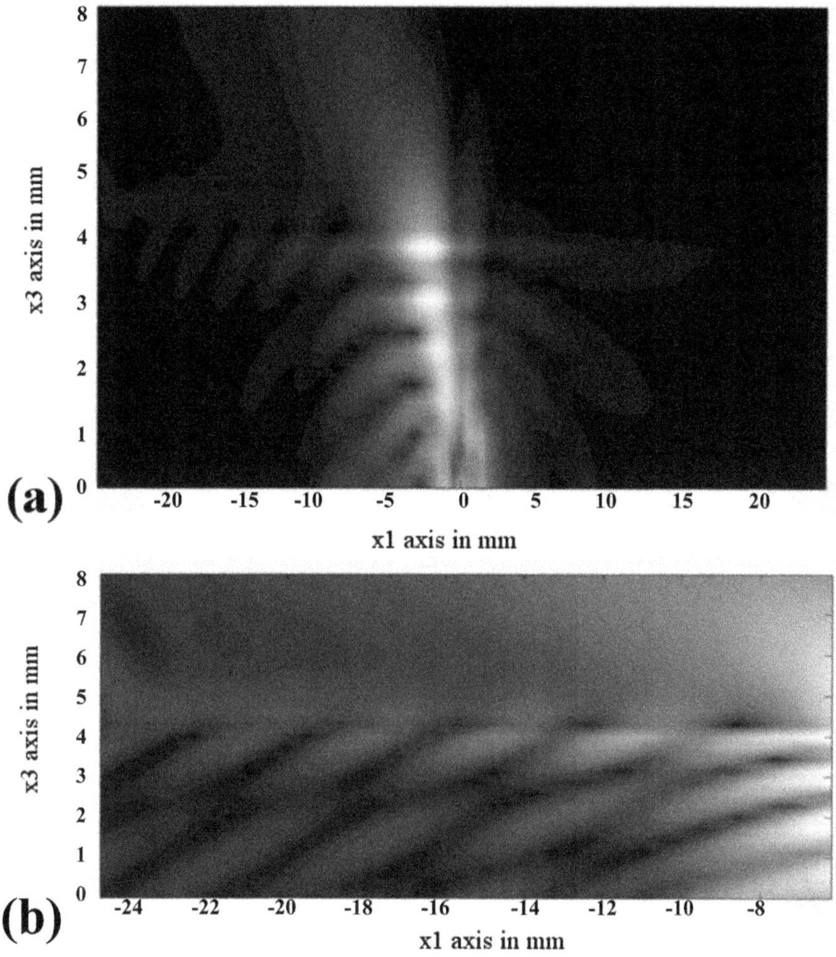

Figure 3.38 (a) Steady-state ultrasonic pressure field in water and normal stress (σ_{33}) in aluminum for incidence angle $\theta = 31.41°$ when x_1 varies (a) from −25 to +25 mm, and (b) from −25 to −6 mm [from Das et al. (2010)].

higher striking angle because the beam is not well collimated. This propagation path is shown by the dotted line 2 in Figure 3.35b. A careful observation of the right portion of the striking beam in Figure 3.38a also reveals the existence of the ray path 3 of Figure 3.35b. However, in this figure all ultrasonic energy appears to die down at the observation point T ($x_1 = -24$ mm, $x_2 = 4$ mm). This is because the energy level at this point is much smaller than the radiating energy from the transducer. However, when the pressure and stress fields are plotted in a different scale for x_1 varying from −25 to −6 mm, as shown in Figure 3.38b, then the ultrasonic energy associated with rays 1 and 2 reaching the observation point T becomes more prominent. The arrival times of rays 1, 2 and 3 of Figure 3.35b can be obtained from the time history plot presented in Figure 3.39. In this figure one can see that the time of arrival of the first noticeable signal is equal to

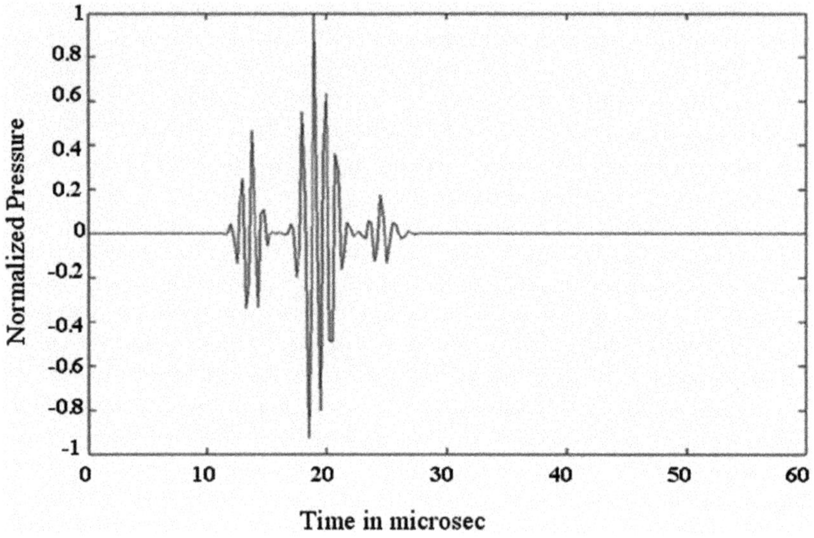

Figure 3.39 Ultrasonic pressure in time domain at the observation point **T** (in Figure 3.35b, L = 24 mm, D = 4 mm and $\theta_c = 31.4°$). The steady-state results for the same problem geometry is shown in Figure 3.38 [from Das et al. (2010)].

10.75 μsec. Theoretical time required by the Rayleigh wave to reach point T can be calculated as

$$\left[\frac{4}{\cos(31.41°)}\right] \times \frac{1}{1.48} + \frac{24 - 4 \times \tan(31.41°)}{2.84} = 10.76\,\mu\text{sec}$$

These two values match very well. A much stronger signal arriving second in Figure 3.39 is the direct signal reaching point T. This signal is denoted as ray 2 and shown by the dotted line in Figure 3.35b. This signal arrives at 16 μsec in Figure 3.39. Theoretical value of the time of arrival of the direct signal is $\frac{\sqrt{4^2 + 24^2}}{1.48} = 16.43\,\mu\text{sec}$; again the matching between these two arrival times is excellent. The smallest signal arriving after 22 μsec corresponds to the ray 3 of Figure 3.35b. It takes longer time to arrive at point T because it travels a much longer path after being reflected by the fluid–solid interface and the transducer face before reaching point T.

3.9 DPSM MODELING FOR ANISOTROPIC MEDIA

Implementation of DPSM requires evaluation of Green's function between many pairs of source and target points. For homogeneous and isotropic media, the Green's function is available as a closed-form analytical expression. But for anisotropic solids, the evaluation of Green's function is more complicated and needs to be done numerically. Nevertheless, important applications such as defect detection in composite materials require anisotropic analysis. Fooladi and Kundu (2017) used DPSM for ultrasonic field modeling in anisotropic materials. This section is taken mostly from that paper which is based on Fooladi's MS thesis and PhD dissertation [Fooladi (2016, 2018)].

Considering the prohibitive computational time of evaluating Green's function numerically for a large number of points, Fooladi and Kundu (2017) used a technique called "windowing" which employs the repetitive pattern of points in DPSM in order to considerably reduce the number of evaluations of Green's function. In addition, they used different resolutions of numerical integration to compute Green's function corresponding to different distances in order to achieve a good balance between time and accuracy. They developed an anisotropic DPSM model equipped with the windowing technique and applied multi-resolution numerical integration to the problem of ultrasonic field modeling in an anisotropic plate immersed in a fluid. The transducers were placed in the fluid on both sides of the plate. They first considered an isotropic plate for verification and rough calibration of the numerical integration. Then they considered an anisotropic composite plate to model the ultrasonic wave field inside the anisotropic plate.

Wang and Achenbach (1994, 1995) formulated elastodynamic time-harmonic Green's functions for anisotropic solids using Radon transform. They applied Radon transform to space variables in order to convert the governing equations from a system of coupled partial differential equations (PDEs) to a system of coupled ordinary differential equations (ODEs). Then, the coupled ODEs were uncoupled by transforming the coordinates to a new set of bases. Next, the uncoupled ODEs were solved, and the solution was transformed back to the original coordinate system. Then, the inverse Radon transform was applied to obtain the Green's function. The result consists of two integrals. One contains the singular term and the other one has the non-singular or the regular term. The singular term is in the form of an integral over an inclined circle on a unit sphere. The regular term is in the form of an integral over the surface of a hemisphere, and is responsible for the majority of the computational time for evaluating the anisotropic Green's function.

The singular term is similar in form to the elastostatic Green's function. It can be reduced to a summation of algebraic terms using the calculus of residue. This method was used by Dederichs and Liebfried (1969) and was later revived by Sales and Gray (1998) who successfully used this method to calculate the Green's function as well as its first and second derivatives. This method assumes that the roots are distinct. The method needs special attention when repeated roots occur for a point. In the case of repeated roots, Sales and Gray (1998) suggested to perturb the point by a small amount in different directions, and then use the average as an approximation.

The solution method developed by Wang and Achenbach (1995) for the elastodynamic time-harmonic Green's function in anisotropic media was adopted by Fooladi and Kundu (2017). The integral representing the singular term was evaluated analytically using the calculus of residue based on the work of Sales and Gray (1998). The integral representing the regular term was computed numerically. As mentioned before, the regular term was responsible for the majority of the computational time associated with the evaluation of the elastodynamic Green's function. For the case of a transversely isotropic material, the integration domain for the regular part of the solution can be reduced from a hemisphere to a quarter-sphere as was done by Fooladi (2018). This improvement is very effective in reducing the computational time, and was used to compute the Green's function of a transversely isotropic material. The Green's function solutions were then used as the building blocks in a DPSM model to simulate the ultrasonic wave propagation in an anisotropic material.

3.9.1 DPSM Modeling of a Solid Plate Immersed in a Fluid

Distribution of point sources for DPSM modeling of a solid plate immersed in a fluid has been discussed earlier in this chapter. It is briefly revisited in this section using the notations used by Fooladi and Kundu (2017) to help the readers understand the subsequent derivations using same notations.

Figure 3.40 shows a solid plate in contact with fluid 1 on the bottom surface, and fluid 2 on the top surface. Two flat square transducers are placed below and above the solid plate, in fluids 1 and 2, respectively. A collection of source points are distributed at the locations of the transducer surfaces and fluid–solid interfaces. In Figure 3.40, only the point sources along the central vertical plane are shown for the sake of clarity. Various sets of point sources are identified by different symbols in this figure. A source strength vector is assigned to each set of point sources, as listed in Table 3.2.

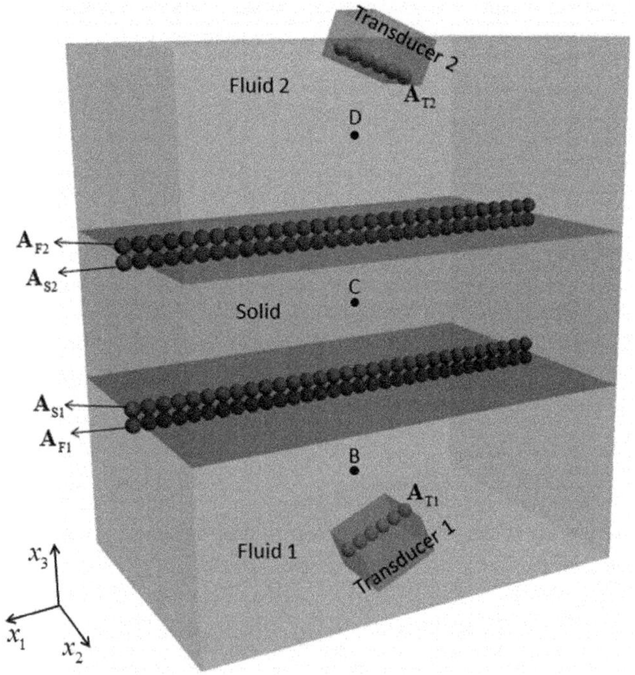

Figure 3.40 Problem description and distribution of point sources [from Fooladi and Kundu (2017)].

Table 3.2: List of Source Strength Vectors

Source Strength Vector	Location of Corresponding Point Sources
$\mathbf{A}_{T1}$	Inside transducer 1 and near its surface
$\mathbf{A}_{T2}$	Inside transducer 2 and near its surface
$\mathbf{A}_{F1}$	Inside fluid 1 and near its interface with the solid
$\mathbf{A}_{F2}$	Inside fluid 2 and near its interface with the solid
$\mathbf{A}_{S1}$	Inside solid and near its interface with fluid 1
$\mathbf{A}_{S2}$	Inside solid and near its interface with fluid 2

Each point source modeling the field in the fluid has a scalar value of source strength corresponding to the pressure at that point. Each point source modeling the field in the solid has three values of source strength corresponding to forces in three different directions in the 3D space. The source points distributed near the transducer surface and the solid–fluid interfaces are used to compute the solution at a target point in the fluid. Three different types of target points are identified based on the problem description shown in Figure 3.40. Let us consider target point B which is placed in fluid 1 below the solid plate. The solution at this point is determined by the strength of source points located inside transducer 1, $\mathbf{A}_{T1}$, and the strength of source points located inside the solid and near the fluid 1–solid interface, $\mathbf{A}_{S1}$. A quantity at this point, q_B, can be written as

$$q_B = \mathbf{M}^q_{B,T1}\ \mathbf{A}_{T1} + \mathbf{M}^q_{B,S1}\ \mathbf{A}_{S1} \tag{3.90}$$

where the array $\mathbf{M}^q_{T,S}$ (here T = B and S = T1 or S1) is constructed from the Green's function of source point(s) S at target point(s) T, for quantity q which can be displacement, velocity, stress or pressure.

At target point C which is placed inside the solid plate, a quantity q_C can be written as

$$q_C = \mathbf{M}^q_{C,F1}\ \mathbf{A}_{F1} + \mathbf{M}^q_{C,F2}\ \mathbf{A}_{F2} \tag{3.91}$$

At target point D which is placed in fluid 2 above the solid plate, a quantity q_D can be written as

$$q_D = \mathbf{M}^q_{D,T2}\ \mathbf{A}_{T2} + \mathbf{M}^q_{D,S2}\ \mathbf{A}_{S2} \tag{3.92}$$

In Equations (3.90–92), if the point is placed in the fluid, then the corresponding quantity q can be displacement vector component, velocity vector component, or pressure. If the point is placed in the solid, then instead of pressure the components of stress tensor are considered.

The boundary and interface conditions are then applied to obtain the values of source strength vectors listed in Table 3.2. The first condition is applied to the velocity vector component normal to the surface of the transducer. Placing a collection of target points on the surface of transducer 1 and denoting them as J1, the first condition can be expressed as [using Eq. (3.90)]

$$\mathbf{V}_{J1} = \mathbf{M}^{V_3}_{J1,T1}\ \mathbf{A}_{T1} + \mathbf{M}^{V_3}_{J1,S1}\ \mathbf{A}_{S1} \tag{3.93}$$

A similar expression can be obtained for a collection of target points placed on the surface of transducer 2. Denoting them as J2, and using equation (3.92), one obtains

$$\mathbf{V}_{J2} = \mathbf{M}^{V_3}_{J2,T2}\ \mathbf{A}_{T2} + \mathbf{M}^{V_3}_{J2,S2}\ \mathbf{A}_{S2} \tag{3.94}$$

At each of the two fluid–solid interfaces, the normal displacement and normal stress are continuous, and the shear stress is zero for a non-viscous fluid. Placing a collection of target points on the lower interface and denoting them by I1, the interface conditions can be obtained using equations (3.90) and (3.91) as

$$\mathbf{M}^{U_3}_{I1,T1}\ \mathbf{A}_{T1} + \mathbf{M}^{U_3}_{I1,S1}\ \mathbf{A}_{S1} = \mathbf{M}^{U_3}_{I1,F1}\ \mathbf{A}_{F1} + \mathbf{M}^{U_3}_{I1,F2}\ \mathbf{A}_{F2} \tag{3.95a}$$

$$\mathbf{M}^{P}_{I1,T1}\ \mathbf{A}_{T1} + \mathbf{M}^{P}_{I1,S1}\ \mathbf{A}_{S1} = -\mathbf{M}^{S_{33}}_{I1,F1}\ \mathbf{A}_{F1} - \mathbf{M}^{S_{33}}_{I1,F2}\ \mathbf{A}_{F2} \tag{3.95b}$$

$$\mathbf{M}_{\mathrm{I1,F1}}^{S_{31}}\,\mathbf{A}_{\mathrm{F1}} + \mathbf{M}_{\mathrm{I1,F2}}^{S_{31}}\,\mathbf{A}_{\mathrm{F2}} = 0 \tag{3.95c}$$

$$\mathbf{M}_{\mathrm{I1,F1}}^{S_{32}}\,\mathbf{A}_{\mathrm{F1}} + \mathbf{M}_{\mathrm{I1,F2}}^{S_{32}}\,\mathbf{A}_{\mathrm{F2}} = 0 \tag{3.95d}$$

where:

- U_i represents the i-th component of the displacement vector
- P is the pressure
- S_{ij} represents the (i, j) component of the stress tensor

Similarly, placing a collection of target points on the upper fluid–solid interface and denoting them as I2, the interface conditions can be obtained using equations (3.91) and (3.92) as

$$\mathbf{M}_{\mathrm{I2,T2}}^{U_3}\,\mathbf{A}_{\mathrm{T2}} + \mathbf{M}_{\mathrm{I2,S2}}^{U_3}\,\mathbf{A}_{\mathrm{S2}} = \mathbf{M}_{\mathrm{I2,F1}}^{U_3}\,\mathbf{A}_{\mathrm{F1}} + \mathbf{M}_{\mathrm{I2,F2}}^{U_3}\,\mathbf{A}_{\mathrm{F2}} \tag{3.96a}$$

$$\mathbf{M}_{\mathrm{I2,T2}}^{P}\,\mathbf{A}_{\mathrm{T2}} + \mathbf{M}_{\mathrm{I2,S2}}^{P}\,\mathbf{A}_{\mathrm{S2}} = -\mathbf{M}_{\mathrm{I2,F1}}^{S_{33}}\,\mathbf{A}_{\mathrm{F1}} - \mathbf{M}_{\mathrm{I2,F2}}^{S_{33}}\,\mathbf{A}_{\mathrm{F2}} \tag{3.96b}$$

$$\mathbf{M}_{\mathrm{I2,F1}}^{S_{31}}\,\mathbf{A}_{\mathrm{F1}} + \mathbf{M}_{\mathrm{I2,F2}}^{S_{31}}\,\mathbf{A}_{\mathrm{F2}} = 0 \tag{3.96c}$$

$$\mathbf{M}_{\mathrm{I2,F1}}^{S_{32}}\,\mathbf{A}_{\mathrm{F1}} + \mathbf{M}_{\mathrm{I2,F2}}^{S_{32}}\,\mathbf{A}_{\mathrm{F2}} = 0 \tag{3.96d}$$

A simultaneous solution of equations (3.93) through (3.96) gives the values of all source strength vectors listed in Table 3.2. After finding the source strengths, the solution at an arbitrary target point inside the fluid or solid can be obtained by application of Equations (3.90), (3.91) and (3.92).

3.9.2 The Windowing Technique

The calculation of array **M** presented in section 3.9.1 requires evaluation of Green's function between many pairs of points. Let us consider the configuration shown in Figure 3.41a, where a collection of source points located on the upper plane are affecting a collection of target points located on the lower plane. Then, in order to build the DPSM model, the Green's function between each pair of source and target points should be evaluated. If for each plane shown in Figure 3.41a the number of points along the two sides or edges are n_1 and n_2, then the total number of points on each plane is $n_1 n_2$, and a total of $(n_1 n_2)^2$ Green's functions need to be evaluated to consider all possible combinations of source and target points.

For isotropic materials, a closed-form expression for Green's function is available, but for anisotropic solids, the Green's function should be evaluated numerically. This makes the anisotropic DPSM computationally more challenging. The "windowing" technique described here can reduce the computation time. The main idea behind the windowing technique is to reduce the number of evaluations for the Green's function by using the repeated patterns of relative positions of target and source points.

Let the source point be located at one of the corners of the plane in Figure 3.41a. The source point chosen in this manner is shown in Figure 3.41b. The goal of the "windowing" technique is to relate every one of the $(n_1 n_2)^2$ combinations of source point/target point combinations to one of the Green's function associated with this particular source point and different target points. To do so, first the collection of target points are extended, as shown in Figure 3.41c, where four sets of initial target points are put next to each other to construct a larger collection of points with $(2n_1 - 1)$ and $(2n_2 - 1)$ points along two edges of the plane. Now, every Green's function between source and target points in

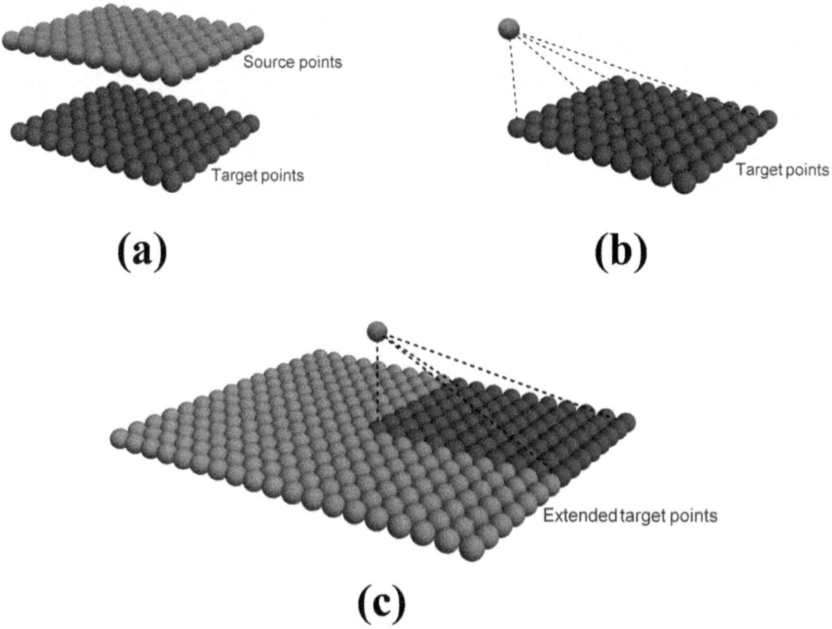

Figure 3.41 Geometrical description of the windowing technique [from Fooladi and Kundu (2017)].

intial configuration shown in Figure 3.41a is equivalent to one of the Green's functions between the chosen source point and the extended collection of target points shown in Figure 3.41c. For example, consider the source point shown in Figure 3.42a. It is one of the n_1n_2 source points of Figure 3.41a. Note that the target points of Figure 3.42a are a subset of the target points of Figure 3.42b. Then, it can be assumed that the source point views the collection of target points through a window. Now, by putting this source point in the position of the source point shown in Figure 3.41c, the corresponding window becomes a subset of the extended plane of target points, as shown in Figure 3.42b. Therefore, the information for the Green's function in Figure 3.42b can be used to obtain the Green's function in Figure 3.42a.

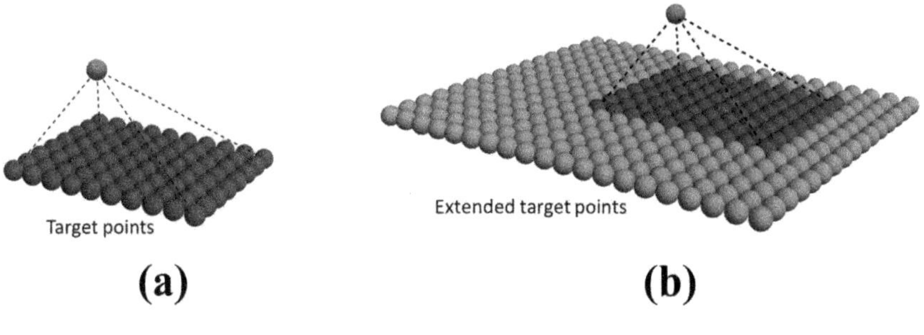

Figure 3.42 A sample source point in the windowing technique [from Fooladi and Kundu (2017)].

Let us define the coordinates (i_1, i_2) of a point in Figures 3.41a–c where i_1 and i_2 are the number of points measured along the plane edges from the chosen corner. Then, the Green's function between a source point with coordinate (i_1, i_2) and a target point with coordinate (j_1, j_2) in Figure 3.41a is equivalent to a Green's function between the chosen source point and a target point with coordinates $(n_1 + j_1 - i_1, n_2 + j_2 - i_2)$ in Figure 3.41c. Therefore, the number of evaluations of Green's function is reduced from $(n_1 n_2)^2$ in Figure 3.41a to $(2n_1 - 1)(2n_2 - 1)$ in Figure 3.41c. For example, for $n_1 = n_2 = 20$, the number of evaluations is reduced from 160,000 to 361.

3.9.3 Elastodynamic Green's Function

3.9.3.1 *General Anisotropic Materials*

The derivation of elastodynamic Green's function for anisotropic materials developed by Wang and Achenbach (1995) is briefly reviewed here. The equations of motion for small deformation in an anisotropic homogenous medium can be written as

$$C_{ijkl} \frac{\partial^2 u_k}{\partial x_j \partial x_l} + f_i = \rho \frac{\partial^2 u_i}{\partial t^2} \tag{3.97}$$

where C_{ijkl} are the components of the stiffness tensor, u_i are the displacement components, f_i are the components of the body force density vector, and ρ is the mass density.

The elastodynamic Green's function $G_{ij}(\mathbf{x}, \mathbf{x}_0)$ gives the displacement at point $\mathbf{x}$ in direction i when a point force is applied at point $\mathbf{x}_0$ in direction j. Time-harmonic force and displacement can be written as

$$f_i(\mathbf{x}, t) = \delta_{ip}\delta(\mathbf{x} - \mathbf{x}_0)e^{-i\omega t} \tag{3.98}$$

and

$$u_i(\mathbf{x}, t) = G_{ip}(\mathbf{x}, \mathbf{x}_0)e^{-i\omega t} \tag{3.99}$$

Substituting equations (3.98) and (3.99) into equation (3.97) one obtains

$$C_{ijkl} \frac{\partial^2 G_{kp}(\mathbf{x}, \mathbf{x}_0)}{\partial x_j \partial x_l} + \rho\omega^2 G_{ip}(\mathbf{x}, \mathbf{x}_0) = -\delta_{ip}\delta(\mathbf{x} - \mathbf{x}_0) \tag{3.100}$$

The above equation has been solved by Wang and Achenbach (1995) using Radon transform. Let $f(\mathbf{x})$ be an arbitrary function in 3D space. The Radon transform of this function $\Re[f(\mathbf{x})]$ is given by

$$\hat{f}(s, \mathbf{n}) = \Re[f(\mathbf{x})] = \int f(\mathbf{x})\delta(s - (\mathbf{n} \cdot \mathbf{x}))d\mathbf{x} \tag{3.101}$$

where $\mathbf{n}$ is a unit vector. The inverse Radon transform is

$$f(\mathbf{x}) = \Re^{-1}[f(\mathbf{x})] = -\frac{1}{8\pi^2} \int_{|\mathbf{n}|=1} \frac{\partial^2 \hat{f}(s, \mathbf{n})}{\partial s^2} dS(\mathbf{n}) \tag{3.102}$$

As can be seen from equation (3.102), the inverse Radon transform involves integration over a unit sphere. The Radon transform of second derivative can be written as

$$\Re\left(\frac{\partial^2 f(\mathbf{x})}{\partial x_i \partial x_j}\right) = n_i n_j \frac{\partial^2 \hat{f}(s, \mathbf{n})}{\partial s^2} \tag{3.103}$$

This is used to convert the PDE into ODE.

Applying Radon transform in equation (3.100) gives

$$\left(K_{ik}(\mathbf{n})\frac{\partial^2}{\partial s^2}+\rho\omega^2\delta_{ik}\right)\hat{G}_{kp}=-\delta_{ip}\delta(s) \tag{3.104}$$

where

$$K_{ik}(\mathbf{n})=C_{ijkl}n_l n_j \tag{3.105}$$

The coupled system of equations (3.104) can be decoupled by transforming the coordinates to a new set of bases formed by eigenvectors of $K_{ik}(\mathbf{n})$. The eigenvalue problem for $K_{ik}(\mathbf{n})$ can be written as

$$K_{ik}E_{km}=\lambda_m E_{im},\quad m=1,2,3\ \left(\text{no sum on } m\right) \tag{3.106}$$

where $\lambda_m(m=1,2,3)$ are eigenvalues of **K**, and **E** is a matrix with columns representing eigenvectors of **K**. In equation (3.106), and in the remainder of this paper whenever the index m refers to the number of eigenvalue or eigenvector or a parameter derived from them, the summation convention does not apply.

By transforming to the new set of bases formed by the eigenspace of **K**, $\hat{\mathbf{G}}$ is converted to $\hat{\mathbf{G}}^*$, and equation (3.106) becomes

$$\left(K_{ik}(\mathbf{n})E_{kl}E_{pn}\frac{\partial^2}{\partial s^2}+\rho\omega^2\delta_{ik}E_{kl}E_{pn}\right)\hat{G}^*_{\mathrm{ln}}=-\delta_{ip}\delta(s) \tag{3.107}$$

By pre-multiplying the above equation by $\mathbf{E}^T$ and post-multiplying by **E**, the above equation can be decoupled as

$$\left(\lambda_m\frac{\partial^2}{\partial s^2}+\rho\omega^2\right)\hat{G}^*_{mq}=-\delta_{mq}\delta(s),\quad m=1,2,3\ \left(\text{no sum on m}\right) \tag{3.108}$$

For each m and q, the solution of the above equation is

$$\hat{G}^*_{mq}=\frac{\mathbf{i}\delta_{mq}}{2\rho c_m^2\alpha_m}e^{\mathrm{i}\alpha_m|s|} \tag{3.109}$$

where, $c_m=\sqrt{\lambda_m/\rho}$ is the phase velocity and $\alpha_m=\omega/c_m$ is the wave number associated with eigenvalue λ_m. By transforming back to the original set of basis, $\hat{\mathbf{G}}$ is obtained as

$$\hat{G}_{kp}=\sum_{m=1}^{3}\frac{\mathbf{i}E_{km}\delta_{mq}E_{pq}}{2\rho c_m^2\alpha_m}e^{\mathrm{i}\alpha_m|s|}=\sum_{m=1}^{3}\frac{\mathbf{i}E_{km}E_{pm}}{2\rho c_m^2\alpha_m}e^{\mathrm{i}\alpha_m|s|} \tag{3.110}$$

Applying the inverse Radon transform to the above expression gives $G_{kp}(\mathbf{x},\mathbf{x}_0)$ as the summation of two parts: the singular part shown by $G^S_{kp}(\mathbf{x},\mathbf{x}_0)$ and the regular part shown by $G^R_{kp}(\mathbf{x},\mathbf{x}_0)$. After some algebraic manipulations, the singular part can be written as

$$G^S_{kp}(\mathbf{x},\mathbf{x}_0)=\frac{1}{8\pi^2 r}\oint_S K^{-1}_{kp}(\xi)\,dS(\xi) \tag{3.111}$$

where $K_{ik}(\xi)=C_{ijkl}\xi_l\xi_j$ and S is an oblique circular path in 3D defined by

$$S=\left\{\xi\in R^3\ \middle|\ \|\xi\|=1,\ \xi\cdot(\mathbf{x}-\mathbf{x}_0)=0\right\} \tag{3.112}$$

The regular part can be written as an integral over a unit sphere as

$$G_{kp}^{R}(\mathbf{x},\mathbf{x}_0)=\frac{\mathrm{i}}{16\pi^2}\int_{|\mathbf{n}|=1}\sum_{m=1}^{3}\frac{\alpha_m E_{km}E_{pm}}{\rho c_m^2}e^{\mathrm{i}\alpha_m|\mathbf{n}\cdot(\mathbf{x}-\mathbf{x}_0)|}dS(\mathbf{n}) \tag{3.113}$$

Using the symmetry properties, the above integral can be written over a unit hemisphere as

$$G_{kp}^{R}(\mathbf{x},\mathbf{x}_0)=\frac{\mathrm{i}}{8\pi^2}\int_{\text{Hemi-sphere}}\sum_{m=1}^{3}\frac{\alpha_m E_{km}E_{pm}}{\rho c_m^2}e^{\mathrm{i}\alpha_m|\mathbf{n}\cdot(\mathbf{x}-\mathbf{x}_0)|}dS(\mathbf{n}) \tag{3.114}$$

3.9.3.2 *Residue Method*

Sales and Gray (1998) used the residue theorem to evaluate the integral for the singular part of the solution shown in equation (3.111). To do so, they used the change of variable $Z=\tan\phi$ where ϕ is the angle from a fixed ray on the circular path of integration. After this change of variables, they wrote equation (3.111) as

$$G_{ij}^{S}(\theta,\psi)=\frac{1}{4\pi^2 r}\int_{-\infty}^{\infty}\frac{P_{ij}(Z)}{Q(Z)}dZ \tag{3.115}$$

where $\mathbf{P}(Z)$ and $Q(Z)$ represent the cofactor and determinant of matrix $\mathbf{K}$, respectively. This integral can be calculated by the residue theorem as

$$G_{ij}^{S}(\theta,\psi)=\frac{\mathrm{i}}{2\pi r}\sum_{n=1}^{3}\text{Residue}\left(\frac{P_{ij}(Z)}{Q(Z)}\right)\Bigg|_{Z=\lambda_n}=\frac{\mathrm{i}}{2\pi r}\sum_{n=1}^{3}\frac{P(\lambda_n)}{Q_n(\lambda_n)} \tag{3.116}$$

where $\lambda_n,\ n=1,2,3$ represent three roots of $Q(Z)$ in the upper half plane, and

$$Q_n(Z)=\frac{Q(Z)}{\lambda-Z_n} \tag{3.117}$$

This application of the residue theorem implies that the roots are distinct. For repeated roots this algorithm is not applicable, but as mentioned by Sales and Gray (1998), it does not appear to be a significant concern since the algorithm performs very well at a very small distance away from the repeated root. Therefore, one can slightly perturb the coordinates of the point corresponding to repeated roots in different directions, and then take an average of the solutions. Repeated roots are expected to occur only at a few isolated points for an anisotropic material. For the particular case of an isotropic material, all points exhibit repeated roots and the perturbation is not applicable. In that case, the integral for the singular part shown in equation (3.111) needs to be evaluated directly using numerical integration.

The derivatives of the displacement Green's function are needed to calculate strain and stress tensors. The derivative of the singular part of the solution can be obtained by first taking derivatives with respect to spherical coordinates and then transforming the result back to a Cartesian coordinate system.

The derivative of equation (3.116) with respect to the radial coordinate r is

$$G_{ij,r}^{S}=-\frac{1}{8\pi^2 r^2}\tilde{G}_{ij}^{S}=\frac{i}{2\pi r^2}\sum_{n=1}^{3}\frac{P(\lambda_n)}{Q_n(\lambda_n)} \tag{3.118}$$

The derivative of equation (3.116) with respect to the polar angle θ is

$$G_{ij,\theta}^{S}=\frac{d}{d\theta}\left(\frac{1}{4\pi^2 r}\int_{-\infty}^{\infty}\frac{P_{ij}(Z)}{Q(Z)}dZ\right)=\frac{1}{4\pi^2 r}\int_{-\infty}^{\infty}\frac{P_{ij,\theta}(Z)Q(Z)-P_{ij}(Z)Q_{,\theta}(Z)}{(Q(Z))^2}dZ \tag{3.119}$$

The residue theorem gives

$$G_{ij,\theta}^{S} = \frac{\mathrm{i}}{2\pi r}\sum_{n=1}^{3}\frac{d}{dZ}\left(\frac{P_{ij,\theta}(Z)Q(Z) - P_{ij}(Z)Q_{,\theta}(Z)}{(Q_n(Z))^2}\right)\Bigg|_{Z=\lambda_n} \tag{3.120}$$

Substituting $Z = \lambda_n$ and considering that $Q(\lambda_n) = 0$, the above expression is reduced to

$$\begin{aligned} G_{ij,\theta}^{S} = \frac{\mathrm{i}}{2\pi r}\sum_{n=1}^{3}\Bigg(&\left(\frac{2Q_{n,Z}(\lambda_n)Q_{,\theta}(\lambda_n) - Q_n(\lambda_n)Q_{,\theta Z}(\lambda_n)}{(Q_n(\lambda_n))^3}\right)P_{jk}(\lambda_n) \\ &+\frac{Q_{,Z}(\lambda_n)}{(Q_n(\lambda_n))^2}P_{jk,\theta}(\lambda_n) - \frac{Q_{,\theta}(\lambda_n)}{(Q_n(\lambda_n))^2}P_{jk,Z}(\lambda_n)\Bigg) \end{aligned} \tag{3.121}$$

The derivative with respect to the azimuthal angle Ψ gives the same result as the above expression with θ replaced by Ψ. One typo in equation (3.20) of the paper by Sales and Gray (1998) was corrected by Fooladi and Kundu (2017), and this corrected version is given in equation (3.121).

3.9.3.3 *Reduction of Integration Domain for Transversely Isotropic Materials*

As mentioned before, numerically computing the integral for the regular part of the solution is responsible for the majority of the computational cost of evaluating the elastodynamic Green's function. The regular part is in the form of an integral over the surface of a unit sphere [equation (3.113)]. Using the symmetry properties, it can be reduced to a unit hemisphere as was shown in equation (3.114). No more simplification seems to be possible for a general anisotropic material. However, for a transversely isotropic material this integration domain can be further reduced from a hemisphere to a quarter-sphere [Fooladi (2016), Fooladi and Kundu (2017)]. This improvement is significant, since it reduces the computational time by almost 50%. This formulation of reduction of the integration domain to a quarter-sphere is briefly reviewed here.

First, the original coordinate system $x_1x_2x_3$ is rotated by transformation matrix $\mathbf{Q}$ to a new coordinate system $x_1'x_2'x_3'$, such that x_3' is the principal direction of the transversely isotropic material and the projection of the vector $\mathbf{x} - \mathbf{x}_0$ on $x_1'x_2'$ plane lies along the x_1' axis. The components of vector $\mathbf{x} - \mathbf{x}_0$ in the rotated coordinate system can be written as

$$(\mathbf{x}' - \mathbf{x}_0') = \mathbf{Q}(\mathbf{x} - \mathbf{x}_0) \tag{3.122}$$

The regular part of the elastodynamic Green's function can be written as an integral over a quarter-sphere in the rotated coordinate system,

$$G_{ij}^{\prime R}(\mathbf{x}', \mathbf{x}_0') = \frac{\mathrm{i}}{8\pi^2}\int_0^{\pi/2}\int_0^{\pi}\frac{\alpha_m}{\rho c_m^2}\,\Sigma_{ij}^{m}(\theta,\phi)\,e^{\mathrm{i}\alpha_m|\mathbf{n}'\cdot(\mathbf{x}'-\mathbf{x}_0')|}\,\sin\phi\, d\theta\, d\phi \tag{3.123}$$

where θ and φ are the polar and azimuthal angles respectively, and

$$\Sigma_{ij}^{m}(\theta,\phi) = 2\begin{bmatrix} (E_{1m}')^2(\theta,\phi) & 0 & (E_{1m}'E_{3m}')(\theta,\phi) \\ 0 & (E_{2m}')^2(\theta,\phi) & 0 \\ (E_{1m}'E_{3m}')(\theta,\phi) & 0 & (E_{3m}')^2(\theta,\phi) \end{bmatrix} \tag{3.124}$$

Similarly, the derivative of the regular part of the solution can be written in the rotated coordinate system as

$$\frac{\partial G_{ij}'^{R}(\mathbf{x}',\mathbf{x}_0')}{\partial x_k'} = -\frac{1}{8\pi^2}\int_0^{\pi/2}\int_0^{\pi}\frac{n_q'\alpha_m^2}{\rho c_m^2}\,\Lambda_{ij}^m(\theta,\phi)\,\mathrm{sign}\left(\mathbf{n}'\cdot(\mathbf{x}'-\mathbf{x}_0')\right)\,e^{i\alpha_m\left|\mathbf{n}'\cdot(\mathbf{x}'-\mathbf{x}_0')\right|}\,\sin\phi\, d\theta\, d\phi \tag{3.125}$$

where

$$\Lambda_{ij}^m = \begin{cases} \Sigma_{ij}^m & \text{for } k = 1, 3 \\ \Pi_{ij}^m & \text{for } k = 2 \end{cases} \tag{3.126}$$

and

$$\Pi_{ij}^m(\theta,\phi) = 2\begin{bmatrix} 0 & (E_{1m}'E_{2m}')(\theta,\phi) & 0 \\ (E_{1m}'E_{2m}')(\theta,\phi) & 0 & (E_{2m}'E_{3m}')(\theta,\phi) \\ 0 & (E_{2m}'E_{3m}')(\theta,\phi) & 0 \end{bmatrix} \tag{3.127}$$

The back transformation of the regular part and its derivatives from the rotated coordinate system to the original coordinate system can be written as

$$G_{ij}^R = Q_{im}Q_{jn}G_{mn}'^R \tag{3.128}$$

and

$$\frac{\partial G_{ij}^R}{\partial x_k} = Q_{kl}Q_{im}Q_{jn}\frac{\partial G_{mn}'^R}{\partial x_l'} \tag{3.129}$$

Once the regular part and its derivatives are computed on the quarter-sphere in the rotated coordinate system, they can be transformed back to the original coordinate system using equations (3.128) and (3.129).

3.9.4 Numerical Examples

3.9.4.1 Isotropic Plate

In the first example, an isotropic plate is considered for which closed-form solutions are available for the Green's functions. This example is solved once with anisotropic DPSM code developed based on the above formulation, and once with isotropic DPSM code. As mentioned before, the numerical evaluation of the Green's function makes anisotropic DPSM simulations considerably slower than their isotropic counterpart. Using more integration points to evaluate the Green's function between each pair of points increases the accuracy, however the computational time can become prohibitive. On the other hand, when the distance between the pair of points is increased, the corresponding Green's function generally needs more integration points to maintain the same level of accuracy. In order to achieve a good balance between the computational time and accuracy, different numbers of integration points can be assigned for different distances. Fooladi and Kundu (2017) suggested four resolutions of integration points based on the distance between the pair of points; this multi-resolution integration scheme was then combined with the windowing technique described in section 3.9.2.2 to reduce the computational time.

The problem geometry is shown in Figure 3.40. The thickness of the solid plate is 3 mm. The length of the plate that was analyzed was 20 mm in the x_1 direction. The number of target and source points placed along the x_1 direction was 53. The source distribution satisfied the convergence criterion that the distance between two adjacent points should be less than the wavelength in the fluid divided by π [Placko and Kundu (2007)]. By further increasing the resolution beyond 53 points, the difference in the results was not noticeable. Along the x_2 direction, nine source and target points were placed such that the distances between two adjacent points along the x_1 and x_2 directions were equal. By increasing the length of the plate in thex_2 direction no noticeable difference in the results on the central x_1x_3 plane was observed.

The transducers had square shape with edge length of 2 mm, and were oriented so that the axis x_3 was normal to their faces. To model each transducer, 81 source points were distributed. A velocity amplitude and a frequency $f = 1\text{MHz}$ were assumed for both transducers. Both upper and lower fluids were assumed to be water with density $\rho = 1000\,\text{kg/m}^3$ and P-wave speed $C_p = 1480\text{m/s}$. The solid plate was assumed to have a density of $\rho = 1600\,\text{kg/m}^3$ and its stiffness properties were defined by the Lame's constants $\lambda = 9.7\,\text{GPa}$ and $\mu = 5\,\text{GPa}$. The material properties for the isotropic solid were chosen close to the graphite-epoxy composite in its transverse direction. The available closed-form solution for isotropic materials allows for a rough calibration of the developed anisotropic model in terms of the resolution of numerical integration in evaluating the Green's function. The selected resolutions were then used to compute the ultrasonic fields for an anisotropic graphite-epoxy composite, which is presented in the next section.

The x_1x_3 plane passing through $x_2 = L_2/2$ was considered. The results for the stress distribution on this plane are shown in Figure 3.43; Figures 3.43a and 3.43b

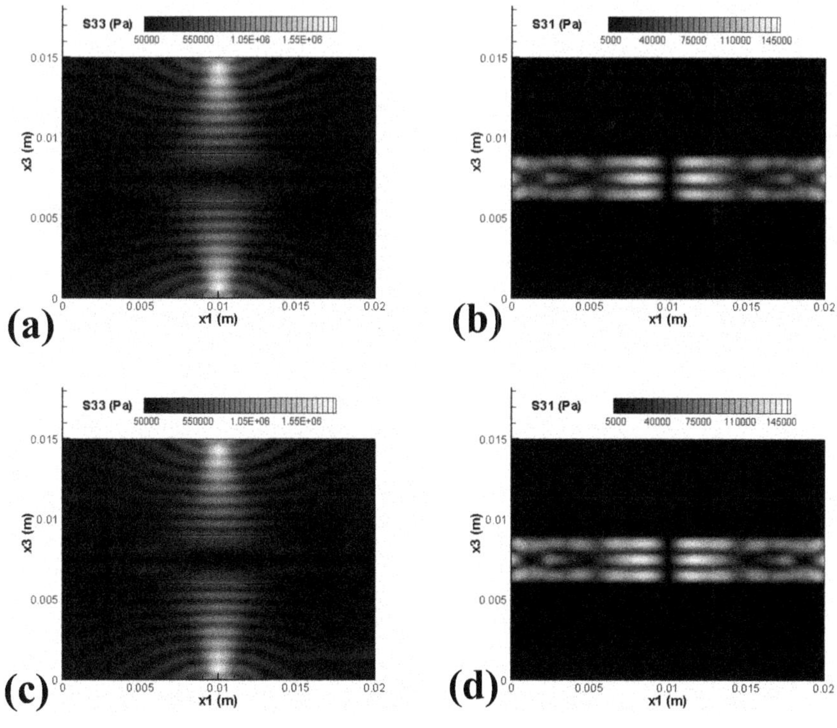

Figure 3.43 Isotropic plate: The results on the central x_1x_3 plane obtained from isotropic DPSM code (a–b) and anisotropic DPSM code (c-d) [from Fooladi and Kundu (2017)].

were obtained from an isotropic DPSM code which uses a closed-form solution for the Green's function; Figures 3.43c and 3.43d were obtained from an anisotropic DPSM code which computes the Green's function numerically. An excellent agreement can be observed between the results produced by the two codes.

3.9.4.2 *Transversely Isotropic Plate*

In the second example, a transversely isotropic plate was considered and the developed DPSM model for anisotropic materials was used. Similar to the previous example, different numbers of integration points were used to evaluate the Green's functions at different distances. This multi-resolution integration scheme was combined with the windowing technique to reduce the computational time.

The problem geometry is the same as the one shown in Figure 3.40. The thickness of the anisotropic solid plate is 3 mm. The transducers are square shaped with an edge length of 2 mm, and were placed parallel to the plate so that the axis x_3 became normal to the transducer faces. A velocity amplitude $V = 1\,\text{m/s}$ and a frequency $f = 1\,\text{MHz}$ were assumed for both transducers. Both upper and lower fluids are water with density $\rho = 1000\,\text{kg/m}^3$ and P-wave speed $C_p = 1480\,\text{m/s}$. The solid plate is a graphite-epoxy composite with a density of $\rho = 1600\,\text{kg/m}^3$. The stiffness tensor was taken from Wang and Achenbach (1995) with $C_{44} = 5\,\text{GPa}$.

$$\mathbf{C} = \begin{bmatrix} 22.73 & 0.9178 & 0.9178 & & & \\ 0.9178 & 1.97 & 0.97 & & & \\ 0.9178 & 0.97 & 1.97 & & & \\ & & & 0.5 & & \\ & & & & 1 & \\ & & & & & 1 \end{bmatrix} \times C_{44} \tag{3.130}$$

Field values were computed on the x_1x_3 plane passing through $x_2 = L_2/2$. The results for stress distribution on this plane are shown in Figure 3.44. A 20 mm length of the plate in the x_1 direction was considered in the analysis. Based on the length and point source spacing in the x_1 direction, the length in the x_2 direction was defined so that the spacing between the neighboring source points along both directions remained the same. The length in the x_2 direction was about 3.4 mm. The reason for having a considerably smaller length in the x_2 direction was to lower the computational time. It was observed that by increasing this length, the solution on the central x_1x_3 plane did not change significantly. The number and distribution of source and target points along the x_1 and x_2 directions are the same for the isotropic and anisotropic plate models.

Next, the x_2x_3 plane passing through $x_1 = L_1/2$ was considered. The results for stress distribution on this plane are shown in Figure 3.45. To obtain the results on the x_1x_3 plane shown in the previous Figure (Figure 3.44), the dimension of the problem was chosen to be large in the x_1 direction and small in the x_2 direction. Next, to show the results on the x_2x_3 plane, the length is chosen to be large in the x_2 direction (20 mm) and small in the x_1 direction (about 3.4 mm). As mentioned before, the reason for such choice of lengths is to lower the computational time. No noticeable change was observed on the plotted planes by increasing the dimensions of that plane or the normal plane.

The anisotropic behavior of the plate is evident when one compares the results on the x_1x_3 plane shown in Figure 3.44 with those on the x_2x_3 plane shown in Figure 3.45. The stress components S_{33}, S_{32}, S_{31}, S_{11} and S_{22} in Figure 3.44 should

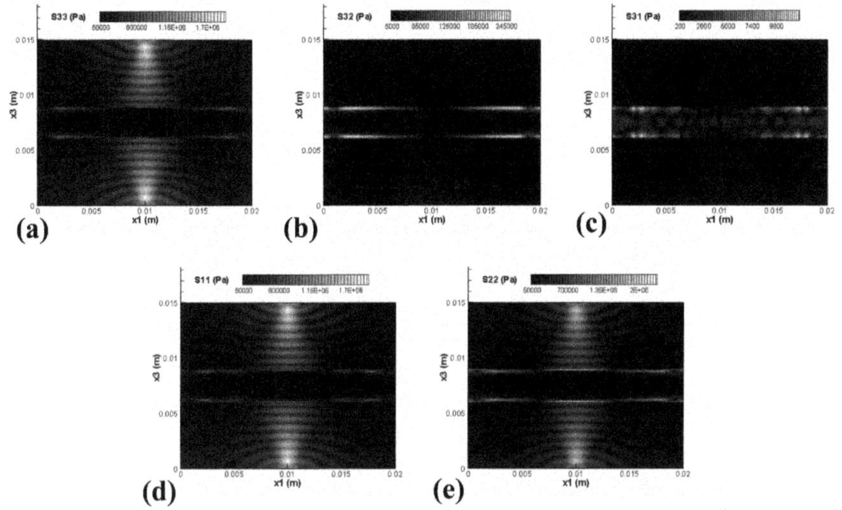

Figure 3.44 Anisotropic plate: Stress distributions on the central x_1x_3 plane [from Fooladi and Kundu (2017)].

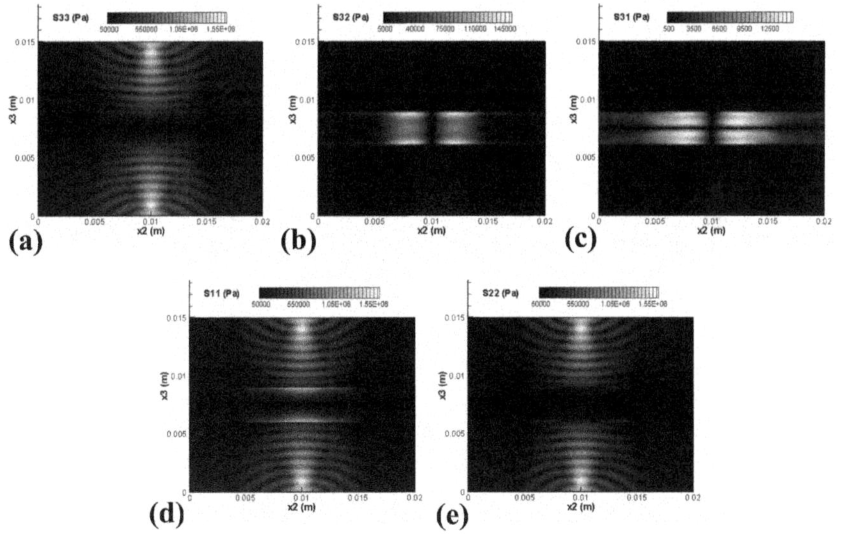

Figure 3.45 Anisotropic plate: Stress distributions on the central x_2x_3 plane [from Fooladi and Kundu (2017)].

be compared with the stress components S_{33}, S_{31}, S_{32}, S_{22} and S_{11} in Figure 3.45, respectively. If the material was isotropic, then these stress components in Figures 3.44 and 3.45 should have been equal.

3.10 CONCLUDING REMARKS

The distributed point source method, or DPSM, was first developed by Placko and Kundu (2004, 2007) for solving engineering problems related to electrostatic, electromagnetic and ultrasonic field modeling. It is a general modeling tool that is capable of solving any scientific or engineering problem using Green's functions or point source solutions in an infinite space as the basic building block. If the Green's function is available, then a problem with complex geometry and boundary conditions can be solved by this technique as illustrated in this chapter and in other publications. To date, a good number of electrostatic, electromagnetic and ultrasonic problems have been solved using DPSM. DPSM has also been used to solve other engineering problems. For example, Wada et al. (2014) modeled viscous fluid layer motion between two vibrating disks using DPSM.

This chapter presents DPSM theory needed for modeling only ultrasonic problems and gives some examples involving isotropic fluids and solids, and anisotropic solids. Both frequency domain and time domain solutions have been presented. Although this chapter presents the ultrasonic problems that have been solved by the author and his research group, it should be noted that other researchers have also used DPSM to solve a variety of other engineering problems. For example, Jarvis and Cegla (2012, 2014) and Benstock et al. (2014) used DPSM and FEM to model ultrasonic wave reflection from a rough surface and concluded that the "[FEM solution is] two orders of magnitude slower than the equivalent DPSM simulation on the same machine". They were interested in measuring the thickness of plates having rough surfaces. For surface crack detection, Kiyasatfar et al. (2011) modeled magnetic flux leakage using DPSM. Wada et al. (2016) combined DPSM and MPS (moving particle semi-implicit) methods to model rotation of levitated fluid droplets with free surface boundary. They used DPSM to calculate Reynolds stress traction force using the idea of effective normal particle velocity through the boundary layer and used it as an input to the MPS surface particles. The droplet was vertically supported by a plane standing wave from an ultrasonic driver and subjected to a rotating acoustic field excited by two acoustic sources on the side-wall with different phases. They successfully reproduced the rotation and acceleration of the droplet numerically and compared their results with available results in the literature. Clearly, DPSM has been used by many investigators to solve a wide range of scientific problems, and only a few ultrasonic problems have been discussed in this chapter.

REFERENCES

Atalar, A., "An Angular-Spectrum Approach to Contrast in Reflection Acoustic Microscopy" *Journal of Applied Physics*, Vol. 49(10), pp. 5130–5139 (1978).

Atalar, A., "A Backscattering Formula for Acoustic Transducers" *Journal of Applied Physics*, Vol. 51(6), pp. 3093–3098 (1980).

Banerjee, S., and T. Kundu, Chapter 4, "Advanced Applications of Distributed Point Source Method-Ultrasonic Field Modeling in Solid Media" in *DPSM for Modeling Engineering, Pub, Problems*, eds. D. Placko and T. Kundu, John Wiley & Sons, pp. 143–229 (2007).

Banerjee, S., and T. Kundu, "Elastic Wave Field Computation in Multilayered Non-Planar Solid Structures: A Mesh-Free Semi-Analytical Approach" *The Journal of the Acoustical Society of America*, Vol. 123(3), pp. 1371–1382 (2008).

Banerjee, S., T. Kundu, and D. Placko, "Ultrasonic Field Modelling in Multilayered Fluid Structures Using DPSM Technique" *Asme Journal of Applied Mechanics*, Vol. 73(4), pp. 598–609 (2006).

Benstock, D., F. Cegla, and M. Stone, "The Influence of Surface Roughness on Ultrasonic Thickness Measurements" *The Journal of the Acoustical Society of America*, Vol. 136(6), p. 3028 (2014).

Cheng, J., W. Lin, and Y. X. Qin, "Extension of the Distributed Point Source Method for Ultrasonic Field Modeling" *Ultrasonics*, Vol. 51(5), pp. 571–580 (2011).

Cobbold, R. S. C., *Foundations of Biomedical Ultrasound*, Oxford University Press, Oxford, New York (2007).

COMSOL AB., "COMSOL Multiphysics: User's Guide" Version 3.5a (2008).

Cooley, J. W., and J. W. Tukey, "An Algorithm for the Machine Computation of the Complex Fourier Series" *Mathematics of Computation*, Vol. 19(90), pp. 297–301 (1965).

Dao, C. M., "Ultrasonic Wave Propagation on an Inclined Solid Half-Space Partially Immersed in a Liquid". Ph.D. Dissertation, Department of Civil Engineering & Engineering Mechanics, University of Arizona (2007).

Dao, C. M., S. Das, S. Banerjee, and T. Kundu, "Wave Propagation in a Fluid Wedge over a Solid Half-Space – Mesh-Free Analysis with Experimental Verification" *International Journal of Solids and Structures*, Vol. 46(11–12), pp. 2486–2492 (2009).

Das, S., S. Banerjee, and T. Kundu, "Modeling of Transient Ultrasonic Wave Propagation in an Elastic Half-Space Using Distributed Point Source Method" *Health Monitoring of Structural and Biological Systems 2010*, SPIE's 17th Annual International Symposium on Smart Structures and Materials & Nondestructive Evaluation and Health Monitoring, San Diego, California, March 7–11, ed. T. Kundu, pp. 76501G–76501 to 76501G-12 (2010).

Dederichs, P. H., and G. Liebfried, "Elastic Green's Function for Anisotropic Cubic Crystals" *Physical Review*, Vol. 188(3), pp. 1175–1183 (1969).

Fooladi, S., "Numerical Implementation of Elastodynamic Green's Function for Anisotropic Media." MS Thesis, Aerospace and Mechanical Engineering Department, University of Arizona, USA (2016).

Fooladi, S., "Distributed Point Source Method for Modeling Wave Propagation in Anisotropic Media." PhD Dissertation, Aerospace and Mechanical Engineering Department, University of Arizona, USA (2018).

Fooladi, S., and T. Kundu, "Ultrasonic field Modeling in Anisotropic Materials by Distributed Point Source Method" *Ultrasonics*, Vol. 78, pp. 115–124 (2017).

Hosten, B., and C. Blateau, "Finite Element Simulation of the Generation and Detection by Air-Coupled Transducers of Guided Waves in Viscoelastic and Anisotropic Materials" *The Journal of the Acoustical Society of America*, Vol. 123(4), pp. 1963–1971 (2008).

Hosten, B., and M. Castaings, "Finite Elements Methods for Modeling the Guided Wave Propagation in Structures with Weak Interfaces" *The Journal of the Acoustical Society of America*, Vol. 117(3), pp. 1108–1113 (2005).

Hosten, B., and M. Castaings, "FE Modeling of Lamb Mode Diffraction by Defects in Anisotropic Viscoelastic Plates" *NDT and E International*, Vol. 39(3), pp. 195–204 (2006).

Jarvis, A. J. C., and F. B. Cegla, "Application of the Distributed Point Source Method to Rough Surface Scattering and Ultrasonic Wall Thickness Measurement" *The Journal of the Acoustical Society of America*, Vol. 132(3), pp. 1325–1335 (2012).

Jarvis, A. J. C., and F. B. Cegla, "Scattering of Near Normal Incidence SH Waves by Sinusoidal and Rough Surfaces in 3-D: Comparison to the Scalar Wave Approximation" *IEEE Transactions on Ultrasonics, Ferroelectrics, and Frequency Control*, IEEE Trans., Vol. 61(7), pp. 1179–1190 (2014).

Kelly, J. F., and R. J. McGough, "An Annular Superposition Integral for Axisymmetric Radiators" *The Journal of the Acoustical Society of America*, Vol. 121(2), pp. 759–765 (2007).

Kiyasatfar, M., M. Golzan, N. Pourmahmoud, and M. Eskandarzade, "Distributed Point Source Technique in Modeling Surface-Breaking Crack in a MFL Test" *Sensors & Transducers*, Vol. 133(10), pp. 108–114 (2011).

Kundu, T., and D. Placko, Chapter 2, "Advanced Theory of DPSM – Modeling Multi-Layered Medium and Inclusions of Arbitrary Shape" in *DPSM for Modeling Engineering, Pub, Problems*, eds. D. Placko and T. Kundu, John Wiley & Sons, pp. 59–96 (2007).

Kundu, T., D. Placko, E. Kabiri Rahani, T. Yanagita, and C. M. Dao, "Ultrasonic Field Modeling: A Comparison between Analytical, Semi-Analytical and Numerical Techniques" *IEEE Transactions on Ultrasonics, Ferroelectric, and Frequency Control*, Vol. 57(12), pp. 2795–2807 (2010).

Kundu, T., D. Placko, T. Yanagita, and S. Sathish, "Micro Interferometric Acoustic Lens: Mesh-Free Modeling with Experimental Verification" *Health Monitoring of Structural and Biological Systems III*, SPIE's 16th Annual International Symposium on Smart Structures and Materials & Nondestructive Evaluation and Health Monitoring, San Diego, California, March 9–12, ed. T. Kundu, Vol. 7295(2), pp. 72951E–72951 to 72950M-11 (2009).

Kundu, T., J. P. Lee, C. Blasé, and J. Bereiter-Hahn, "Acoustic Microscope Lens Modeling and Its Application in Determining Biological Cell Properties from Single and Multi-Layered Cell Models" *The Journal of the Acoustical Society of America*, Vol. 120(3), pp. 1646–1654 (2006).

Lobkis, O. I., and P. V. Zinin, "Acoustic Microscopy of Spherical Objects Theoretical Approach" *Acoustic Letters*, Vol. 14, pp. 168–172 (1990).

Mast, T. D., and F. Yu, "Simplified Expansions for Radiation from a Baffled Circular Piston" *The Journal of the Acoustical Society of America*, Vol. 118(6), pp. 3457–3464 (2005).

Mellow, T. J., "On the Sound Field of a Resilient Disk in an Infinite Baffle" *The Journal of the Acoustical Society of America*, Vol. 120(1), pp. 90–101 (2006).

Mellow, T. J., "On the Sound Field of a Resilient Disk in Free Space" *The Journal of the Acoustical Society of America*, Vol. 123(4), pp. 1880–1891 (2008).

Moreau, L., M. Castaings, B. Hosten, and M. V. Predoi, "An Orthogonality Relation-Based Technique for Post-Processing Finite Element Predictions of Waves Scattering in Solid Waveguides" *The Journal of the Acoustical Society of America*, Vol. 120(2), pp. 611–620 (2006).

O'Neil, H. T., "Theory of Focusing Radiators" *The Journal of the Acoustical Society of America*, Vol. 21(5), pp. 516–526 (1949).

Placko, D., and T. Kundu, Chapter 2, "Modeling of Ultrasonic Field by Distributed Point Source Method". In *Ultrasonic Nondestructive Evaluation: Engineering and Biological Material Characterization*, ed. T. Kundu, CRC Press, pp. 143–202 (2004).

Placko, D., and T. Kundu, eds., *DPSM for Modeling Engineering Problems*, Wiley and Sons (2007).

Placko, D., T. Yanagita, E. K. Kabiri Rahani, and T. Kundu, "Mesh-Free Modeling of the Interaction between a Point Focused Acoustic Lens and a Cavity" *IEEE Transactions on Ultrasonics, Ferroelectrics and Frequency Control*, Vol. 57(6), pp. 1396–1404 (2010).

Sales, M. A., and L. J. Gray, "Evaluation of the Anisotropic Green's Function and Its Derivatives" *Computers and Structures*, Vol. 69(2), pp. 247–254 (1998).

Schmerr, L. W., *Fundamentals of Ultrasonic Nondestructive Evaluation – A Modeling Approach*, Plenum Press, New York (1998).

Wada, Y., K. Yuge, H. Tanaka, and K. Nakamura, "Analysis of Ultrasonically Rotating Droplet Using Moving Particle Semi-Implicit and Distributed Point Source Methods" *Japanese Journal of Applied Physics*, Vol. 55(7S1), pp. 07KE06-1 to 9 (2016).

Wada, Y., T. Kundu, and K. Nakamura, "Mesh-Free Distributed Point Source Method for Modeling Viscous Fluid Layer Motion between Disks Vibrating at Ultrasonic Frequency" *The Journal of the Acoustical Society of America*, Vol. 136(2), pp. 466–474 (2014).

Wang, C.-Y., and J. D. Achenbach, "Elastodynamic Fundamental Solutions for Anisotropic Solids" *Geophysical Journal International*, Vol. 118(2), pp. 384–392 (1994).

Wang, C.-Y., and J. D. Achenbach, "Three-Dimensional Time-Harmonic Elastodynamic Green's Functions for Anisotropic Solids" *Proceedings of the Royal Society A: Mathematical, Physical and Engineering Sciences*, Vol. 449(1937), pp. 441–458 (1995).

4 Nonlinear Ultrasonic Techniques for Nondestructive Evaluation

4.1 INTRODUCTION

By "nonlinear ultrasonic technique for damage detection" one means sensing and recording the damage induced material nonlinearity by appropriate use of ultrasonic waves. Ultrasonic waves generally produce small amplitude stresses in a material. This minute and almost negligible nonlinearity induced by an ultrasonic wave is very difficult to detect by traditional mechanics of materials approach. To investigate if a material is behaving linearly or nonlinearly, in the traditional mechanics of materials approach the material is loaded to check if the stress–strain relation is linear or nonlinear.

Many materials, such as metals, behave linearly at low stress levels while the nonlinear behavior is observed when it is subjected to high stresses. One may argue that a metal when loaded by a low amplitude ultrasonic wave should only exhibit a linear response since the stress induced by a low amplitude ultrasonic wave is very small. Therefore, it might seem logical to think that there is no need to discuss nonlinear ultrasonic techniques for low amplitude ultrasonic waves. However, it should be noted that a stress–strain curve that looks linear may not be linear. Figure 4.1 shows three stress–strain curves – one perfectly linear (line A), one moderately nonlinear that looks almost linear (line B) and one highly nonlinear (line C). The slightly nonlinear curve B has a very large radius of curvature. Such nonlinearity is difficult to detect from its stress–strain plot because for small stress or strain increments this curve looks like a perfect straight line.

An apparently linear elastic material, such as steel below its yield point, can also exhibit moderate nonlinearity (as depicted by curve B in Figure 4.1) below its yield point if the material contains distributed micro-damage. In such a material, even a small amplitude wave can cause a very small nonlinear response. This nonlinearity can be detected by various nonlinear ultrasonic methods like the higher harmonic (HH) technique, frequency modulation (FM) technique, nonlinear impact resonant acoustic spectroscopy (NIRAS) and the sideband peak count (SPC) technique. These techniques and a number of other techniques have been discussed in detail in a book on nonlinear ultrasonic and vibro-acoustic techniques for nondestructive evaluation (NDE) [Kundu (2019)]. Readers are referred to this book for detailed descriptions of various nonlinear techniques. In this chapter, fundamental concepts of nonlinear ultrasonic techniques and some applications of these techniques in the NDE of materials are presented with a few examples.

In previous chapters it has been discussed how relatively large defects whose dimensions are in the order of, or preferably greater than, the wavelength of the interrogating waves can be detected by analyzing the scattered ultrasonic fields. Scattering of ultrasonic fields by a defect can be modeled following the linear analysis – assuming a linear stress–strain relation, as considered in first three chapters. Large damages having dimensions in the order of ultrasonic signal wavelength or larger affect the linear ultrasonic parameters – reflection and transmission coefficients, wave velocity and attenuation – and therefore can be detected by monitoring these parameters. However, very small defects and dislocations with dimensions much smaller than the wavelength do not significantly affect these linear parameters. It is shown in this chapter that nonlinear acoustic parameters are noticeably affected by very small defects, such as fatigue generated dislocations at the grain boundaries, and micro-cracks that are much

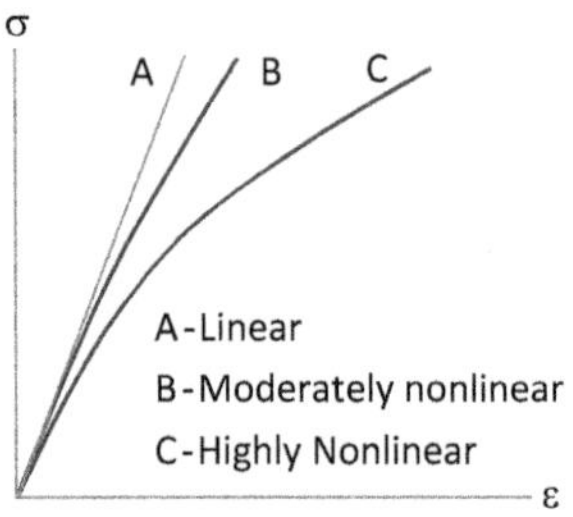

Figure 4.1 Stress–strain relation for perfectly linear (A), moderately nonlinear (B) and highly nonlinear (C) materials.

smaller than the wavelength of the interrogating ultrasonic waves. Since in comparison to the linear parameters the nonlinear parameters are more sensitive to such small defects, the nonlinear parameters are used to monitor dislocations, micro-cracks and other changes in the material that are hard to monitor by linear ultrasonic techniques.

The nonlinear ultrasonic analysis assumes nonlinear stress–strain curve while the linear analysis considers linear stress–strain relation. Nonlinearity in the stress–strain relation can appear due to various effects – one popular belief is that as waves propagate through the damaged material crack surfaces near the crack tips open and close causing variations in stiffness and temperature of the material in that region. This variation in stiffness and temperature causes nonlinear material response.

When waves of two different frequencies propagate through a linear material their frequencies do not change; they propagate independently without having any interaction between them. In the following section it is shown that if the material is nonlinear then the interaction between these two waves of frequencies, f_1 and f_2, produce waves of several other frequencies $(mf_1 \pm nf_2)$ where m and n are integer values. Waves having these new frequencies can be analyzed to detect very small defects. This technique is called the *Nonlinear Wave Modulation Spectroscopy* (NWMS) technique.

When a wave of single frequency f_1 propagates through a nonlinear material new waves of frequencies mf_1, where m is an integer, are generated in the material. These waves are called *higher harmonics*. Higher harmonics and wave modulation spectroscopy can be efficiently used to sense a very small degree of nonlinearity in the material generated by micro-damage.

It is now well-established that linear ultrasonic parameters are not sufficiently sensitive to the microscopic degradation that affect a material's integrity. On the other hand, micro-defects, although too small to be detected by linear ultrasonic techniques, can significantly alter the nonlinear acoustic parameters of the material. One interesting distinction between linear and nonlinear ultrasonic NDE is that in nonlinear ultrasonic NDE, the defects are characterized by analyzing acoustic signals whose frequencies are different from those of the input signals while in the linear ultrasonic technique no signal exists whose frequency is different from the frequencies of the input signals.

In the following section the basic mechanics responsible for the nonlinear ultrasonic response of materials is presented using simple one-dimensional models. It is followed by a few examples of nonlinear ultrasonic NDE.

4.2 ONE-DIMENSIONAL ANALYSIS OF WAVE PROPAGATION IN A NONLINEAR MATERIAL

4.2.1 Stress–Strain Relations of Linear and Nonlinear Materials

The one-dimensional stress–strain relation for a linear elastic material is given by Hooke's law,

$$\sigma = E\varepsilon = E_0\varepsilon \tag{4.1}$$

where $E = E_0$ is the Young's modulus, which is a material constant, independent of applied stress (or strain) as discussed in Chapter 1. For a classical nonlinear material, the stress–strain relation can be expressed as

$$\begin{aligned}\sigma = E\varepsilon &= E_0\left(1-\beta\varepsilon-\gamma\varepsilon^2-\delta\varepsilon^3-\eta\varepsilon^4-\ldots\right)\varepsilon\\ &=E_0\varepsilon-E_1\varepsilon^2-E_2\varepsilon^3-E_3\varepsilon^4-E_4\varepsilon^5+\ldots\end{aligned} \tag{4.2}$$

Equation (4.2) shows that for a nonlinear material the stress–strain relation can be modeled as the superposition of linear, quadratic, cubic, fourth, fifth and even higher order terms. Higher order nonlinear behavior can be modeled simply by taking more terms on the right-hand side of the equation.

In Figure 4.1 the curve A is modeled by Eq. (4.1) while both curves B and C can be modeled by Eq. (4.2). The number of terms on the right-hand side of this equation depends on whether curves B and C are quadratic, cubic or even higher order polynomials. The nonlinear behavior of the material can be defined by two terminologies – *degree of nonlinearity* and *order of nonlinearity*. The *degree of nonlinearity* (how large the curvature of the stress–strain curve is) is dependent on the values of the nonlinear material constants – β, γ, δ, η… or E_1, E_2, E_3, E_4 … etc. of Eq. (4.2) – while the *order of nonlinearity* defines up to which order polynomial terms are kept in Eq. (4.2). A material can have a low order but high degree of nonlinearity when it has a large value of β but its $\gamma = \delta = \eta = \cdots = 0$, on the other hand, another material can have a higher order of nonlinearity but of low degree. For such a material the nonlinear constants β, γ, δ, η are non-zero but should have very small values.

Figure 4.1 shows line C has a higher degree of nonlinearity than curve B. Whether curve B or C has a higher order of nonlinearity (quadratic, cubic, … etc.) is difficult to say just by looking at these two curves. Although curve C has a higher degree of nonlinearity than curve B the two curves can have the same order of nonlinearity or any one can have a higher order of nonlinearity than the other one. The materials that show a higher degree of nonlinearity generally also have a higher or same order of nonlinearity. In the following subsections it is discussed how these nonlinear materials respond when excited by a wave of single frequency and multiple frequencies.

4.2.2 Nonlinear Material Excited by a Wave of Single Frequency

Consider a nonlinear material excited by a harmonic wave of frequency f_1 propagating in the x direction. This wave generates the following displacement field u in the material,

$$\begin{aligned}u(x,t) = Ae^{i(kx-\omega_1 t)} &= A\cos(kx-\omega_1 t)+iA\sin(kx-\omega_1 t)\\ &= A_1(x)\sin(\omega_1 t)+B_1(x)\cos(\omega_1 t)\end{aligned} \tag{4.3}$$

where the angular frequency $\omega_1 = 2\pi f_1$. Functions $A_1(x)$ and $B_1(x)$ can be expressed in terms of sin(kx), cos(kx) and constant A after some trigonometric manipultion.

The strain field for this input displacement for one-dimensional problem is given by

$$\varepsilon(x,t) = \frac{\partial u(x,t)}{\partial x} = A_1'(x)\sin(\omega_1 t) + B_1'(x)\cos(\omega_1 t) \tag{4.4}$$

where $A_1'(x)$, $B_1'(x)$ denote the first derivatives of $A_1(x)$ and $B_1(x)$, respectively.

For a classical nonlinear quadratic stress–strain relation

$$\sigma(x,t) = E\varepsilon(x,t) = \left[E_0 - E_1(\varepsilon)\right]\varepsilon(x,t) = E_0\varepsilon(x,t) - E_1\,\varepsilon^2(x,t) \tag{4.5}$$

The harmonic stress field generated in this nonlinear material by the input displacement field of Eq. (4.3) is then given by

$$\begin{aligned}\sigma(x,t) &= E_0\varepsilon(x,t) - E_1\,\varepsilon^2(x,t) = E_0\left[A_1'(x)\sin(\omega_1 t) + B_1'(x)\cos(\omega_1 t)\right] \\ &\quad - E_1\left\{\left[A_1'(x)\right]^2\sin^2(\omega_1 t) + \left[B_1'(x)\right]^2\cos^2(\omega_1 t) \right. \\ &\quad \left. + 2A_1'(x)B_1'(x)\sin(\omega_1 t)cos(\omega_1 t)\right\}\end{aligned} \tag{4.6}$$

or,

$$\begin{aligned}\sigma(x,t) &= E_0\left[A_1'(x)\sin(\omega_1 t) + B_1'(x)\cos(\omega_1 t)\right] \\ &\quad - E_1\left\{\frac{\left[A_1'(x)\right]^2\left[1-\cos(2\omega_1 t)\right]}{2} + \frac{\left[B_1'(x)\right]^2\left[1+\cos(2\omega_1 t)\right]}{2} + A_1'(x)B_1'(x)\sin(2\omega_1 t)\right\}\end{aligned} \tag{4.7}$$

In equation (4.7) one can clearly see that the linear term that is multiplied by E_0 contains waves of angular frequency ω_1 only while the nonlinear term which is multiplied by E_1 contains waves of frequency $2\omega_1$ (higher harmonic wave).

Next let us assume the stiffness modulus E to be a quadratic function of strain. Then the stress–strain relation becomes cubic:

$$E = E_0 - E_1\varepsilon = E_2\varepsilon^2 \tag{4.8}$$

$$\sigma = E\varepsilon = E_0\varepsilon - E_1\varepsilon^2 - E_2\varepsilon^3 \tag{4.9}$$

For a harmonic excitation of such a material, the input displacement field can be assumed to be

$$u(x,t) = A(x)\sin(\omega t) \tag{4.10}$$

To keep the analysis simple the cosine term is not considered in the above expression but can easily be considered if necessary, as done in the previous derivation [Eqs. (4.3–7)].

From the one-dimensional strain–displacement relation the strain field for the above displacement field is obtained as

$$\varepsilon(x,t) = \frac{\partial u(x,t)}{\partial x} = A'(x)\sin(\omega t) = a(x)\sin(\omega t) \tag{4.11}$$

where $a(x) = A'(x)$.

Therefore, the stress field for this material is

$$\begin{aligned}\sigma(x,t) &= E_0\varepsilon(x,t) - E_1\,\varepsilon^2(x,t) - E_2\,\varepsilon^3(x,t)\\ &= E_0 a(x)\sin(\omega t) - E_1\,a^2(x)\sin^2(\omega t) - E_2\,a^3(x)\sin^3(\omega t)\\ &= E_0 a(x)\sin(\omega t) - E_1\,a^2(x)\frac{\{1-\cos(2\omega t)\}}{2}\\ &\quad -E_2 a^3(x)\frac{\{3\sin(\omega t)-\sin(3\omega t)\}}{4}\end{aligned} \tag{4.12}$$

From Eq. (4.12) one can clearly see that if this material is excited by a signal of frequency ω then the stress fields of frequency ω, 2ω, and 3ω are generated. Similarly, when the stress–strain relation contains fourth and fifth order terms, then the nonlinear material response should contain waves of frequency 4ω and 4ω also when it is excited by frequency ω.

The above analysis can be summarized in this manner: when a nonlinear material is excited by a wave of frequency ω, then the material response produces a wave of frequency ω as well as its higher harmonics of frequency 2ω, 3ω, 4ω, 5ω,...etc., depending on the order of nonlinearity of the material. These higher harmonics are not generated in a linear material.

Eq. (4.12) also shows that the wave amplitudes associated with the fundamental frequency ω and the higher harmonic frequencies 2ω and 3ω are $E_0 a(x)$, $E_1 a^2(x)$ and $E_2 a^3(x)$, respectively. When the stiffness modulus is expressed as

$$E = E_0 - E_1\varepsilon - E_2\varepsilon^2 = E_0\left(1-\beta\varepsilon-\gamma\varepsilon^2\right) \tag{4.13}$$

then the amplitudes A_1, A_2 and A_3 of the fundamental frequency (ω) and the two higher harmonics (2ω and 3ω) are given by

$$\begin{aligned}A_1 &= E_0 a(x)\\ A_2 &= E_1\,a^2(x) = E_0\,\beta a^2(x)\\ A_3 &= E_2\,a^3(x) = E_0\,\gamma a^3(x)\end{aligned} \tag{4.14}$$

From Eq. (4.14),

$$\frac{A_2}{A_1^2} = \frac{\beta}{E_0} \tag{4.15}$$

and

$$\frac{A_3}{A_1^3} = \frac{\gamma}{E_0^2} \tag{4.16}$$

Since E_0 is a constant, from Eqs. (4.15) and (4.16) one can write

$$\beta \propto \frac{A_2}{A_1^2} \tag{4.17}$$

and

$$\gamma \propto \frac{A_3}{A_1^3} \tag{4.18}$$

Similarly, if more higher order terms in the stress–strain relation are kept, as given in Eq. (4.2),

$$\sigma = E\varepsilon = E_0\left(1-\beta\varepsilon-\gamma\varepsilon^2-\delta\varepsilon^3-\eta\varepsilon^4\right)\varepsilon \tag{4.2}$$

then more higher order harmonics of frequency 2ω, 3ω, 4ω and 5ω are generated and those higher order material constants are related to the fundamental wave amplitude and higher harmonic amplitudes in the following manner:

$$\delta \propto \frac{A_4}{A_1^4} \tag{4.19}$$

$$\eta \propto \frac{A_5}{A_1^5} \tag{4.20}$$

Figure 4.2 illustrates the higher harmonic generation. Figure 4.2a shows three stress–strain relations – linear (A), quadratic (B) and cubic (C). When a monochromatic wave of frequency ω passes through these materials then the output signal contains waves of frequency ω only for material A, waves of frequency ω and 2ω for material B and waves of frequency ω, 2ω and 3ω for material C, as shown in Figure 4.2b. Since for most materials the nonlinear material constants β and γ are very small one can easily see from equations (4.15) and (4.16) that amplitudes A_2 and A_3 of higher harmonics must be much smaller than A_1, the fundamental wave amplitude.

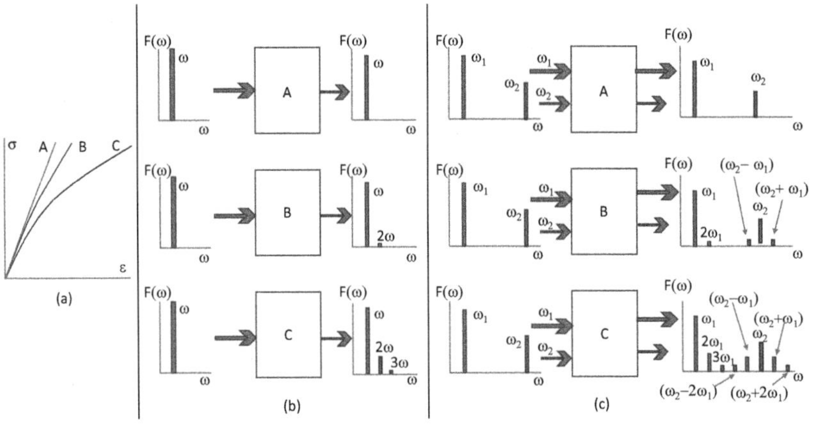

Figure 4.2 (a) Same as Figure 4.1 – Stress–strain relations for linear and nonlinear materials; (b) input and output spectra for linear and nonlinear materials for monochromatic wave excitation – linear material produces no higher harmonic (top row), while nonlinear quadratic material generates one higher harmonic at frequency 2ω (middle row) and cubic material generates two higher harmonics at frequencies 2ω and 3ω (bottom row); (c) input and output spectra when linear and nonlinear materials are excited by waves of two different frequencies ω_1 and ω_2 – linear material shows no modulation and no higher harmonic generation (top row), while nonlinear quadratic material generates one higher harmonic at frequency $2\omega_1$ and shows modulated waves at frequencies $(\omega_2+\omega_1)$ and $(\omega_2-\omega_1)$ (middle row) and cubic material generates two higher harmonics at $2\omega_1$ and $3\omega_1$ and shows more modulated waves (bottom row).

4.2.3 Nonlinear Material Excited by Waves of Two Different Frequencies

Let waves of two different frequencies excite the nonlinear material simultaneously – a high amplitude low frequency wave and a low amplitude high frequency wave are applied together. These two waves are known as a pumping wave (high amplitude low frequency wave) and a probing wave (low amplitude high frequency wave). The harmonic displacement generated by the pumping wave of frequency f_1 is expressed as

$$u_1(\mathrm{x},\mathrm{t}) = A_1(\mathrm{x})\sin(\omega_1 t) \tag{4.21}$$

where angular frequency $\omega_1 = 2\pi f_1$. To keep the expressions simple only the sine term is considered. Similarly, the probing wave of frequency f_2 gives a displacement field

$$u_2(\mathrm{x},t) = A_2(\mathrm{x})\sin(\omega_2 t) \tag{4.22}$$

Then the total displacement field is

$$u(\mathrm{x},\mathrm{t}) = u_1(\mathrm{x},\mathrm{t}) + u_2(\mathrm{x},\mathrm{t}) = A_1(\mathrm{x})\sin(\omega_1 t) + A_2(\mathrm{x})\sin(\omega_2 t) \tag{4.23}$$

The strain field for one-dimensional problem is given by

$$\varepsilon(\mathrm{x},\mathrm{t}) = \frac{\partial u(\mathrm{x},\mathrm{t})}{\partial \mathrm{x}} = A_1'(\mathrm{x})\sin(\omega_1 t) + A_2'(\mathrm{x})\sin(\omega_2 t) \tag{4.24}$$

or,

$$\varepsilon(\mathrm{x},\mathrm{t}) = a_1(\mathrm{x})\sin(\omega_1 t) + a_2(\mathrm{x})\sin(\omega_2 t) \tag{4.25}$$

where $a_i(\mathrm{x}) = A_i'(\mathrm{x})$ denotes the first derivative of $A_i(\mathrm{x})$, $i = 1$ and 2. For a classical nonlinear quadratic stress–strain relation the total stress can be expressed as

$$\begin{aligned}\sigma(\mathrm{x},\mathrm{t}) &= E_0\varepsilon(\mathrm{x},\mathrm{t}) - E_1\,\varepsilon^2(\mathrm{x},\mathrm{t}) = E_0\left[a_1(\mathrm{x})\sin(\omega_1 t) + a_2(\mathrm{x})\sin(\omega_2 t)\right] \\ &\quad - E_1\left[a_1^2(\mathrm{x})\sin^2(\omega_1 t) + a_2^2(\mathrm{x})\sin^2(\omega_1 t) + 2a_1(\mathrm{x})a_2(\mathrm{x})\sin(\omega_1 t)\sin(\omega_2 t)\right]\end{aligned} \tag{4.26}$$

or,

$$\begin{aligned}\sigma(\mathrm{x},\mathrm{t}) &= E_0\left[a_1\sin(\omega_1 t) + a_2\sin(\omega_2 t)\right] \\ &\quad - E_1\left[\frac{a_1^2}{2}\{1-\cos(2\omega_1 t)\} + \frac{a_2^2}{2}\{1-\cos(2\omega_2 t)\}\right. \\ &\quad \left. + a_1a_2\left(\cos\{(\omega_1-\omega_2)t\} - \cos\{(\omega_1+\omega_2)t\}\right)\right]\end{aligned} \tag{4.27}$$

In equation (4.27) one can clearly see that the linear term which is multiplied by E_0 only contains waves of angular frequency ω_1 and ω_2 while the nonlinear term produces waves of frequencies $2\omega_1$ and $2\omega_2$ (higher harmonic waves) as well as $\omega_1 \pm \omega_2$ (modulated waves). Small peaks at frequencies $\omega_1 \pm \omega_2$ are called modulated wave peaks, or simply *sideband peaks* or *sidebands*. If cubic terms are also kept in the stress–strain relation, then following similar derivations one can show that in addition to waves of frequency $2\omega_1$ and $2\omega_2$ and $\omega_1 \pm \omega_2$, waves of frequency $3\omega_1, 3\omega_2, \omega_1 \pm 2\omega_2$ and $2\omega_1 \pm \omega_2$ should be also generated. Therefore, more higher harmonic peaks and sidebands are generated for a material having a cubic stress–strain relation in comparison to a material having a quadratic stress–strain relation. Figure 4.2(c) illustrates sideband and higher harmonic generation when the nonlinear material is excited simultaneously by waves of two different frequencies.

4.2.4 Detailed Analysis of One-Dimensional Wave Propagation in a Nonlinear Rod

Let us take a rod whose quadratic stress–strain relation is given by

$$\sigma = E\varepsilon = (E_0 - E_1\varepsilon)\varepsilon = E_0(1-\beta\varepsilon)\varepsilon \tag{4.28}$$

In this one-dimensional problem let us assume the displacement u and the wave propagation direction both to be in x direction. For this problem displacement, strain and stress fields are functions of time t and position x only.

The one-dimensional governing equation in the absence of body force is

$$\sigma_{,x} = \rho u_{,tt} \tag{4.29}$$

Substitution of Eq. (4.28) into Eq. (4.29) gives

$$\begin{aligned} \sigma_{,x} &= \left[E_0(1-\beta\varepsilon)\varepsilon\right]_{,x} = E_0\varepsilon_{,x} - E_0\beta\varepsilon\varepsilon_{,x} - E_0\beta\varepsilon_{,x}\,\varepsilon = E_0\varepsilon_{,x} - 2E_0\beta\varepsilon\varepsilon_{,x} \\ &= E_0 u_{,xx} - 2E_0\beta u_{,x}\,u_{,xx} = \rho u_{,tt} \end{aligned} \tag{4.30}$$

or,

$$u_{,tt} = \frac{E_0}{\rho}\left(u_{,xx} - 2\beta u_{,x}\,u_{,xx}\right) = c^2\left(u_{,xx} - 2\beta u_{,x}\,u_{,xx}\right) \tag{4.31}$$

where

$$c = \sqrt{\frac{E_0}{\rho}} \tag{4.32}$$

Since the amplitude of the second harmonic wave is much less than that of the primary wave, the perturbation method can be applied to solve the nonlinear governing equation (4.31). In the perturbation approach the displacement field u is expressed as a summation of two components, u_1 and u_2, that can be labeled as linear and nonlinear components:

$$u = u_1 + u_2 \tag{4.33}$$

The nonlinear wave equation (4.31) can be decomposed into two wave equations:

$$u_{1,tt} = c^2 u_{1,xx} \tag{4.34}$$

and

$$u_{2,tt} = c^2 u_{2,xx} - 2\beta c^2 u_{1,x} u_{1,xx} \tag{4.35}$$

where:

- u_1 is the primary wave having the fundamental frequency
- u_2 is the second harmonic wave with double frequency

The general solution for u_1 of the wave equation (4.34) is given by

$$u_1 = A_1\sin(kx - \omega t) \tag{4.36}$$

where, A_1 can be a real or complex constant. Therefore, equation (4.35) can be written as

$$\begin{aligned} u_{2,tt} &= c^2 u_{2,xx} - 2\beta c^2\left[kA_1\cos(kx-\omega t)\right]\left[-k^2A_1\sin(kx-\omega t)\right] \\ &= c^2u_{2,xx} + 2\beta c^2k^3A_1^2\sin(kx-\omega t)\cos(kx-\omega t) \\ &= c^2u_{2,xx} + \beta c^2k^3A_1^2\sin\{2(kx-\omega t)\} \end{aligned} \tag{4.37}$$

To solve the above differential equation assume

$$u_2 = xf\left(\frac{x}{c}-t\right) \tag{4.38}$$

Then

$$\begin{aligned} u_{2,x} &= f\left(\frac{x}{c}-t\right) + \frac{x}{c}f'\left(\frac{x}{c}-t\right) \\ u_{2,xx} &= \frac{2}{c}f'\left(\frac{x}{c}-t\right) + \frac{x}{c^2}f''\left(\frac{x}{c}-t\right) \\ u_{2,tt} &= xf''\left(\frac{x}{c}-t\right) \end{aligned} \tag{4.39}$$

Substituting equations (4.38) and (4.39) into equation (4.37) one obtains

$$xf''\left(\frac{x}{c}-t\right) = c^2\left[\frac{2}{c}f'\left(\frac{x}{c}-t\right) + \frac{x}{c^2}f''\left(\frac{x}{c}-t\right)\right] + \beta c^2k^3A_1^2\sin\{2(kx-\omega t)\} \tag{4.40}$$

or,

$$f'\left(\frac{x}{c}-t\right) = -\frac{1}{2c}\beta c^2k^3A_1^2\sin\{2(kx-\omega t)\} = -\frac{1}{2}\beta ck^3A_1^2\sin\left\{2\omega\left(\frac{x}{c}-t\right)\right\} \tag{4.41}$$

If one assumes

$$f\left(\frac{x}{c}-t\right) = a_2\cos\left\{2\omega\left(\frac{x}{c}-t\right)\right\} \tag{4.42}$$

then substituting Eq. (4.42) into Eq. (4.41) gives

$$f'\left(\frac{x}{c}-t\right) = -2\omega a_2\sin\left\{2\omega\left(\frac{x}{c}-t\right)\right\} = -\frac{1}{2}\beta ck^3A_1^2\sin\left\{2\omega\left(\frac{x}{c}-t\right)\right\} \tag{4.43}$$

Therefore,

$$a_2 = \frac{c\beta k^3A_1^2}{4\omega} = \frac{\omega\beta k^3A_1^2}{k4\omega} = \frac{\beta k^2A_1^2}{4} \tag{4.44}$$

From Eqs. (4.38), (4.42) and (4.44) one then obtains

$$\begin{aligned} u_2 &= xf\left(\frac{x}{c}-t\right) = xa_2\cos\left\{2\omega\left(\frac{x}{c}-t\right)\right\} \\ &= \frac{x}{4}A_1^2k^2\beta\cos\{2(kx-\omega t)\} = A_2\cos\{2(kx-\omega t)\} \end{aligned} \tag{4.45}$$

Clearly, the amplitude of the propagating second harmonic wave whose frequency is twice the frequency of the fundamental wave is given by

$$A_2 = \frac{x}{4}A_1^2k^2\beta \tag{4.46}$$

where A_1 is the amplitude of the fundamental wave, and A_2 is the amplitude of the second harmonic. In equation (4.46) k is the wave number. The material nonlinearity can be quantified by the nonlinear parameter β, which is related to the amplitude of the second harmonic and the square of the fundamental wave amplitude, as given in Eq. (4.46). This equation can be rewritten in the following form:

$$\beta = \frac{4A_2}{A_1^2k^2x} \tag{4.47}$$

Thus, the material nonlinearity parameter β can be evaluated by measuring the amplitudes of the fundamental wave and the second harmonic wave generated in an ultrasonic test.

If the rod is replaced by a half-space and one-dimensional wave propagation in the x direction in this non-linear half-space is considered, then the only change will be in the stress–strain relation. The one-dimensional stress–strain relation $\sigma = E\varepsilon$ of Eq. (4.28) for the rod will be replaced by the stress–strain relation $\sigma = (\lambda + 2\mu)\varepsilon$ which is valid for a half-space when only one normal strain component is non-zero while all other strain components are zero. In this relation the Lame's constants λ and μ are independent of strain for a linear material but are strain dependent for a nonlinear material. Such analysis for the longitudinal wave propagation in a bulk material gives similar relations as given in Eqs. (4.46) and (4.47) (see Kundu et al. (2019) for detail derivation.)

One interesting property of the second harmonic for the longitudinal wave propagation in a bulk material (half-space or full space) or in a rod is evident from Eq. (4.46). Equation (4.46) clearly shows that the second harmonic amplitude increases linearly with x. This is known as the *accumulative* or *cumulative effect* of the second harmonic. This property can improve the signal-to-noise ratio of the second harmonic simply by letting the wave propagate a longer distance as evident from equation (4.46). From equations (4.46) and (4.47) one can write

$$\frac{A_2}{A_1^2} \propto \beta x \tag{4.48}$$

Thus, the material nonlinearity can be evaluated by measuring the ratio of the second harmonic wave amplitude to the square of the fundamental wave amplitude as a function of the distance between the transmitter and the receiver, as shown in Figure 4.3.

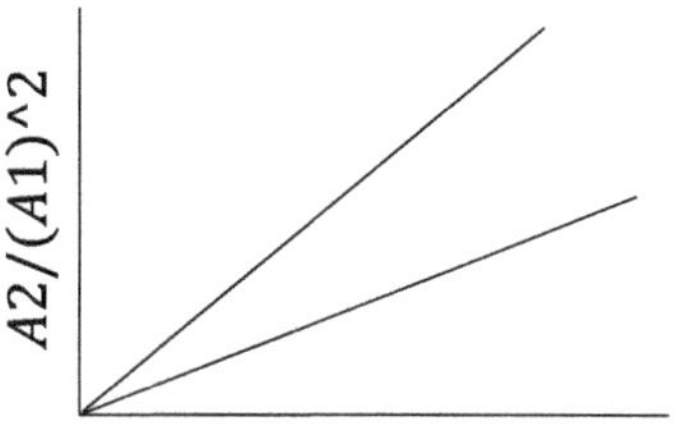

Figure 4.3 Variations of $A_2/(A_1)^2$ with propagation distance; A_2 is the amplitude of the second harmonic and A_1 is the amplitude of the primary wave. Two lines correspond to two nonlinear materials. Higher slope of the line implies higher nonlinearity.

4.2.5 Higher Harmonic Generation for Other Types of Wave

4.2.5.1 Transverse Wave Propagation in a Nonlinear Bulk Material

When a transverse bulk wave propagates in a nonlinear material then a higher order longitudinal wave is generated, but not any higher order transverse wave. The displacement field for the higher harmonic longitudinal wave is given by Kundu et al. (2019):

$$u_2 = \frac{-\beta_t k_S^3 A_1^2}{2\left(k_S^2 - k_P^2\right)} \sin\left[\left(k_S - k_P\right)x\right] \exp\left[i\left(k_P + k_S\right)x - 2i\omega t\right] \tag{4.49}$$

Therefore, the amplitude of the second harmonic is

$$A_2 = \frac{-\beta_t k_S^3 A_1^2}{2\left(k_S^2 - k_P^2\right)} \sin\left\{\left(k_S - k_P\right)x\right\} \tag{4.50}$$

From Eq. (4.56) the material nonlinearity parameter β_t can be expressed as

$$\beta_t = \frac{A_2}{A_1^2} \frac{-2\left(k_S^2 - k_P^2\right)}{k_S^3 \sin\left\{\left(k_S - k_P\right)x\right\}} \tag{4.51}$$

As shown in Figure 4.4, the amplitude of the second harmonic longitudinal wave generated by the incident shear wave becomes zero at certain propagation distances. Since in this case the second harmonic wave amplitude does not grow with the propagation distance one can say that the second order longitudinal wave component generated by the primary transverse wave does not have the cumulative effect. Therefore, it is not very useful for nondestructive testing since its signal-to-noise ratio cannot be improved by simply letting the second harmonic wave propagate a longer distance.

4.2.5.2 Guided Wave Propagation in a Nonlinear Wave-guide

The second harmonic generation is more difficult for the guided wave due to its dispersive nature and multi-mode propagation characteristics. The multi-mode feature of guided waves makes it difficult to generate an experimentally single pure mode. The second harmonic generation for guided waves in half-spaces, isotropic plates and pipes have been discussed in detail in Kundu et al. (2019). Interested readers are referred to that book for detailed derivations of various

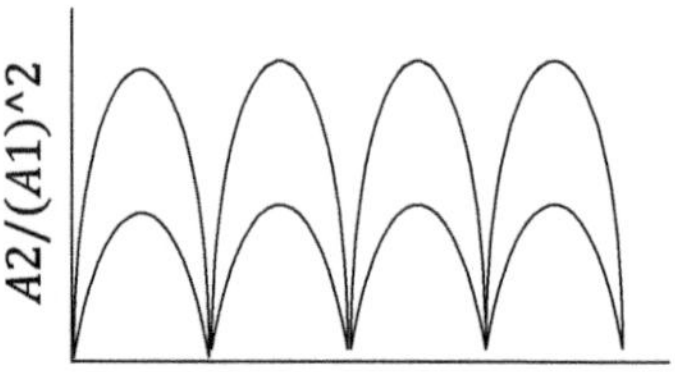

Figure 4.4 Variations of $\frac{A_2}{A_1^2}$ with propagation distance when a higher harmonic longitudinal wave is generated by the primary transverse wave; A_2 is the amplitude of the higher harmonic longitudinal wave and A_1 is the amplitude of the primary transverse wave. Two lines are for two nonlinear materials. The curve with higher peak values correspond to a material with higher nonlinearity.

equations. Only the final results without any derivation are presented in the following subsections.

4.2.5.2.1 Acoustic Nonlinear Parameter for Surface Wave Propagation

The nonlinear parameter for the surface wave propagation can be expressed as (Herrmann, et al., 2006; Li and Cho, 2016; Kundu et al., 2019)

$$\beta = \frac{A_2}{A_1^2} \frac{8ip}{k_R k_P^2 x} \left(1 - \frac{2k_R^2}{k_R^2 + q^2} \right) D_\alpha \tag{4.52}$$

where A_1 and A_2 are the displacement amplitudes of the fundamental wave and the second harmonic wave, respectively measured in the direction perpendicular to the propagating surface; k_R is the Rayleigh wave number $\left(k_R = \frac{\omega}{c_R} \right)$, k_P is the P-wave number $\left(k_P = \frac{\omega}{c_P} \right)$, $p = \sqrt{k_R^2 - k_P^2}$, $q = \sqrt{k_R^2 - k_S^2}$,

The attenuation correction factor D_α is given by

$$D_\alpha = \left(\frac{m}{1 - e^{-m}} \right), \quad m = (\alpha_2 - 2\alpha_1) x \tag{4.53}$$

where α_1 and α_2 are the attenuation coefficients of the fundamental and the second harmonic waves, respectively.

One can clearly see in Eq. (4.52) that the second harmonic amplitude A_2 has cumulative effect, i.e. it linearly increases with the propagation distance x when other parameters (fundamental wave amplitude, P-wave number, Rayleigh wave number and attenuation coefficient) remain unchanged.

4.2.5.2.2 Acoustic Nonlinear Parameter for Lamb Wave Propagation

As mentioned before the measurement of the second harmonic amplitude for characterization of material nonlinearity is typically aimed at determining the value of the nonlinear parameter β. The nonlinear parameter is related to the amplitudes of the second harmonic and fundamental waves. The nonlinear parameter for the Lamb wave propagation can be written as [Kundu et al. (2019)]

$$\beta = \frac{8}{k^2 x}\frac{A_2}{A_1^2}F \tag{4.54}$$

where:

- A_1 and A_2 are the amplitudes of the fundamental and the second harmonic waves, respectively
- k is the wave number $\left(k = \frac{\omega}{c}\right)$ associated with the propagating wave mode
- x is the wave propagation distance
- F is the feature function of the guided wave nonlinear parameter
- F is defined as a function of the frequency, mode type, material properties and dimensions of the wave-guide. If the wave mode, frequency and dimensions, such as thickness of the wave-guide, do not change, then F is a constant. If a relative nonlinear parameter $\bar{\beta}$ is defined as $\bar{\beta} = A_2 / A_1^2$, then from equation (4.54)

$$\beta \propto \bar{\beta} / x \tag{4.55}$$

For both a Lamb wave in a plate and a cylindrical guided wave in a pipe the ratio of the amplitude of the second harmonic and the fundamental wave grows with the propagation distance because of the cumulative effect. The amplitude grows up to a certain distance after which the material attenuation effect dominates and it starts to decay. Amplitudes of the higher harmonic waves induced by the material nonlinearity are functions of wave propagation distance as shown in Eq. (4.55), but the harmonics induced by the instrumental nonlinearity are not. Therefore, if one keeps the experimental setup and conditions unchanged, as the guided wave propagation distance increases, the strength of the second harmonic wave generated by the nonlinearity of the specimen increases due to the cumulative effect; however, the second harmonic generated by the nonlinearity in the experimental setup or ambient noise does not increase with the propagation distance. Demonstration of this cumulative effect is necessary to ensure that measurements taken from the specimens are truly due to the damage induced nonlinearity and not the nonlinearity in the measurement system.

4.3 USE OF NONLINEAR BULK WAVES FOR NONDESTRUCTIVE EVALUATION

The second harmonic generated by propagating longitudinal waves has been widely used to evaluate material nonlinearity. In this section, an example from Li et al. (2013) is given that shows that nonlinear longitudinal ultrasonic waves can be used for monitoring micro-structural changes in the material from heat treatment.

4.3.1 Nonlinear Acoustic Parameter Measurement

The nonlinear parameter β for longitudinal wave propagation in a bulk material is obtained simply by replacing the wave number k for the rod wave or bar wave in Eq. (4.47) by P-wave number k_P,

$$\beta = \frac{4A_2}{A_1^2 k_P^2 x} \tag{4.56}$$

where:

- A_1 and A_2 are the amplitudes of the fundamental wave and the second harmonic wave, respectively
- x is the wave propagation distance

During this experimental study, the β parameter refers to the expression $\frac{A_2}{A_1^2}$ and is denoted as $\bar{\beta} = \frac{A_2}{A_1^2}$. Note that

$$\bar{\beta} = \frac{A_2}{A_1^2} \propto \beta x \tag{4.57}$$

where the proportionality constant is $\frac{k_P^2}{4}$. Alternately, one can also write from Eq. (4.57) $\bar{\beta} \propto \beta$ when the proportionality constant is $\frac{xk_P^2}{4}$. Eq. (4.56) is sometimes written as

$$\beta = \frac{4\bar{\beta}}{k_P^2 x} \tag{4.58}$$

The β parameter $\bar{\beta}$ is obtained experimentally from the ratio $\frac{A_2}{A_1^2}$.

4.3.2 Experimental Results

Four rectangular plate specimens of dimensions 200 mm × 30 mm × 5 mm made of Inconel X-750 were investigated by Li et al. (2013). One specimen was left untreated at room temperature, and the other three were subjected to different heat treatment conditions.

A 5 MHz signal was sent through the specimen. Figures 4.5a, b show a typical time history and its FFT plot after the ultrasonic signal propagated through the specimen.

As shown in equations (4.56) and (4.57), the nonlinear parameter of the ultrasonic longitudinal waves can be expressed as a function of the ratio between the second harmonic amplitude (A_2) and the square of the primary wave amplitude (A_1^2) for a fixed wave number and propagation distance. This value helps us to correlate the acoustic nonlinearity parameter $\bar{\beta}$ with the material nonlinearity β of the specimens for different heat treatment conditions. The measured parameters are normalized with respect to their initial values to show relative changes.

The variation of the nonlinear acoustic parameter for the specimens with different heat treatment conditions is shown in Figure 4.6. Compared to the pristine

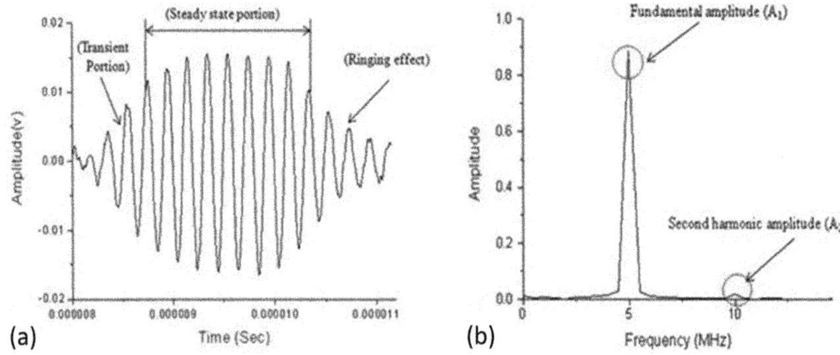

Figure 4.5 (a) Time domain signal and (b) frequency spectrum (FFT) of the signal [from Li et al. (2013)].

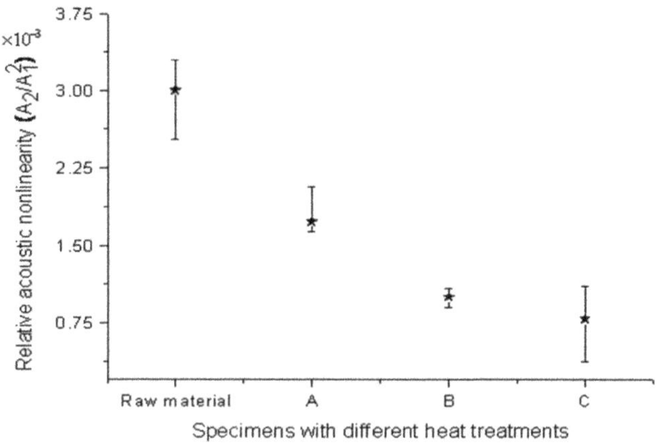

Figure 4.6 Variation of acoustic nonlinearity parameter $\bar{\beta}$ in specimens after different heat treatments [from Li et al. (2013)].

specimen that did not go through any heat treatment, all three heat treated specimens showed distinctly lower nonlinear acoustic parameter value. The acoustic nonlinearity decreased monotonically as the heat treatment condition changed from A to C. See Li et al. (2013) for details on heat treatment conditions.

Figure 4.7 shows the sensitivities of different linear and nonlinear acoustic parameters to heat treatment conditions. All acoustic parameters in the heat treated specimens have been normalized with respect to their initial values for the pristine material, to show the relative changes for each parameter. This figure shows that the wave velocity increased by 0.8% and the attenuation decreased by 16% after heat treatment C. However, the acoustic nonlinearity parameter showed the most noticeable change, a 40% drop after heat treatment

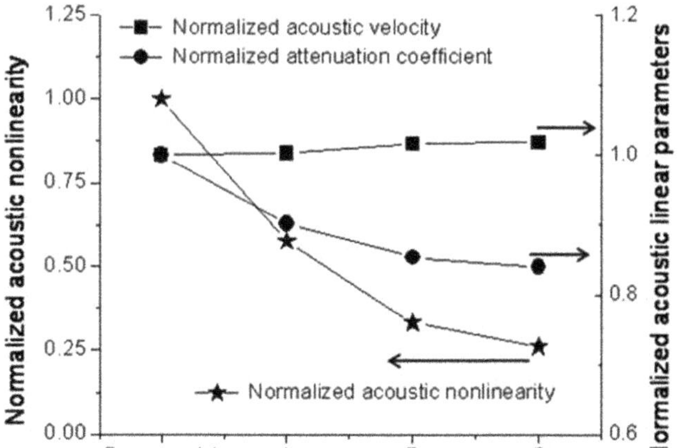

Figure 4.7 Comparison of sensitivity of acoustic linear and nonlinear parameters to heat treatment; the linear parameters (acoustic wave velocity and attenuation coefficient) and acoustic nonlinear parameter are normalized with respect to their values at the pristine state [from Li et al. (2013)].

A, a drop of over 60% after heat treatment B, and a drop of over 70% after heat treatment C [Li et al. (2013)].

Clearly, in comparison to the linear parameters, the nonlinear ultrasonic parameter is more sensitive to the small variations of the mechanical properties of the material induced by different heat treatment conditions.

4.4 USE OF NONLINEAR LAMB WAVES FOR NONDESTRUCTIVE EVALUATION

4.4.1 Phase Matching for Nonlinear Lamb Wave Experiments

Single primary modes in a wave-guide can generate multiple secondary modes. However, attention is given on the Lamb modes that have same phase and group velocities as the primary mode. This is because only the wave mode with proper phase matching and group velocity matching with the primary mode survives after it propagates a certain distance, while all other modes decay due to destructive interference with one another. Cumulative effect grows the phase matched higher harmonic modes up to a point, then it starts to decay due to the material attenuation. Some primary and secondary Lamb wave modes which satisfy the phase matching condition are shown in Figure 4.8. Two sets of modes, marked by circular and star markers in Figure 4.8, satisfy the phase matching conditions and have the potential to generate cumulative higher harmonic waves. For example, at 2.8 km/s phase velocity the S_0 mode has a frequency close to 2 MHz while for the A_1, S_1 and A_2 modes these frequencies are close to 4, 6 and 8 MHz, respectively. Similarly, at 4.4 km/s phase velocity the S_1 mode has a frequency close to 3 MHz while for the S_2 and S_3 modes these frequencies are close to 6 and 9 MHz, respectively. Therefore, if A_1 is used as the primary wave, then the second harmonic will be generated as the A_2 wave mode. Note that the slope of the phase velocity dispersion curves of the A_1 and A_2 modes at the marked positions will also have to be the same to have the same group velocities

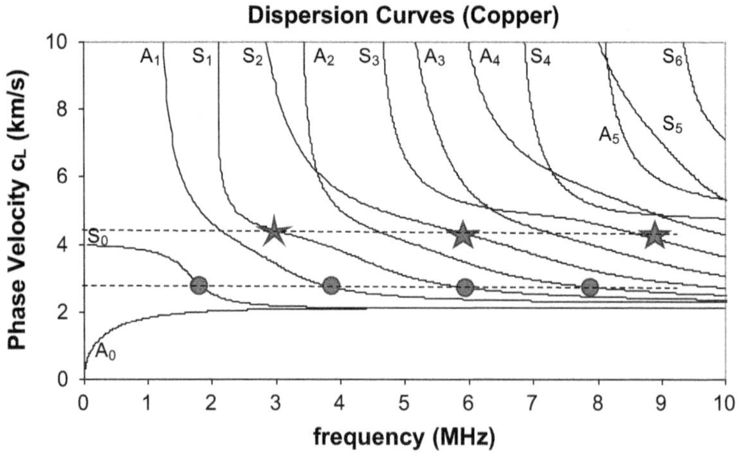

Figure 4.8 Some phase matching guided wave modes that can serve as primary and higher harmonic modes in a copper plate are shown by circular and star markers on the dispersion curve plot. Slopes of the dispersion curves at the marked positions will have to be the same to have matching group velocity for these waves to propagate together and generate measurable higher harmonic amplitudes utilizing the cumulative effect.

of the primary and secondary waves to have the cumulative property of the generated second harmonic. Similarly, the S_1 mode with 4.4 km/s phase velocity can be used as the primary mode when measurable higher harmonics are observed in the S_2 and S_3 modes, provided these three modes have the same group velocity at the marked positions.

4.4.2 Experimental Results

Li et al. (2012) investigated effects of thermal fatigue on unidirectional and symmetric quasi-isotropic carbon/epoxy laminates with stacking sequences $[0]_6$. Specimens were 1 mm thick. The two other dimensions were 400 mm×400 mm. The specimens were subjected to thermal fatigue loading between 70°C and −55°C with a constant cooling and heating time of 15 minutes to cause thermal degradation. The test specimens were subjected to 100, 200 and 1000 thermal cycles.

The S_1 mode at frequency 2.25 MHz and the S_2 mode at 4.5 MHz had the same phase and group velocities (9.6 km/s) and were used as the primary mode and the higher harmonic mode. The ratio of the second harmonic amplitude and the square of the primary wave mode amplitude (A_2 / A_1^2) were then plotted as a function of the wave propagation distance x in Figure 4.9. The normalized second harmonic amplitude grows with the propagation distance because of the cumulative effect.

Nonlinear parameter $\hat{\beta}$ is obtained from the slope of the line shown in Figure 4.9 as [also see Eq. (4.57)]

$$\hat{\beta} = \frac{\bar{\beta}_2 - \bar{\beta}_1}{x_2 - x_1} \tag{4.59}$$

When the slope of the acoustic nonlinearity with wave propagation distance is used to represent the relative acoustic nonlinearity parameter it minimizes the effect of other nonlinearity arising from the couplant and the instrument. Figure 4.10 shows

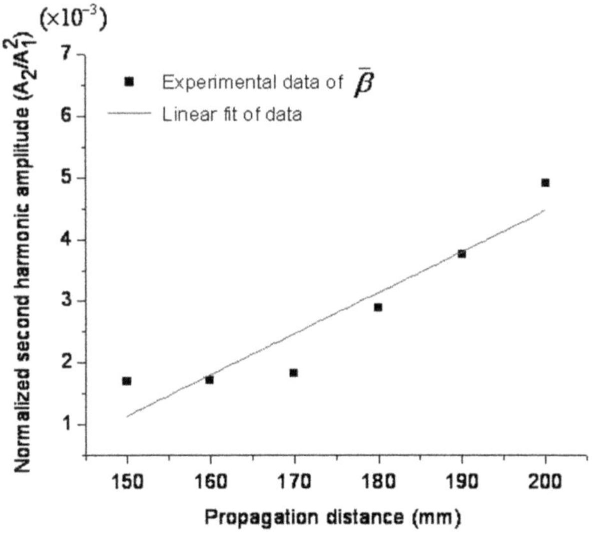

Figure 4.9 Variation of the normalized second harmonic amplitude as a function of the propagation distance of the Lamb wave [from Li et al. (2012)].

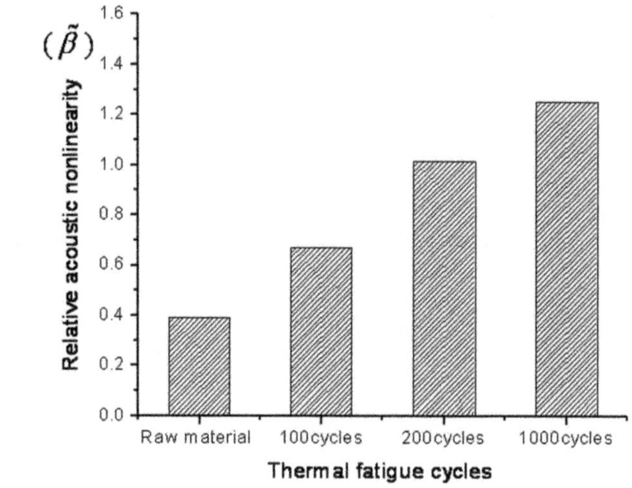

Figure 4.10 Relative acoustic nonlinearity [$\hat{\beta}$, see equation (4.59)] as a function of thermal cycles [from Li et al. (2012)].

how the thermal fatigue cycle number is related to the relative nonlinear parameter. As the number of thermal cycles increases, the values of the acoustic nonlinearity change significantly.

The sensitivities of the linear and nonlinear acoustic parameters to the thermal cycles are shown in Figure 4.11. All measured parameters are normalized

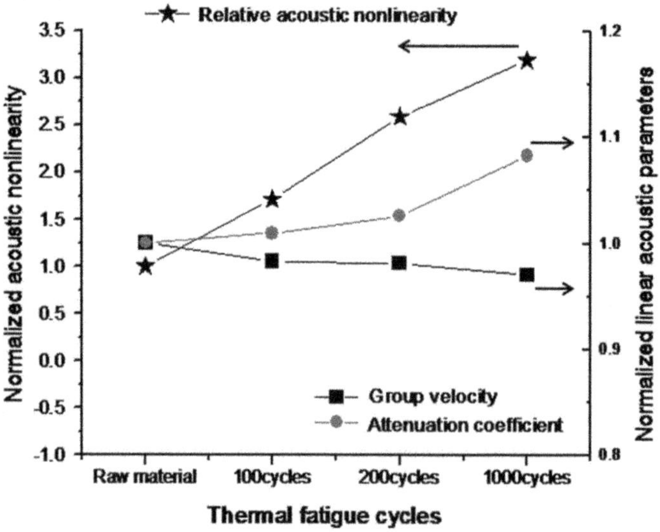

Figure 4.11 Comparison of the sensitivity of acoustic linear and nonlinear ultrasonic parameters to thermal fatigue damage in composite specimens; the acoustic linear parameters (group velocity and attenuation coefficient) and nonlinearity parameter are normalized with respect to their values in the pristine or undamaged condition [from Li et al. (2012)].

with respect to their values in the pristine condition to display only the relative changes. Note that with the increasing number of thermal cycles, the wave velocity decreases slightly while the attenuation and the nonlinearity parameter increase. The changes of the linear parameter values are not as significant as those changes for the nonlinear parameter value. For example, after one hundred cycles of thermal fatigue the micro-damage accumulated in the specimen shows insignificant change on the linear parameters (velocity and attenuation) but noticeable change on the nonlinear parameter (over 50%). These results show that in comparison to the linear parameters the nonlinear ultrasonic parameter has a significantly higher sensitivity to early stages of damage.

4.5 NONLINEAR RESONANCE TECHNIQUE

For nonlinear materials the resonance frequency and attenuation depend on the excitation amplitude, unlike linear materials for which these two parameters are independent of the excitation amplitude. The resonance frequency shift and attenuation variations with increasing amplitude of excitation are monitored in the nonlinear resonance techniques. A downward resonance peak shift and increase of attenuation with increasing excitation amplitude are observed in nonlinear materials. The resonance frequencies of a specimen can be determined by hitting the specimen with an impactor which can be, for example, an instrumented hammer. Multiple resonance spectra are recorded by continuously increasing the excitation amplitude in consecutive strikes. This technique is known as Nonlinear Impact Resonance Acoustic Spectroscopy, or NIRAS (Van Den Abeele et al., 2000).

Eiras et al. (2013) used this technique to monitor randomly distributed fiber reinforced mortar specimens under an accelerated ageing process. Fibers give the composite mortar specimens higher fracture toughness and also introduce some nonlinearity in its response, probably due to the additional interfaces between the fibers and the mortar matrix. After accelerated ageing, many fibers were damaged and weakened, and some even disappeared, making the composite specimen less tough, but its response became more linear because the interface area between the fibers and the mortar matrix was reduced – the interfaces were effectively replaced by cavities. Figure 4.12 shows five different vibrational modes – FLEX1, TOR1, FLEX2, TOR2 and TOR3 – that were generated when the mortar composite plate specimens were hit by an instrumented hammer. FLEX and TOR stand for flexural and torsional modes, while the numbers 1, 2 and 3 indicate first, second and third vibrational modes, respectively. Figure 4.13 shows recorded resonance peaks for a pristine or unaged specimen (continuous lines) and after it was aged by soaking in a thermal bath for 150 hours (dotted lines). Five peaks that can be seen in this figure correspond to five vibrational modes shown in Figure 4.12. Since the specimen was hit by an instrumented hammer multiple times with increasing intensity, multiple curves

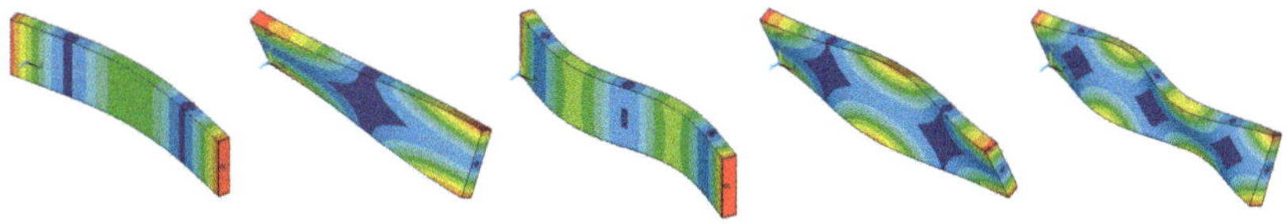

Figure 4.12 Five different vibrational mode shapes generated in a plate specimen when it is hit by an instrumented hammer – mode shapes are shown from left to right FLEX1, TOR1, FLEX2, TOR2 and TOR3. FLEX and TOR correspond to flexural and torsional modes, respectively [from Eiras et al. (2013)].

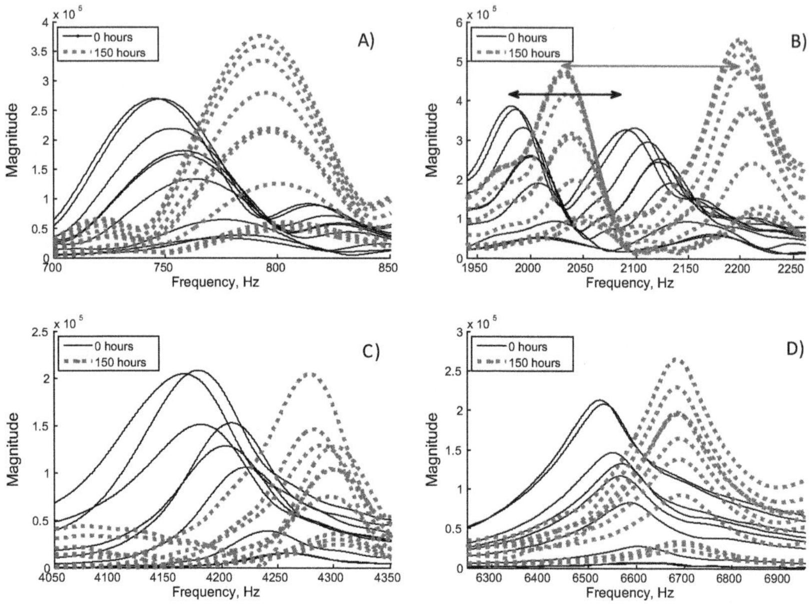

Figure 4.13 Resonance peaks for plate specimens in pristine condition and after 150 hours of accelerated ageing for 5 vibrational modes shown in Figure 4.12 which are A) FLEX1, B) TOR1 and FLEX2, C) TOR2 and D) TOR3 [from Eiras et al. (2013)].

with increasing peak values can be seen in this figure for every vibrational mode. Note that for the pristine specimen (continuous lines) the resonance frequency goes down significantly as the amplitude of excitation increases, while for the aged specimen it goes down only slightly. Figure 4.14 shows the shifts in the resonance frequency of the first flexural mode as a function of the amplitude of excitation. Experimental points and best fitted straight lines are plotted for ageing times – after 0, 40, 80, 120 and 150 hours of soaking in the thermal bath. In this plot one can clearly see that as the fiber reinforced mortar composite specimens age their response becomes more linear.

Eiras et al. (2014) simplified the NIRAS technique by hitting the specimen only once and collecting and analyzing the data from different regions of the time history plots as the vibration amplitude decays with time. Since the intensity of vibration goes down with time naturally, multiple impacts are not necessary for recording the material response for different intensities of excitation. One can call this technique Single Impact NIRAS, or SI-NIRAS. Figure 4.15 illustrates the SI-NIRAS technique. The left column of this figure shows the vibration signals generated by hitting a Portland cement mortar specimen in its pristine condition (top row) and after it was subjected to 10 freezing–thawing cycles (middle row) and 15 cycles (bottom row). All time signals show the natural vibration amplitude going down with time. Fourier transforms of the part of the signal shown within the dotted window were taken and plotted in the middle column. Then the window was gradually slided from left to right and Fourier transforms of the part of the signal within every position of the window were taken and plotted in the middle column. This process was continued until the recorded signal within the window became negligibly small. Final positions of the window

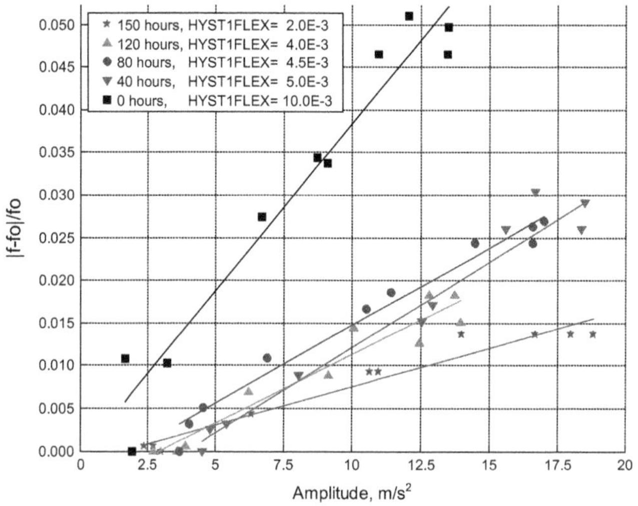

Figure 4.14 Resonance frequency shift for the first flexural mode (FLEX1) as a function of the impact energy level for different ageing times [from Eiras et al. (2013)].

where the calculations were stopped are shown in the left column by the right windows (drawn with continuous lines). A total of 50 Fourier spectra were taken from 50 segments of every vibration signal as the window stopped at 50 positions during its slide from left to right. The FFT spectra for all 50 window positions are shown in the middle column. The right column shows the plot of the peak frequency as a function of the window position. Since the 50 positions of the window 1 to 50 corresponds to increasing time, the time window positions plotted along the horizontal axis of the third column can simply be considered as the time axis as well. Middle and right columns show that the resonance frequency almost remains unchanged with time for the pristine specimen but for the damaged specimens the resonance frequency increases with time as the intensity of vibration goes down. The specimen subjected to 15 cycles shows more nonlinear response compared to the specimen subjected to 10 cycles of freezing and thawing.

Two examples presented in section 4.5 show that in the first case the pristine sample has higher nonlinear response and in the second case it has lower nonlinear response compared to the damaged samples. Therefore, depending on what specimen is being investigated and how it is being damaged, the pristine specimen can show a higher or lower level of nonlinearity compared to the damaged specimen.

4.6 PUMP WAVE AND PROBE WAVE BASED TECHNIQUE

Material nonlinearity can be also monitored by exciting the specimen simultaneously by acoustic waves of two different frequencies – a pump frequency and a probe frequency. Different variations of this technique exist and have been discussed in Kundu et al. (2019). The most popular version is known as the Nonlinear Wave Modulation Spectroscopy (NWMS). This technique excites the specimen using a high frequency of low amplitude probe wave (of frequency f_{probe}), and a low frequency of high amplitude pump wave (of frequency f_{pump}). The pump wave is typically generated by vibrating a specimen at one of its

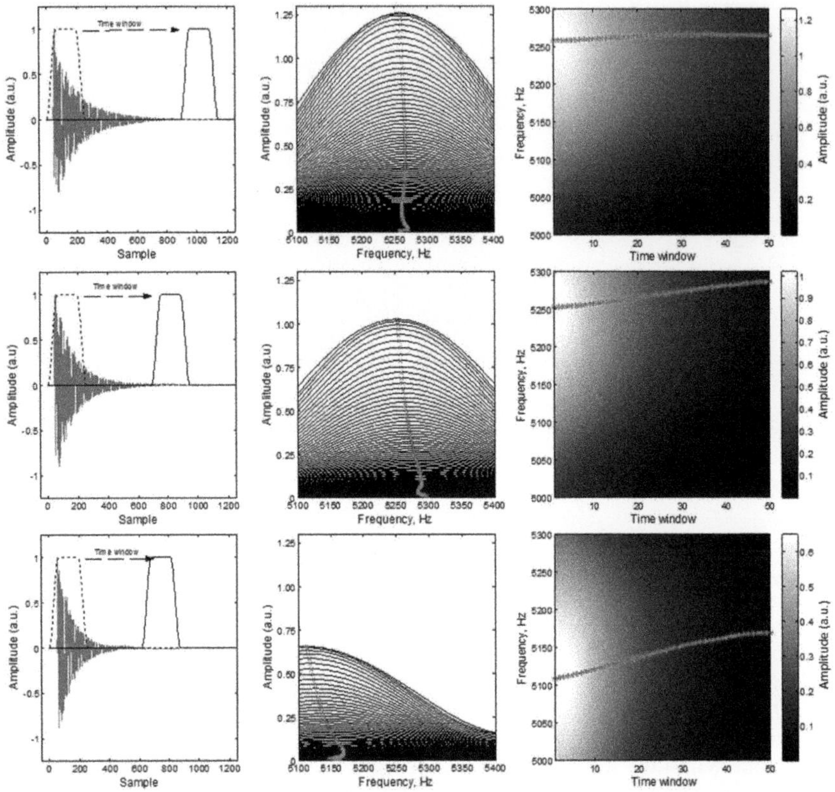

Figure 4.15 Illustration of the SI-NIRAS technique – time signals (left column), extracted windowed spectra (center column) and stacked spectra plotted in gray scale as a function of the window position (right column) for 0 cycle (top row), 10 cycles (middle row) and 15 cycles (bottom row) of freezing–thawing. Points on the right column indicate the frequencies corresponding to the peak amplitude positions [from Eiras et al. (2014)].

resonance modes. The pump vibration induces stresses within the sample, while the probe wave senses the variation of the elastic modulus produced by the pump vibration. For linear materials, the frequency of the probe wave remains unchanged while propagating through the vibrating material. However, in a nonlinear material the frequency of the probe wave is modulated by the pump wave excitation. The interaction of the two waves – with frequencies f_{probe} and f_{pump} – produces sidebands in the frequency spectra of the received signal, as explained analytically in section 4.2.3.

Compared to the higher harmonic generation technique, the modulation technique offers some advantages [Donskoy and Sutin (1998)]. First, higher harmonic generation requires a homogeneous path between the transmitter and the receiver to take advantage of the cumulative effect. It might be difficult to satisfy this requirement in the presence of reflecting boundaries and other structural inhomogeneities. Second, high voltages are needed to generate high amplitude fundamental modes and detectable higher harmonic modes. High voltage excitation can add some nonlinear background signal, which may affect the sensitivity of the higher harmonic technique.

4.7 SIDEBAND PEAK COUNT (SPC) TECHNIQUE

Instead of generating sidebands by nonlinear interaction between a pump wave and a probe wave, in the SPC technique interaction between multiple waves propagating at different frequencies are used to generate multiple sidebands. Peaks of these sidebands are then monitored. These multiple waves can be either bulk waves or guided waves propagating at different frequencies. In a plate if the Lamb wave is generated by a broadband excitation [such as a laser beam striking a plate (see Figure 4.16)] then multiple Lamb modes are generated in the plate over a wide frequency range. In a nonlinear plate the interaction between various Lamb modes can generate multiple sidebands. Instead of measuring the amplitudes of these sidebands, in the SPC technique the number of frequency peaks of the sidebands above a threshold level are counted. The SPC technique was first introduced in Eiras et al. (2013).

Figure 4.17 illustrates this technique. If waves of multiple frequency pass through a linear material then their frequency values do not change. The left spectral plot of Figure 4.17 shows three distinct peaks that correspond to three different frequencies of the input signal. The input signal can be bulk waves of three different frequencies in a bulk material or three different guided wave modes having three different frequencies. In a linear material peaks at the spectral plot of the output signal appear at the same frequencies as those of the input signal spectrum, but peak amplitudes vary due to material attenuation and scattering as shown in the top-right plot. In a nonlinear material, however, interactions between these waves of different frequency produce additional peaks due to the frequency modulation effect as shown in the bottom-right spectral plot of Figure 4.17. Typically, amplitudes of the additional peaks generated by the frequency modulation effect are much smaller than the major peaks, often less than 1 to 2% of the highest peak value. The SPC plot is generated by moving a threshold line, shown by the horizontal dashed line in the bottom-right plot. The threshold line is moved vertically between two pre-set values, for example,

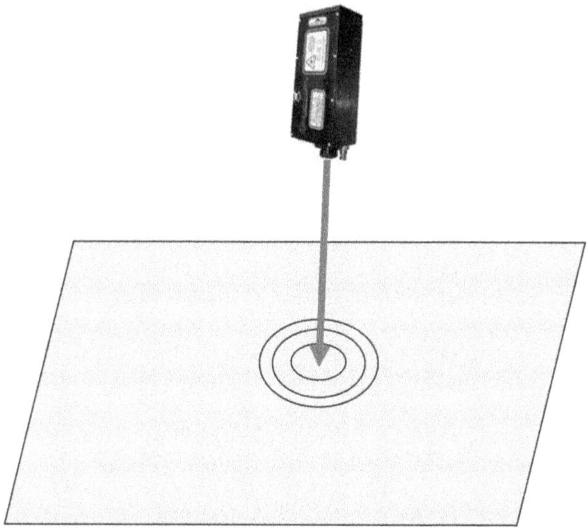

Figure 4.16 A laser gun striking a plate generates multiple Lamb modes in the plate from this broadband excitation.

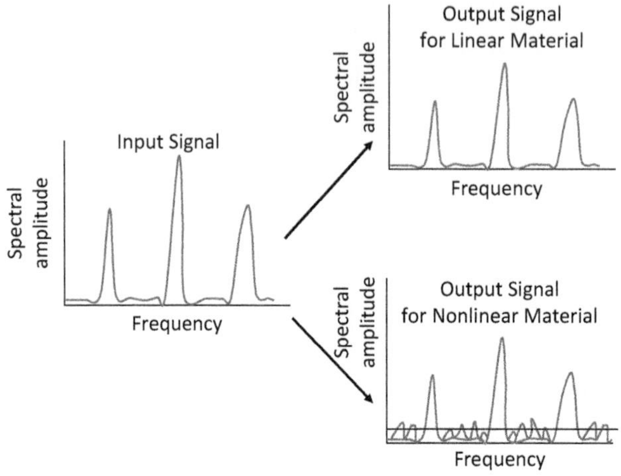

Figure 4.17 Left diagram shows a typical spectral plot of the input signal. Top-right diagram shows a typical spectral plot of the output signal for a linear elastic material. Spectral plots for both input and output signals show peaks at the same frequencies when the material is linear. Although peak positions do not change, their amplitudes may vary due to attenuation and scattering of waves as the waves propagate through the material; bottom-right figure shows additional small peaks (called sideband peaks) appearing in the spectral plot of the output signal for a nonlinear material. Peaks appearing above the threshold line (the horizontal dashed line shown in the bottom-right figure) are counted in the SPC technique.

between 0 and 2% of the highest peak value, and all peaks that are above this threshold line but below the 2% line are counted and plotted against the moving threshold value. This SPC plot (number of peaks as a function of the threshold value) indicates the degree of material nonlinearity. A nonlinear material should give a higher value of SPC compared to a linear material.

Hafezi et al. (2017) have theoretically generated SPC plots using a peri-ultrasound modeling technique, which is based on a peridynamic modeling principle applied to ultrasonic problems. They showed that when elastic waves propagate through cracked and un-cracked structures, the SPC plot for a structure containing a thin crack is greater than that for a structure containing a thick crack or no crack, as shown in Figure 4.18. This is because compared to a thick crack more areas of the opposing surfaces of a thin crack come into contact when elastic waves pass through the crack. For a thick crack, it's possible that only the regions near the crack tips can come into contact and generate a slight nonlinear response of the material as the wave passes through it, but for a thin crack opposing crack surfaces near the crack tip as well as away from the crack tip can come into contact, thus generating a nonlinear response. A linear elastic material having no crack should show no nonlinear behavior and ideally the SPC values generated for a crack-free linear elastic material should be zero.

4.7.1 Experimental Evidence of SPC Measuring Material Nonlinearity

The SPC technique was first successfully applied to monitor the material nonlinearity variation in glass-fiber reinforced cement (GRC) composite specimens subjected to an accelerated ageing process by Eiras et al (2013). Liu et al. (2014)

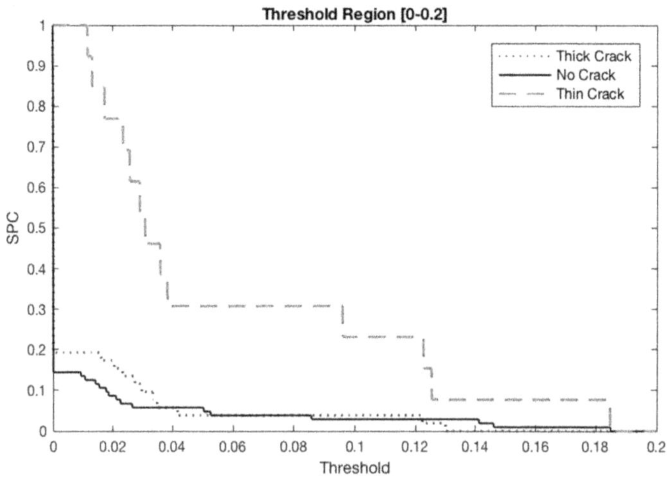

Figure 4.18 SPC plots for cracked and un-cracked structures. The structure containing a thin crack shows the highest SPC values indicating maximum nonlinearity. The diagram shows the SPC plot for a structure containing a thin crack much greater than those for structures containing a thick crack or no crack [from Hafezi et al. (2017)].

later applied this technique to monitor fatigue cracks in aluminum plates and aircraft fitting-lugs with complex geometries.

For non-contact monitoring of the damage induced nonlinearity in a plate, a pulse laser beam was shot on the target structure to create ultrasonic waves and the response was measured by a laser Doppler vibrometer (LDV) [Liu et al. (2014)]. When a pulse laser beam strikes a small area of the specimen, a localized heating of the surface causes a thermo-elastic expansion of the material and creates ultrasonic waves [Scruby and Drain (1990)].

The SPC technique requires a broadband excitation of the structure. Broadband excitation can generate multiple Lamb modes in a plate and interaction between these modes can produce multiple sidebands in a nonlinear material. As the material nonlinearity increases the number of measurable sideband peaks above a threshold value also increase. Therefore, from the sideband peak counts the degree of material nonlinearity can be estimated in a qualitative manner.

Spectral plots from the recorded time history show multiple peaks, as shown in Figure 4.17. These peaks correspond to different Lamb modes propagating in a plate. In a plate damaged due to the nonlinearity introduced in the plate by internal damages, multiple sidebands, as shown in the bottom-right plot of Figure 4.17, are generated. If the number of peaks above a threshold value is counted, then this number should be higher for a nonlinear material compared to a linear material as long as the threshold value is small enough to count the sideband peaks. When the SPC values are plotted as a function of the threshold value, then they continuously decrease, as shown in Figures 4.18 (theoretical plot) and 4.19 (experimental plot). Figure 4.19 shows how the SPC decreases with increasing threshold value for three different plate specimens made of GRC composites before and after ageing. These specimens were aged artificially by soaking them in hot water for several hours [Eiras et al. (2013)]. Although there are significant variations in the experimental results for the three specimens, the

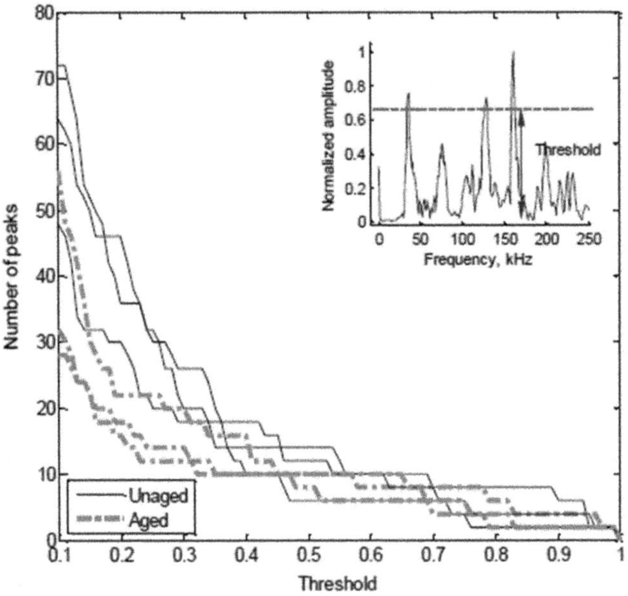

Figure 4.19 SPC technique applied to glass-fiber reinforced cement (GRC) composite specimens. For a specific value of the threshold the peak count is higher for the unaged material indicating the unaged material is more nonlinear than the aged material. Other nondestructive and destructive tests confirmed this observation. Three curves for aged and unaged samples were obtained from three different specimens [Eiras et al. (2013)].

SPC values clearly show a noticeable decrease as the three specimens were aged. Mechanical destructive tests and NIRAS tests (discussed in the previous section) confirmed that with ageing these samples became more brittle and weaker but their stress–strain behavior became more linear. Experimental results of Figure 4.19 and theoretical plots of Figure 4.18 look qualitatively similar.

Instead of simply counting the side band peaks above a threshold value, Liu et al. (2014) defined normalized *SPC* as the ratio of the number of peaks (N_{peak}) above a moving threshold (*th*) to the total number of peaks (N_{total}). While counting the peaks, they did not distinguish the small and large peaks, arguing that most of the peaks are small sideband peaks, as shown in the bottom-right plot of Figure 4.17, therefore a few large peaks corresponding to the primary wave modes should not significantly change the total peak counts. Thus, according to Liu et al. (2014), the normalized *SPC* count above a threshold value is defined as

$$SPC = \frac{N_{\text{peak}}(th)}{N_{\text{total}}} \tag{4.60}$$

where all peaks – the dominant peaks as well as the sideband peaks above the threshold level – are counted, as shown in Figure 4.20a. Note that this new definition of normalized *SPC* is different from the earlier definition of SPC. To highlight this difference, normalized *SPC* is written in italics as "*SPC*". In this definition, the value of *SPC* should be 1 for threshold value equal to 0, since the total number of peaks $[N_{\text{total}}]$ and the number of peaks above the threshold $[N_{\text{peak}}(th)]$ should be the same. When the material is degraded or damaged,

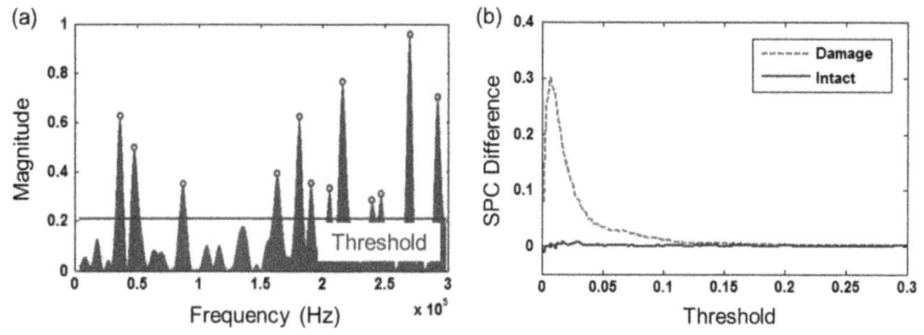

Figure 4.20 Illustration of normalized *SPC* and *SPC*-difference plots: (a) normalized *SPC* is defined as the ratio of the number of frequency peaks above a moving threshold to the total number of peaks in the spectral plot; (b) plot of the *SPC*-difference, which is defined as the difference between the *SPC* values obtained from the current damaged and initial intact stages over a varying threshold. It clearly shows a distinct bump for the damaged material when the threshold is relatively low, indicating the presence of more small peaks in the spectral plot of the recorded signal for the damaged specimen [Liu et al. (2014)].

it often becomes more nonlinear, and as a result more sideband peaks are observed in the spectral plots provided the defect sizes are small. Figure 4.18 shows how small defects increase the degree of nonlinearity but big defects like wide open cracks and cavities do not. Therefore, the *SPC* value for the damaged case should be larger than that for the intact case in general when the threshold value is small but not zero. For a threshold value equal to zero the *SPC* is 1 for both damaged and undamaged specimens. For a large threshold value, on the other hand, the *SPC* should have a very small value, close to 0, for both damaged and undamaged specimens. For intermediate threshold positions *SPC* values should be between 0 and 1 and these values should be higher for the damaged specimen than the pristine specimen in general. Therefore, if one plots the *SPC*-difference values for damaged and undamaged specimens, then it should show 0 for normalized threshold value equal to 0 as well as 1 (threshold is normalized with respect to the peak spectral value). In between these two extreme values (0 and 1) the *SPC*-difference plot should show a bump. The bigger the bump, the more nonlinear the specimen is.

Figure 4.20b shows a typical *SPC*-difference plot obtained experimentally for intact and fatigued specimens. In this figure one can see that the *SPC*-difference is positive for the damaged material and the maximum *SPC*-difference appears when the normalized threshold value is around 1.5% of the largest peak in the spectral plot.

4.8 CONCLUDING REMARKS

As mentioned before, a stress–strain curve that looks linear to the naked eye may not be linear. A nonlinear curve that has a very large radius of curvature looks linear when a small part of the nonlinear curve is plotted. In such a material even a small amplitude wave can cause a very small nonlinear response of the material. The challenge is how to detect such a small nonlinear response in the material. Various nonlinear acoustic/ultrasonic techniques discussed in this chapter (higher harmonic generation, frequency modulation, sideband peak count and resonance acoustic spectroscopy, to name a few) can detect minute nonlinear responses in a material. The SPC technique is relatively new and simple to use.

Detecting the nonlinearity in the material response is of interest because in this manner one can detect the early stages of material degradation, that is, when the cracks and dislocations start to form and introduce nonlinearity in the material behavior. Nonlinear ultrasonic techniques are of little use after the cracks have been well developed. For fully developed macro-cracks the crack surfaces are generally well-separated and remain separated when the waves propagate through the material. It should be noted that simply the presence of cracks does not make the material nonlinear. It is the clapping effect (opposing surfaces of the cracks coming into contact with each other and then moving away as the waves propagate) and other nonlinear effects such as heat generation due to crack surfaces rubbing against each other that make the material nonlinear and generate nonlinear ultrasonic responses. Closing and opening of the crack tips for macro-cracks when the elastic waves pass through the material can also create a nonlinear response in the material, but this response is not as strong as that observed for distributed micro-cracks.

REFERENCES

Donskoy, D. M. and A. M. Sutin, "Vibro-acoustic modulation nondestructive evaluation technique", *Journal of Intelligent Material Systems and Structures*, Vol. 9(9), pp. 765–771 (1998).

Eiras, J. N., J. Monzó, J. Payá, T. Kundu, and J. S. Popovics, "Non-classical nonlinear feature extraction from standard resonance vibration data for damage detection", *The Journal of the Acoustical Society of America*, Vol. 135(2), pp. EL82–ELEL87 (2014).

Eiras, J. N., T. Kundu, M. Bonilla, and J. Payá, "Nondestructive monitoring of ageing of alkali resistant Glass Fiber Reinforced Cement (GRC)," *Journal of Nondestructive Evaluation*, Vol. 32(3), pp. 300–314 (2013).

Hafezi, M. H., R. Alebrahim, and T. Kundu, "Peri-ultrasound for modeling linear and nonlinear ultrasonic response," *Ultrasonics*, Vol. 80, pp. 47–57 (2017).

Herrmann, J., J. Kim, L. J. Jacobs, J. Qu, J. W. Littles, and M. F. Savage, "Assessment of material damage in a nickel-based superalloy using nonlinear rayleigh surface waves", *Journal of Applied Physics*, Vol. 99(12), p. 124913 (2006).

Kundu, T., ed., *Nonlinear Ultrasonic and Vibro-Acoustic Techniques for Nondestructive Evaluation*, Springer (2019).

Kundu, T., J. N. Eiras, W. Li, P. Liu, H. Sohn, and J. Payá, Chapter 1 "Fundamentals of nonlinear acoustical techniques and sideband peak count", In *Nonlinear Ultrasonic and Vibro-Acoustical Techniques for Nondestructive Evaluation*, ed. T. Kundu, ASA Press, Springer Nature, Switzerland (2019).

Li, W., and Y. Cho,"Combination of nonlinear ultrasonics and guided wave tomography for imaging the micro-defects," *Ultrasonics*, Vol. 65, pp. 87–95 (2016).

Li, W., Y. Cho, and J. D. Achenbach, "Detection of thermal fatigue in composites by second harmonic lamb waves", *Smart Materials and Structures*, Vol. 21(8), p. 085019 (2012).

Li, W., Y. Cho, J. D. Achenbach, and J. D. Achenbach, "Assessment of heat treated inconel X-750 alloy by nonlinear ultrasonics", *Experimental Mechanics*, Vol. 53(5), pp. 775–781 (2013).

Liu, P., H. Sohn, T. Kundu, and S. Yang, "Noncontact detection of fatigue cracks by Laser Nonlinear Wave Modulation Spectroscopy (LNWMS)", *NDT and E International*, Vol. 66, pp. 106–116 (2014).

Scruby, C. B. and L. E. Drain, *Laser Ultrasonics: Techniques and Applications*, Taylor and Francis, London, UK (1990).

Van Den Abeele, K. E. -A., J. Carmeliet, J. A. Ten Cate, and P. A. Johnson, "Nonlinear Elastic Wave Spectroscopy (NEWS) techniques to discern material damage, part II: Single-mode nonlinear resonance acoustic spectroscopy" *Research in Nondestructive Evaluation*, Vol. 12(1), pp. 31–42 (2000).

5 Acoustic Source Localization

5.1 INTRODUCTION

Localization of acoustic source is necessary in pinpointing the region of crack initiation in a structure, or identifying the point where a structure may have been hit by a foreign object. Both crack formation and strike of an object generate acoustic waves that propagate through the structure in different directions as bulk waves and/or guided waves. These propagating waves are recorded and analyzed by various acoustic source localization (ASL) techniques to identify the source of the propagating waves. Therefore, ASL is important for any autonomous system built for structural health monitoring. Developing more efficient ASL techniques is a topic of active research which is continuously evolving. This chapter discusses different ASL techniques developed in last few decades.

To localize the acoustic source one needs to record the time of arrivals (TOA) of the elastic waves at sensor locations and calculate the time difference of arrivals (TDOA) between various sensors. Different techniques for localizing an acoustic source, their advantages as well as limitations, are discussed in the following. This chapter is written based on a review article published earlier by the author (Kundu, 2014). Newer developments since the publication of that article are also included here.

Different ASL techniques have been proposed in the literature. Some techniques are restricted to isotropic materials and some can be applied to both isotropic and anisotropic materials. Some techniques require precise knowledge of the direction dependent velocity profiles in the anisotropic body while other techniques do not need this information. Some methods require accurate values of the TOAs of the acoustic waves at the receiver locations while other techniques can function without that information. Most published papers that introduce a new ASL technique emphasize the advantages of the introduced technique while hardly mentioning the limitations of the new technique. For this reason a comprehensive review and comparisons of the available techniques were given by Kundu (2014) in his review article.

One can encounter acoustic sources of different types. The source type depends on how the acoustic waves are generated. For example, acoustic waves can be generated by (1) impact of a foreign object, (2) crack formation, such as matrix cracking and delamination in a composite material, or (3) structural element failure, such as cable failure in a bridge, failure of rebars in reinforced pre-stressed concrete structures, or fiber breakage in a composite material. The process of locating the source of the generated acoustic waves, by recording and analyzing the propagating acoustic signals is commonly known as the *acoustic source localization* technique. It is an important step of *structural health monitoring* (SHM) because it pinpoints the region of the structure that should be inspected carefully for any possibility of damage initiation or propagation at that location.

Two components of SHM that have received significant attention from the research community are *Diagnosis* and *Prognosis*. Diagnosis is the characterization of the damage – measuring its size, location and orientation. It is necessary to get a clear idea about the severity of the damage. Nondestructive testing and evaluation (NDT&E) is used for damage diagnosis. Prognosis estimates the remaining life of the structure. Knowledge of fracture mechanics and fatigue crack growth is necessary for predicting the remaining life of the structure under certain loading conditions and it is known as prognosis.

For large structures it is neither possible nor necessary to inspect every region of the structure with same level of attention; therefore, one needs to focus on certain regions that are more susceptible to damage initiation. These regions are commonly known as *hot spots*. Regions that are subjected to relatively high level of stress constitute the hot spot regions. However, sometimes the hot spot regions in a large plate or shell type structure cannot be predetermined. For example, the wing and the fuselage of an airplane or the outer surface of a space shuttle can have damage initiated at any point if it is struck by a foreign object, such as a piece of debris striking a space shuttle flying in space, a flying bird hitting an airplane fuselage, or a tool dropping on an airplane wing during regular maintenance and repair. After any such hit the structure should be inspected near the impact region. Therefore, this region must be identified by the ASL technique. Then more careful inspection of that region can be conducted to conclude if any significant damage occurred there.

Critical structures made of composite materials need to be monitored continuously, not only for detecting the point of impact of a foreign object but also for monitoring matrix cracking, fiber breakage and inter-laminar delamination. Ultrasonic sensors are used for this purpose. Ultrasonic transducers can work in two modes – and *passive* (Jata et al., 2007). For active mode inspection acoustic actuators generate ultrasonic signals that are then recorded by ultrasonic sensors (Giurgiutiu, 2003). However, in passive mode inspection the impacting foreign objects, fiber breaks or crack formation and propagation act as the acoustic source (Mal et al., 2003a,b). Ultrasonic sensors are placed on or inside the structure to efficiently receive ultrasonic signals from the acoustic source and monitor its condition (Wang and Chang, 2000; Kessler et al., 2002; Park and Chang, 2003; Manson et al., 2003; Köhler et al., 2004; Mal et al., 2005).

The passive monitoring techniques are discussed in this chapter. Early works on source localization were carried out by Tobias (1976) for isotropic structures and by Sachse and Sancar (1987) for anisotropic structures. Early attempts of locating the acoustic emission sources in anisotropic plates required the measurement of two dominant pulses in a waveform whose speeds of propagation, c_1 and c_2 were known, and the receiving sensors were to be placed as a sensor-array on the periphery of a circle or on two orthogonal lines (Sachse and Sancar (1987). Other restrictions of these early analyses were (Castagnede et al., 1989), (1) the order of the elastic symmetry of the solid is to be orthorhombic or higher, (2) the principal axes of the solid are to be known *a priori* and to be oriented along the coordinate axes of the specimen, (3) the sensors comprising the receiving array must be placed on the principal planes of the material. Although the first constraint condition is approximately satisfied for most engineering materials, it may not be true in some cases. Even widely used engineering materials like fiber reinforced composite solids may violate this condition. Note that although fiber reinforced composite solids are often assumed to be orthotropic or transversely isotropic materials, the non-uniform distribution of fibers may not make xz, yz or xy planes to be the planes of symmetry.

Since those early works many techniques for source localization have been proposed. These newer methods are reviewed in this chapter – starting with simple isotropic structures and ending with highly anisotropic structures having complex geometry.

5.2 SOURCE LOCALIZATION IN ISOTROPIC PLATES

5.2.1 Triangulation Technique for Isotropic Plates with Known Wave Speed

The triangulation technique (Tobias, 1976) is the most popular source localization technique for isotropic and homogeneous structures. This technique is illustrated with the help of Figure 5.1. Three sensors (denoted by 1, 2 and 3 in Figure 5.1) are attached to a plate. If the plate has an acoustic source at point P then the wave generated by the acoustic source propagates through the plate and strikes the sensors at different times.

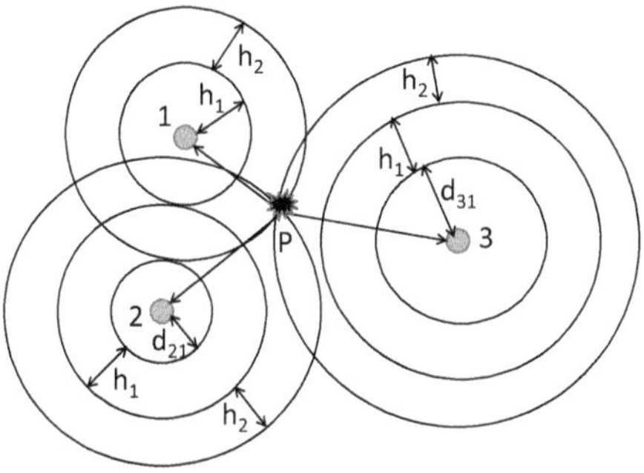

Figure 5.1 Triangulation technique – Three sensors placed at positions 1, 2 and 3 receive the acoustic waves generated by the source at position P. The radius of each circle corresponds to the distance traveled by the wave from the source to the sensor (Kundu, 2014).

For plate type structures the generated wave is the Lamb wave. Let us denote the times of travel of the wave to sensors 1, 2 and 3 as t_1, t_2 and t_3, respectively. Since the exact clock time (T_0) of the acoustic event (impact or crack formation) generating the acoustic waves is unknown it is impossible to obtain travel times t_1, t_2 and t_3 from the recorded arrival times or times of detection (clock times) T_1, T_2 and T_3 of the arriving waves at the three sensors. However, it is easy to see that

$$t_{ij} = t_i - t_j = (T_i - T_0) - (T_j - T_0) = T_{ij} \tag{5.1}$$

Thus, although clock times T_i and T_j are not equal to travel times t_i and t_j, respectively their differences T_{ij} and t_{ij} are same. If the wave speed in the plate is denoted by c then the distance traveled (d_i) by the wave to reach the i-th sensor is given by

$$d_i = c \times t_i \tag{5.2}$$

Since t_i is unknown, d_i cannot be obtained from Eq. (5.2). However, the difference between the distances d_i and d_j traveled by the waves going from the source point to the sensors i and j can be calculated as

$$d_{ij} = c \times t_{ij} \tag{5.3}$$

If the time of detection T_1 for sensor 1 is smaller than both T_2 and T_3, then one can say that additional distances traveled by the wave to go to sensors 2 and 3 compared to its path to sensor 1 are given by d_{21} and d_{31}, respectively. These values can be calculated from Eq. (5.3).

To obtain the acoustic source location P two circles with radii d_{21} and d_{31} whose centers coincide with the sensor locations 2 and 3, respectively, are first drawn (shown by dotted lines in Figure 5.1). Then the radii of these two circles are increased by h_1 and a third circle of radius h_1 is drawn about sensor 1. These three circles are shown by dashed lines in Figure 5.1. If the three circles do not intersect at a common point, then their radii are continuously increased by the same amount until all three circles intersect at a common point, as shown in Figure 5.1 by three solid circles. The radii of these three circles are equal to the distances (d_i) traveled by the acoustic waves propagating from the source to the sensors.

It should be noted here that if the wave velocity in the plate is increased from c to $2c$ then d_{21} and d_{31} is increased by a factor of 2, as shown in Figure 5.2, and

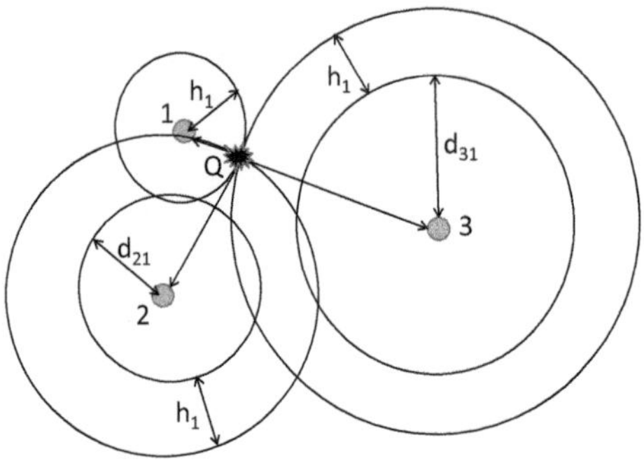

Figure 5.2 Triangulation technique – Similar to Figure 5.1, but in this figure the wave speed in the plate is twice of that for Figure 5.1 resulting a different location of the acoustic source for the same difference of time of flights for the three receiving sensors (Kundu, 2014).

following the same procedure as described above the predicted source point will move from point P in Figure 5.1 to point Q in Figure 5.2. Clearly, in this approach one needs to know accurately the wave speed in the plate to correctly localize the acoustic source from three sensors. Small variations in the wave speed estimation can significantly alter the predicted source location. This ambiguity in the source location prediction can be reduced by increasing the number of receiving sensors.

5.2.2 Triangulation Technique for Isotropic Plates with Unknown Wave Speed

When the wave speed (c) in the isotropic plate is unknown then also the triangulation technique works; however, then the following system of nonlinear equations must be solved to locate the acoustic source. In Figure 5.3 if the acoustic source is P then (Liang et al., 2013)

$$\begin{aligned} c \times t_{12} &= d_1 - d_2 \\ c \times t_{23} &= d_2 - d_3 \\ d_1 \sin\theta_1 &= d_2 \sin\theta_2 \\ d_1 \cos\theta_1 + d_2 \cos\theta_2 &= \sqrt{(x_1 - x_2)^2 + (y_1 - y_2)^2} = D_{12} \end{aligned} \tag{5.4}$$

From the acoustic source at P the distances d_1, d_2 and d_3 of the receiving sensors S_1, S_2 and S_3, respectively, and the angles θ_1 and θ_2 are shown in Figure 5.3. D_{12} is the distance between sensors 1 and 2. t_{12} and t_{23} are defined in Eq. (5.1). In the above equation set, although many unknowns appear – c, d_1, d_2, d_3, θ_1 and θ_2 – there are only three independent unknowns – the wave velocity c and the

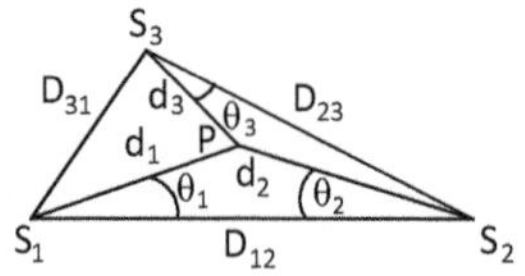

Figure 5.3 Three sensors at positions S_1, S_2 and S_3 form a triangle from which the term triangulation technique came (Kundu, 2014).

coordinates of the acoustic source x_0 and y_0. This is because unknowns d_1, d_2, d_3, θ_1 and θ_2 can be expressed in terms of x_0, y_0 and the known coordinates of the receiving sensors x_i, y_i (i=1,2, 3). This technique is the called triangulation technique because three sensors form a triangle, as shown in Figure 5.3.

5.2.3 Optimization Based Technique for Isotropic Plates with Unknown Wave Speed

The triangulation technique discussed in Section 5.2.1 works well for a plate having a known wave speed. If the wave speed is also unknown then a system of nonlinear equations needs to be solved to localize the acoustic source as described in Section 5.2.2. It makes the triangulation technique less attractive. An alternative formulation was presented by Kundu et al. (2008) that works equally well for isotropic and anisotropic plates and hence appears to be more attractive than the triangulation technique. This formulation presented below was obtained by specializing the general formulation given by Kundu et al. (2008) for source localization in anisotropic plate, to the isotropic case.

If the coordinates of the i-th receiving sensor S_i, are (x_i, y_i) and coordinates of the source are (x_0, y_0), then the distance of the i-th sensors from the impact point is given by

$$d_i = \sqrt{(x_i - x_0)^2 + (y_i - y_0)^2} \tag{5.5}$$

The time of travel t_i of the generated wave from the source to the i-th sensor can be related to the distance d_i of Eq. (5.5) in the following manner:

$$d_i = \sqrt{(x_i - x_0)^2 + (y_i - y_0)^2} = ct_i \tag{5.6}$$

In Eq. (5.6) c is the unknown wave speed. Eliminating the wave speed c from Eq. (5.6) one obtains

$$\begin{aligned} \frac{d_2}{d_1} = \frac{t_2}{t_1} = \frac{\sqrt{(x_2 - x_0)^2 + (y_2 - y_0)^2}}{\sqrt{(x_1 - x_0)^2 + (y_1 - y_0)^2}} = r_{21} \\ \frac{d_3}{d_1} = \frac{t_3}{t_1} = \frac{\sqrt{(x_3 - x_0)^2 + (y_3 - y_0)^2}}{\sqrt{(x_1 - x_0)^2 + (y_1 - y_0)^2}} = r_{31} \end{aligned} \tag{5.7}$$

Therefore,

$$\begin{aligned} &\left(\frac{d_2}{d_1}\right)^2 + \left(\frac{d_3}{d_1}\right)^2 = \left(\frac{t_2}{t_1}\right)^2 + \left(\frac{t_3}{t_1}\right)^2 = \frac{(x_2 - x_0)^2 + (y_2 - y_0)^2}{(x_1 - x_0)^2 + (y_1 - y_0)^2} + \frac{(x_3 - x_0)^2 + (y_3 - y_0)^2}{(x_1 - x_0)^2 + (y_1 - y_0)^2} \\ &\Rightarrow r_{21}^2 + r_{31}^2 = \frac{(x_2 - x_0)^2 + (x_3 - x_0)^2 + (y_2 - y_0)^2 + (y_3 - y_0)^2}{(x_1 - x_0)^2 + (y_1 - y_0)^2} \\ &\Rightarrow r_{21}^2 + r_{31}^2 = \frac{2\left(x_0^2 + y_0^2 - x_0x_2 - x_0x_3 - y_0y_2 - y_0y_3\right) + x_2^2 + x_3^2 + y_2^2 + y_3^2}{x_0^2 + y_0^2 - 2\left(x_0x_1 + y_0y_1\right) + x_1^2 + y_1^2} \end{aligned} \tag{5.8}$$

From Eq. (5.8) it is possible to define an error function $E(x_0, y_0)$, also known as the objective function in the optimization literature, in the following manner:

$$E(x_0,y_0)=\Big[2\big(x_0^2+y_0^2-x_0x_2-x_0x_3-y_0y_2-y_0y_3\big)+x_2^2+x_3^2+y_2^2+y_3^2-\big(r_{21}^2+r_{31}^2\big)\big(x_0^2+y_0^2-2(x_0x_1+y_0y_1)+x_1^2+y_1^2\big)\Big]^2 \tag{5.9}$$

It should be noted here that in Eq. (5.9) only the coordinates (x_0, y_0) of the source point are unknowns. If the assumed coordinate values (x_0, y_0) are different from that of the true source point, then a positive value of the error function $E(x_0, y_0)$ of Eq. (5.9) should be obtained. However, if the assumption is correct, then the error function should give a zero value. Therefore, one can obtain (x_0, y_0) by minimizing the error function expression given in Eq. (5.9). The expression given in Eq. (5.9) has the terms r_{21} and r_{31} that require the knowledge of the time of arrival $t_i = T_i - T_0$ [see Eqs. (5.1) and (5.7)] of the acoustic wave from the source to the i-th sensor. However, this value is not known since the clock time (T_0) of the acoustic event is unknown. For this reason it is desirable that the objective function be expressed in terms of relative times of arrival as defined in Eq. (5.1).

The relative times of arrival and their ratio is obtained from Eqs. (5.1) and (5.7):

$$\frac{t_{21}}{t_{31}}=\frac{\left\{\sqrt{(x_2-x_0)^2+(y_2-y_0)^2}-\sqrt{(x_1-x_0)^2+(y_1-y_0)^2}\right\}}{\left\{\sqrt{(x_3-x_0)^2+(y_3-y_0)^2}-\sqrt{(x_1-x_0)^2+(y_1-y_0)^2}\right\}} \tag{5.10}$$

Then the objective function $E(x_0, y_0)$ can be defined as

$$E(x_0,y_0)=\left(\frac{\left\{\sqrt{(x_2-x_0)^2+(y_2-y_0)^2}-\sqrt{(x_1-x_0)^2+(y_1-y_0)^2}\right\}}{\left\{\sqrt{(x_3-x_0)^2+(y_3-y_0)^2}-\sqrt{(x_1-x_0)^2+(y_1-y_0)^2}\right\}}-\frac{t_{21}}{t_{31}}\right)^2 \tag{5.11}$$

Ideally, for the correct values of (x_0, y_0) the objective function should give a zero value while for wrong values of (x_0, y_0) it should give a positive value. Therefore, one needs to minimize the value of this objective function. Note that in the above definition of the objective function the time t_1 has been used twice in computing t_{21} and t_{31} while t_2 and t_3 have been used only once. To give equal importance or bias to the three measured arrival times at the three sensor locations the error function can be defined in a different manner as shown below. With this definition of the objective function, the source point location prediction should not be biased toward any sensor.

$$\begin{aligned}E(x_0,y_0)=&\left(\left\{\frac{\sqrt{(x_1-x_0)^2+(y_1-y_0)^2}-\sqrt{(x_2-x_0)^2+(y_2-y_0)^2}}{\sqrt{(x_2-x_0)^2+(y_2-y_0)^2}-\sqrt{(x_3-x_0)^2+(y_3-y_0)^2}}\right\}-\frac{t_{12}}{t_{23}}\right)^2\\&+\left(\left\{\frac{\sqrt{(x_2-x_0)^2+(y_2-y_0)^2}-\sqrt{(x_3-x_0)^2+(y_3-y_0)^2}}{\sqrt{(x_3-x_0)^2+(y_3-y_0)^2}-\sqrt{(x_1-x_0)^2+(y_1-y_0)^2}}\right\}-\frac{t_{23}}{t_{31}}\right)^2\\&+\left(\left\{\frac{\sqrt{(x_3-x_0)^2+(y_3-y_0)^2}-\sqrt{(x_1-x_0)^2+(y_1-y_0)^2}}{\sqrt{(x_1-x_0)^2+(y_1-y_0)^2}-\sqrt{(x_2-x_0)^2+(y_2-y_0)^2}}\right\}-\frac{t_{31}}{t_{12}}\right)^2\end{aligned} \tag{5.12}$$

By some optimization scheme (such as simplex algorithm; Nelder and Mead (1965) or genetic algorithm (Barricelli, 1957, Fraser and Burnell, 1970) the acoustic source point (x_0, y_0) can be localized by minimizing the above objective function.

One shortcoming of Eq. (5.12) is that for certain values of the unknown (x_0, y_0) the denominator $\sqrt{(x_j - x_0)^2 + (y_j - y_0)^2} - \sqrt{(x_i - x_0)^2 + (y_i - y_0)^2}$ can vanish. For those values of (x_0, y_0) the objective function becomes infinity and special care must be taken during the computation of the objective function to avoid these singular points. This problem can be easily avoided by modifying the definition of the error function as given below:

$$E(x_0, y_0) = \begin{pmatrix} t_{23}\left\{\sqrt{(x_1 - x_0)^2 + (y_1 - y_0)^2} - \sqrt{(x_2 - x_0)^2 + (y_2 - y_0)^2}\right\} \\ -t_{12}\left\{\sqrt{(x_2 - x_0)^2 + (y_2 - y_0)^2} - \sqrt{(x_3 - x_0)^2 + (y_3 - y_0)^2}\right\} \end{pmatrix}^2$$
$$+ \begin{pmatrix} t_{31}\left\{\sqrt{(x_2 - x_0)^2 + (y_2 - y_0)^2} - \sqrt{(x_3 - x_0)^2 + (y_3 - y_0)^2}\right\} \\ -t_{23}\left\{\sqrt{(x_3 - x_0)^2 + (y_3 - y_0)^2} - \sqrt{(x_1 - x_0)^2 + (y_1 - y_0)^2}\right\} \end{pmatrix}^2 \quad (5.13)$$
$$+ \begin{pmatrix} t_{12}\left\{\sqrt{(x_3 - x_0)^2 + (y_3 - y_0)^2} - \sqrt{(x_1 - x_0)^2 + (y_1 - y_0)^2}\right\} \\ -t_{31}\left\{\sqrt{(x_1 - x_0)^2 + (y_1 - y_0)^2} - \sqrt{(x_2 - x_0)^2 + (y_2 - y_0)^2}\right\} \end{pmatrix}^2$$

or,

$$E(x_0, y_0) = \left(t_{23}\{d_1 - d_2\} - t_{12}\{d_2 - d_1\}\right)^2 + \left(t_{31}\{d_2 - d_3\} - t_{23}\{d_3 - d_1\}\right)^2 + \left(t_{12}\{d_3 - d_1\} - t_{31}\{d_1 - d_2\}\right)^2 \quad (5.14)$$

The predictions can be improved further by placing more receiving sensors on the structure. For n number of sensors there are $n(n-1)/2$ unique sensors pairs. To be unbiased, the error function needs to include information from every unique sensor pair and can be constructed in the following manner.

Considering all possible sensor pair combinations the objective function for a plate monitoring system with n receiving sensors is obtained as

$$E(x_0, y_0) = \sum_{i=1}^{n-1}\sum_{j=i+1}^{n}\sum_{k=1}^{n-1}\sum_{l=k+1}^{n}\left[t_{ij}(d_k - d_l) - t_{kl}(d_i - d_j)\right]^2 \quad (5.15)$$

5.2.4 Beamforming Technique for Isotropic Plates

The beamforming technique for ASL was introduced by McLaskey et al. (2010). It requires a small array of four to eight sensors. He et al. (2012) followed the beamforming technique for source localization in a thin steel plate. The original beamforming method proposed by McLaskey et al. (2010) and used by He et al.

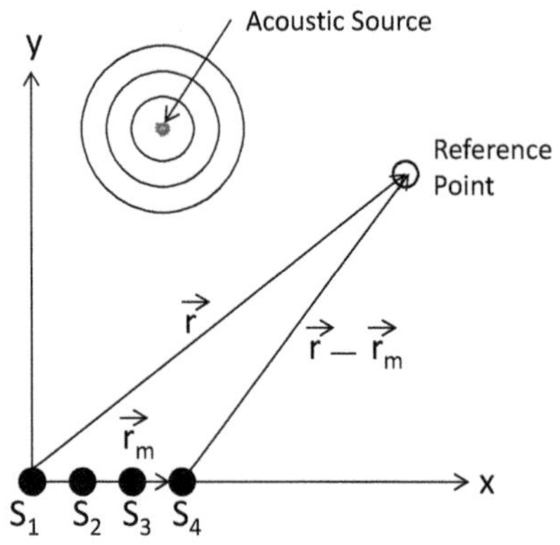

Figure 5.4 Four sensors at positions S_1, S_2 and S_3 and S_4 are used for the beamforming technique. The parameter b given in Eq. (5.16) attains the maximum value when the reference point coincides with the acoustic source point (Kundu, 2014).

(2012) assumed constant wave speed in all directions and therefore worked for an isotropic medium only. The basic principle of the beamforming technique is based on the delay-and-sum algorithm and is illustrated in Figure 5.4. This figure shows four sensors receiving acoustic signals generated by an acoustic event such as an impact. The array $b(\vec{r},t)$. is generated from M signals recorded by M receivers ($M=4$ in Figure 5.4), after applying time delays $\Delta_m(\vec{r})$ to the recorded signals and multiplying them with the weight factor w_m, as shown below.

$$b(\vec{r},t)=\frac{1}{M}\sum_{m=1}^{M}w_m x_m\left(t-\Delta_m(\vec{r})\right) \tag{5.16}$$

$$\Delta_m(\vec{r})=\frac{|\vec{r}|-|\vec{r}-\vec{r}_m|}{c} \tag{5.17}$$

Here $\vec{r}$ represents the distance of a reference point (also called the focal point) from the first sensor, which is denoted as S_1 in Figure 5.4. The variable x_m(t) represents the signal acquired by the m-th sensor S_m. $\Delta_m(\vec{r})$ is the individual time delay for the m-th sensor and c. is the propagation speed of the acoustic wave. For different reference points (or focal points) the time delays $\Delta_m(\vec{r})$ are different. When the reference point and the acoustic source point coincide then the delay-and-sum algorithm of Eq. (5.16) gives the maximum value of b because then the recorded signals are added in phase. All weight factors w_m can be taken as 1, or w_m can be equated to the normalizing factors so that the peak values recorded by every sensor be 1. The advantage of the beamforming technique is that it does not require the exact time of arrival of a specific wave mode and thus can handle noisy signals if the noise is Gaussian white noise.

5.2.5 Strain Rossette Technique for Isotropic Plates with Unknown Wave Speed

Matt and Lanza di Scalea (2007) and Salamone et al. (2010, 2011) exploited the directivity properties of MFC (Micro-Fiber Composite) sensors for ASL. Using three such sensors arranged as rosettes the principal strain directions were obtained. Betz et al. (2007) also obtained the principal strain directions from the strain rosettes constructed from the optical fiber Bragg grating (FBG) sensors. The wave propagation direction can be predicted from this strain rosette arrangement when that direction coincides with the principal strain direction, as is the case for an isotropic plate. The acoustic source in the isotropic plate can be localized from the intersection point of two wave propagation directions obtained from two rosettes made of a total of six MFC sensors. This technique is not influenced by the noise and the dispersion of waves propagating through the structure since the time of arrival of the wave is not required. However, since this technique requires that the principal strain direction must coincide with the group velocity (or energy velocity) direction obtained by connecting the source and the receiver with a straight line, it is not applicable for anisotropic plates.

5.2.6 Source Localization by Modal Acoustic Emission

Modal analysis of the propagation of elastic waves in a thin isotropic plate gives the dispersion characteristics of the propagating modes. From the group velocity dispersion curves one can calculate the time difference between the arrival times of the extensional and flexural plate wave modes at a sensor position for a certain distance between the acoustic source and the receiver (Jiao et al. (2004)). Therefore, in a thin plate it is possible to localize the source from the time difference between the arrival times of these two wave modes recorded by a single receiving sensor for a one-dimensional wave propagation problem (Surgeon and Wevers, 1999; Jiao et al., 2004) or by two sensors for two-dimensional wave propagation in a plate (Toyama et al., 2001). In this technique, since the waveform generated by the acoustic source is obtained theoretically, it is possible to obtain some information about the acoustic source in addition to its location. Gorman (1991) has shown that pencil lead breaking on a plate surface at 0, 30, 60 and 90 degrees generate wave forms having different in-plane and out-of-plane components of motion. Advantages of this technique are that the acoustic source can be localized with fewer sensors, and in addition to the source localization some of its characteristics can be predicted. The limitation of this approach is that the plate properties and its thickness must be known for the theoretical analysis *a priori*. For an isotropic plate it might not be a problem; however, for an anisotropic plate getting all material properties accurately is very difficult. Another restriction of this technique is that the mechanics problem of wave propagation in a plate satisfying appropriate governing equations and boundary conditions must be solved first. For other techniques described in Sections 5.2.1 to 5.2.5 the solution of this mechanics problem is not required.

5.3 SOURCE LOCALIZATION IN ANISOTROPIC PLATES

5.3.1 Beamforming Technique for Anisotropic Structure

Nakatani et al. (2012a,b) extended the beamforming technique to anisotropic structures. For an anisotropic plate (or shell) Eq. (5.17) takes the following form:

$$\Delta_m(\vec{r}) = \frac{|\vec{r}|}{c(\theta_1)} - \frac{|\vec{r} - \vec{r}_m|}{c(\theta_m)} \tag{5.18}$$

where θ_1 and θ_m correspond to the propagation directions to the reference point from the first sensor (S_1) and the m-th sensor (S_m), respectively. It is obvious that this technique requires *a priori* knowledge of the direction dependent velocity profile. By adjusting the time delay $\Delta_m(\vec{r})$, the recorded signals can be aligned in time before they are added. When the reference point coincides with the acoustic source then all recorded signals going through the delay-and-sum operation of Eq. (5.16) are added in phase and thus the value of b becomes the maximum.

If the coordinates of the reference point are (x_0, y_0) and the coordinates of the receiving sensor S_m are (x_m, y_m) then the distance d_m of the m-th sensor from the reference point is

$$d_m = |\vec{r} - \vec{r}_m| = \sqrt{(x_m - x_0)^2 + (y_m - y_0)^2}\, m = 1,2,3,4 \tag{5.19}$$

From Eqs. (5.18) and (5.19) one obtains

$$\Delta_m(x_0, y_0) = \frac{\sqrt{(x_1 - x_0)^2 + (y_1 - y_0)^2}}{c(\theta_1)} - \frac{\sqrt{(x_m - x_0)^2 + (y_m - y_0)^2}}{c(\theta_m)} \tag{5.20}$$

Substituting Eq. (5.20) into Eq. (5.16) one gets

$$b(x_0, y_0) = \frac{1}{M}\sum_{m=1}^{M} w_m x_m \left[t - \frac{\sqrt{(x_1 - x_0)^2 + (y_1 - y_0)^2}}{c(\theta_1)} + \frac{\sqrt{(x_m - x_0)^2 + (y_m - y_0)^2}}{c(\theta_m)} \right] \tag{5.21}$$

(x_0, y_0) values that give the maximum value of b are the coordinates of the acoustic source. Therefore, the problem is reduced to an optimization problem of finding (x_0, y_0) values for maximizing the b value.

5.3.2 Optimization Based Technique for Source Localization in Anisotropic Plates

Various versions of the optimization based source localization technique in an anisotropic plate are available in the literature (Kundu et al. (2007, 2009); Kundu et al. (2008); Hajzargerbashi et al., 2011; Koabaz et al., 2012. The derivation presented here is based on the work of Kundu et al. (2007, 2008) which is the generalized version of the formulation presented in Section 5.2.3. Eq. (5.5) of Section 5.2.3 remains same for isotropic and anisotropic plates. However, Eq. (5.6) is changed when the wave speed c is direction dependent.

$$d_i = \sqrt{(x_i - x_0)^2 + (y_i - y_0)^2} = c(\theta_i)t_i \tag{5.22}$$

From Eq. (5.22) one obtains

$$\begin{aligned} \frac{d_2}{d_1} &= \frac{c(\theta_2)t_2}{c(\theta_1)t_1} = \frac{\sqrt{(x_2 - x_0)^2 + (y_2 - y_0)^2}}{\sqrt{(x_1 - x_0)^2 + (y_1 - y_0)^2}} = r_{21} \\ \frac{d_3}{d_1} &= \frac{c(\theta_3)t_3}{c(\theta_1)t_1} = \frac{\sqrt{(x_3 - x_0)^2 + (y_3 - y_0)^2}}{\sqrt{(x_1 - x_0)^2 + (y_1 - y_0)^2}} = r_{31} \end{aligned} \tag{5.23}$$

Therefore,

$$\left(\frac{d_2}{d_1}\right)^2+\left(\frac{d_3}{d_1}\right)^2=\left(\frac{c(\theta_2)t_2}{c(\theta_1)t_1}\right)^2+\left(\frac{c(\theta_3)t_3}{c(\theta_1)t_1}\right)^2$$

$$=\frac{(x_2-x_0)^2+(y_2-y_0)^2}{(x_1-x_0)^2+(y_1-y_0)^2}+\frac{(x_3-x_0)^2+(y_3-y_0)^2}{(x_1-x_0)^2+(y_1-y_0)^2}$$

$$\Rightarrow r_{21}^2+r_{31}^2=\frac{(x_2-x_0)^2+(x_3-x_0)^2+(y_2-y_0)^2+(y_3-y_0)^2}{(x_1-x_0)^2+(y_1-y_0)^2} \tag{5.24}$$

$$\Rightarrow r_{21}^2+r_{31}^2=\frac{2\left(x_0^2+y_0^2-x_0x_2-x_0x_3-y_0y_2-y_0y_3\right)+x_2^2+x_3^2+y_2^2+y_3^2}{x_0^2+y_0^2-2(x_0x_1+y_0y_1)+x_1^2+y_1^2}$$

Note that Eq. (5.24) is very similar to Eq. (5.8); the only difference between the two is that r_{21} and r_{31} are now functions of direction dependent wave speed. The objective function $E(x_0,y_0)$ can be constructed from Eq. (5.24) in the same manner as Eq. (5.9) was formed from Eq. (5.8). The constructed objective function for the anisotropic plate has identical form as Eq. (5.9). Coordinates (x_0,y_0) of the acoustic source are obtained by minimizing this objective function.

If the wave speed is $c(\theta)$ then the relative times of arrival (or difference between two arrival times) and their ratios can be defined as

$$t_{21}=\frac{d_2}{c(\theta_2)}-\frac{d_1}{c(\theta_1)}=\frac{c(\theta_1)d_2-c(\theta_2)d_1}{c(\theta_2)c(\theta_1)}$$

$$=\frac{1}{c(\theta_2)c(\theta_1)}\left\{c(\theta_1)\sqrt{(x_2-x_0)^2+(y_2-y_0)^2}\right.$$

$$\left.-c(\theta_2)\sqrt{(x_1-x_0)^2+(y_1-y_0)^2}\right\}$$

$$t_{31}=\frac{c(\theta_1)d_3-c(\theta_3)d_1}{c(\theta_3)c(\theta_1)}=\frac{1}{c(\theta_3)c(\theta_1)}\left\{c(\theta_1)\sqrt{(x_3-x_0)^2+(y_3-y_0)^2}\right. \tag{5.25}$$

$$\left.-c(\theta_3)\sqrt{(x_1-x_0)^2+(y_1-y_0)^2}\right\}$$

$$\Rightarrow\frac{t_{21}}{t_{31}}=\frac{c(\theta_3)}{c(\theta_2)}\frac{\left\{c(\theta_1)\sqrt{(x_2-x_0)^2+(y_2-y_0)^2}-c(\theta_2)\sqrt{(x_1-x_0)^2+(y_1-y_0)^2}\right\}}{\left\{c(\theta_1)\sqrt{(x_3-x_0)^2+(y_3-y_0)^2}-c(\theta_3)\sqrt{(x_1-x_0)^2+(y_1-y_0)^2}\right\}}$$

Objective function $E(x_0,y_0)$ is then defined as

$$E(x_0,y_0)=\left(\frac{c(\theta_3)}{c(\theta_2)}\frac{\left\{c(\theta_1)\sqrt{(x_2-x_0)^2+(y_2-y_0)^2}-c(\theta_2)\sqrt{(x_1-x_0)^2+(y_1-y_0)^2}\right\}}{\left\{c(\theta_1)\sqrt{(x_3-x_0)^2+(y_3-y_0)^2}-c(\theta_3)\sqrt{(x_1-x_0)^2+(y_1-y_0)^2}\right\}}-\frac{t_{21}}{t_{31}}\right)^2 \tag{5.26}$$

Ideally, for the correct values of (x_0,y_0) the objective function should give a zero value, while for wrong values of (x_0,y_0) it should have a positive value. Therefore, this objective function needs to be minimized. Note that in the above definition of the objective function the time t_1 has been used twice in computing t_{21} and t_{31}, while t_2 and t_3 have been used only once. Similar to Section 5.2.3, to give equal importance or bias to the three measured arrival times at the three sensor locations, the objective function can be defined in a different manner as shown below. With this definition of the objective function any bias is eliminated and as a result the acoustic source prediction should not be strongly influenced by the experimental error in any one measurement of the arrival time.

$$E(x_0,y_0)=\left(\frac{c(\theta_3)}{c(\theta_1)}\left\{\frac{c(\theta_2)\sqrt{(x_1-x_0)^2+(y_1-y_0)^2}-c(\theta_1)\sqrt{(x_2-x_0)^2+(y_2-y_0)^2}}{c(\theta_3)\sqrt{(x_2-x_0)^2+(y_2-y_0)^2}-c(\theta_2)\sqrt{(x_3-x_0)^2+(y_3-y_0)^2}}\right\}-\frac{t_{12}}{t_{23}}\right)^2$$
$$+\left(\frac{c(\theta_1)}{c(\theta_2)}\left\{\frac{c(\theta_3)\sqrt{(x_2-x_0)^2+(y_2-y_0)^2}-c(\theta_2)\sqrt{(x_3-x_0)^2+(y_3-y_0)^2}}{c(\theta_1)\sqrt{(x_3-x_0)^2+(y_3-y_0)^2}-c(\theta_3)\sqrt{(x_1-x_0)^2+(y_1-y_0)^2}}\right\}-\frac{t_{23}}{t_{31}}\right)^2 \quad (5.27)$$
$$+\left(\frac{c(\theta_2)}{c(\theta_3)}\left\{\frac{c(\theta_1)\sqrt{(x_3-x_0)^2+(y_3-y_0)^2}-c(\theta_3)\sqrt{(x_1-x_0)^2+(y_1-y_0)^2}}{c(\theta_2)\sqrt{(x_1-x_0)^2+(y_1-y_0)^2}-c(\theta_1)\sqrt{(x_2-x_0)^2+(y_2-y_0)^2}}\right\}-\frac{t_{31}}{t_{12}}\right)^2$$

As before, following some optimization scheme (simplex algorithm or genetic algorithm) the impact point (x_0, y_0) can be obtained by minimizing the above objective function.

One shortcoming of Eq. (5.27) is that for certain values of the unknown (x_0, y_0) the denominator $c(\theta_i)\sqrt{(x_j-x_0)^2+(y_j-y_0)^2}-c(\theta_j)\sqrt{(x_i-x_0)^2+(y_i-y_0)^2}$ can vanish. For those values of (x_0, y_0) the objective function becomes infinity and special care must be taken during the computation of the objective function to avoid these singular points. However, this problem can be easily avoided by modifying the definition of the objective function, as shown below.

$$E(x_0,y_0)=\left(\begin{array}{l} t_{23}c(\theta_3)\left\{c(\theta_2)\sqrt{(x_1-x_0)^2+(y_1-y_0)^2}-c(\theta_1)\sqrt{(x_2-x_0)^2+(y_2-y_0)^2}\right\} \\ -t_{12}c(\theta_1)\left\{c(\theta_3)\sqrt{(x_2-x_0)^2+(y_2-y_0)^2}-c(\theta_2)\sqrt{(x_3-x_0)^2+(y_3-y_0)^2}\right\}\end{array}\right)^2$$
$$+\left(\begin{array}{l} t_{31}c(\theta_1)\left\{c(\theta_3)\sqrt{(x_2-x_0)^2+(y_2-y_0)^2}-c(\theta_2)\sqrt{(x_3-x_0)^2+(y_3-y_0)^2}\right\} \\ -t_{23}c(\theta_2)\left\{c(\theta_1)\sqrt{(x_3-x_0)^2+(y_3-y_0)^2}-c(\theta_3)\sqrt{(x_1-x_0)^2+(y_1-y_0)^2}\right\}\end{array}\right)^2 \quad (5.28)$$
$$+\left(\begin{array}{l} t_{12}c(\theta_2)\left\{c(\theta_1)\sqrt{(x_3-x_0)^2+(y_3-y_0)^2}-c(\theta_3)\sqrt{(x_1-x_0)^2+(y_1-y_0)^2}\right\} \\ -t_{31}c(\theta_3)\left\{c(\theta_2)\sqrt{(x_1-x_0)^2+(y_1-y_0)^2}-c(\theta_1)\sqrt{(x_2-x_0)^2+(y_2-y_0)^2}\right\}\end{array}\right)^2$$

or,

$$E(x_0,y_0)=\left(t_{23}c(\theta_3)\{c(\theta_2)d_1-c(\theta_1)d_2\}-t_{12}c(\theta_1)\{c(\theta_3)d_2-c(\theta_2)d_1\}\right)^2$$
$$+\left(t_{31}c(\theta_1)\{c(\theta_3)d_2-c(\theta_2)d_3\}-t_{23}c(\theta_2)\{c(\theta_1)d_3-c(\theta_3)d_1\}\right)^2 \quad (5.29)$$
$$+\left(t_{12}c(\theta_2)\{c(\theta_1)d_3-c(\theta_3)d_1\}-t_{31}c(\theta_3)\{c(\theta_2)d_1-c(\theta_1)d_2\}\right)^2$$

Where d_j is the distance between the impact point (x_0, y_0) and the j-th receiving sensor location (x_j, y_j),

$$d_j = \sqrt{(x_j - x_0)^2 + (y_j - y_0)^2} \tag{5.30}$$

The predictions can be improved further by placing more receiving sensors on the structure. For n number of sensors there are $\frac{n(n-1)}{2}$ unique sensor pairs.

To be unbiased, the objective function needs to include information from every unique sensor pair and can be constructed in the following manner.

From Eq. (5.25) it is possible to obtain the following equations for sensor pairs i-j and k-l

$$t_{ij} = \frac{d_i}{c(\theta_i)} - \frac{d_j}{c(\theta_j)} = \frac{c(\theta_j)d_i - c(\theta_i)d_j}{c(\theta_i)c(\theta_j)}$$

$$t_{kl} = \frac{d_k}{c(\theta_k)} - \frac{d_l}{c(\theta_l)} = \frac{c(\theta_l)d_k - c(\theta_k)d_l}{c(\theta_k)c(\theta_l)}$$

$$\Rightarrow \frac{t_{ij}}{t_{kl}} = \frac{c(\theta_k)c(\theta_l)}{c(\theta_i)c(\theta_j)} \frac{\{c(\theta_j)d_i - c(\theta_i)d_j\}}{\{c(\theta_l)d_k - c(\theta_k)d_l\}} \tag{5.31}$$

$$\Rightarrow \frac{t_{ij}}{t_{kl}} - \frac{c(\theta_k)c(\theta_l)}{c(\theta_i)c(\theta_j)} \frac{\{c(\theta_j)d_i - c(\theta_i)d_j\}}{\{c(\theta_l)d_k - c(\theta_k)d_l\}} = 0$$

$$\Rightarrow t_{ij}c(\theta_i)c(\theta_j)\{c(\theta_l)d_k - c(\theta_k)d_l\} - t_{kl}c(\theta_k)c(\theta_l)\{c(\theta_j)d_i - c(\theta_i)d_j\} = 0$$

Considering all possible sensor pair combinations the objective function for a plate monitoring system with n receiving sensors is obtained as

$$E(x_0, y_0) = \sum_{i=1}^{n-1}\sum_{j=i+1}^{n}\sum_{k=1}^{n-1}\sum_{l=k+1}^{n}\left[t_{ij}c(\theta_i)c(\theta_j)\big(d_kc(\theta_l) - d_lc(\theta_k)\big) - t_{kl}c(\theta_k)c(\theta_l)\big(d_ic(\theta_j) - d_jc(\theta_i)\big)\right]^2 \tag{5.32}$$

Note that the angle θ_i of the wave propagation direction from the source (x_0, y_0) to the station (x_i, y_i) is measured from the horizontal axis and can be obtained from the following equation:

$$\theta_i = \tan^{-1}\left(\frac{y_i - y_0}{x_i - x_0}\right) \tag{5.33}$$

The above equation is valid for all possible combinations of (x_0, y_0) and (x_i, y_i) for which the computed θ_i values should vary between $-\pi/2$ and $+\pi/2$. Since the wave velocity in θ_i and $(\theta_i + \pi)$ directions should be the same it is not necessary to consider any angle beyond the boundaries $\theta = -\pi/2$ and $+\pi/2$ for computing the wave velocity in all possible directions between $-\pi/2$ and $+3\pi/2$.

Hajzargerbashi et al. (2011) introduced another objective function for source localization as shown below:

$$\begin{aligned} E(x_0, y_0) = {} & \left(c(\theta_1)c(\theta_2)t_{12} - \sqrt{(x_1 - x_0)^2 + (y_1 - y_0)^2}\,c(\theta_2) + \sqrt{(x_2 - x_0)^2 + (y_2 - y_0)^2}\,c(\theta_1)\right)^2 \\ & + \left(c(\theta_1)c(\theta_3)t_{13} - \sqrt{(x_1 - x_0)^2 + (y_1 - y_0)^2}\,c(\theta_3) + \sqrt{(x_3 - x_0)^2 + (y_3 - y_0)^2}\,c(\theta_1)\right)^2 \\ & + \left(c(\theta_2)c(\theta_3)t_{23} - \sqrt{(x_2 - x_0)^2 + (y_2 - y_0)^2}\,c(\theta_3) + \sqrt{(x_3 - x_0)^2 + (y_3 - y_0)^2}\,c(\theta_2)\right)^2 \end{aligned} \tag{5.34}$$

If four receiving sensors a used, then the objective function expression proposed by Hajzargerbashi et al. (2011) takes the following form,

$$\begin{aligned}E(x_0,y_0)=&\left(c(\theta_1)c(\theta_2)t_{12}-\sqrt{(x_1-x_0)^2+(y_1-y_0)^2}c(\theta_2)+\sqrt{(x_2-x_0)^2+(y_2-y_0)^2}c(\theta_1)\right)^2+\\&\left(c(\theta_1)c(\theta_3)t_{13}-\sqrt{(x_1-x_0)^2+(y_1-y_0)^2}c(\theta_3)+\sqrt{(x_3-x_0)^2+(y_3-y_0)^2}c(\theta_1)\right)^2+\\&\left(c(\theta_1)c(\theta_4)t_{14}-\sqrt{(x_1-x_0)^2+(y_1-y_0)^2}c(\theta_4)+\sqrt{(x_4-x_0)^2+(y_4-y_0)^2}c(\theta_1)\right)^2+\\&\left(c(\theta_2)c(\theta_3)t_{23}-\sqrt{(x_2-x_0)^2+(y_2-y_0)^2}c(\theta_3)+\sqrt{(x_3-x_0)^2+(y_3-y_0)^2}c(\theta_2)\right)^2+\\&\left(c(\theta_2)c(\theta_4)t_{24}-\sqrt{(x_2-x_0)^2+(y_2-y_0)^2}c(\theta_4)+\sqrt{(x_4-x_0)^2+(y_4-y_0)^2}c(\theta_2)\right)^2+\\&\left(c(\theta_3)c(\theta_4)t_{34}-\sqrt{(x_3-x_0)^2+(y_2-y_0)^2}c(\theta_4)+\sqrt{(x_4-x_0)^2+(y_4-y_0)^2}c(\theta_3)\right)^2\end{aligned} \tag{5.35}$$

The predictions can be improved further by placing more receiving sensors on the structure. To be unbiased, the objective function needs to include information from every possible sensor pair the same number of times. An increase in the number of sensors results in an increase in the number of terms in the objective function and an increase in the run time of the computer code. With three sensors the number of terms in the objective function is three; by changing the number of sensors to four, the number of terms increases to six. For n sensors the general form of this objective function is given by

$$\begin{aligned}E(x_0,y_0)=\sum_{i=1}^{n-1}\sum_{j=i+1}^{n}\Big[&c(\theta_i)c(\theta_j)t_{ij}-\sqrt{(x_i-x_0)^2+(y_i-y_0)^2}c(\theta_j)\\&+\sqrt{(x_j-x_0)^2+(y_j-y_0)^2}c(\theta_i)\Big]^2\end{aligned} \tag{5.36}$$

Note that for n number of sensors there are $n(n-1)/2$ unique pairs of sensor and as a result $n(n-1)/2$ terms appear in Eq. (5.36). However, in Eqs. (5.15) and (5.32) the number of terms in the summation series is

$$\sum_{m=1}^{\left[\frac{n(n-1)}{2}-1\right]} m$$

Note that for three, four and five receiving sensors the number of terms in Eq. (5.36) should be three, six and ten, respectively whereas in Eqs. (5.15) and (5.32) these numbers are 3, 15 and 45, respectively. Clearly, the computational efficiency increases significantly for greater than three receiving sensors when the objective function given in Eqs. (5.34–36) is used instead of the earlier expressions Eqs. (5.15) and (5.32).

5.3.3 Source Localization in Anisotropic Plates without Knowing Their Material Properties

Two techniques described in Sections 5.3.1 and 5.3.2 require the knowledge of the direction dependent velocity profile in the anisotropic plate for source localization. Only for isotropic plates could acoustic sources be localized without having any knowledge of the wave speed in the plate, as described in Sections 5.2.2 to 5.2.5. Kundu et al. (2012) proposed a technique for acoustic source localization in an anisotropic plate with the help of only six receiving sensors without knowing the direction dependent velocity profile in the plate and without any need to solve a system of nonlinear equations. Kundu et al. (2012) then experimentally verified this technique for plates made of both isotropic (requiring four sensors) and anisotropic materials (requiring six sensors). It should be noted that Ciampa and Meo (2010) and Ciampa et al. (2012) also proposed a technique for source localization in anisotropic plates without knowing the plate properties. However, their technique required the solution of a system of nonlinear equations, and therefore was computationally more demanding. The technique proposed by Kundu (2012) and experimentally verified by Kundu et al. (2012) is briefly described in the following.

Three receiving sensors S_1, S_2 and S_3 are mounted on the plate as shown in Figure 5.5. If the coordinates of three receiving sensors S_1, S_2 and S_3 are (x_1, y_1), (x_2, y_2) and (x_3, y_3), respectively, then from Figure 5.5 it is clear that $x_2 = x_1 + d$, $x_3 = x_1$, $y_2 = y_1$ and $y_3 = y_1 + d$. The coordinate of the acoustic source (A) is given by (x_A, y_A). The distance d between the sensors is much smaller than the distance D between the acoustic source A and the i-th sensor S_i. Therefore, the inclination angle θ of lines AS_1, AS_2 and AS_3 (see Figure 5.5) can be assumed to be approximately the same. Because of this assumption the received signals at these three sensors should be almost identical but slightly time-shifted and the wave velocity in the direction of the sensors S_1, S_2 and S_3 from the source point A should be almost the same even for an anisotropic plate. Angle θ can be expressed as

$$\theta = \tan^{-1}\left(\frac{y_1 - y_A}{x_1 - x_A}\right) \approx \tan^{-1}\left(\frac{y_2 - y_A}{x_2 - x_A}\right) \approx \tan^{-1}\left(\frac{y_3 - y_A}{x_3 - x_A}\right) \tag{5.37}$$

After arriving at sensor S_1 the time taken by the wave front to reach sensors S_2 and S_3 can be denoted as $t_{21} = t_2 - t_1$ and $t_{31} = t_3 - t_1$, respectively. These two time delays are given by

$$t_{21} = \frac{d\cos\theta}{c(\theta)} \tag{5.38}$$

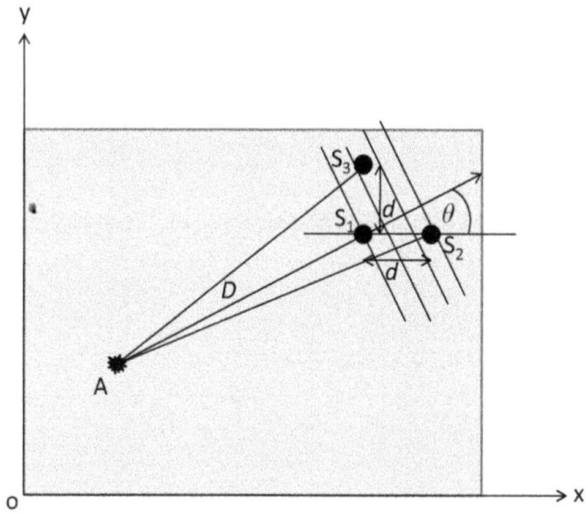

Figure 5.5 Three sensors at positions S_1, S_2 and S_3 are needed to get the direction of the source by the method described in Section 5.3.3 (Kundu et al., 2012).

$$t_{31} = \frac{d\sin\theta}{c(\theta)} \tag{5.39}$$

where $c(\theta)$ is the wave velocity in the θ direction. From Eqs. (5.38) and (5.39) one can easily obtain

$$\theta = \tan^{-1}\left(\frac{t_{31}}{t_{21}}\right) \tag{5.40}$$

From Eq. (5.38)

$$c(\theta) = \frac{d \times \cos\theta}{t_{21}} = \frac{d \times t_{21}}{t_{21}\sqrt{t_{21}^2 + t_{31}^2}} = \frac{d}{\sqrt{t_{21}^2 + t_{31}^2}} \tag{5.41}$$

In the above equation $\cos\theta = \frac{t_{21}}{\sqrt{t_{21}^2 + t_{31}^2}}$ is obtained from the following consideration.

From Figure 5.5 it is clear that $(AS_2 - AS_1) = c(\theta)t_{21}$ and $(AS_3 - AS_1) = c(\theta)t_{31}$. Three lines AS_1, AS_2 and AS_3 are assumed parallel, which should be the case when the source is far away from the sensors. Note that the two triangles S_1S_2P and S_1S_3Q are similar triangles when AS_1, AS_2 and AS_3 are parallel. The lines AS_2 and AS_3 intersect the wave front going through sensor S_1 at points P and Q, respectively. Therefore,

$$\cos\theta = \frac{PS_2}{S_1S_2} = \frac{PS_2}{\sqrt{PS_2^2 + PS_1^2}} = \frac{PS_2}{\sqrt{PS_2^2 + QS_3^2}} = \frac{c(\theta)t_{21}}{\sqrt{c(\theta)^2 t_{21}^2 + c(\theta)^2 t_{31}^2}} = \frac{t_{21}}{\sqrt{t_{21}^2 + t_{31}^2}} \tag{5.42}$$

From Eqs (5.40) and (5.41) the wave propagation direction and the wave speed in that direction are obtained in terms of experimentally measured values t_{21} and t_{31}. If three more sensors S_4, S_5 and S_6 are mounted near another corner of the plate as shown in Figure 5.6 then the wave propagation direction θ_4 from the

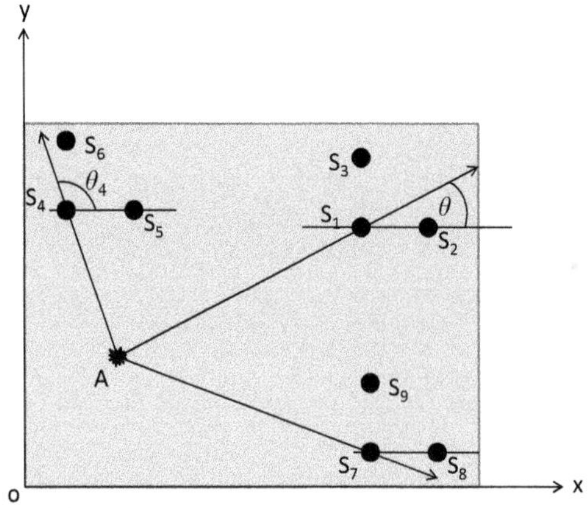

Figure 5.6 Acoustic source can be localized from the intersection point of two direction lines generated by two clusters of sensors. A third cluster can be used to investigate if the third direction line also goes through the intersection point of the other two lines to reconfirm the prediction (Kundu et al., 2012).

acoustic source to sensor S_4 and the wave speed in that direction $c(\theta_4)$ can be obtained in the same manner from t_{54} and t_{64} from the following equations:

$$\theta_4 = \tan^{-1}\left(\frac{t_{64}}{t_{54}}\right) \tag{5.43}$$

$$c(\theta_4) = \frac{d}{\sqrt{t_{54}^2 + t_{64}^2}} \tag{5.44}$$

From Eqs. (5.37) and (5.40) of the S_1, S_2, S_3 sensor system, and from similar two equations for the S_4, S_5, S_6 sensor system, one can write

$$\tan\theta = \frac{y_1 - y_A}{x_1 - x_A} = \frac{t_{31}}{t_{21}} \tag{5.45}$$

$$\tan\theta_4 = \frac{y_4 - y_A}{x_4 - x_A} = \frac{t_{64}}{t_{54}} \tag{5.46}$$

Equations (5.45) and (5.46) give a system of two linear equations with two unknowns x_A and y_A that can be uniquely solved. In other words, two straight lines with inclination angles θ and θ_4 going through sensors S_1 and S_4 intersect at a point that should be the acoustic source point, as shown in Figure 5.6.

5.3.3.1 *Determination of t_{ij}*

It should be noted that all derived values – acoustic wave propagation direction (θ and θ_4 in Figure 5.6), acoustic source location (A in Figures 5.5 and 5.6) and the direction dependent wave speed $c(\theta)$ – are obtained from t_{ij}. Therefore, it is necessary to measure t_{ij} accurately. Since the distance d between the sensors is small the time difference t_{ij} between two recorded signals by i-th and j-th sensors placed in close proximity is expected to be small. However, this small time difference can still be measured accurately in the following manner.

Let the recorded transient signals by i-th and j-th sensors be expressed as two arrays $\mathbf{I}(t) = [I_1, I_2, I_3, \ldots. I_N]$ and $\mathbf{J}(t) = [J_1, J_2, J_3, \ldots. J_N]$. Here I_n and J_n represent the signal values at time t_n. Note that the time increment δt between two successive points in the transient signal is given by $\delta t = T / N - 1$ where T is the total recorded time and N is the total number of points in the transient signal. These two arrays can be added or multiplied after giving a small time shift in one of the two arrays as shown below.

$$U(\Delta t) = \sum_{n=1}^{N-m} \left[\left| I_n + J_{n+m} \right| \right] \tag{5.47}$$

$$V(\Delta t) = \sum_{n=1}^{N-m} \left[I_n \times J_{n+m} \right] \tag{5.48}$$

where

$$\Delta t = m \times \delta t \tag{5.49}$$

If $U(\Delta t)$ and $V(\Delta t)$ are plotted, then they should reach their maximum values at $\Delta t = t_{ji}$ because then these two arrays are in phase. If two arrays in phase are added and all negative terms are made positive after addition by taking their magnitudes as shown in Eq. (5.47), then that value should be higher than the value obtained by adding the same two arrays when they are not in phase. The same thing can be said for the two arrays when they are multiplied as in Eq. (5.48). In this manner t_{ji} can be measured very accurately with precision equal to δt, the time increment of the recorded transient signal.

5.3.3.2 Improving and Checking the Accuracy of Prediction

The accuracy of the prediction can be checked and improved by introducing a third sensor cluster made of three more receiving sensors (S_7, S_8 and S_9) placed at another location of the plate as shown in Figure 5.6. If the third line generated by this sensor cluster goes through the intersection point of the first two lines generated by sensor clusters (S_1, S_2, S_3) and (S_4, S_5, S_6), as shown in Figure 5.6, then it can be concluded that the prediction is accurate and reliable. Otherwise, if the three lines form a triangle instead of coinciding at one point then there is some uncertainty associated with this prediction. In that case the intersection point of two longer lines (connecting the source and the cluster) should be considered as the acoustic source point and the shortest line should be ignored. This is because the sensor cluster closest to the acoustic source is expected to have the maximum error since for a short distance between the sensor cluster and the acoustic source the assumption that the lines connecting the source point and the three sensors are almost parallel is violated.

5.3.3.3 Experimental Verification

Experimental verification of the method described in Section 5.3.3 was provided by Kundu et al. (2012). They conducted the experiment on a curved anisotropic plate or shell made of fiber reinforced composite material having non-uniform thickness as shown in Figure 5.7. The unwrapped dimension of the composite

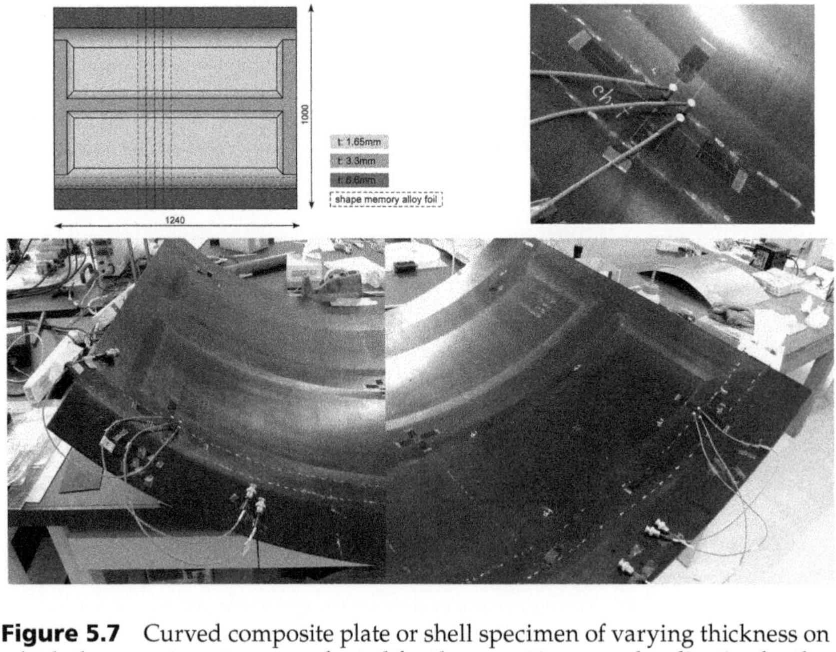

Figure 5.7 Curved composite plate or shell specimen of varying thickness on which the experiment was conducted for the acoustic source localization by the method described in Section 5.3.3 (Kundu et al., 2012).

plate was 1000 mm × 1240 mm. Two sensor clusters (S_1, S_2, S_3) and (S_4, S_5, S_6) were attached on the concave side of the curved plate near its two corners, as shown in Figure 5.7. Three sensors in each cluster were placed at the three vertices of an isosceles right triangle whose side lengths were 15 mm, 15 mm and $15\sqrt{2} = 21.21$ mm. The distance between the two clusters was 900 mm. The acoustic source (the pencil lead break location) was then predicted following the formulation described in Sections 5.3.3 and 5.3.3.1. Predictions of the acoustic wave propagation directions from the acoustic source to the sensor clusters near the bottom-left and bottom-right corners of the plate are shown by the straight lines in Figure 5.8. In this figure one can see ten different lines emitting from each cluster (five dashed and five solid lines). Since some of these lines coincide one cannot see ten distinct lines. Ideally all these lines should coincide since although they were obtained from five different experiments at five different times the pencil lead was broken at the same position. Directions of these lines were obtained from Eqs. (5.45) and (5.46) while Δt_{ij} in those equations were obtained from either Eq. (5.47) that gave dashed lines or from Eq. (5.48) that gave solid lines. Intersection points of these two sets of lines emitting from the two clusters are the predicted locations of the acoustic source. The actual source location is also shown by a solid diamond marker in Figure 5.8.

Although predicted points in Figure 5.8 were found to be close to the true location of the acoustic source for this particular experiment, in some other experiments the predicted points were found to be inconsistent and far away from the true source location, as shown in the left plot of Figure 5.9. This figure was generated by Nakatani et al. (2014) for a flat composite plate specimen. They investigated the reason for this discrepancy and found that although it was anticipated that the signals recorded by the three sensors in a cluster should be slightly time-shifted but otherwise almost identical, in reality they were not. The differences in the later parts of the recorded signals were found to be more

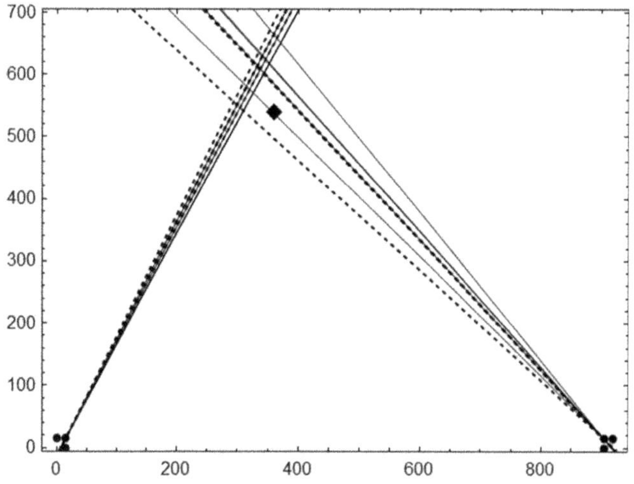

Figure 5.8 For the composite specimen shown in Figure 5.7 the predicted acoustic source locations are obtained from the intersections of the two sets of straight lines drawn from the two clusters of sensors (shown by solid circles) placed near the bottom-left and bottom-right corners of the specimen. Actual source position is shown by the diamond marker (Kundu et al., 2012).

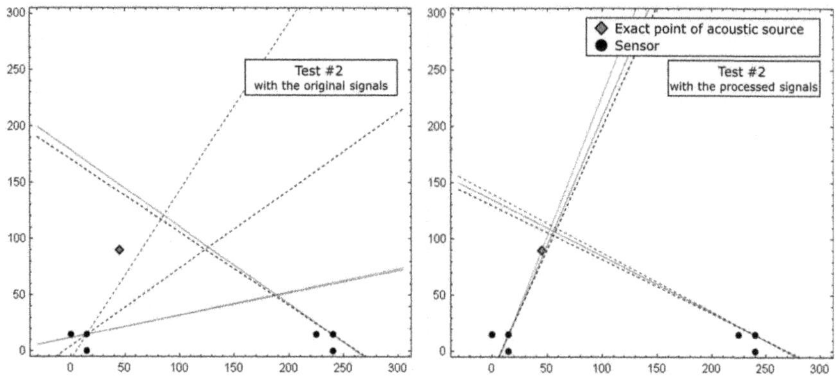

Figure 5.9 Predicted acoustic source locations in a composite plate are the intersection points of the straight lines drawn from the two clusters of sensors (shown by solid circles). The left figure is obtained by considering the complete signals and the right figure is obtained from the initial parts of the recorded signals. True acoustic source locations are shown by the diamond marker (Nakatani et al., 2014).

prominent than the initial parts. Their suggestion was to use only the initial parts (up to first dip and peak) of the recorded signals. Only this initial part of the signal should be substituted in Eqs. (5.47) and (5.48) to avoid any inconsistency in prediction. This modification improved the predictions significantly as illustrated in Figure 5.9 where the left plot shows the predictions without this modification while the right plot shows the predictions with the modification. Analyzing the experimental data one gets three distinct predictions, as shown in the left plot of Figure 5.9, considering the entire signal. Only the initial parts of the signals were considered to generate the right plot predicting the acoustic source location. Clearly, the right plot gives better source location.

Another improvement was suggested by Yin et al. (2018). They proposed to use a *Z*-shaped sensor cluster made of four sensors instead of an *L*-shaped cluster made of three sensors and showed that two such *Z*-shaped sensor clusters made of a total of eight sensors can accurately predict the acoustic source.

It should be pointed out that *L* and *Z*-shaped sensor cluster techniques work well for weakly anisotropic plates. However, for highly anisotropic plates these techniques fail and wave front shape based techniques discussed in Section 5.8 should be used.

5.3.4 Source Localization and Its Strength Estimation without Knowing the Plate Material Properties by Poynting Vector Technique

Guyomar et al. (2011) proposed a technique based on the Poynting vector analysis from which one can predict if the acoustic source is inside or outside a specific region and can give an estimate of the strength of the source. This technique is useful for monitoring a critical region, to predict if that region has been hit by a foreign object, or if a crack has been formed in that region. The working principle of this technique is demonstrated with the help of Figure 5.10. Let the critical region (sometimes called "hot spot") in a large plate be within a rectangular region PQRS. For automatic monitoring of this region (to predict if the acoustic source is within this region), piezoelectric sensors are attached to the plate along the boundary of this rectangular region PQRS. After every

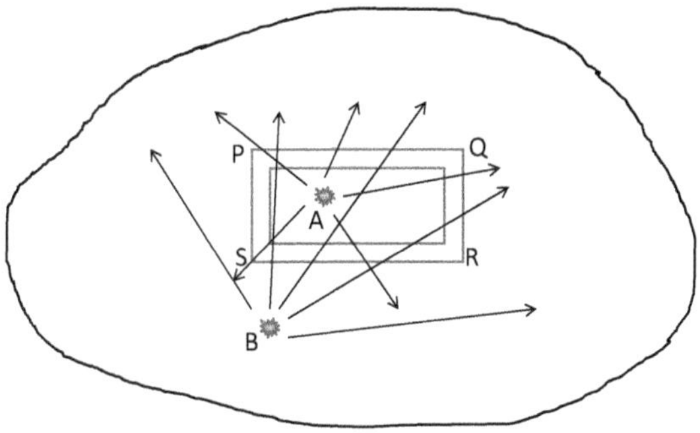

Figure 5.10 Sensors are placed along a pre-defined boundary PQRS for acoustic source detection by the Poynting vector technique (Kundu, 2014).

acoustic event the net acoustic energy leaving the monitored region is obtained using Poynting Vectors. It is done by subtracting the total energy entering the monitored region from the total energy leaving the region. If the impact point A is inside the region then the net energy leaving the region should be positive but if the impact point B is outside the region then it should be either zero (if the plate material has zero attenuation and no energy is lost within the monitored region) or negative (if some energy is absorbed in the monitored region because of the material attenuation). In this manner, without knowing the elastic properties of the plate material, the acoustic source location can be predicted – whether outside or inside a monitored region. However, the exact location of the acoustic source cannot be predicted by this technique. If the source is inside the monitored region, then the severity of the source can be predicted by calculating the amount of energy leaving the monitored region. The stronger the acoustic source, the more energy that leaves the monitored region. In other techniques described in Sections 5.3.2 and 5.3.3 the strength of the acoustic source can be indirectly estimated from the strength of the signal recorded by the receiving sensors, since powerful sources generate stronger acoustic signals. It should be noted that the Poynting vector technique was also used for electromagnetic source localization by Nehorai and Paldi (1994)

5.4 SOURCE LOCALIZATION IN COMPLEX STRUCTURES

For a simple three-dimensional solid medium or a two-dimensional plate-like structure where the acoustic waves propagate along a straight line from the acoustic source to the receiving sensors, the techniques that have been discussed above work well. These methods also work for a shell type structure with the simplifying assumption that when the shell is unwrapped to form a flat plate the acoustic wave propagates along a straight line in that unwrapped flat plate from the source to the receiver. However in complex three-dimensional or two-dimensional structures, as shown in Figure 5.11, straight line propagation of the generated wave from the acoustic source to the receiving sensors is not possible because of the presence of internal cavities or other inhomogenities. The structures' complex boundary geometry may also restrict the abilities of the above described techniques to localize the acoustic source. In these situations alternate techniques for source localization must be followed. In the two geometries shown in Figure 5.11 what one can see is that from the acoustic source point A the generated wave can propagate in a straight line to one of the sensors (S_1) but not to all sensors (for example S_2 and S_3) because of the internal cavities

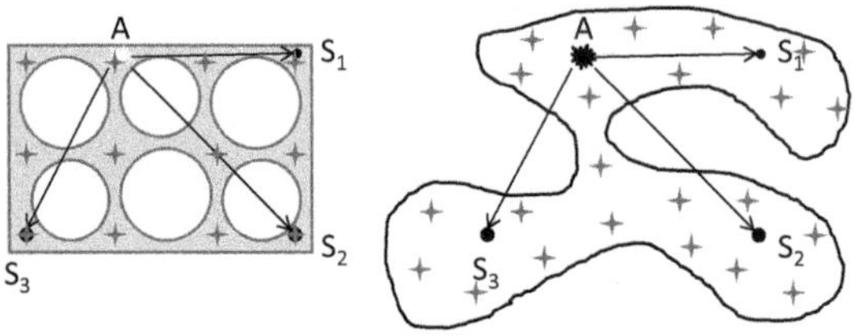

Figure 5.11 Geometries of two complex structures where the elastic waves cannot propagate along a straight line from the source to the sensor (Kundu, 2014).

or inhomogenities on its path (left figure) or because of a complex boundary geometry (right figure) that obstructs the straight line propagation of the generated wave.

For source localization in these complex structures one has two options – (1) to follow a more labor intensive and computationally demanding artificial neural network (ANN) or time reversal technique based on the impulse response function (IRF) or (2) to place a dense distribution of acoustic sensors on the structure. These two techniques are discussed in the following sections.

5.4.1 Source Localization in Complex Structures by Time Reversal and Artificial Neural Network Techniques

Ing et al. (2005) and Ribay et al. (2007) proposed the ASL by the time reversal technique. This method requires three steps – (1) collecting a training data set by mechanically impacting multiple points of a structure and measuring corresponding IRFs by one or more surface mounted sensors; (2) recording the signals at the sensors for an actual impact event; and (3) identifying the IRF in the training data set, which gives the maximum correlation with the IRF generated by the actual impact event, and thus determining the corresponding mechanical impact point. A similar approach for source localization by inverting the impact response functions and avoiding the wave mechanics analysis was proposed by Grabec and Sachse (1989), Sribar and Sachse (1993) and Kosel et al. (2000), among others. They used ANN and pattern recognition techniques to solve this inverse problem. The advantages of these techniques are that they do not require the knowledge of the wave velocity or the structural geometry and can estimate the strength of the source in addition to its location; however, it is labor intensive and computationally demanding. It requires repeated training, which can take a long time to cover a large area, and it needs to be conducted by manual impacts or by employing robotic devices.

Park et al. (2012) automated and expedited the training by using a scanning laser Doppler vibrometer (SLDV), which can measure out-of-plane velocity based on the Doppler effect (Staszewski et al., 2004; Scruby and Drain, 1990). They generated IRF by exciting the surface mounted PZT transducers and sensed the signal by SLDV based on the reciprocity between these IRFs. In principle, this technique can be applied for source localization in any complex structure, however, the degree of complexity of the structures so far studied has been rather limited. For example, Park et al. (2012) have considered aircraft wing and aluminum fuselage where waves can travel directly from the acoustic source to the sensor without encountering any obstacle on its path, and therefore some of the techniques described in Sections 5.3.2 and 5.3.3 that do not require wave velocity information are also applicable for source localizations in these structures.

More complex structures having circular holes on the path between the acoustic source and the sensor have been considered by Baxter et al. (2007) and Hensman et al. (2010). Their approach is briefly described here. For waves propagating from the acoustic source to the two sensors i and j along unobstructed paths, the ideal time difference between the arrival times at the two sensors can be written as

$$\Delta t_{ij} = \frac{|E - S_i| - |E - S_j|}{c} \tag{5.50}$$

where $|E - S_i|$ is the Euclidean distance between the source location and the i-th sensor. However, in presence of the obstacles as shown in Figure 5.11, the experimentally recorded time difference ΔT_{ij} is expected to be different from Δt_{ij}. Therefore, minimization of the following objective function can localize the source in a structure where waves propagate unobstructed.

$$Z = \sum_{i,j} \left[\Delta T_{ij} - \frac{|E - S_i| - |E - S_j|}{c} \right] = \sum_{i,j} \left[\Delta T_{ij} - \Delta t_{ij} \right] \tag{5.51}$$

However, such minimization cannot localize the acoustic source in complex structures, as shown in Figure 5.11. For source localization in such complex structures Baxter et al. (2007) proposed the "Delta T" method for locating AE events based on a set of artificial training data, generated by a pencil lead break (Hsu, 1977). This Delta T method generated ten artificial sources at each grid point in the problem that they considered. It involved taking the mean of the ten arrival times at each sensor, and constructing a map of the expected ΔT information between each pair of sensors across the structure using linear interpolation between grid points. This map then replaced the Δt_{ij} expression in the above objective function, so that the test events could match the training data. Hensman et al. (2010) improved this technique. They introduced probabilistic interpretation and thus required less training data.

5.4.2 Source Localization by Densely Distributed Sensors

The second alternative for source localization in complex structures is to place multiple sensors on the structure, as shown by the crosses in Figure 5.11. If the acoustic source A happens to be very close to one such sensor, as in the left figure, then the strain recorded by that sensor will be the highest; therefore one can say with certainty that the acoustic source is located very close to that sensor. If the acoustic source is near several sensors, as in the right figure, then the strain measured by several sensors will be high. In that case one can only say that the impact point is located close to those sensors.

For composite plate inspection thin piezoelectric material can be deposited on a flexible layer called a SMART™ layer to fabricate multiple sensors on a flexible surface (Tracy and Chang, 1996; Lin, 1998; Lin and Chang, 1998; Seydel and Chang, 1999; Wang, 1999; Wang and Chang, 2000; Park and Chang, 2003). Then this layer can be placed inside the composite plate to protect the sensors from the adverse environment and from the direct hit of the impacting object. The SMART™ layer can be also attached on the surface of a complex structure to be monitored.

Instead of the piezoelectric sensors, multiple optical fiber Bragg grating (FBG) sensors have been also attached on the surface of a structure, and the point of impact has been predicted by estimating where the impact generated strain should be maximum when the point of impact is near multiple sensors but not necessarily hitting one sensor (Hiche et al., 2011). Although techniques based on distributed sensors should in principle work for complex structures like those shown in Figure 5.11, most of the studies reported in the literature have been limited to flat or curved composite panels for which the techniques reported in Section 5.3 should also work and will require fewer sensors.

5.5 SOURCE LOCALIZATION IN THREE-DIMENSIONAL STRUCTURES

It should be noted that Eq. (5.51) is a general objective function that can be minimized for source localization in one-, two- or three-dimensional structures. Minimization of this objective function requires some optimization technique. Ting et al. (2012) localized acoustic sources in cylindrical coal samples in this manner. Dong and Li (2012) simplified the source localization in three-dimensional bodies by placing sensors at selected corner points of a cube and then deriving an analytical expression for source localization.

5.6 AUTOMATIC DETERMINATION OF TIME OF ARRIVAL

A number of source localization techniques such as triangulation technique, modal acoustic emission and optimization based techniques discussed in Sections 5.2.1–3, 5.2.6 and 5.3.2 require accurate determination of the exact time of arrival of the acoustic signal. For automatic source localization (without human intervention) it is very important to have this parameter determined accurately by proper signal processing applied to the recorded acoustic signals. Several approaches have been proposed to determine the exact time of arrival of the acoustic signal in addition to the standard threshold crossing technique, such as wavelet transform, analysis of the LTA/STA (long-time average/short-time average), and higher order statistics (HOS) (Lokajicek and Klima, 2006), use of special characteristic functions based on Akaike information criterion (AIC) (Sedlak et al., 2009) and Maeda's equation (Sedlak et al., 2013). However, for a good number of source localization techniques, such as those described in Sections 5.2.4 and 5.3.3 where the time difference of arrival of acoustic signals at two different sensors is needed, the cross-correlation technique as discussed earlier is sufficient.

5.7 UNCERTAINTY IN ACOUSTIC SOURCE PREDICTION

All experimental techniques have some uncertainties associated with their predictions because of the unavoidable errors in the experimental measurements of various parameters such as the time of flight and the strain value that are often used for the ASL and its strength estimation. Niri et al. (2012) and Niri and Salamone (2012) proposed a probabilistic approach for ASL. They took into account the systematic errors in the time of flight measurements due to Heisenberg uncertainty principle then used an extended Kalman filter to iteratively estimate the acoustic source location and the wave velocity in an isotropic plate. They took into account the uncertainties in the wave velocity and the time of flight measurements and filtered out the uncertainty using an extended Kalman filter by fusing multi-sensor data.

5.8 SOURCE LOCALIZATION IN ANISOTROPIC PLATES BY ANALYZING PROPAGATING WAVE FRONTS

The ASL techniques in anisotropic plates without knowing the plate properties, as described in Section 5.3.3, work only for weakly anisotropic materials. This is because all techniques discussed in that section assume that wave energy propagates along a straight line from the acoustic source to the receiver. However, for anisotropic plates this is not true. For a highly anisotropic plate the wave path deviates significantly from a straight line. Therefore, extending the direction of wave propagation at the receiver position along a straight line and assuming the acoustic source is located on that line is not correct, especially for a highly anisotropic plate.

Any reliable method should assume that waves in anisotropic plates propagate along curved lines and form non-circular wave fronts. As a result, for a highly anisotropic solid the ASL techniques that assume straight line propagation of the wave energy from the source to the receiver are bound to produce some error.

Park et al. (2017) introduced a new technique for ASL in an anisotropic plate considering non-circular wave front shapes. In a highly anisotropic plate they considered two common wave front shapes – rhombus and ellipse – that are typically observed for an orthotropic plate. They showed how the acoustic source could be successfully localized without knowing all material properties of the anisotropic plate. In the following this technique is described in detail step by step as was presented in Park et al. (2017). Their technique was verified by both numerical and experimental results (Park, 2016).

5.8.1 Wave Propagation Direction Vector Measurement by Sensor Clusters

The wave propagation direction at the cluster position is measured first. Park et al. (2017) followed the technique introduced by Kundu (2012) where an L-shaped sensor cluster of three sensors was attached to the plate. Two corner sensors at P_1 and P_2 were placed at the end of the cluster at a distance (d) from the middle sensor (O), as shown in Figure 5.12. The sensor cluster was mounted on the surface of the plate to record incoming wave signals synchronously. Figure 5.13a shows signals recorded by sensors placed at O and P_1 that were generated by an acoustic source far away from the sensor cluster. Since the acoustic source was located far away from the cluster the wave front passing through the cluster can be assumed to be a plane wave front, as depicted in Figure 5.12. Note that the first one or two peaks of each signal of Figure 13a display good agreement in shape but their shapes deviate significantly in the later parts. The time difference of arrival (TDOA) between these two signals can

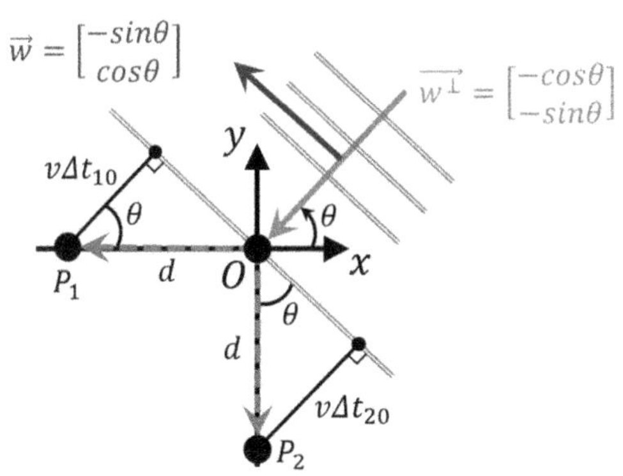

Figure 5.12 Plane wave front at the sensor cluster location, sensors P_1 and P_2 are located at a distance d from sensor O. Direction vector ($\overrightarrow{w^{\perp}}$) and parallel vector ($\vec{w}$) of the plane wave front are obtained by measuring TDOAs (time difference of arrivals) between two pairs of sensors (P_1-O and P_2-O) and then computing their ratio, $\tan\theta = \Delta t_{20} / \Delta t_{10}$ (Park et al., 2017).

be accurately computed from the cross-correlation of the first one or two peaks of the two signals.

The direction vector ($\overrightarrow{w^{\perp}}$) and the parallel vector ($\vec{w}$) of the plane wave front toward the point O from the first quadrant of the local x–y coordinates can be described as

$$\left[\overrightarrow{w^{\perp}}\ \vec{w}\right] = \begin{bmatrix} -\cos\theta & -\sin\theta \\ -\sin\theta & \cos\theta \end{bmatrix} \tag{5.52}$$

where θ follows the sign convention: counter-clock-wise to be positive, and it is obtained from four-quadrant inverse tangent with two TDOAs

$$\tan\theta = \frac{\sin\theta}{\cos\theta} = \frac{v\Delta t_{20} / d}{v\Delta t_{10} / d} = \frac{\Delta t_{20}}{\Delta t_{10}} = \frac{t_2 - t_0}{t_1 - t_0} \tag{5.53}$$

$$\theta = \tan^{-1}\left(\frac{t_2 - t_0}{t_1 - t_0}\right) \tag{5.54}$$

With the plane wave front approximation, one can assume that the three sensors receive identical signal patterns near the first arrival and the signals are dispersed afterwards. TDOA is a time shift between two received signals which is computed by the cross-correlation technique, by plotting the product of two signals when for one signal the time shift is continuously changed. Two signals recorded by two sensors are displayed together in Figure 5.13a. Similar signals with a small time shift were found in the range between 1100 μs and 1500 μs. The time shift between them can be easily found by the cross-correlation technique. This method examines the similarity between the two given signals by computing

$$\left[F(t) \star G(t)\right](\tau) = \int F(t)G(t+\tau)dt \tag{5.55}$$

The maximum value of the cross-correlation plot corresponds to the time shift as shown in Figure 5.13b.

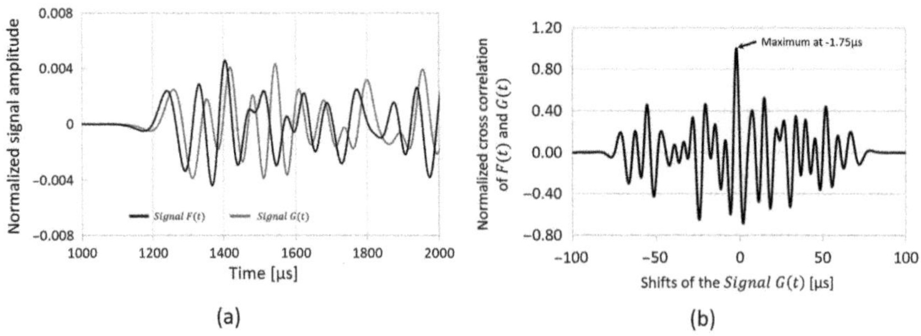

Figure 5.13 (a) Signal $F(t)$ is measured at sensor O, and signal $G(t)$ is measured at sensor P_1 (see Figure 5.12 for sensor locations). Note that the signal patterns are very similar in the time range 1100 μs to 1500 μs. (b) Cross-correlation between the two signals in the time range (1100 to 1500 μs) is computed and displayed. The maximum correlation exists at −1.75 μs which corresponds to the time shift between the two signals (Park et al., 2017).

Table 5.1: Orthotropic Material Properties Used in the Numerical Simulation

Orthotropic Material Properties	Values	
Mass density	1.5×10^{-9} tonnes / mm^3 = 1.5 gm/cc	
Elastic moduli	E_1	66400 MPa
	E_2 and E_3	6000 MPa
Poisson's ratios	ν_{12}	0.2
	ν_{23} and ν_{31}	0.25
Shear moduli	G_{12}	1400 MPa
	G_{23} and G_{31}	2100 MPa

5.8.2 Numerical Simulation of Wave Propagation in an Anisotropic Plate

A $500\text{ mm} \times 500\text{ mm} \times 2\text{ mm}$ thin anisotropic plate was modeled by cuLISA3D software for numerical simulation with orthotropic plate material properties, as given in Table 5.1 (Park et al., 2017). In the model, eight elements through the plate thickness – i.e. $\Delta z = 0.25$ mm, with in-plane element size $\Delta x = \Delta y = 0.5$ mm, were used. The total number of elements in the model was 8 million. The time step, Δt, was taken to be 0.025 μs to ensure stability of the explicit time integration scheme.

The acoustic source was located at the center of the plate (250, 250) and was excited by a two-period sine signal modulated by the Gauss window. Free boundary conditions were used in the model not only to avoid computational complexity of the simulation model with respect to the model size but also to maintain high accuracy on the massively parallel computation. Figure 5.14a shows resultant wave fronts of the numerical simulation at $t = 100\ \mu s$. Different sensor clusters were placed at various positions as shown in Figure 5.14b. Sensors in each cluster were 15 mm apart in both horizontal and vertical directions. Time histories recorded by the sensors were obtained from the numerically simulated data, as shown in Figures 5.13 and 5.15.

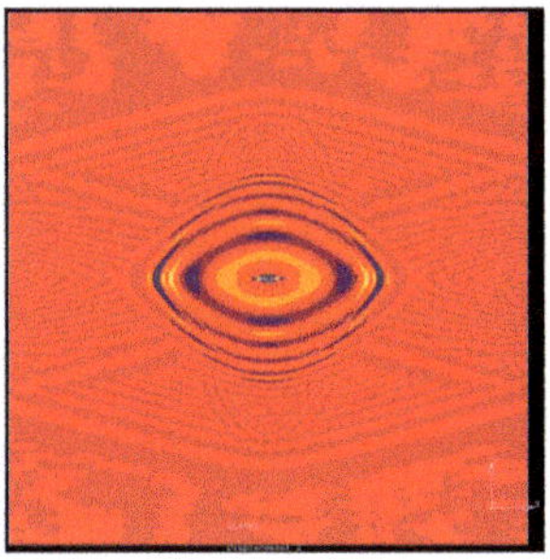

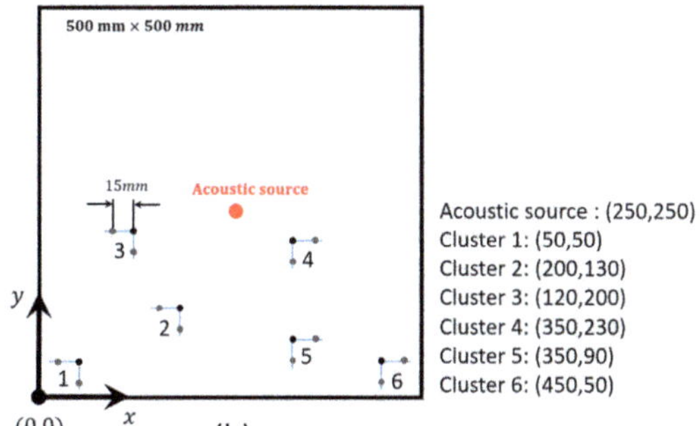

(a) (b)

Figure 5.14 (a) Numerically simulated propagating elastic waves in an anisotropic plate. This image was obtained at $t = 100\,\mu s$. (b) Six clusters at different locations recorded the wave signal. Sensors in each cluster are 15 mm apart along both horizontal and vertical directions. For example, three sensors of cluster 1 are located at (35,50), (50,50), and (50,35) (Park et al., 2017).

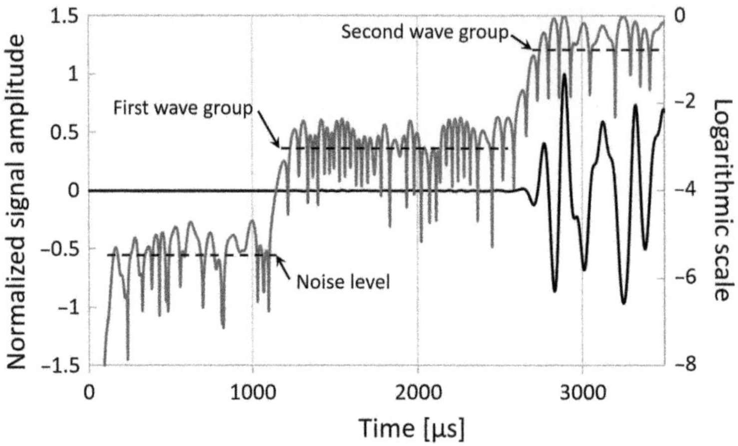

Figure 5.15 A signal recorded by one sensor is plotted in the dark black solid line. When absolute value of the signal is plotted in logarithmic scale as shown in gray solid line, three significantly different levels of signal energy are noticed that represent white noise or numerical error, first wave group arrival and second wave group arrival (Park et al., 2017).

All 18 sensors recorded incoming signals and stored them until the simulation ended. One received signal (dark black solid line) is plotted in Figure 5.15. When one takes the absolute value of the signal and plots it in logarithmic scale then three different magnitude levels are clearly revealed. The lowest level indicates numerical errors. The first wave group hits the sensor carrying a relatively small amount of energy but propagating faster while the second wave group arrives later with higher level of energy (gray solid line in Figure 5.15).

All $\overrightarrow{w^{\perp}}$ described in Section 5.8.1 are computed by measuring TDOAs at every cluster for the first and second wave groups. Then lines along each $\overrightarrow{w^{\perp}}$ were drawn in Figure 5.16. The lines should go through the acoustic source if the wave front is circular, as explained earlier. However, for the anisotropic plate considered here the lines do not go through a common point. In Figure 5.16a, almost two sets of parallel lines are observed on two sides of the acoustic source location. In Figure 5.16b, multiple crossing points at various locations can be observed. This figure underlines the need for a more accurate source localization technique without knowing the plate's material properties. The new technique developed by Park et al. (2017) is described in the following section.

5.8.3 Wave Front Based Source Localization Technique

Earlier techniques as described in Section 5.3.3 assumed wave energy propagating along a straight line from the source to the receiver. However, this assumption is not true for anisotropic plates. Park et al. (2017) proposed to localize the acoustic source from the shape of the propagating wave front instead of trying to trace the wave energy path. They considered two wave front shapes – rhombus and ellipse.

5.8.3.1 Rhombus Wave Front

In anisotropic plates propagating waves form non-circular wave fronts – typically close to a rhombus or an ellipse. When a rhombus wave front is formed then

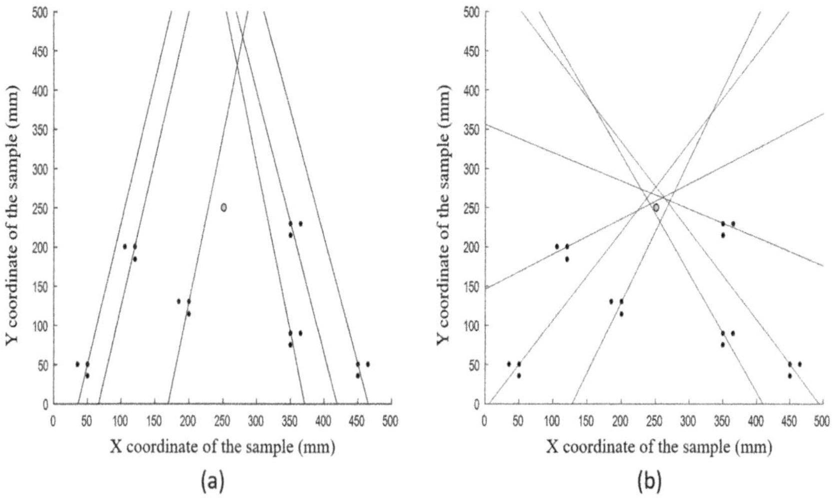

Figure 5.16 Lines parallel to every $\overrightarrow{w^{\perp}}$ (see Figure 5.12) are displayed where $\overrightarrow{w^{\perp}}$ are obtained from the measured TDOAs for every cluster using Eq. (5.54). Predictions are from (a) the first wave group with rhombus wave front and (b) the second wave group with non-circular (almost elliptic) wave front. Note that the intersection points which are predictions for the acoustic source locations are quite far from the actual source location denoted by the single dot in the middle at coordinate (250, 250). The prediction error is more in Figure 5.16a (Park et al., 2017).

measured wave propagation direction vectors ($\overrightarrow{w^{\perp}}$) for two adjacent clusters become parallel regardless of the cluster's location, as shown in Figure 5.16a. In this example, the rhombus wave front propagates faster than the other wave front so the sensors detect it first without any interference with the reflected waves from the boundary.

For ASL the concentric rhombus wave fronts depicted in Figure 5.17 are analyzed. Geometric properties of the concentric rhombuses are utilized to determine the acoustic source location. All concentric rhombuses share a vertical diagonal and a horizontal diagonal, and the intersection of these two diagonals should be the acoustic source location as shown in Figure 5.17. It should be noted that the shape velocity or the velocity of the rhombus wave front is constant. The distance between two sensor clusters S_r and S_1 along the wave propagation direction is denoted by d_{r1}. TDOA between these two sensor clusters is $(t_r - t_1)$. Then the shape velocity $\mu = \frac{d_{r1}}{(t_r - t_1)}$.

In order to compute the shape velocity, Park et al. (2017) placed two clusters S_r and S_I in the same quadrant. According to them, the minimum number of required sensor clusters would be four to localize the acoustic source in an anisotropic plate that generates a rhombus wave front. However, the number of required clusters can be reduced further to three because it is also possible to calculate the wave front shape velocity from a single cluster (either S_r or S_I) also. After obtaining the wave propagation direction through that cluster one can use Eq. (5.41) to calculate the speed of the wave front passing through the cluster and avoid the need to use two clusters in the same quadrant.

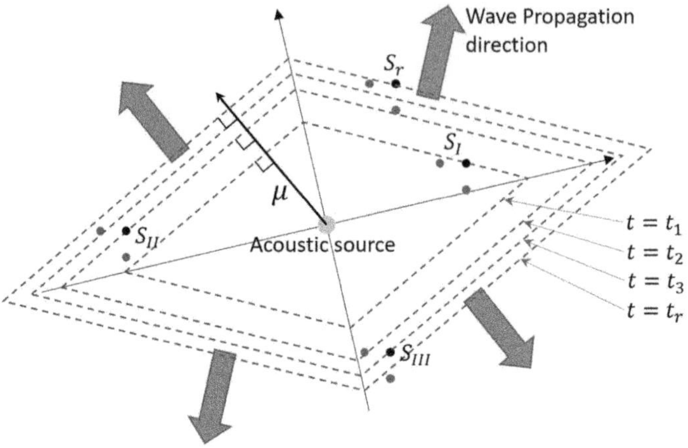

Figure 5.17 Rhombus wave front is generated by an acoustic source and expands with a shape velocity (or wave front velocity) μ. At $t = t_1$, sensor cluster S_I captures the wave front first. Then clusters S_{II}, S_{III}, and S_r receive signals sequentially at t_2, t_3, and t_r, respectively. Simple vector analyses with direction vectors at all clusters and TDOAs allow to form the diagonals of the concentric rhombuses. The intersection point of the two diagonals gives the final acoustic source location (Park et al., 2017).

One diagonal of the rhombus can be obtained by simple vector analysis with direction vectors and TDOA $(t_2 - t_1)$ between clusters S_I and S_{II}. Similarly, the second diagonal of the rhombus can be obtained from direction vectors and TDOA $(t_3 - t_1)$ between clusters S_I and S_{III}. These two diagonals are denoted as vertical and horizontal diagonals. Detail derivation is given below. After two diagonals are obtained from the concentric rhombuses, the acoustic source location is identified from the intersection point of the two diagonals. Therefore, only geometric properties of the rhombus shape and direction vector measurements at three cluster positions are needed to predict the exact source location without knowing any material properties or direction dependent velocity profile of the anisotropic plate.

Direction vectors and parallel vectors at the sensor clusters were described in Section 5.8.1. How to obtain TDOAs between any two sensors by applying cross-correlation technique to the received signals at those two sensors has been also discussed earlier. In the following the vector analysis for localizing the acoustic source is introduced. Let us denote the direction vector at S_r and S_I as $\overrightarrow{u^{\perp}}$ and the parallel vector at S_I as $\vec{u}$. If the distance between wave front L_1 and L_r is d_{1r}, then the shape velocity or the wave front velocity can be obtained in the following manner:

$$\mu = \frac{d_{1r}}{t_r - t_1} = \frac{\left\| \text{proj}_{\overrightarrow{u^{\perp}}}\left(\overrightarrow{S_R} - \overrightarrow{S_I}\right) \right\|}{t_r - t_1} = \left\| \frac{\left(\overrightarrow{S_R} - \overrightarrow{S_I}\right) \cdot \overrightarrow{u^{\perp}}}{\overrightarrow{u^{\perp}} \cdot \overrightarrow{u^{\perp}}} \overrightarrow{u^{\perp}} \right\| / (t_r - t_1) \tag{5.56}$$

where $\text{proj}_{\vec{b}}\vec{a}$ represents vector projection of $\vec{a}$ onto $\vec{b}$.

Another direction vector and parallel vector at S_{II} are denoted as $\overrightarrow{v^{\perp}}$ and $\vec{v}$, respectively. Next, the parametric representation (κ_1: some number in real space) for line L_1 on 2D plane is introduced in the following manner:

$$L_1 = \left\{ \overrightarrow{S_I} + \kappa_1 \vec{u} \mid \kappa_1 \in \mathbb{R} \right\} \tag{5.57}$$

where $\mathbb{R}$ is real number space and all vectors have two coordinate components, $(x,y)^{\mathrm{T}}$. If TDOA between S_I and S_{II} is zero, the bisector V_{bi} in Figure 5.18a becomes the vertical diagonal of the rhombus directly. Otherwise, the line L_1 should be shifted by d_{12} to align the rhombus at $t = t_2$ in order to find the true vertical bisector, V_{tr}. Using the given shape velocity, the shift is simply

$$d_{12} = \mu(t_2 - t_1) \tag{5.58}$$

The shifted line (L_1') may be derived from the following parametric representation (κ_1': some number in real space) involving the shift:

$$L_1' = \left\{ \overrightarrow{S_I} + d_{12}\frac{\overrightarrow{u^{\perp}}}{\left\|\overrightarrow{u^{\perp}}\right\|} + \kappa_1'\vec{u} \,\middle|\, \kappa_1' \in \mathbb{R} \right\} \tag{5.59}$$

and the line (L_2) passing through S_{II} is given in the same manner:

$$L_2 = \left\{ \overrightarrow{S_{II}} + \kappa_2\vec{v} \,\middle|\, \kappa_2 \in \mathbb{R} \right\} \tag{5.60}$$

Then $L_1' = L_2$ is solved to find the intersection point (P_V) of L_1' and L_2, denoted as

$$\overrightarrow{S_I} + d_{12}\frac{\overrightarrow{u^{\perp}}}{\left\|\overrightarrow{u^{\perp}}\right\|} + \kappa_1'\vec{u} = \overrightarrow{S_{II}} + \kappa_2\vec{v} \tag{5.61}$$

By rearranging Eq. (5.61) as follows, one can solve the simultaneous equations of two unknowns, κ_1' and κ_2:

$$\begin{bmatrix} \kappa_1' \\ \kappa_2 \end{bmatrix} = \begin{bmatrix} \underline{u} & -\underline{v} \end{bmatrix}^{-1} \left(\underline{S_{II}} - \underline{S_I} - d_{12}\frac{\underline{u^{\perp}}}{\left\|\underline{u^{\perp}}\right\|} \right) \tag{5.62}$$

P_V is calculated by either substituting κ_1' into Eq. (5.59) or substituting κ_2 into Eq. (5.60) in the parametric formulation of the true bisector V_{tr} (solid line in Figure 5.18a)

$$V_{tr} = \left\{ \overrightarrow{P_V} + \kappa_V\overrightarrow{b_V} \,\middle|\, \kappa_V \in \mathbb{R} \right\} \tag{5.63}$$

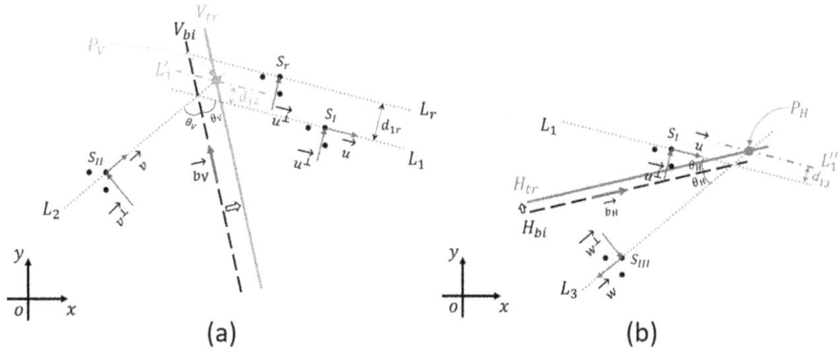

Figure 5.18 (a) V_{tr} and (b) H_{tr}, are obtained from direction vectors and TDOAs at four clusters (Park et al., 2017).

where $\overrightarrow{b_V}$ is parallel to the bisector and is obtained by adding two unit vectors:

$$\overrightarrow{b_V} = \frac{\overrightarrow{u^\perp}}{\left\|\overrightarrow{u^\perp}\right\|} + \frac{\overrightarrow{v^\perp}}{\left\|\overrightarrow{v^\perp}\right\|} \tag{5.64}$$

To obtain the true horizontal diagonal (H_{tr}) of the rhombus wave front, one can follow the same procedure as described in Eqs. (5.58) to (5.64). The new shift (d_{13}) and the new shifted line (L_1'') are computed based on the TDOAs between S_I and S_{III}, as shown in Figure 5.18b

$$d_{13} = \mu(t_3 - t_1) \tag{5.65}$$

and

$$L_1'' = \left\{\overrightarrow{S_I} + d_{13}\frac{\overrightarrow{u^\perp}}{\left\|\overrightarrow{u^\perp}\right\|} + \kappa_1''\vec{u} \,\middle|\, \kappa_1'' \in \mathbb{R}\right\} \tag{5.66}$$

The other line passing through S_{III} is expressed using the direction vector ($\vec{w}$) as

$$L_3 = \left\{\overrightarrow{S_{III}} + \kappa_3\vec{w} \,\middle|\, \kappa_3 \in \mathbb{R}\right\} \tag{5.67}$$

To obtain the intersection point P_H, one needs to solve

$$L_1'' = L_3 \tag{5.68}$$

and denote Eq. (5.68) as

$$\overrightarrow{S_I} + d_{13}\frac{\overrightarrow{u^\perp}}{\left\|\overrightarrow{u^\perp}\right\|} + \kappa_1''\vec{u} = \overrightarrow{S_{III}} + \kappa_3\vec{w} \tag{5.69}$$

By rearranging Eq. (5.69) one gets the following system of simultaneous equations from which two unknowns, κ_1'' and κ_3, can be solved:

$$\begin{bmatrix} \kappa_1'' \\ \kappa_3 \end{bmatrix} = \begin{bmatrix} \underline{u} & -\underline{w} \end{bmatrix}^{-1} \left(\underline{S_{III}} - \underline{S_I} - d_{13}\frac{\underline{u^\perp}}{\left\|\underline{u^\perp}\right\|} \right) \tag{5.70}$$

P_H is calculated by either substituting κ_1'' into Eq. (5.66) or substituting κ_3 into Eq. (5.67) and is used in the parametric formulation of the bisector H_{tr} (solid line in Figure 5.18b)

$$H_{tr} = \left\{\overrightarrow{P_H} + \kappa_H\overrightarrow{b_H} \,\middle|\, \kappa_H \in \mathbb{R}\right\} \tag{5.71}$$

where $\overrightarrow{b_H}$ is parallel to the bisector:

$$\overrightarrow{b_H} = \frac{\overrightarrow{u^\perp}}{\left\|\overrightarrow{u^\perp}\right\|} + \frac{\overrightarrow{w^\perp}}{\left\|\overrightarrow{w^\perp}\right\|} \tag{5.72}$$

Since the two true bisectors are diagonals of the rhombus wave front, we can conclude that the final intersection point of V_{tr} and H_{tr} must be the acoustic source location. The final step of predicting the source position (P_S) is to solve the following relation:

$$V_{tr} = H_{tr} \tag{5.73}$$

After substituting Eq. (5.63) and Eq. (5.71) into Eq. (5.73) one gets

$$\overrightarrow{P_V} + \kappa_V \overrightarrow{b_V} = \overrightarrow{P_H} + \kappa_H \overrightarrow{b_H} \tag{5.74}$$

After some algebraic manipulation,

$$\begin{bmatrix} \kappa_V \\ \kappa_H \end{bmatrix} = \begin{bmatrix} \underline{b_V} & -\underline{b_H} \end{bmatrix}^{-1} \left(\underline{P_H} - \underline{P_V} \right) \tag{5.75}$$

Estimated acoustic source location $\left(P_S^*\right)$ shown in Figure 5.19 is finally determined by either substituting κ_V into Eq. (5.63) or substituting κ_H into Eq. (5.71).

5.8.3.2 Elliptical Wave Front

Another common non-circular wave front for anisotropic plates is an ellipse or close to an ellipse, as shown in Figure 5.14a. Figure 5.16 shows how the source localization technique with circular wave front assumption fails for anisotropic plates when the actual wave front is close to a rhombus (see Figure 5.16a) or an ellipse (see Figure 5.16b). In the example considered here the elliptic wave front propagates behind the rhombus wave front (see Figure 5.14a) but the propagating energy is much higher for the elliptic wave front and therefore gives higher signal to noise ratio in the recorded signal for this wave front. The analysis for the elliptical wave front and near elliptical wave front that is presented below is based on the work of Sen and Kundu (2018).

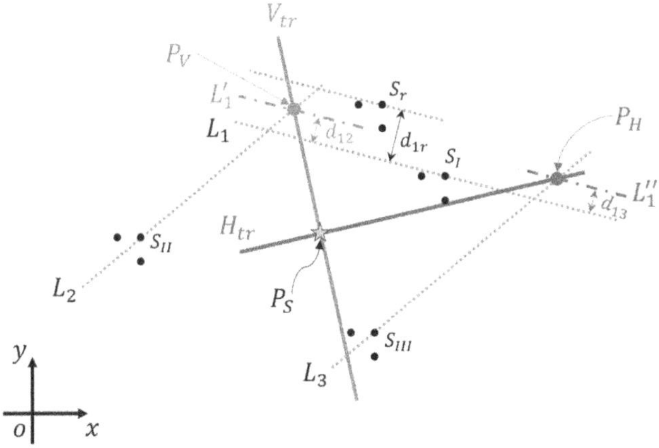

Figure 5.19 Two computed bisectors (V_{tr} and H_{tr}) are drawn together. The intersection point of the two lines (marked by a star symbol) indicates the final acoustic source position obtained after a series of vector analyses involving direction vectors and TDOAs (Park et al., 2017).

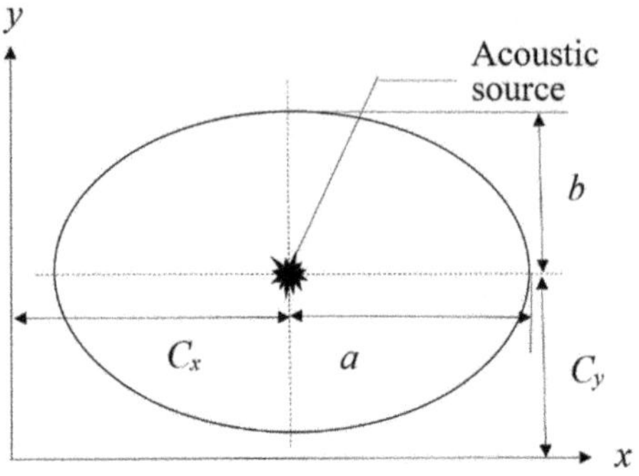

Figure 5.20 Schematic of an elliptical wave front (Sen and Kundu, 2018).

The equation of an elliptical wave front as shown in Figure 5.20 is given by

$$\frac{(x-C_x)^2}{a^2}+\frac{(y-C_y)^2}{b^2}=1 \tag{5.76}$$

where (x,y) are the coordinates of a point on the ellipse, (C_x,C_y) are the coordinates of the acoustic source, and a and b are the lengths of the semi-major and semi-minor axes, respectively. Taking derivative of Eq. (5.76) with respect to x leads to

$$\frac{dy}{dx}=-\frac{b^2}{a^2}\frac{x-C_x}{y-C_y} \tag{5.77}$$

An acoustic event generates wave fronts that are concentric in nature. Therefore, the ratio b^2/a^2 can be replaced by a single (positive) unknown parameter λ. Thus, Eq. (5.77) may be rewritten as

$$\lambda(x-C_x)+m(y-C_y)=0 \tag{5.78}$$

where $m=dy/dx$. Since the wave fronts pass through the sensor clusters, for the ith sensor cluster, Eq. (5.78) gives

$$\lambda(x_i-C_x)+m_i(y_i-C_y)=0 \tag{5.79}$$

with (x_i,y_i) denoting the coordinates of the ith sensor cluster and m_i is the slope m measured at that cluster. It may be noted that m_i physically represents the slope of the tangent line to the wave front at (x_i,y_i), i.e.,

$$m_i=\tan\psi_i \tag{5.80}$$

where ψ_i is the angle ψ (shown in Figure 5.21) for the ith sensor cluster.

It should be noted that Park et al. (2017) considered two different orientations of L-shaped sensor cluster as shown in Figures 5.21a and 5.21b, denoted as Type-a and Type-b sensor clusters, respectively. The key difference between these two orientations lies in the relative positions of the three sensors A, O and B within a cluster. As can be seen from Figure 5.21, the Type-a sensor cluster receives the incident wave-angle in the form of an acute angle, while the Type-b sensor cluster receives the incident wave-angle in the form of an obtuse angle. It can be shown that this basic difference between the two types of sensor clusters

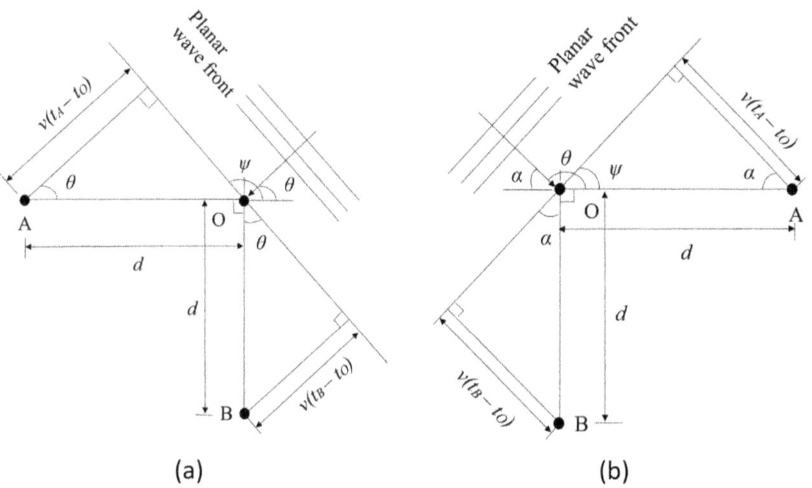

Figure 5.21 (a) Type-a and (b) Type-b sensor clusters (Sen and Kundu, 2018).

leads to tangents of the incident wave-angles given by the opposite signs of the same measured quantity. Both these types of sensor clusters were mounted on a plate such that the lines AO and OB are parallel to the x and y-axes of the assumed coordinate system, respectively.

The angle θ is the incident angle of the planar wave front striking the sensor cluster. The wave front first arrives at sensor O, then it strikes either sensor A or B. If t_O, t_A and t_B denote the times of arrival (TOAs) of the signal from the acoustic source to the sensors O, A and B, respectively, then for the Type-a configuration [see Figure 5.21a], $\tan\theta$ can be expressed as

$$\tan\theta = \frac{\sin\theta}{\cos\theta} = \frac{v(t_B - t_O)/d}{v(t_A - t_O)/d} = \frac{t_B - t_O}{t_A - t_O} \tag{5.81}$$

Similarly, for the Type-b configuration (see Figure 5.21b),

$$\tan\theta = -\tan\alpha = -\frac{\sin\alpha}{\cos\alpha} = -\frac{v(t_B - t_O)/d}{v(t_A - t_O)/d} = -\frac{t_B - t_O}{t_A - t_O} \tag{5.82}$$

In Eqs. (5.81) and (5.82) v is the velocity of the wave. In order to obtain a sufficiently accurate estimation of $\tan\theta$, TOAs (i.e., t_O, t_A and t_B) or TDOAs (i.e., t_B–t_O and t_A–t_O) need to be measured as accurately as possible. This can be achieved by either applying the cross-correlation technique to the two signals [see Eq. (5.55)] or using the Akaike information criterion (AIC) (Park, 2016; Ebrahimkhanlou and Salamone, 2017) given by

$$\text{AIC}(k) = k\ln[\text{var}\{U(1:k)\}] + (K - k - 1)\ln[\text{var}\{U(k+1:K)\}] \tag{5.83}$$

where k is the sample number corresponding to time t in the recorded signal, K is the total number of samples recorded, U is the recorded signal amplitude and $\text{var}\{U(r:s)\}$ denotes the variance of all signal amplitude values between (and including) rth and sth samples. If AIC is plotted against k, two distinct valleys are observed. The first valley represents the arrival of the first wave group, while the second valley represents the arrival of the second wave group (Park, 2016). The values of t corresponding to the first and second valleys are the required

TOAs at the sensor for the first and second wave groups, respectively. The second wave group approximately forms the elliptic wave front while the first wave group forms the rhombus wave front.

It follows from Figure 5.21 that ψ_i is equal to $\theta_i + \frac{\pi}{2}$ and $\theta_i - \frac{\pi}{2}$ for Type-a and Type-b sensor clusters, respectively. Therefore, Eq. (5.80) can be rewritten as

$$m_i = -\frac{1}{\tan\theta_i} \tag{5.84}$$

where $\tan\theta_i$ can be obtained using Eq. (5.81) or (5.82) for the *ith* sensor cluster. Thus, Eq. (5.79) consists of three unknowns: λ, C_x and C_y. If n number of sensor clusters are used then Eq. (5.79) leads to n simultaneous nonlinear equations. These equations can be solved using the Levenberg–Marquardt algorithm which minimizes the following objective function

$$\Phi = \sum_{i=1}^{n}\left[\lambda(x_i - C_x) + m_i(y_i - C_y)\right]^2 \tag{5.85}$$

It should be noted that to obtain the unknowns λ, C_x and C_y, a minimum of three clusters are needed. Therefore, the minimum value for n should be 3. The Levenberg–Marquardt algorithm (LMA) to solve the three unknowns by minimizing Φ is described in the following:

Let $\beta = [\lambda \quad C_x \quad C_y]^T$, $f_i(\beta) = \lambda(x_i - C_x) + m_i(y_i - C_y)$ and $\mathbf{f}(\beta) = [f_1 \quad f_2 \quad \ldots \quad f_n]^T$. To start the Levenberg–Marquardt algorithm, an initial guess of $\beta = \beta_0$ is needed. Then the following steps are performed to solve $\boldsymbol{\beta}$ iteratively:

1. Step 1: Compute the Jacobian $\mathbf{J}$ as

$$\mathbf{J} = \begin{bmatrix} \frac{\partial f_1}{\partial \lambda} & \frac{\partial f_1}{\partial C_x} & \frac{\partial f_1}{\partial C_y} \\ \frac{\partial f_2}{\partial \lambda} & \frac{\partial f_2}{\partial C_x} & \frac{\partial f_2}{\partial C_y} \\ \ldots & \ldots & \ldots \\ \frac{\partial f_n}{\partial \lambda} & \frac{\partial f_n}{\partial C_x} & \frac{\partial f_n}{\partial C_y} \end{bmatrix} = \begin{bmatrix} x_1 - C_x & -\lambda & -m_1 \\ x_2 - C_x & -\lambda & -m_2 \\ \ldots & \ldots & \ldots \\ x_n - C_x & -\lambda & -m_n \end{bmatrix} \tag{5.86}$$

2. Step 2: Evaluate $\mathbf{f}(\beta)$.
3. Step 3: Find the increment of $\boldsymbol{\beta}$ as

$$\Delta\boldsymbol{\beta} = -[\mathbf{J}^T\mathbf{J}]^{-1}\mathbf{J}^T\mathbf{f} \tag{5.87}$$

4. Step 4: Calculate a new $\boldsymbol{\beta}$ by adding $\Delta\boldsymbol{\beta}$ to the current $\boldsymbol{\beta}$.
5. Step 5: Calculate the Euclidean norm of $\Delta\beta$. If the result is more than a predefined tolerance value, then go back to step 1. Otherwise, exit the algorithm and report the latest $\boldsymbol{\beta}$ as the final solution. Thus, λ, C_x and C_y can be obtained systematically and the predicted location of the acoustic source is given by (C_x, C_y).

5.8.3.3 Numerical Validation for Rhombus Wave Front

Wave fronts generated by the LISA numerical simulation (Packo et al., 2015) are utilized to validate the wave front based technique for source localization. As the recorded signal at all sensor clusters have two distinct arrivals as shown in Figure 5.15, the method appropriate for the rhombus wave front is employed first, and then the technique for the elliptic wave front is employed. The following results were obtained by analyzing different parts of the time signals, namely the first arrival for the rhombus wave front and the second arrival for the elliptic wave front. In Figure 5.22, the estimated source location P_S^* for the rhombus wave front is denoted by a solid circle and the relevant straight lines obtained from the vector analysis given in Eqs. (5.6) to (5.75) are plotted. Three sensor clusters (S_I, S_{II}, S_{III}) were placed in three different quadrants and the reference sensor cluster S_r was placed in the third quadrant with S_I. The analysis starts with the calculation of direction vectors, $\overrightarrow{u^\perp}$, $\overrightarrow{v^\perp}$ and $\overrightarrow{w^\perp}$ at the three sensor cluster positions. The rhombus wave front velocity or shape velocity (μ) is obtained from $\overrightarrow{u^\perp}$, and TDOA value t_{r1} between S_r and S_I. Alternately, it could also be obtained from the time difference of arrivals of the signal between two sensors in the same cluster – either S_r or S_I. TDOA value t_{21} and $\overrightarrow{u^\perp}$ allows us to obtain the first shifted line (L_1'), then P_V is obtained, which is the intersection point of lines L_1' and L_2. The vertical diagonal (the vertical solid line V_{tr} in Figure 5.22) of the rhombus wave front can also be obtained by drawing a straight line which is parallel to the bisecting vector $\overrightarrow{b_V}$ and going through point P_v. In a similar manner, with two sensors, S_I and S_{III}, the horizontal diagonal (the solid line H_{tr} in Figure 5.22) can be drawn. Finally, the intersecting point of the two diagonals is the estimated acoustic source location. All computed vectors, points from the numerical simulation and vector analysis are listed in Table 5.2. Note that the prediction error [or the distance of the predicted point (P_S^*) from the

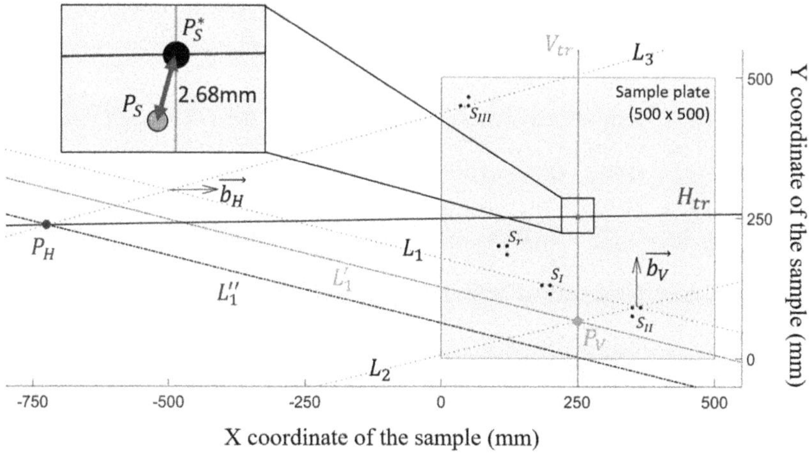

Figure 5.22 Computed lines, vectors and points obtained from the vector analysis for the rhombic wave front using numerically simulated results. The estimated location (P_s^*) of the acoustic source (250.714, 252.581) is only 2.68 mm away from the true source location (P_S). The gray square depicts the 500 mm × 500 mm plate (Park et al., 2017).

Table 5.2: Computed Vectors and Points from the Numerical Simulation and the Vector Analysis

	Coordinates (mm)	Computed Vectors		Computed Vectors or Points
S_I	(200, 130)	$\overrightarrow{u^\perp} = \begin{pmatrix}-1 & -4.1429\end{pmatrix}^T$	$\overrightarrow{b_V}$	$\begin{pmatrix}0 & -1.9442\end{pmatrix}^T$
S_{II}	(350, 90)	$\overrightarrow{v^\perp} = \begin{pmatrix}1 & -4.1429\end{pmatrix}^T$	$\overrightarrow{b_H}$	$\begin{pmatrix}-0.5006 & -0.0081\end{pmatrix}^T$
S_{III}	(50, 450)	$\overrightarrow{w^\perp} = \begin{pmatrix}-1 & 3.6250\end{pmatrix}^T$	P_V	(250.714, 66.034)
S_r	(120, 200)	$\overrightarrow{u^\perp} = \begin{pmatrix}-1 & -4.1429\end{pmatrix}^T$	P_H	(−722.667, 236.851)
Rhombus shape velocity			μ	2.0112 km / s
Actual location of acoustic source			$\mathbf{P_S}$	**(250.00, 250.00)**
Final estimation of acoustic source			$\mathbf{P_S^*}$	**(250.71, 252.58)**

Note that the presented vectors are not normalized.

actual acoustic source location (P_S)] is only 2.68 mm, while the distances between the source and four clusters vary from 130 to 283 mm.

Source localizations obtained from the elliptic wave front assumption was not very accurate (see Park et al., 2017). This inaccuracy for the elliptic wave front is due to the fact that the simulated wave front was not perfectly ellipse. Sen and Kundu (2018) tackled this problem by considering a non-elliptic wave front as described in the following. Sections 5.8.3.4 and 5.8.3.5 are taken from their work.

5.8.3.4 Wave Front Modeled by Non-Elliptical Parametric Curve

The wave front in an anisotropic plate generally deviates from an exact elliptical shape, as can be seen in the above example. Hence, a curve that can represent a more general shape than an ellipse is needed and was introduced by Sen and Kundu (2018). They modified Eq. (5.76) to obtain the following general expression for the parametric curve:

$$\gamma \frac{|x - C_x|^{p+1}}{p+1} + \frac{|y - C_y|^{q+1}}{q+1} = c \tag{5.88}$$

where γ, p and q are three unknown positive numbers and c is an arbitrary constant. Taking a derivative of Eq. (5.88) with respect to x leads to

$$m = (-1)^{quad} \gamma \frac{|x - C_x|^p}{|y - C_y|^q} \tag{5.89}$$

where *quad* denotes the quadrant number of the point (x, y) with respect to the straight lines $x = C_x$ and $y = C_y$. Thus,

$$\begin{aligned} quad &= 1 \text{ if } x > C_x \text{ and } y \geq C_y \\ &= 2 \text{ if } x \leq C_x \text{ and } y > C_y \\ &= 3 \text{ if } x < C_x \text{ and } y \leq C_y \\ &= 4 \text{ if } x \geq C_x \text{ and } y < C_y \end{aligned} \tag{5.90}$$

It should be noted that the parametric curve of Eq. (5.88) is symmetric with respect to the straight lines $x = C_x$ and $y = C_y$. Further, from Eqs. (5.89) and (5.90), it follows that the derivatives of the curve at four points on these two lines (two points along $x = C_x$ and two along $y = C_y$) are same as those for an ellipse at the corresponding points. Eq. (5.88) indicates that the parametric curve becomes an ellipse when both p and q are equal to 1. Hence, ellipse is a special case of the family of curves represented by Eq. (5.88). By assuming (C_x, C_y) as (0,0) along with different combinations of γ as 1.0 and 1.5, p and q as 0.6, 1.1 and 1.6, and c as 10, 15 and 20, a wide range of curves can be generated from Eq. (5.88), as shown in Figure 5.23. Each of the 18 plots of Figure 5.23 considers a constant set of values for p, q and γ, and shows three concentric curves corresponding to three different values of c but for the same values of the other parameters p, q and γ.

In order to demonstrate the generality of Eq. (5.88) for modeling a realistic wave front, the numerically generated wave front shown in Figure 5.14a is modeled by this equation. The parameters of Eq. (5.88) are adjusted by trial and

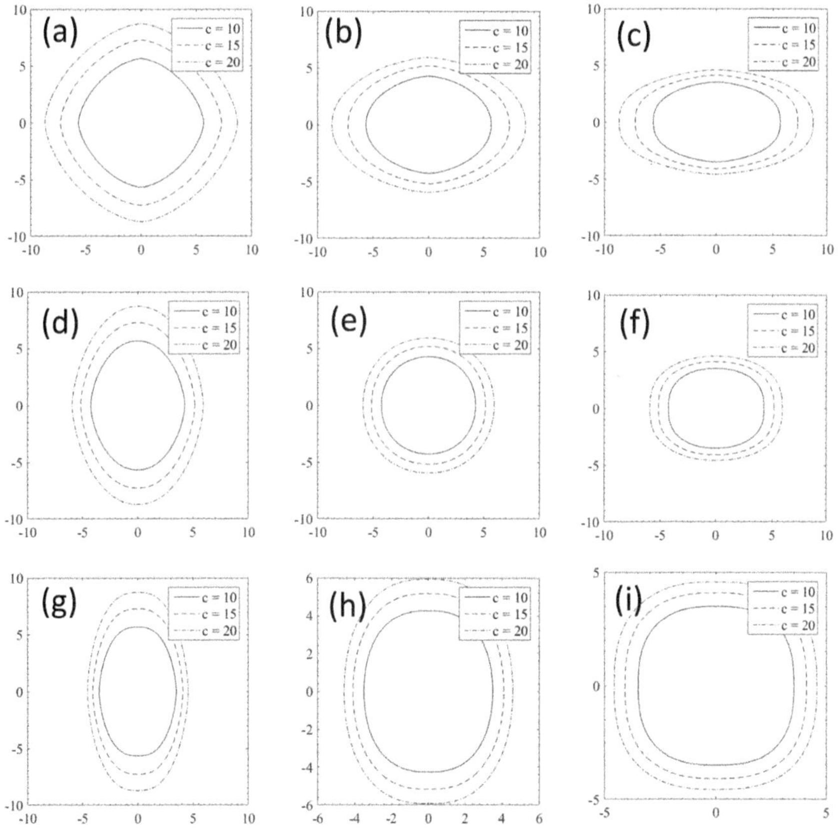

Figure 5.23 Generated parametric curves given by Eq. (5.88) considering different values of c, p, q and γ (a) $p = 0.6$, $q = 0.6$ and $\gamma = 1.0$, (b) $p = 0.6$, $q = 1.1$ and $\gamma = 1.0$, (c) $p = 0.6$, $q = 1.6$ and $\gamma = 1.0$, (d) $p = 1.1$, $q = 0.6$ and $\gamma = 1.0$, (e) $p = 1.1$, $q = 1.1$ and $\gamma = 1.0$, (f) $p = 1.1$, $q = 1.6$ and $\gamma = 1.0$, (g) $p = 1.6$, $q = 0.6$ and $\gamma = 1.0$, (h) $p = 1.6$, $q = 1.1$ and $\gamma = 1.0$, (i) $p = 1.6$, $q = 1.6$ and $\gamma = 1.0$, (j) $p = 0.6$, $q = 0.6$ and $\gamma = 1.5$, (k) $p = 0.6$, $q = 1.1$ and $\gamma = 1.5$, (l) $p = 0.6$, $q = 1.6$ and $\gamma = 1.5$, (m) $p = 1.1$, $q = 0.6$ and $\gamma = 1.5$, (n) $p = 1.1$, $q = 1.1$ and $\gamma = 1.5$, (o) $p = 1.1$, $q = 1.6$ and $\gamma = 1.5$, (p) $p = 1.6$, $q = 0.6$ and $\gamma = 1.5$, (q) $p = 1.6$, $q = 1.1$ and $\gamma = 1.5$ and (r) $p = 1.6$, $q = 1.6$ and $\gamma = 1.5$ (Sen and Kundu, 2018).

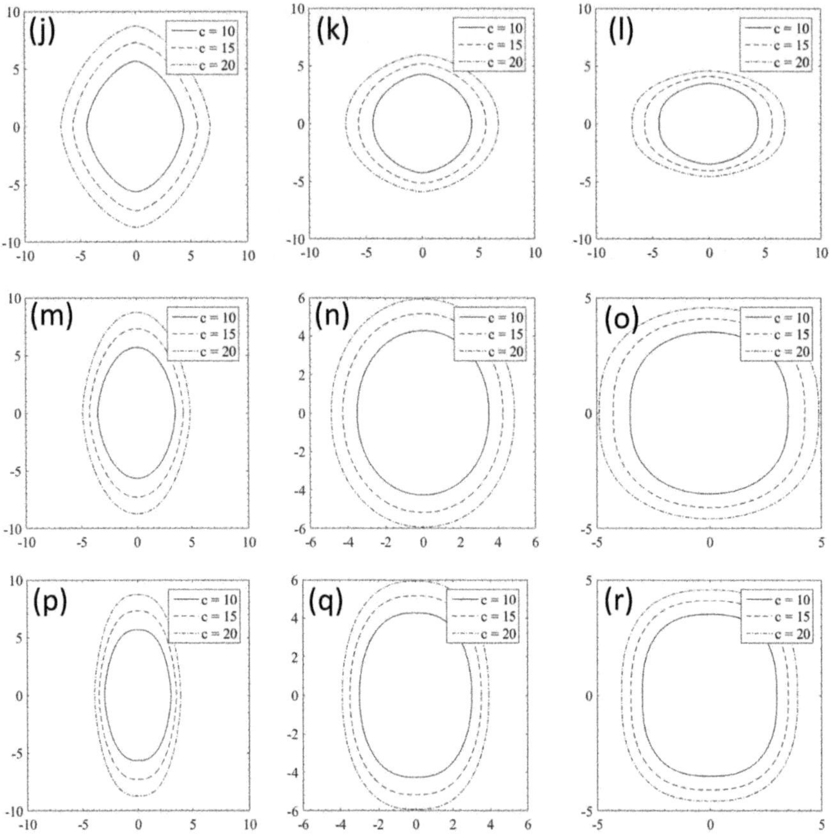

Figure 5.23 (Continued)

error so that the generated curve matches closely with the wave front shape. This matching is shown in Figure 5.24. The figure shows the best-fitted ellipse and the best-fitted parametric curve plotted on the numerically generated wave front. The parameters chosen for the parametric curve of Eq. (5.88) to generate Figure 5.24 are: $C_x, C_y = 250,\ 250$, $p = 0.6$, $q = 0.6$, $\gamma = 0.46$ and $c = 550$, whereas for the elliptical curve the chosen parameters are: $(C_x, C_y) = (250, 250)$, $p = 1.0$, $q = 1.0$, $\gamma = 0.4$ and $c = 2200$. Although these two curves look almost identical in Figure 5.24, a close observation of the two curves reveals some differences. For example, compared to the ellipse the parametric curve is extended more along the major and minor axes. Sen and Kundu (2018) showed that even this small difference can produce a noticeable improvement in the source localization.

As in the case of an elliptical wave front, n sensor clusters can be considered to solve the unknowns. For the ith sensor cluster, Eq. (5.89) may be rewritten as

$$(-1)^{quad_i}\gamma\,|\,x_i - C_x\,|^p - m_i\,|\,y_i - C_y\,|^q = 0 \tag{5.91}$$

where $quad_i$ denotes the quad value [1, 2, 3 or 4, as defined in Eq. (5.90)] for the ith cluster. Eq. (5.91) consists of five unknowns: p, q, γ, C_x and C_y. Thus, n

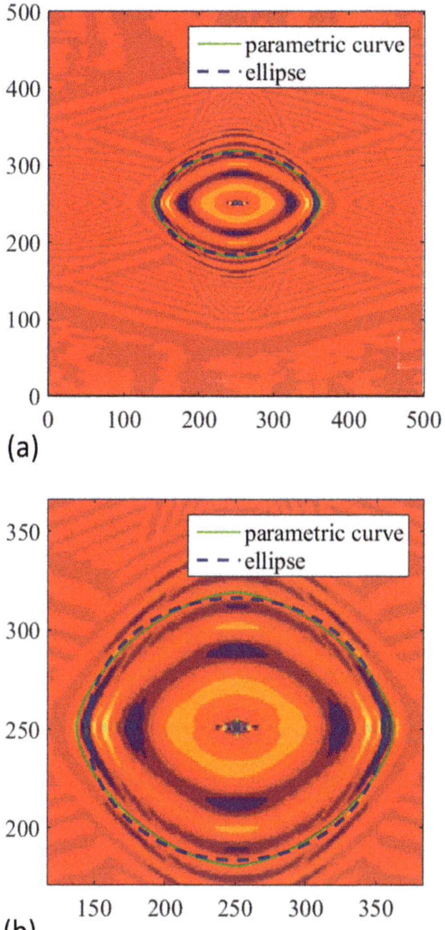

Figure 5.24 (a) A comparison between a numerically generated wave front and wave fronts modeled by the parametric curve (thin continuous line) and ellipse (thick dashed line), (b) zoomed-in view. It should be noticed that compared to the ellipse the parametric curve is extended more along both major and minor axes (Sen and Kundu, 2018).

simultaneous nonlinear equations are obtained from Eq. (5.91). The following objective function S can be minimized to solve the unknowns.

$$S = \sum_{i=1}^{n} [(-1)^{\text{quad}_i} \gamma \,|\, x_i - C_x \,|^p - m_i \,|\, y_i - C_y \,|^q]^2 \tag{5.92}$$

It was observed that if both exponents p and q were considered to be unknown variables, the minimization process often failed to converge. To avoid this, p was considered to be an unknown parameter while q was set as a known constant value. Different constant values for q were considered to arrive at the best fit between the parametric curve and the numerically generated wave front. The simplex algorithm (Nelder and Mead, 1965) can be used to minimize S. The specific version of this algorithm that was used in this example for minimizing this objective function was given by Lagarias et al. (1998) and is described below.

Let $\mathbf{x} = [p \quad \gamma \quad C_x \quad C_y]^T$. Five initial vertices for $\mathbf{x}$ (i.e., five different initial guesses for $\mathbf{x}$) are considered and S is calculated using Eq. (5.92) for all five vertices. Then the following steps are performed to iteratively solve the unknowns.

1. Step 1: Order the five vertices so that $S(\mathbf{x}_1) \le S(\mathbf{x}_2) \le S(\mathbf{x}_3) \le S(\mathbf{x}_4) \le S(\mathbf{x}_5)$.
2. Step 2: Compute the reflection point $\mathbf{x}_r$ by

$$\mathbf{x}_r = \bar{\mathbf{x}} + \rho(\bar{\mathbf{x}} - \mathbf{x}_5) \tag{5.93}$$

where $\bar{\mathbf{x}} = \sum_{i=1}^{4} \mathbf{x}_i / 4$ is the centroid of the four best points (i.e., all vertices except for $\mathbf{x}_5$), and ρ is the coefficient of reflection chosen as 1. Compute $S(\mathbf{x}_r)$. If $S(\mathbf{x}_1) \le S(\mathbf{x}_r) < S(\mathbf{x}_4)$, replace $\mathbf{x}_5$ by $\mathbf{x}_r$ and go to step 1. Otherwise, go to step 3.

3. Step 3: If $S(\mathbf{x}_r) < S(\mathbf{x}_1)$, compute the expansion point $\mathbf{x}_e$ as given below, otherwise go to step 4.

$$\mathbf{x}_e = \bar{\mathbf{x}} + \chi(\mathbf{x}_r - \bar{\mathbf{x}}) \tag{5.94}$$

where χ is the coefficient of expansion chosen as 2. Evaluate $S(\mathbf{x}_e)$. If $S(\mathbf{x}_e) < S(\mathbf{x}_r)$, replace $\mathbf{x}_5$ by $\mathbf{x}_e$ and go to step 1. Otherwise, for $S(\mathbf{x}_e) \ge S(\mathbf{x}_r)$, replace $\mathbf{x}_5$ with $\mathbf{x}_r$ and go to step 1.

4. Step 4: If $S(\mathbf{x}_r) \ge S(\mathbf{x}_4)$, go to step 4a.

Step 4a: If $S(\mathbf{x}_4) \le S(\mathbf{x}_r) < S(\mathbf{x}_5)$, perform an outside contraction by calculating x_c as shown in Eq. (5.95). Otherwise, if $S(\mathbf{x}_r) \ge S(\mathbf{x}_5)$, go to step 4b.

$$\mathbf{x}_c = \bar{\mathbf{x}} + \eta(\mathbf{x}_r - \bar{\mathbf{x}}) \tag{5.95}$$

where η is the coefficient of contraction chosen as 1/2. Evaluate $S(\mathbf{x}_c)$. If $S(\mathbf{x}_c) \le S(\mathbf{x}_r)$, replace $\mathbf{x}_5$ with $\mathbf{x}_c$ and go to step 1. Otherwise, go to step 5.

Step 4b: Perform an inside contraction by calculating $\mathbf{x}_{cc}$ as

$$\mathbf{x}_{cc} = \bar{\mathbf{x}} - \eta(\bar{\mathbf{x}} - \mathbf{x}_5) \tag{5.96}$$

Evaluate $S(\mathbf{x}_{cc})$. If $S(\mathbf{x}_{cc}) < S(\mathbf{x}_5)$, replace $\mathbf{x}_5$ with $\mathbf{x}_{cc}$ and go to step 1. Otherwise, go to step 5.

5. Step 5: Perform a shrink by replacing all points except $\mathbf{x}_1$ by $\mathbf{x}_i = \mathbf{x}_1 + \sigma(\mathbf{x}_i - \mathbf{x}_1)$, $i = 2, 3, 4$ and 5. σ is the coefficient of shrinkage chosen as 1/2.

The above-mentioned steps can be performed until the standard deviation of the computed values of S at the current vertices becomes less than a pre-defined tolerance level. At this stage, the algorithm is terminated and the latest $\mathbf{x}_1$ is reported as the final solution of the unknown vector $\mathbf{x}$. The predicted location of the acoustic source is then given by (C_x, C_y).

5.8.3.5 *Numerical Validation for Non-Elliptical Wave Fronts*

To illustrate the effectiveness of the proposed approach for acoustic source localization, the same numerical example is considered for ASL assuming elliptical and non-elliptical wave front shapes, as shown in Figure 5.24. The schematic diagram of the plate and the locations of the source and the six sensor clusters are shown in Figure 5.14. The bottom-left corner of the plate is the origin (0.0, 0.0) of the coordinate system. It may be observed that the sensor clusters 1, 2 and 3 are of Type-a, whereas 4, 5 and 6 are of Type-b (see Figure 5.21). The TOAs of the second wave front are estimated at all sensor clusters using AIC as discussed earlier [see Eq. (5.83)], and then from Eqs. (5.81) and (5.82) $\tan\theta$ is estimated at these cluster locations. m_i is then computed from Eq. (5.84) for $i = 1, 2, \ldots$ and 6. These computed values are −1.50, −0.47, −4.00, 2.88, 0.60 and 1.50 for sensor clusters 1, 2, 3, 4, 5 and 6, respectively. The acoustic source is then estimated following the proposed techniques for elliptical wave front Eq. (5.76) and non-elliptical wave front represented by parametric curves of Eq. (5.88). For the parametric curve representation of the wave front, it was observed after trying out several constant positive values for the parameter q that $q = 1.0$ leads to a sufficiently close match between the wave front and the parametric curve resulting an accurate estimation of the acoustic source location without causing any problem in the convergence of the solution algorithm.

If only three sensor clusters (i.e., $n = 3$) are used to predict the acoustic source, a total of 20 combinations of three sensor clusters are possible from the six clusters shown in Figure 5.14. The acoustic source location was predicted for all these combinations using both wave front models – elliptic and non-elliptic. Disregarding some unrealistic predictions (Sen and Kundu (2018), the predicted results are shown graphically in Figure 5.25. The average of the predicted source coordinates from the two models are shown in Figure 5.26. It can be seen that the average location of the source predicted by the parametric curve-based approach is closer to the actual source location.

Similarly, four sensor clusters ($n = 4$) can also be considered for predicting the acoustic source leading to 15 possible sensor cluster combinations. The average

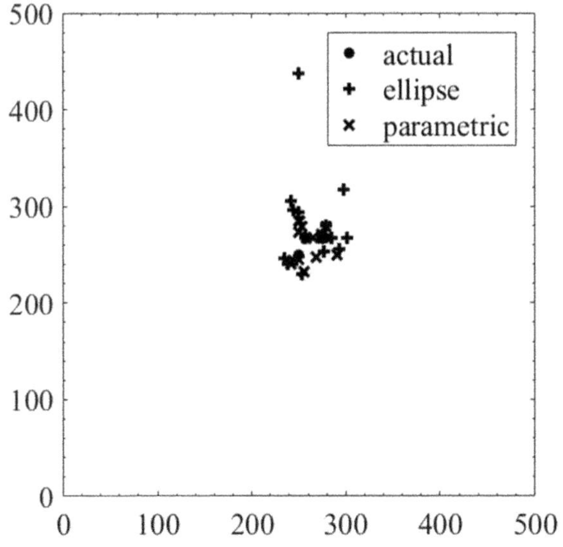

Figure 5.25 Coordinates of the true acoustic source point (denoted by the solid circle) and source locations predicted from various sensor clusters when the wave front is assumed to be elliptical (denoted by the "+" marker), and non-elliptical parametric curve (denoted by the "x" marker) for $n=3$ (Sen and Kundu, 2018).

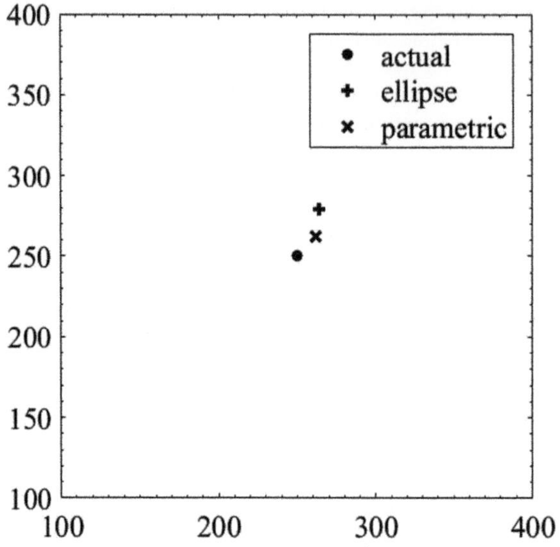

Figure 5.26 Coordinates of the true acoustic source location (denoted by the solid circle) and the average of the source locations predicted when the wave front is assumed to be elliptical (denoted by the "+" marker), and non-elliptical parametric curve (denoted by the "x" marker) for $n = 3$ (Sen and Kundu, 2018).

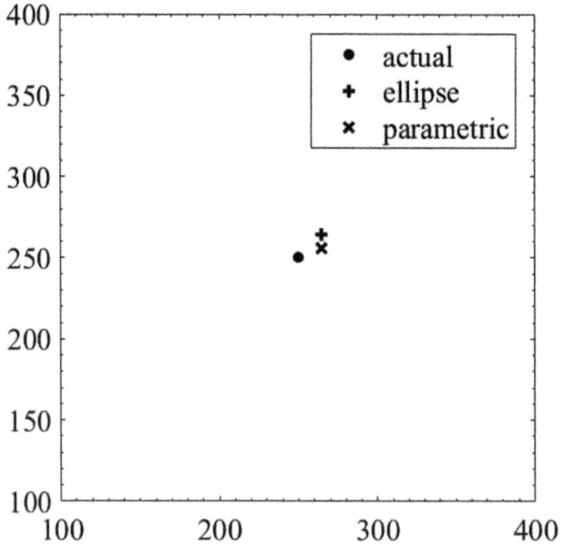

Figure 5.27 Coordinates of the true acoustic source location (denoted by the solid circle) and the average of the source locations predicted when the wave front is assumed to be elliptical (denoted by the "+" marker), and non-elliptical parametric curve (denoted by the "x" marker) for $n = 4$ (Sen and Kundu, 2018).

source coordinates are calculated and plotted in Figure 5.27. This figure shows that for $n=4$ also, the parametric curve-based approach performs better than the ellipse-based approach.

When five sensor clusters ($n = 5$) are considered for the acoustic source prediction six different sensor cluster combinations are possible. For all these combinations the acoustic source is localized using the two models. The average source coordinates are evaluated and shown in Figure 5.28.

Finally, when all six sensor clusters are considered simultaneously ($n = 6$), only one combination of sensor clusters is possible that includes all clusters 1 to 6. In this case, the ellipse-based technique predicts the source coordinates as (265.80 mm, 262.24 mm), giving an error of 19.98 mm, whereas the coordinates predicted by the parametric curve-based approach is (260.39 mm, 251.70 mm), giving an error of 10.53 mm. Thus, both models give satisfactory results, however the parametric curve-based model localizes the source more accurately than the elliptical wave front model. These results are shown in Figure 5.29.

To graphically show how the means and standard deviations of the errors vary with the number of sensor clusters used for these predictions, these values are plotted against the number of sensor clusters in Figures 5.30 and 5.31, respectively. One can see that as the number of clusters increases, the mean error in source localization decreases for both elliptical and non-elliptical wave fronts. Also, note that the parametric curve-based non-elliptical wave front gives a lower mean error than the elliptical wave front. The standard deviation of errors is also lower for the non-elliptical wave front than the elliptical wave front when three and five clusters are considered, and is marginally more for the four clusters.

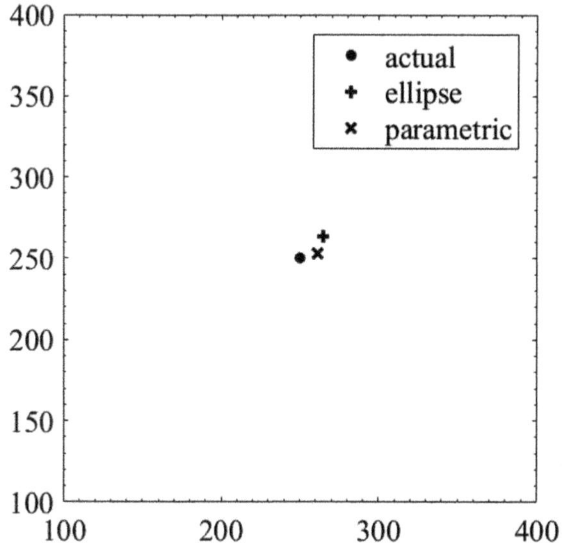

Figure 5.28 Coordinates of the true acoustic source location (denoted by the solid circle) and the average of the source locations predicted when the wave front is assumed to be elliptical (denoted by the "+" marker), and non-elliptical parametric curve (denoted by the "x" marker) for $n=5$ (Sen and Kundu, 2018).

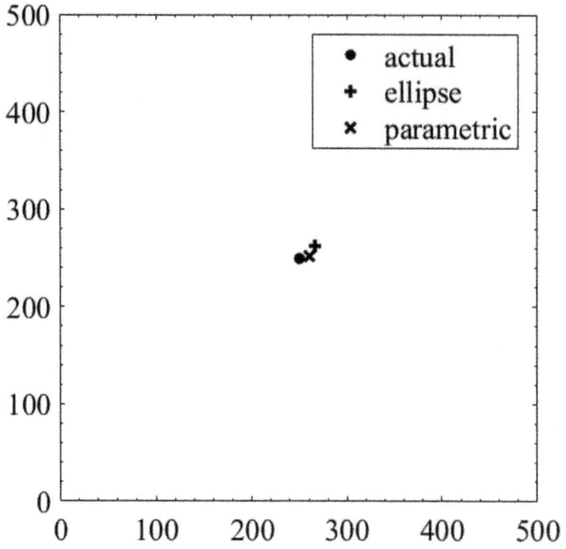

Figure 5.29 Coordinates of the true acoustic source location (denoted by the solid circle) and the source locations predicted when the wave front is assumed to be elliptical (denoted by the "+" marker), and non-elliptical parametric curve (denoted by the "x" marker) for $n = 6$ (Sen and Kundu, 2018).

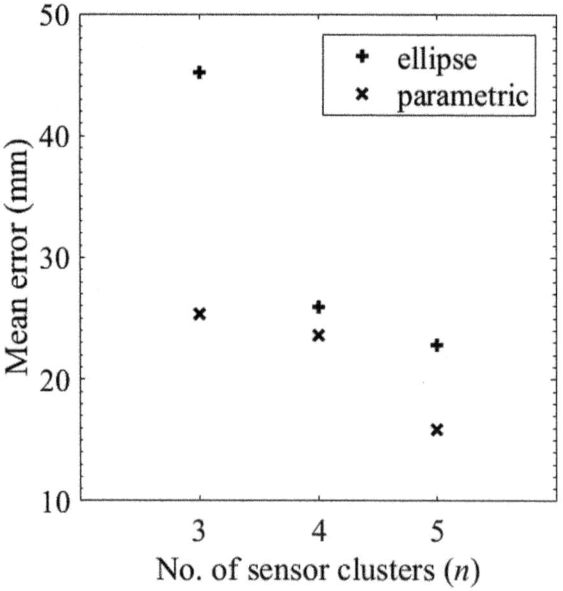

Figure 5.30 Mean error for the predicted source coordinates for elliptical (denoted by the "+" markers) and parametric curve-based non-elliptical wave fronts (denoted by the "x" markers) for $n = 3, 4$ and 5 (Sen and Kundu, 2018).

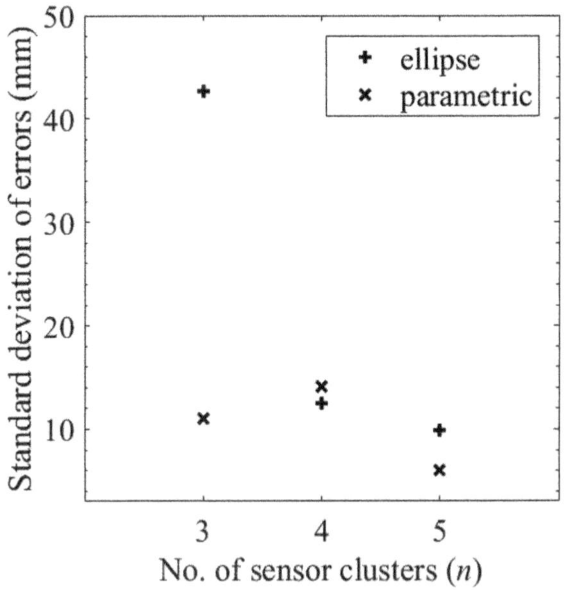

Figure 5.31 Standard deviation of the errors for the predicted source coordinates for elliptical (denoted by the "+" markers) and parametric curve-based non-elliptical wave fronts (denoted by the "x" markers) for $n = 3$, 4 and 5 (Sen and Kundu, 2018).

5.9 CONCLUDING REMARKS

For about the last five decades the ASL technique in isotropic and anisotropic structures has been a popular topic of research. Many techniques have been proposed. While the advantages of a new technique have been routinely highlighted by the proposer, often its limitations have not been mentioned. While describing various source localization techniques, this chapter discussed both advantages and disadvantages of these techniques.

Many ASL techniques work very well for isotropic plates. Some of these techniques work for anisotropic plates as well. Most techniques require known properties of the plate and its dimensions, especially if the plate is anisotropic. Only a handful techniques are available for predicting the acoustic source location in an anisotropic plate with the help of a few receiving sensors without knowing the material properties of the plate. For plates having weak anisotropy the use of *L*- and *Z*-shaped sensor clusters (discussed in Section 5.3.3) is probably the most efficient technique for ASL for which absolutely no information on the plate material and its dimensions is necessary. For highly anisotropic plates however, the only way one can predict the acoustic source location with a handful sensors and without knowing the exact material properties of the plate is to get an approximate idea about the wave front shape in the plate (whether the wave front should be rhombus, elliptical or nearly elliptical in shape) and then localizing the acoustic source based on that shape, as discussed in Section 5.8.

Acoustic source localization in more complex three-dimensional structures requires more research and development especially if one wants to do it without knowing the material properties of the structure and with the help of a few sensors, as was done for the plate. Until such an advanced new technique is developed, the only option available to us is to place a large number of sensors over the entire structure and then to localize the acoustic source by identifying which sensor received the acoustic event generated signal first. One can then conclude that the acoustic source must be close to that sensor since it recorded the signal first. Therefore, to predict the acoustic source with high accuracy one needs to place a large number of sensors distributed over the entire structure. One may also follow a more labor intensive and computationally demanding artificial neural network or time reversal technique based on the impulse response function that has been also discussed in this chapter.

REFERENCES

Barricelli, N. A., *Symbiogenetic Evolution Processes Realized by Artificial Methods*, pp. 143–182 (1957). https://books.google.com/books?id=FVZxGwAACAAJ&dq=inauthor:%22Nils+Aall+Barricelli%22&hl=en&sa=X&ved=0ahUKEwiw0c3W7_fdAhVFwMQHHevoANcQ6AEINTAD

Baxter, M. G., R. Pullin, K. M. Holford, and S. L. Evans, "Delta T source location for acoustic emission" *Mechanical Systems and Signal Processing*, Vol. 21(3), pp. 1512–1520 (2007).

Betz, D. C., G. Thursby, B. Culshaw, and W. J. Staszewski, "Structural damage location with fiber bragg grating rosettes and lamb waves" *Structural Health Monitoring: an International Journal*, Vol. 6(4), pp. 299–308 (2007).

Castagnede, B., W. Sachse, and K. Y. Kim, "Location of point like acoustic emission sources in anisotropic plates" *The Journal of the Acoustical Society of America*, Vol. 86(3), pp. 1161–1171 (1989).

Ciampa, F., and M. Meo, "A New algorithm for acoustic emission localization and flexural group velocity determination in anisotropic structures" *Composites Part A: Applied Science and Manufacturing*, Vol. 41(12), pp. 1777–1786 (2010).

Ciampa, F., M. Meo, and E. Barbieri, "Impact localization in composite structures of arbitrary cross section" *Structural Health Monitoring: an International Journal*, Vol. 11(6), pp. 643–655 (2012).

Dong, L., and X. Li, "Three-dimensional analytical solution of acoustic emission or microseismic source location under cube monitoring network" *Transactions of Nonferrous Metals Society of China*, Vol. 22(12), pp. 3087–3094 (2012).

Ebrahimkhanlou, A., and S. Salamone, "Acoustic emission source localization in thin metallic plates: A single-sensor approach based on multimodal edge reflections" *Ultrasonics*, Vol. 78, pp. 134–145 (2017).

Fraser, A., and D. Burnell, *Computer Models in Genetics*, McGraw-Hill, New York (1970).

Giurgiutiu, V., "Lamb wave generation with piezoelectric wafer active sensors for structural health monitoring" *Proceedings of SPIE*, Vol. 5056, pp. 111–122 (2003).

Gorman, M. R., "AE source orientation by plate wave analysis" *Journal of Acoustic Emission*, Vol. 9(4), pp. 283–288 (1991).

Grabec, I., and W. Sachse, "Application of an intelligent signal processing system to acoustic emission analysis" *The Journal of the Acoustical Society of America*, Vol. 85(3), pp. 1226–1235 (1989).

Guyomar, D., M. Lallart, L. Petit, and X.-J. Wang, "Impact localization and energy quantification based on the power flow: A low-power requirement approach" *Journal of Sound and Vibration*, Vol. 330(13), pp. 3270–3283 (2011).

Hajzargerbashi, T., T. Kundu, and S. Bland, "An improved algorithm for detecting point of impact in anisotropic inhomogeneous plates" *Ultrasonics*, Vol. 51(3), pp. 317–324 (2011).

He, T., Q. Pan, Y. Liu, X. Liu, and D. Hu, "Near-field beamforming analysis for acoustic emission source localization" *Ultrasonics*, Vol. 52(5), pp. 587–592 (2012).

Hensman, J., R. Mills, S. G. Pierce, K. Worden, and M. Eaton, "Locating acoustic emission sources in complex structures using gaussian processes" *Mechanical Systems and Signal Processing*, Vol. 24(1), pp. 211–223 (2010).

Hiche, C., C. K. Coelho, and A. Chattopadhyay, "A strain amplitude algorithm for impact localization on composite laminates" *Journal of Intelligent Material Systems and Structures*, Vol. 22(17), pp. 2061–2067 (2011).

Hsu, N. N., *"Acoustic Emissions Simulator"*, US Patent 4,018,084 (1977).

Ing, R. K., N. Quieffin, S. Catheline, and M. Fink, "In solid localization of finger impacts using acoustic time-reversal process" *Applied Physics Letters*, Vol. 87(20), p. 2004104 (2005).

Jata, K. V., T. Kundu, and T. Parthasarathy, "An introduction to failure mechanisms and ultrasonic inspection", in *Advanced Ultrasonic Methods for Material and Structure Inspection*, T. Kundu, ed. ISTE Ltd., London, UK, and Newport Beach, CA, USA (2007), Chapter 1, pp. 1–42 (2007).

Jiao, J., C. He, B. Wu, R. Fei, and X. Wang, "Application of wavelet transform on modal acoustic emission source location in thin plates with one sensor" *International Journal of Pressure Vessels and Piping*, Vol. 81(5), pp. 427–431 (2004).

Kessler, S. S., S. M. Spearing, and C. Soutis, "Damage detection in composite materials using lamb wave methods" *Smart Materials and Structures*, Vol. 11, pp. 795–803 (2002).

Koabaz, M., T. Hajzargarbashi, T. Kundu, and M. Deschamps, "Locating the acoustic source in an anisotropic plate" *Structural Health Monitoring: an International Journal*, Vol. 11(3), pp. 315–323 (2012).

Köhler, B., F. Schubert, and B. Frankenstein, "Numerical and experimental investigation of lamb wave excitation, propagation and detection for SHM", Proceedings of the Second European Workshop on Structural Health Monitoring, pp. 993–1000 (2004).

Kosel, T., I. Grabec, and P. Muzic, "Location of acoustic emission sources generated by air flow" *Ultrasonics*, Vol. 38(1–8), pp. 824–826 (2000).

Kundu, T., "A new technique for acoustic source localization in an anisotropic plate without knowing its material properties", *6th European Workshop on Structural Health Monitoring*, Dresden, Germany, July 3–6 (2012).

Kundu, T., "Acoustic source localization" *Ultrasonics*, Vol. 54(1), pp. 25–38 (2014).

Kundu, T., H. Nakatani, and N. Takeda, "Acoustic source localization in anisotropic plates" *Ultrasonics*, Vol. 52(6), pp. 740–746 (2012).

Kundu, T., S. Das, S. A. Martin, and K. V. Jata, "Locating point of impact in anisotropic fiber reinforced composite plates" *Ultrasonics*, Vol. 48(3), pp. 193–201 (2008).

Kundu, T., S. Das, and K. V. Jata, "Point of impact prediction in isotropic and anisotropic plates from the acoustic emission data", *Journal of the Acoustical Society of America*, Vol. 122(4), pp. 2057–2066 (2007).

Kundu, T., S. Das, and K. V. Jata "Impact point detection in stiffened plates by acoustic emission technique", *Smart Materials and Structures*, Vol. 18, Article 035006 (2009).

Lagarias, J. C., J. A. Reeds, M. H. Wright, and P. E. Wright, "Convergence properties of the nelder–mead simplex method in low dimensions" *SIAM Journal on Optimization*, Vol. 9(1), pp. 112–147 (1998).

Liang, D., S. Yuan, and M. Liu, "Distributed coordination algorithm for impact location of preciseness and real-time on composite structures" *Measurement*, Vol. 46(1), pp. 527–536 (2013).

Lin, M., "Manufacturing of composite structures with a built-in network of piezoceramics" Ph.D. Dissertation, Department of Mechanical Engineering, Stanford University (1998).

Lin, M., and F. K. Chang, "Development of SMART Layers for Built-In Diagnostics for Composite Structures", *The 13th Annual ASC Technical Conference on Composite Materials*, Baltimore, MD, September, 1998 (1998).

Lokajicek, T., and K. Klima, "A first arrival identification system of Acoustic Emission (AE) Signals by means of a high-order statistics approach" *Measurement Science and Technology*, Vol. 17(9), pp. 2461–2466 (2006).

Mal, A. K., F. Ricci, S. Banerjee, and F. Shih, "A conceptual structural health monitoring system based on wave propagation and modal data" *Structural Health Monitoring: an International Journal*, Vol. 4(3), pp. 283–293 (2005).

Mal, A. K., F. Ricci, S. Gibson, and S. Banerjee "Damage detection in structures from vibration and wave propagation data" *Proceedings of SPIE*, Vol. 5047, pp. 202–210 (2003a).

Mal, A. K., F. Shih, and S. Banerjee, "Acoustic emission waveforms in composite laminates under low velocity impact" *Proceedings of SPIE*, Vol. 5047, pp. 1–12 (2003b).

Manson, G.K.. Worden, K., and D.J. Allman, "Experimental validation of structural health monitoring methodology II: Novelty detection on an aircraft wing" *Journal of Sound and Vibration*, Vol. 259, pp. 345–363 (2003).

Matt, H. M., and F. Lanza di Scalea, "Macro-fiber composite piezoelectric rosettes for acoustic source location in complex structures" *Smart Matererials and Structures*, Vol. 16, pp. 1489–1499 (2007).

McLaskey, G. C., S. D. Glaser, and C. U. Grosse, "Beamforming array techniques for acoustic emission monitoring of large concrete structures" *Journal of Sound and Vibration*, Vol. 329(12), pp. 2384–2394 (2010).

Nakatani, H., T. Hajzargarbashi, K. Ito, T. Kundu, and N. Takeda, "Impact localization on a cylindrical plate by near-field beamforming analysis,", *Sensors and Smart Structures Technologies for Civil, Mechanical, and Aerospace Systems, SPIE's 2012 Annual International Symposium on Smart Structures and Nondestructive Evaluation*, San Diego, California, March 12–15, 2012, ed. M. Tomizuka, Vol. 8345 (2012a).

Nakatani, H., T. Hajzargarbashi, K. Ito, T. Kundu, and N. Takeda, "Locating point of impact on an anisotropic cylindrical surface using acoustic beamforming technique" *Key Engineering Materials*, 4th Asia-Pacific Workshop on Structural Health Monitoring, Melbourne, Australia, December 5–7, 2012, Vol. 558, pp. 331–340 (2012b).

Nakatani, H., T. Kundu, and N. Takeda, "Improving accuracy of acoustic source localization in anisotropic plates" *Ultrasonics*, Vol. 54(7), pp. 1776–1788 (2014).

Nehorai, A., and E. Paldi, *"Method for electromagnetic source localization"*, US Patent Number 5315308 (filed Nov 4, 1992), awarded May 24, 1994.

Nelder, J. A., and R. Mead, "A simplex method for function minimization" *The Computer Journal*, Vol. 7(4), pp. 308–313 (1965).

Niri, E. D., and S. Salamone, "A probabilistic framework for acoustic source localization in plate-like structures" *Smart Materials and Structures*, Vol. 21, pp. 035009–1:16 (2012).

Niri, E. D., S. Salamone, and P. Singla, ""Acoustic Emission (AE) Source Localization using Extended Kalman Filter (EKF)", *Health Monitoring of Structural and Biological Systems" Proceedings of SPIE*, ed. T. Kundu, Vol. 8348, pp. 834804–1:15 (2012).

Packo, P., T. Bielak, A. B. Spencer, T. Uhl, W. J. Staszewski, K. Worden, T. Barszcz, P. Russek, and K. Wiatr, "Numerical simulations of elastic wave propagation using graphical processing units—comparative study of high-performance computing capabilities" *Computer Methods in Applied Mechanics and Engineering*, Vol. 290, pp. 98–126 (2015).

Park, B., H. Sohn, S. E. Olson, M. P. DeSimio, K. S. Brown, and M. M. Derriso, "Impact localization in complex structures using laser based time reversal" *Structural Health Monitoring: An International Journal*, Vol. 11(5), pp. 577–588 (2012).

Park, J., and F.K. Chang, "Built-In Detection of Impact Damage in Multi-Layered Thick Composite Structures" *Proceedings of the Fourth International Workshop on Structural Health Monitoring*, pp. 1391–1398 (2003).

Park, W. H., "Acoustic source localization in an anisotropic plate without knowing its material properties" PhD dissertation, Department of Aerospace and Mechanical Engineering, The University of Arizona. (2016)

Park, W. H., P. Packo, and T. Kundu, "Acoustic source localization in an anisotropic plate without knowing its material properties—a new approach" *Ultrasonics*, Vol. 79, pp. 9–17 (2017).

Ribay, G., S. Catheline, D. Clorennec, R. K. Ing, N. Quieffin, and M. Fink, " Acoustic impact localization in plates: Properties and stability to temperature variation" *IEEE Transactions on Ultrasonics, Ferroelectrics, and Frequency Control*, Vol. 54(2), pp. 378–385 (2007).

Sachse, W. H., and S. Sancar, "Acoustic emission source location on plate-like structures using a small array of transducers" *The Journal of the Acoustical Society of America*, U.S. Patent No. 4,592,034, Vol. 81(1), pp. 206–206 (1987).

Salamone, S., I. Bartoli, J. Rhymer, F. Lanza di Scalea, and H. Kim, ""Validation of the piezoelectric rosette technique for locating impacts in complex aerospace panels", Health Monitoring of Structural and Biological Systems", *Pub. Bellingham, WASH, 18th Annual International Symposium on Smart Structures/NDE*, March 7–10, 2011, *San Diego, California*, ed. T. Kundu, Vol. 7984, pp. 79841E1–79811 (2011).

Salamone, S., I. Bartoli, P. di Leo, F. Lanza di Scalea, A. Ajovalasit, L. D'Acquisto, J. Rhymer, and H. Kim, "High-velocity impact location on aircraft panels using macro-fiber composite piezoelectric rosettes" *Journal of Intelligent Material Systems and Structures*, Vol. 21(9), pp. 887–896 (2010).

Scruby, C. B., and L. E. Drain, *Laser Ultrasonics: Techniques and Applications*, Taylor & Francis, New York, NY, pp. 76–85 (1990).

Sedlak, P., Y. Hirose, and M. Enoki, "acoustic emission localization in thin multilayer plates using first-arrival determination" *Mechanical Systems and Signal Processing*, Vol. 36(2), pp. 636–649 (2013).

Sedlak, P., Y. Hirose, S. A. Khan, M. Enoki, and J. Sikula, "new automatic localization technique of acoustic emission signals in thin metal plates" *Ultrasonics*, Vol. 49(2), pp. 254–262 (2009).

Sen, N., and T. Kundu, "A new wave front shape-based approach for acoustic source localization in an anisotropic plate without knowing its material properties" *Ultrasonics*, Vol. 87, pp. 20–32 (2018).

Seydel, R., and F. K. Chang, "Impact load identification of stiffened composite plates with built-in piezo-sensors" *Proceedings of the SPIE Smart Structures and Materials Conference*, Newport Book Company, Beach, CA, (March 1999).

Sribar, R., and W. Sachse, "An experimental investigation of the ae source location and magnitude on 2-d frame structures using intelligent signal processing" *The Journal of the Acoustical Society of America*, Vol. 93(4), part 2, pp. 2279–2279 (1993).

Staszewski, W. J., B. C. Lee, L. Mallet, and F. Scarpa, "Structural health monitoring using scanning laser vibrometry: I. lamb wave sensing" *Smart Materials and Structures*, Vol. 13(2), pp. 251–260 (2004).

Surgeon, M., and M. Wevers, "One sensor linear location of acoustic emission events using plate wave theories" *Materials Science and Engineering: A*, Vol. 265(1–2), pp. 254–261 (1999).

Ting, A., Z. Ru, L. Jianfeng, and R. Li, "Space-time evolution rules of acoustic emission location of unloaded coal sample at different loading rates" *International Journal of Mining Science and Technology*, Vol. 22(6), pp. 847–854 (2012).

Tobias, A., "Acoustic emission source location in two dimensions by an array of three sensors" *Non-Destructive Testing*, Vol. 9(1), pp. 9–12 (1976).

Toyama, N., J. -H. Koo, R. Oishi, M. Enoki, and T. Kishi, "Two-dimensional AE source location with two sensors in thin CFRP Plates" *Journal of Materials Science Letters*, Vol. 20(19), pp. 1823–1825 (2001).

Tracy, M., and F. K. Chang, "Identifying impact load in composite plates based on distributed piezo-sensors," *The Proceedings of SPIE Smart Structures and Materials Conference*, San Diego, CA" (1996).

Wang, C., "Built-In impact damage detection for composite plates" Ph.D. Dissertation, Department of Aeronautics and Astronautics, Stanford University (1999).

Wang, S. C., and F. K. Chang, "Diagnosis of impact damage in composite structures with built-in piezoelectrics network" *Proceedings of SPIE*, Vol. 3990, pp. 13–19 (2000).

Yin, S., Z. Cui, and T. Kundu, "Acoustic source localization in anisotropic plates with "Z" shaped sensor clusters" *Ultrasonics*, Vol. 84, pp. 34–37 (2018).

Index

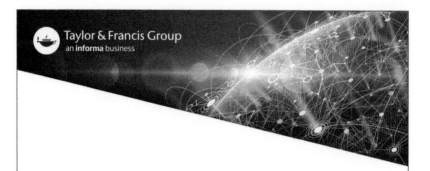